The Chemistry and Biology of Nitroxyl (HNO)

The Chemistry and Biology of Nitroxyl (HNO)

Edited by

Fabio Doctorovich
Universidad de Buenos Aires and CONICET, Buenos Aires, Argentina

Patrick J. Farmer
Baylor University, Waco, TX, United States

Marcelo A. Marti
Universidad de Buenos Aires and CONICET, Buenos Aires, Argentina

ELSEVIER

AMSTERDAM • BOSTON • HEIDELBERG • LONDON
NEW YORK • OXFORD • PARIS • SAN DIEGO
SAN FRANCISCO • SINGAPORE • SYDNEY • TOKYO

Elsevier
Radarweg 29, PO Box 211, 1000 AE Amsterdam, Netherlands
The Boulevard, Langford Lane, Kidlington, Oxford OX5 1GB, United Kingdom
50 Hampshire Street, 5th Floor, Cambridge, MA 02139, United States

British Library Cataloguing-in-Publication Data
A catalogue record for this book is available from the British Library

Library of Congress Cataloging-in-Publication Data
A catalog record for this book is available from the Library of Congress

ISBN: 978-0-12-800934-5

Publisher: Cathleen Sether
Acquisition Editor: Katey Birtcher
Editorial Project Manager: Jill Cetel
Production Project Manager: Anitha Sivaraj
Designer: Vicky Pearson Esser

Typeset by MPS Limited, Chennai, India

Contents

List of Contributors

L. Álvarez Universidad de Buenos Aires and CONICET, Buenos Aires, Argentina

S.E. Bari Universidad de Buenos Aires, Buenos Aires, Argentina; CONICET, Buenos Aires, Argentina

D. Basudhar National Cancer Institute, Frederick, MD, United States

G. Bharadwaj University of California, Davis School of Medicine, Sacramento, CA, United States

D.E. Bikiel Universidad de Buenos Aires and CONICET, Buenos Aires, Argentina

S. Bruce King Wake Forest University, Winston-Salem, NC, United States

T.A. Chavez Johns Hopkins University, Baltimore, MD, United States

F. Doctorovich Universidad de Buenos Aires and CONICET, Buenos Aires, Argentina

B. Dong University of Jinan, Jinan, Shandong, People's Republic of China

P.J. Farmer Baylor University, Waco, TX, United States

Milos Filipovic Université de Bordeaux, IBGC, UMR 5095, Bordeaux, France; CNRS, IBGC, UMR 5095, Bordeaux, France

J.M. Fukuto Sonoma State University, Rohnert Park, CA, United States

L.E. Goodrich University of Michigan, Ann Arbor, MI, United States

D.A. Guthrie Johns Hopkins University, Baltimore, MD, United States

M. Hamer Universidad de Buenos Aires and CONICET, Buenos Aires, Argentina

X. Han Baylor University, Waco, TX, United States

I. Ivanović-Burmazović Friedrich-Alexander University Erlangen-Nuremberg, Erlangen, Germany

D.A. Kass Johns Hopkins Medical Institutions, Baltimore, MD, United States

G. Keceli Johns Hopkins Medical Institutions, Baltimore, MD, United States

M.R. Kumar Baylor University, Waco, TX, United States

N. Lehnert University of Michigan, Ann Arbor, MI, United States

W. Lin University of Jinan, Jinan, Shandong, People's Republic of China

M.A. Martí Universidad de Buenos Aires and CONICET, Buenos Aires, Argentina

Z. Miao Wake Forest University, Winston-Salem, NC, United States

R.J. Millikin Sonoma State University, Rohnert Park, CA, United States

K.M. Miranda University of Arizona, Tucson, AZ, United States

M.A. Morales Vásquez Universidad de Buenos Aires and CONICET, Buenos Aires, Argentina

M. Muñoz Universidad de Buenos Aires and CONICET, Buenos Aires, Argentina

N.I. Neuman Universidad de Buenos Aires and CONICET, Buenos Aires, Argentina; Universidad Nacional del Litoral, Santa Fe, Argentina

S. Nourian Johns Hopkins University, Baltimore, MD, United States

J.A. Olabe Universidad de Buenos Aires, Buenos Aires, Argentina; CONICET, Buenos Aires, Argentina

N. Paolocci Johns Hopkins Medical Institutions, Baltimore, MD, United States; University of Perugia, Perugia, Italy

J. Pellegrino Universidad de Buenos Aires and CONICET, Buenos Aires, Argentina

M. Ren University of Jinan, Jinan, Shandong, People's Republic of China

D.J. Salmon University of Minnesota, Minneapolis, MN, United States

L.D. Slep Universidad de Buenos Aires, Buenos Aires, Argentina; CONICET, Buenos Aires, Argentina

S.A. Suarez Universidad de Buenos Aires and CONICET, Buenos Aires, Argentina

J.P. Toscano Johns Hopkins University, Baltimore, MD, United States

C. Van Stappen University of Michigan, Ann Arbor, MI, United States

M.A. Morales Vásquez Universidad de Buenos Aires and CONICET, Buenos Aires, Argentina

D.A. Wink National Cancer Institute, Frederick, MD, United States

A.L. Zapata Baylor University, Waco, TX, United States

Introduction: A Bit of History and General Facts About Nitroxyl: From Interstellar Molecule to Biological Gasotransmitter

Chemical and physical properties of small di-/tri-atomic species and reaction pathways for their generation lay the foundation of chemical science and set the principles for chemical evolution of the universe. Although basic, the chemistry of small molecules can often pose great scientific challenges. One such challenging species is HNO (nitroxyl, also known as nitrosyl hydride, nitroso hydrogen or azanone) and its conjugated base NO^- that is a product of one electron reduction of nitric oxide ($NO^\bullet$), isoelectronic with molecular oxygen (O_2).

Due to the high cosmic abundances of hydrogen, oxygen, and nitrogen atoms[1] a search for their simple molecular assemblies in outer space has started in the sixties,[2] and in the seventies resulted in the detection of nitroxyl as the first interstellar molecule which contained the NO-bond.[3] Later its existence was confirmed in the Sagittarius Dwarf Galaxy, a satellite galaxy of our Milky Way.[4] However, very low temperatures, highly diluted (10^3-10^5 molecules/cm^3)[5] and rather reducing interstellar medium is not comparable to the terrestrial environment, where nitroxyl appears to be unstable, with the reported lifetime about milliseconds to 40 s depending on experimental circumstances.[6,7] Reaction conditions similar to those in interstellar space can be mimicked by matrix isolation methods or simulated in silico, which enables trapping and spectroscopic characterization of unstable species and their theoretical analysis, respectively. Therefore, the first reported attempt (by Paul Harteck at Kaiser-Wilhelm-Institut für Physikalische Chemie und Elektrochemie in Berlin-Dahlem in 1933) to prepare HNO was by the addition reaction of $H^\bullet$ atom to $NO^\bullet$ at cryogenic temperatures of liquid air (c. − 196°C).[8] A transparent, yellow, solid reaction product, described as a $[HNO]_n$ aggregate, was hygroscopic, relatively stable at the temperature of liquid air, but explosive upon very tiny stimulus resulting in H_2 and $NO^\bullet$ (different from expected H_2O and N_2 as products). By slow warming it transformed predominantly into hyponitrous acid (HON=NOH ↔ HO–NH–N=O) and maybe partly in its isomer nitrylamid (nitramide, H_2N-NO_2), finally resulting in N_2O (nitrous oxide) at room temperature (Scheme I.1). Harteck was not sure whether in that experiment discrete HNO was formed; however, he reported that in the reaction of O atoms with ammonia (NH_3), where oxygen formally substituted H_2, "the HNO as a radical species was generated

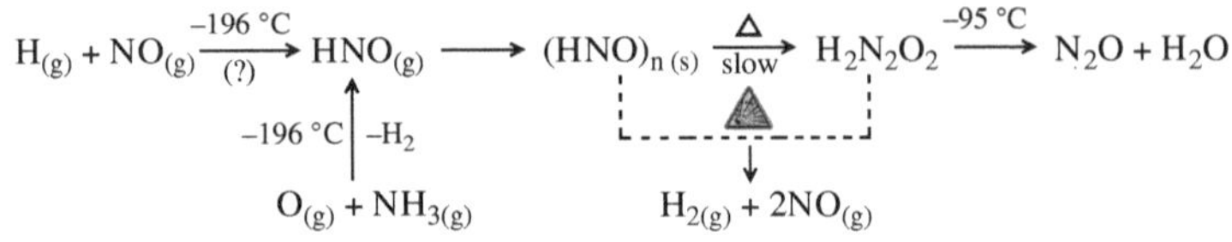

Scheme I.1 Reactions/transformations described by Harteck in "The preparation of HNO or $[HNO]_n$."[8]

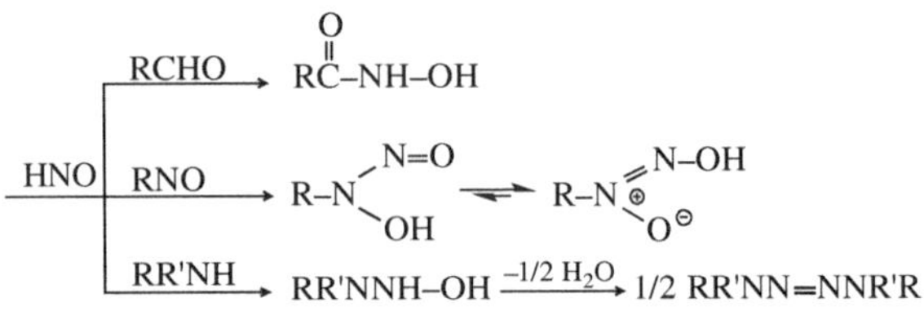

Scheme I.2 Some of the nitroxyl reactions described by Angeli.[12]

in the gas phase" (translated from German).[8] With that evidence he confirmed intermediate existence of HNO that was earlier proposed within the catalytic oxidation of NH_3 to $NO^\bullet$.[9]

In the same period, Smallwood suggested formation of a quite short-lived molecule of the type H–NO to explain catalytic effect of $NO^\bullet$ on the recombination of $H^\bullet$ atoms to form H_2 at room temperature conditions (Eq. I.1 and Eq. I.2).[10]

$$H^\bullet + NO^\bullet \rightarrow \textit{H}\text{-NO} \tag{I.1}$$

$$\textit{H}\text{-NO} + H^\bullet \rightarrow H_2 + NO \tag{I.2}$$

Different from those gas-phase reactions, Angelo Angeli had even earlier, in 1901, described a way to obtain HNO in aqueous solution by decomposition of "salts of nitrohydroxylaminic acid" ($Na_2N_2O_3$, known as Angeli's salt, i.e. trioxodinitrate), which resulted in the corresponding nitrites and, what he called, "the unsaturated residue nitroxyl: $N^\bullet OH$."[11] In 1905 he even studied the reactivity of nitroxyl toward aldehydes, nitroso-derivatives, and secondary amines (Scheme I.2). He was not sure whether it had "the constitution of dihydroxyammonia, $NH(OH)_2$, or the corresponding anhydride, NHO." However, based on its generation from hydroxylamine by oxidation with Caro's acid (peroxymonosulfuric acid, H_2SO_5) he properly positioned nitroxyl in "the series of oxidation products of ammonia: ammonia, hydroxylamine, dihydroxyammonia, nitrous acid."[12] Nowadays application of Angeli's salt experiences a renaissance as a widely used reagent for the in situ production of HNO, which has also inspired development of other nitroxyl donors with a particular emphasis on their utilization as molecular tools for elucidation of biochemical/physiological role of related nitrogen species and as promising therapeutic agents (for more on HNO donors' applications and corresponding HNO release mechanisms see other chapters in this book).

But the first exact characterization of HNO (and DNO) was reported some 30 years later by Dalby, who determined its molecular structure based on gas phase absorption spectra in the region 6500–7700 Å.[13] He obtained HNO by the flash photolysis of nitromethane, nitroethane, isoamyl nitrite, and mixtures of nitric oxide and ammonia. Under those experimental conditions the reported lifetime of HNO was about 1/10 s. In the same year Brown and Pimentel generated HNO by photolysis of methyl nitrite in an argon matrix and for the first time obtained its infrared spectrum, which supported "the structure HNO over the possible arrangement HON."[14] In several follow-up investigations, flash photolysis of a mixture of H_2 and $NO^•$ was used for the preparation of HNO by the reaction of $H^•$ and $NO^•$.[15]

Thanks to all these experiments, it was clear that HNO is not an elusive species but rather a reactive intermediate, which, depending on reaction conditions, can decay according to different mechanisms (i.e., dimerization to N_2O, oxidation to $NO^•$, peroxynitrite, nitrite or nitroso-species, reduction to hydroxylamine, binding to metal centers, etc.; for more about these processes see other chapters in this book). With time its implication has been recognized in the variety of processes: combustion of nitrogen-containing fuels, oxidation of atmospheric nitrogen, reaction of NH_2 radical with O_3 (ozone) and O_2 in troposphere,[16] hydrocarbon combustion,[17] and relatively late in the biological nitrogen cycle, particularly denitrification by nitric oxide reductase cytochrome P450nor,[18] in ecologically related physicochemical and biological techniques for NO_x (consisting of 95% nitric oxide and 5% nitrogen dioxide) removal from industrial flue gases,[19] as well as in mammalian physiology.[20] In that last context, based on the above-mentioned earlier physicochemical investigations of the gas-phase reactions, it seems that HNO is a gaseous molecule and as such could be classified as a gasotransmitter in biological systems together with $NO^•$, CO, and H_2S.

Giving such broad significance of nitroxyl, elucidation of its electronic nature, its acid–base and redox properties became quite attractive goal for physicochemical and theoretical studies demonstrated by a number of publications on this topic in the last five decades.[21] Especially intriguing is a fact that its protonated, deprotonated, and tautomeric forms can exist in singlet and triplet states, which makes interconversion between them kinetically demanding and cause a difference in nitroxyl reactivity depending on its multiplicity, i.e. reaction conditions that stabilize its particular forms. Therefore, it is not surprising that in early days nitroxyl was often referred to as a radical.[8,21a] Nowadays it is well established that the bent ^{1}HNO form represents the global minimum in the potential energy surfaces of both singlet and triplet sates of nitroxyl.[22] The linear HNO structure is by ca. 67 kcal/mol less stable than the bent form.[23] Although Dalby had already predicted the probable existence of a triplet $^3A''$ state,[13] which was expected to lay 18.0 to 19.0 kcal above the ground state,[24] ^{3}HNO has never been directly observed.[21a] Recent high level ab initio quantum chemical studies confirmed the adiabatic singlet–triplet energy splitting for HNO of 18.45 ± 0.2 kcal/mol.[21g] On the other hand, the triplet state of ^{3}HON isomer (isonitroso hydrogen or hydroxy nitrene) was isolated in argon matrix at 10 K,[21f] though it should be for 26.18 kcal/mol higher in energy relative to the ^{1}HNO global minimum, whereas its singlet ^{1}HON state is disfavored at any

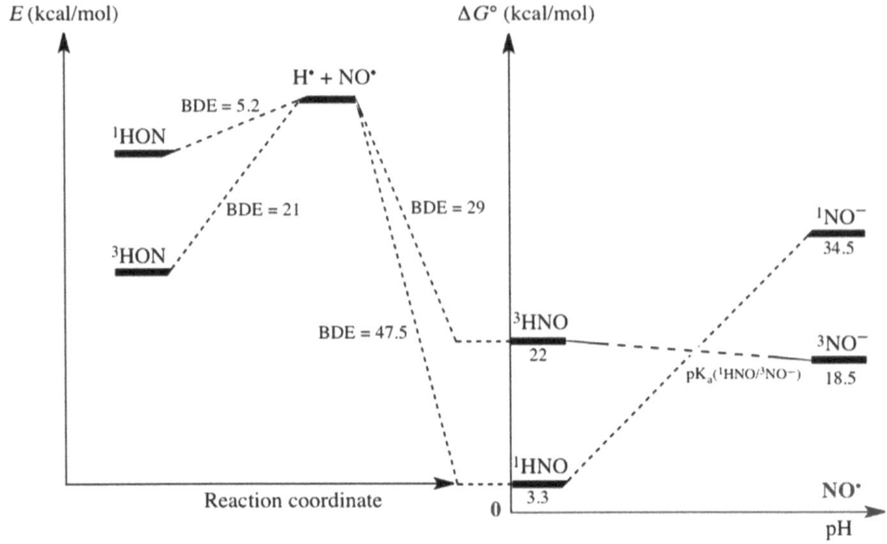

Figure I.1 Schematic relative energy profile for the HON–HNO–NO^- system. Approximate energies (either enthalpy based on BDE=bond dissociation energy,[21g] in the left panel, or Gibbs free energy in correlation to redox potential, in the right panel) are given relative to $NO^{\bullet}$.

temperature, laying 42.23 kcal and 16.05 kcal above ^{1}HNO and ^{3}HON, respectively (Figure I.1).[21g] Thus, the observed photochemical equilibrium between HNO and HON is presumably reached by the dissociation of the two isomers into $NO^{\bullet}$ radicals and H atoms, involving a change of multiplicity.[21f,g] Importantly, in the anionic NO^- form the triplet state is more stable than the singlet one by ca. 16 kcal/mol,[21e,25] and the (de)protonation triggered spin crossover equilibrium is reached at pH equal to pK_a of the $^1HNO/^3NO^-$ conjugated acid–base pair (Figure I.1). This multiplicity change makes the determination of the related pK_a value of nitroxyl, as well as a pH dependent redox potential of the $NO^{\bullet}$/HNO couple quite challenging. While in the older literature the pK_a value of 4.7[26] and the redox potentials of +0.39 V and −0.35 V for the $NO^{\bullet}/^3NO^-$ and $NO^{\bullet}/^1NO^-$ couples,[27] respectively, were reported, the substantially revised values are $pK_a(^1HNO/^3NO^-) = 11.4$, -0.8 ± 0.2 V (for $NO^{\bullet}/^3NO^-$) and -1.7 ± 0.2 V ($NO^{\bullet}/^1NO^-$) *versus* NHE.[21e,h,i] These pure electron transfer processes operate at pH $\geq pK_a$ (Eq. I.3), whereas at lower pH a proton coupled electron transfer (PCET) dominates (Eq. I.4). The PCET mechanism makes the $NO^{\bullet}$ reduction thermodynamically more favorable and the corresponding standard redox potential (at pH = 0) for the $NO^{\bullet}/^1HNO$ couple is only −0.15 V versus SHE,[28] being −0.55 V at pH = 7.[20a,d]

$$NO^{\bullet} + e^- \leftrightarrow {}^3NO^- \quad E^{\circ}(NO^{\bullet}/3NO^-) = -0.8 \pm 0.2V^{21i} \tag{I.3}$$

$$NO^{\bullet} + e^- + H^+ \leftrightarrow {}^1HNO \quad E^{\circ}(NO^{\bullet}, H^+/{}^1HNO) = -0.150 \pm 0.021\ V^{28} \qquad (I.4)$$

Despite all that knowledge collected for more than a century of HNO chemistry investigations, a number of open questions remain. Some of them are related to the relatively new field of HNO biological redox signaling.[20] In that context an intriguing question is how HNO is generated in living systems and what are its molecular targets in the biological milieu. A breakthrough in that field has required a development of advanced methodologies for selective in situ detection of short-lived nitroxyl both in vitro and in vivo, which we have witnessed in the last decade. For the detection of intracellular nitroxyl, fluorescent probes were designed, in particular those based on the Cu(II)/Cu(I) redox couple,[29] whereas for the time resolved monitoring of the HNO production, i.e., kinetics of its generation/decomposition in vitro and in extracellular in vivo surroundings, an electrochemical methodology has been established that uses an HNO specific electrode based on cobalt porphyrin (Por) attachment. (More about HNO trapping and detection you can read in other chapters in this book.)

In general, hydroxylamine (NH_2OH), L-arginine and its derivative N-hydroxy-L-arginine, as well as $NO^{\bullet}$, could serve as probable substrates for the physiological production of HNO, whereas *S*-nitrosothiols (RSNO) might also be considered. Number of enzymatic and nonenzymatic processes that utilize these substrates, among others, and result in HNO will be also reviewed in other chapters in this book. Oxidative pathways for heme protein-mediated HNO production from various nitrogen-containing substrates have been previously summarized in the literature[30] and will be tackled here as well, whereas a special attention will be paid to reductive pathways.

Ivana Ivanović-Burmazović
Friedrich-Alexander University Erlangen-Nuremberg,
Erlangen, Germany

References

1. Allen CW. *Astrophysical quantities*. 2nd ed. London: Athlone Press; 1963.
2. Vardya MS. March with depth of molecular abundances in the outer layers of K and M stars. *Mon Notic Roy Astron Soc* 1966;**134**:347–70.
3. (a) Ulich BL, Hollis JM, Snyder LE. Radio detection of nitroxyl (HNO): the first interstellar NO bond. *Astrophys J Lett* 1977;**217**:L105–8.
   (b) Pickles JB, Williams DA. Interstellar nitroxyl. *Nature* 1978;**271**:335–6.
4. Turner BE. A molecular line survey of Sagittarius B2 and Orion-KL from 70 to 115 GHz. II - Analysis of the data. *Astrophys J Suppl Ser* 1991;**76**:617–86.
5. Herbst E. Die Chemie des interstellaren Raumes. *Angew Chem* 1990;**102**:627–41.
6. Hughes MN, Cammack R. Synthesis, chemistry, and applications of nitroxyl ion releasers sodium trioxodinitrate or Angeli's salt and Piloty's acid. *Meth Enzymol* 1999; **301**:279–87.
7. Takagi K, Saito S. Microwave spectrum of DNO. *J Mol Spectrosc* 1972;**44**:81–7.

8. Harteck P. Die Darstellung von HNO bzw. $[HNO]_n$ (The preparation of HNO or $[HNO]_n$). *Berichte der Deutschen Chemischen Gesellschaft* 1933;**66B**:423–6.
9. (a) Nagel E. Über den Verlauf der katalytischen Oxydation von Ammoniak. *Ztschr Elektrochem* 1930;**36**:754–6.
   (b) Bodenstein M. Versuch einer Theorie der katalytischen Ammoniakverbrennung. *Ztschr Angew Chem* 1927;**40**:174–7.
   (c) Andrussow I. Über die katalytische Ammoniakoxydation.III. *Ztschr Angew Chem* 1927;**40**:166–74.
10. Smallwood HM. The rate of recombination of atomic hydrogen. *J Am Chem Soc* 1929;**51**:1985–99.
11. Angeli A, Angelico F. Reactions of nitroxyl [NOH]. *Atti della Accademia Nazionale dei Lincei, Classe di Scienze Fisiche, Matematiche e Naturali, Rendiconti* 1901;**10**(v):164–8.
12. Angeli A, Angelico F. New reactions of nitroxyl (dihydroxyammonia). *Gazz Chim Ital* 1905;**35**(i):152–9.
13. Dalby FW. The spectrum and structure of the HNO molecule. *Can J Phys* 1958;1336–71.
14. Brown HW, Pimentel GC. Photolysis of nitromethane and of methyl nitrite in an argon matrix; Infrared detection of nitroxyl, HNO. *J Chem Phys* 1958;**29**:883–8.
15. (a) Clyne MAA, Thrush BA. Reaction of hydrogen atoms with nitric oxide. *Trans Faraday Soc* 1961;**57**:1305–14.
   (b) Clyne MAA, Thrush BA. Mechanism of chemiluminescent reactions involving nitric oxide-the H + NO reaction. *Disc Faraday Soc* 1962;**33**:139–48.
   (c) Ibaraki T, Kusunoki I, Kodera K. Study of chemiluminescence by means of crossed beams: nitric oxide-hydrogen atom system. *Chem Letters* 1973;**2**:317–20.
16. Patrick R, Golden DM. Kinetics of the reactions of NH_2 Radicals with O_3 and O_2. *J Phys Chem* 1984;**88**:491–5.
17. Keçeli M, Shiozaki T, Yagi K, Hirata S. Anharmonic vibrational frequencies and vibrationally-averaged structures of key species in hydrocarbon combustion: HCO^+, HCO, HNO, HOO, HOO^-, CH_3^+, and CH_3. *Mol Phys* 2009;**107**:1283–301.
18. (a) Shoun H, Fushinobu S, Jiang L, Kim S-W, Wakagi T. Fungal denitrification and nitric oxide reductase cytochrome $P450_{nor}$. *Phil Trans R Soc B* 2012;**367**:1186–94.
   (b) Doctorovich F, et al. Azanone (HNO) interaction with hemeproteins and metalloporphyrins. In: van Eldik R, Olabe J, editors. *Adv Inorg Chem*, vol. 64. Amsterdam: Elsevier; 2012. p. 97–139.
   (c) Daiber A, et al. *Nitric Oxide reductase ($P450_{no}r$) from* Fusarium oxysporum. *Chapter 14. The Smallest biomolecules: diatomics and their interactions with heme proteins*. Amsterdam: Elsevier; 2008. p. 354–77.
   (d) Sulc F, et al. Bioinorganic chemistry of the HNO ligand. In: Ghosh A, editor. *The smallest biomolecules: diatomics and their interactions with heme proteins.* Amsterdam: Elsevier; 2008. p. 429–62 (Chapter 16).
   (e) Lehnert N, et al. *Structure and bonding in heme–nitrosyl complexes and implications for biology. Structure and bonding*, vol. 154. Berlin, Heidelberg: Springer-Verlag; 2014. p. 155–224.
19. van der Maas P, van de Sandt T, Klapwijk B, Lens P. Biological reduction of nitric oxide in aqueous Fe(II)EDTA solutions. *Biotechnol Prog* 2003;**19**:1323–8.
20. (a) Flores-Santana W, Salmon DJ, Donzelli S, Switzer CH, Basudhar D, Ridnour L, et al. The specificity of nitroxyl chemistry is unique among nitrogen oxides in biological systems. *Antioxid Redox Signal* 2011;**14**(9):1659–74.

(b) Paolocci N, Wink DA. The shy Angeli and his elusive creature: the HNO route to vasodilation. *Am J Physiol Heart Circ Physiol* 2009;**296**:H1217–20.

(c) Bullen ML, Miller AA, Andrews KL, Irvine JC, Ritchie RH, Sobey CG, et al. Nitroxyl (HNO) as a vasoprotective signaling molecule. *Antioxid Redox Signal* 2011;**14**(9):1675–86.

(d) Ellis A, Li CG, Rand MJ. Differential actions of L-cysteine on responses to nitric oxide, nitroxyl anions and EDRF in the rat aorta. *Br J Pharmacol* 2000;**129**(2):315–22.

(e) Fukuto JM, Carrington SJ. HNO signalling mechanisms. *Antioxid Redox Signal* 2011;**14**(9):1649–57.

(f) Tocchetti CG, Stanley BA, Murray CI, Sivakumaran V, Donzelli S, Mancardi D, et al. Playing with cardiac "redox switches": the "HNO way" to modulate cardiac function. *Antioxid Redox Signal* 2011;**14**(9):1687–98.

21. (a) Dombek MG. Multiconfigurational electronic wavefunctions in the full optimized reaction space: the isomerization of nitrosyl hydride to nitrogen hydroxide in the lowest singlet and triplet states. *Retrospective Theses and Dissertations* 1977;6065.

(b) Guadagnini R, Schatz GC, Walch SP. Global potential energy surfaces for the lowest $^{1}A'$, $^{3}A'$, and $^{1}A'$ states of HNO. *J Chem Phys* 1995;**102**:774–6783.

(c) Mordaunt DH, Flöthmann H, Stumpf M, Keller H-M, Beck C, Schinke R, et al. The dissociation of HNO. I. Potential energy surfaces for the $\tilde{X}^{1}A'$, $\tilde{A}1A'$, and $\tilde{a}3A'$ states. *J. Chem. Phys.* 1997;**107**:6603–15.

(d) Shafirovich V, Lymar SV. Nitroxyl and its anion in aqueous solutions: spin states, protic equilibria, and reactivities toward oxygen and nitric oxide. *Proc Natl Acad Sci USA* 2002;**99**:7340–5.

(e) Maier G, Reisenauer HP, De Marco M. Isonitroso hydrogen (hydroxy nitrene, HON). *Angew Chem Int Ed* 1999;**38**:108–10.

(f) Bozkaya U, Turney JM, Yamaguchi Y, Schaefer 3rd HF. The lowest-lying electronic singlet and triplet potential energy surfaces for the HNO–NOH system: energetics, unimolecular rate constants, tunneling and kinetic isotope effects for the isomerization and dissociation reactions. *J Chem Phys* 2012;**136**:164303–15.

(g) Bartberger MD, Fukuto JM, Houk KN. On the acidity and reactivity of HNO in aqueous solution and biological systems. *PNAS* 2001;**98**:2194–8.

(h) Bartberger MD, Liu W, Ford E, Miranda KM, Switzer C, Fukuto JM, et al. The reduction potential of nitric oxide (NO) and its importance to NO biochemistry. *Proc Natl Acad Sci USA* 2002;**99**:10958–63.

22. Zhang Y. Computational investigations of HNO in biology. *J Inorg Biochem* 2013;**118**:191–200.

23. Jursic BS. Complete basis set and gaussian computational study of bond dissociation energies, enthalpy of formation and rearrangement barriers for the XNO nitric oxide derivatives. *J Mol Struct (THEOCHEM)* 1999;**492**:35–43.

24. (a) Ishiwata T, Akimoto H, Tanaka I. Chemiluminescent spectra of HNO and DNO in the reaction of O (^{3}P)/O_2 with NO and hydrocarbons or aldehydes. *Chem Phys Lett* 1973;**21**:322–5.

(b) Ishiwata T, Akimoto H, Tanaka I. *Chem Phys Lett* 1974;**27**:260–3.

25. Tronc M, Huetz A, Landau M, Pichou F, Reinhardt J. Resonant vibrational excitation of the NO ground state by electron impact in the 0.1–3 eV energy range. *J Phys B* 1975;**8**:1160–9.

26. Gratzel M, Taniguchi S, Henglein A. A pulse radiolytic study of short-lived byproducts on nitric oxide reduction in aqueous solution. *Ber Bunsen-Ges Phys Chem* 1970;**74**:1003–10.

27. Stanbury DM. Reduction potentials involving inorganic free radicals in aqueous solution. In: Sykes AG, editor. *Adv Inorg Chem*, Vol. 33. Amsterdam: Elsevier; 1989. p. 69–138.
28. Armstrong DA, Huie RE, Koppenol WH, Lymar SV, Merényi G, Neta P, et al. Standard electrode potentials involving radicals in aqueous solution: inorganic radicals. *Pure App Chem* 2015;**87**:1139–50.
29. (a) Wrobel AT, Johnstone TC, Deliz Liang A, Lippard SJ, Rivera-Fuentes P. A fast and selective near-infrared fluorescent sensor for multicolor imaging of biological nitroxyl (HNO). *J Am Chem Soc* 2014;**136**:4697–705.
    (b) Rosenthal J, Lippard SJ. Direct detection of nitroxyl in aqueous solution using a tripodal copper(II) bodipy complex. *J Am Chem Soc* 2010;**132**:5536–7.
    (c) Loas A, Radford RJ, Deliz Liang A, Lippard SJ. Solid-phase synthesis provides a modular, lysine-based platform for fluorescent discrimination of nitroxyl and biological thiols. *Chem Sci* 2015;**6**:4131–40.
30. Reisz JA, Bechtold E, King SB. Oxidative heme protein-mediated nitroxyl (HNO) generation. *Dalton Trans* 2010;**39**:5203–12.

HNO: Redox Chemistry and Interactions With Small Inorganic Molecules

1

M. Hamer, M.A. Morales Vásquez and F. Doctorovich
Universidad de Buenos Aires and CONICET, Buenos Aires, Argentina

Chapter Outline

1.1 Introduction

In this chapter, we will briefly describe the fundamental chemistry of ^1HNO and $^3NO^-$ making a comparison with NO reactivity. Rate constants for the reactions mentioned below are summarized in Table 1.1.[1,2]

Despite being a three atoms molecule, the chemical characteristics/properties of HNO are complex and less straightforward as compared to NO.

One of the intriguing properties of HNO involves its acid–base chemistry. HNO is a weak acid with an accepted pK_a of 11.4[3,11] and a singlet ground state.[12,13] However, the deprotonation of ^1HNO generates the corresponding anion $^3NO^-$, which has a triplet ground state,[14,15] analogous to the isoelectronic O_2 molecule. Consequently, the loss of a proton from HNO is not a simple acid/base equilibrium but a spin-forbidden slow deprotonation (Table 1.1, Eq. 1.1)[3]:

$$^1HNO + OH^- \rightleftarrows {}^3NO^- + H_2O \tag{1.1}$$

Thus, as HNO acid–base equilibrium species have different electronic spin states, both deprotonation of ^1HNO (Table 1.1, Eq. 1.1) and reprotonation of $^3NO^-$ (Table 1.1; Eq. 1.2) are very slow. Taking this into account, in biological media (pH = 7.4) ^1HNO is likely to be the exclusive species present. And in case that $^3NO^-$ is produced, it may have a long enough lifetime to be chemically relevant (milliseconds), even under acidic conditions, since the protonation rate for this anion is slow.

The Chemistry and Biology of Nitroxyl (HNO). DOI: http://dx.doi.org/10.1016/B978-0-12-800934-5.00001-3

Table 1.1 Rate constants for reactions of azanone, azanone anion and nitric oxide with biologically relevant small molecules

Eq. #	Reaction	Rate constant[a]	References
1.1	$^1HNO + OH^- \rightarrow {}^3NO^- + H_2O$	$4.9 \times 10^4\ M^{-1}\ s^{-1}$	[3]
1.2	$^3NO^- + H_2O \rightarrow {}^1HNO + OH^-$	$1.2 \times 10^2\ s^{-1}$	[3]
1.3	$^1HNO + {}^1HNO \rightarrow HONNOH$	$8.0 \times 10^6\ M^{-1}\ s^{-1}$	[3]
	$HONNOH \rightarrow N_2O + H_2O$	$5.0 \times 10^{-4}\ s^{-1}$	[4]
1.4	$^3NO^- + {}^1HNO \rightarrow N_2O + OH^-$	$6.6 \times 10^9\ M^{-1}\ s^{-1}$	[5]
1.6	$O_2 + HNO \rightarrow ONOOH$	$1.8 \times 10^4\ M^{-1}\ s^{-1}$	[3,6]
1.7	$^3NO^- + O_2 \rightarrow ONOO^-$	$2.7 \times 10^9\ M^{-1}\ s^{-1}$	[3]
1.8	$NO + O_2^- \rightarrow ONOO^-$	$4–7 \times 10^9\ M^{-1}\ s^{-1}$	[7,8]
1.9	$2\ NO + O_2 \rightarrow 2\ NO_2$	$2.54 \times 10^6\ M^{-2}\ s^{-1}$	[9]
1.12	$^1HNO + NO \rightarrow N_2O_2^- + H^+$	$5.8 \times 10^6\ M^{-1}\ s^{-1}$	[10]
1.13	$^3NO^- + NO \rightarrow N_2O_2^-$	$3.0 \times 10^9\ M^{-1}\ s^{-1}$	[10]
1.14	$H_2S + HNO \rightarrow RS\text{-}NHOH$	$1.2 \times 10^6\ M^{-1}\ s^{-1}$	[6]

[a]Rate constants are given at room temperature and pH = 7.

1.2 Dimerization

HNO chemistry is significantly difficult to describe due to its high reactivity. HNO spontaneously dimerizes with a second-order rate constant of c.10^7 to produce hyponitrous acid which finally dehydrates to give nitrous oxide (Table 1.1, Eq. 1.3).[3,11,16,17] As a result, detection of N_2O as an end product might serve as a marker for the presence of HNO.

$$2\ HNO \rightarrow HONNOH \rightarrow N_2O + H_2O \tag{1.3}$$

From pH 7.5 to 10.5, where $HN_2O_2^-$ ($pK_a = 10.9$) is the major species in solution, the decomposition rate of $HN_2O_2^-$ exhibits a plateau with $k_{(25°)} = 5.0 \times 10^{-4}\ s^{-1}$ (half-life 23 min).[4] Outside this range of pH, the decomposition rate becomes slower up to values near $10^{-6}\ s^{-1}$. Below pH = 5, in absence of radical scavengers, hyponitrous acid can decompose by a radical chain mechanism producing N_2 and NO_3^-.[18] Therefore, it must be taken in mind that, below pH = 5, ethanol or other radical scavengers should be added to the reaction mixtures in order to avoid complications involved with the radical chain mechanism.

Recently, a new mechanism for HNO dimerization in aqueous solution has been described.[19] The developed DFT studies showed a detailed initial stage of the reaction mechanism suggesting that N_2O formation is dominated by initial formation of the *cis*-HNO dimer, followed by rapid proton transfer reactions and decomposition of the *cis*-hyponitrite anion. Based on the calculated pK_a values for the acid–base equilibrium of the *cis*-hyponitrous acid, under physiological

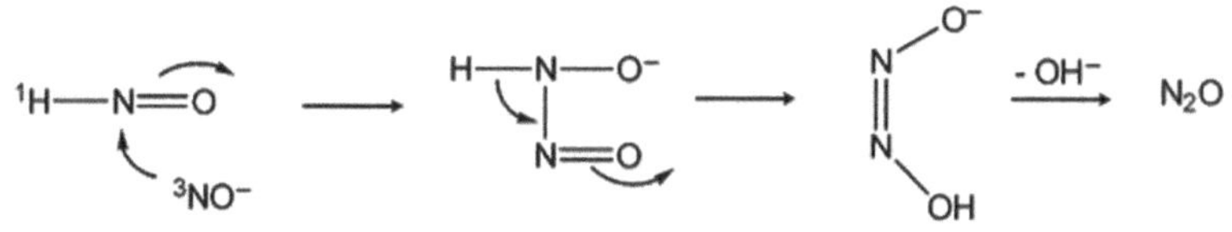

Scheme 1.1 Nucleophilic addition of nitroxyl anion to the N atom of ^{1}HNO.

conditions N_2O formation is not determined by the generally thought decomposition of the *cis*-HONNOH but by the decomposition of the *cis*-$N_2O_2H^-$ anion.

Concerning $^3NO^-$, although it has been suggested that it dimerizes with $k > 8 \times 10^6\ M^{-1}\ s^{-1}$,[19] there is no experimental evidence to support this claim. According to Bonner and coworkers, "Coulomb barrier considerations lead one to expect inhibition of the dimerization reaction with increasing deprotonation of HNO."[20] However, it has been recently reported by Lymar and coworkers that the spin-forbidden reaction of $^3NO^-$ and ^{1}HNO takes place with a rate constant of $6.6 \times 10^9\ M^{-1}\ s^{-1}$ (Eq. 1.4).[5]

$$^3NO^- + {}^1HNO \rightarrow N_2O + OH^- \tag{1.4}$$

It should be noted that this result is based on indirect kinetic observations. However, it is a surprising fact—even for the authors—that a spin-forbidden reaction would be so fast. According to Lymar, this could be due to the very large driving force for this reaction (with an estimated ΔG of $-80\ kcal\ mol^{-1}$). Although $^3NO^-$ is isoelectronic with O_2, it is expected to be more nucleophilic due to its negative charge. Consequently, reaction 1.4 could be thought of as a nucleophilic addition of nitroxyl anion to the N atom of ^{1}HNO to produce the intermediate $HONNO^-$ (after electronic rearrangement and a 1,2 H shift), which decomposes to produce N_2O (Scheme 1.1).

Although $^1HN=O$ could be expected to suffer this type of nucleophilic attack in a similar way to carbonyl ($R_2C=O$) compounds, this particular reaction is complicated by the different spin states of azanone. Possibly, further theoretical studies are needed to understand the reaction details. Contrasting ^{1}HNO, NO has little tendency to dimerize to $(NO)_2$ with a small equilibrium constant $K = 8.360 \times 10^{-2}$ (120 K), and therefore is rather stable as a monomer in solution.[20]

1.3 Reaction with molecular oxygen

The reaction of ^{1}HNO with O_2 has been studied in the gas phase[21] and is rather slow, mainly due to their different spin states, $k \approx 10^3\ M^{-1}s^{-1}$ (Table 1.1).

Initially, the product of the reaction between ^{1}HNO and O_2 was experimentally unknown. Therefore, it was proposed that it proceeded through Eq. (1.5), leading to

NO and a radical hydroperoxy species with a rate constant $k = 3\text{–}8 \times 10^3\ M^{-1}\ s^{-1}$,[3] since this reaction is not only energetically favorable but is also spin allowed, while the direct formation of peroxynitrite is a spin-forbidden reaction (Eq. 1.6).

$$^1HNO + O_2 \rightarrow NO^\bullet + HO_2^\bullet \quad (1.5)$$

$$O_2 + HNO \rightarrow ONOOH \quad (1.6)$$

A number of studies on the reactivity of HNO with molecular oxygen have been reported[22–24] and the peroxynitrite formation remained skeptical and understudied. Recently, $ONOO^-$ formation by reaction of HNO with O_2 was identified and evidenced with a $1.8 \times 10^4\ M^{-1}s^{-1}$ second-order rate constant, by Sikora and collaborators.[6] However, peroxynitrite can be formed from the products of reaction 5.

Besides, $^3NO^-$ also reacts with O_2 at a second-order rate to produce peroxynitrite (Eq. 1.7),[3]

$$^3NO^- + O_2 \rightarrow ONOO^- \quad (1.7)$$

in a reaction isoelectronic with the second-order reaction 1.8.[7,8,19]

$$NO + O_2^- \rightarrow ONOO^- \quad (1.8)$$

On the other hand, nitric oxide reacts with O_2 following third-order kinetics, and at a slower rate. The mechanism of this reaction is shown in a simplified manner in Eqs. (1.9–1.11).[9]

$$2\ NO + O_2 \rightarrow\rightarrow\rightarrow 2\ NO_2 \quad (1.9)$$

$$NO_2 + NO \rightarrow N_2O_3 \quad (1.10)$$

$$N_2O_3 + H_2O \rightarrow 2\ NO_2^- + 2\ H^+ \quad (1.11)$$

1.4 Reaction with NO

Both ^1HNO and $^3NO^-$ react with NO, with quite distinct second-order rate constants which differ by three orders of magnitude favoring reaction with $^3NO^-$ (Table 1.1, Eqs. 1.12 and 1.13)[10]:

$$^1HNO + NO \rightarrow N_2O_2^- + H^+ \quad (1.12)$$

$$^3NO^- + NO \rightarrow N_2O_2^- \quad (1.13)$$

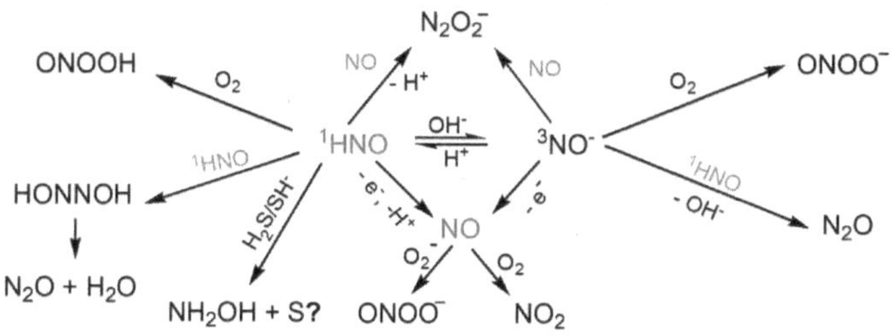

Scheme 1.2 Reactions of nitroxyl and nitric oxide.

The resulting $N_2O_2^-$ radical extremely rapidly acquires another NO molecule, producing the closed shell $N_3O_3^-$ anion, which decays to the final products $N_2O + NO_2^-$ with a rate constant of ca. 300 s^{-1}.

A summary of the reactions of NO and HNO/NO^- with small molecules is shown in Scheme 1.2.[1,2]

1.5 Reaction with H_2S

The electrophilic reactivity of HNO[25,26] is selective to the nucleophiles it reacts with. Theoretical analysis foretells that it does not react either with water or alcohols[26] but it is very reactive toward thiols.

One of the earliest reports related to the reactivity of thiols (R-SH) with HNO was by Doyle et al.[25] They suggested that the reaction of thiophenol with HNO generates N-hydroxysulfenamide (Eq. 1.14), as an unstable intermediate, via attack of a nucleophilic sulfur atom on the electrophilic nitrogen atom of HNO. Depending on the disposal of other reactive thiols, N-hydroxysulfenamide can react along two different paths. In absence of another thiol, this intermediate suffers rearrangement to the sulfinamide (Eq. 1.16). However, in presence of excess of thiol, the final products are disulfide and NH_2OH (Eq. 1.15). Later, Wong et al.[27] and Nagasawa and coworkers[28] supported the hypothesis of this intermediate, although this species has been yet isolated and characterized. It is worth mentioning that HNO can oxidize the sulfur atom by four electrons in a single sequence when the sulfinamide is formed.[29]

$$\text{RSH} + \text{HNO} \rightarrow \text{RS-NHOH} \quad (1.14)$$

$$\text{RSNH} + \text{R'SH} \rightarrow \text{RSSR'} + \text{NH}_2\text{OH} \quad (1.15)$$

$$\text{RS-NHOH} \rightarrow \text{RS(O)NH}_2 \quad (1.16)$$

Although there are many examples of HNO reactivity with thiols, little is known about the reaction between HNO and H_2S. Recently, Sikora and coworkers[6] have reported the kinetic parameters and a k value of $1.2 \times 10^6\ M^{-1}\ s^{-1}$ was postulated for this reaction. The products were not identified, but NH_2OH and sulfur or polysulfides could be expected as the reaction products.

1.6 Redox-related reactions

The reduction of NO to nitroxyl was also studied and the accepted standard reduction potential for the $NO/^3NO^-$ couple was assigned as -0.8 V (Table 1.2, Eq. 1.17).[11] At physiological pH, ^1HNO is expected to be the main nitroxyl-related species, which has an estimated $E^0_{(NO,H^+/^1HNO)} \approx -0.14$ V (Table 1.2, Eq. 1.18), becoming -0.55 V at pH 7 (all values against NHE).[3] This single-electron potential implicate that the one-electron oxidation process for HNO is feasible under biological conditions, although close to the lower limit of the biological reductants. On the other hand, the one-electron potential for the $HNO,H^+/H_2NO\cdot$ couple (Table 1.2, Eq. 1.19) is within the physiological range[30] and it is a possible fate for HNO.

Liochev and Fridovich had discussed the possibility of the production of NO^-/HNO within living cells.[32] They argued that the redox potential could be much less negative because the ratio [NO]:[HNO] could be high and the log of this ratio influences the actual redox potential, according with the Nernst electrode equation. The calculated redox potential $E^{+1}_{(NO,H/HNO)}$ at millimolar and picomolar concentrations of NO and HNO, respectively, is -0.05 V at pH = 7.4 and $+0.10$ V at pH = 5.

However, it is currently a hot matter of discussion if NO could be reduced to $^3NO^-$ or ^1HNO under these conditions, since the reduction potentials mentioned above (-0.8 and -0.55 V) are in the limit of biological reducing agents and other species which are present in higher concentrations, such as O_2, are more favorable to be reduced. It is important to stress that the above-mentioned reduction potential of -0.55 V is the result of an estimation, since it cannot be measured directly due to the irreversibility of the NO•/HNO redox couple. This assessment was done by Lymar and coworkers,[3] assuming $\Delta G_f^\circ{}_{(HNO(aq))} \sim 115\ kJ\ mol^{-1}$. However, it is important to note that if this value is overestimated by just a few $kJ\ mol^{-1}$ (20–30), the NO•/HNO redox couple would be much closer to the thermodynamic range of many biological reductants. Moreover, it has been recently shown that in

Table 1.2 Reduction potentials for one-electron reductions of NO to azanone and its anion

Eq. #	Reaction	Potential vs NHE	References
1.17	$NO + e^- \rightarrow {}^3NO^-$	$E^\circ < -0.8$ V	[31]
1.18	$NO + H^+ + e^- \rightarrow {}^1HNO$	$E^\circ \approx -0.14$ V	[3]
1.19	$NOH + H^+ + e^- \rightarrow H_2NO\cdot$	$E^\circ \approx 0.52$ V	[30]

reducing environments, such as those containing high concentrations of ascorbate (vitamin C), HNO can be produced from endogenous NO.[33,34] In such environments, the concentration of O_2 or other species prone to be reduced by ascorbate should be at a minimum. On the other hand, in oxidizing environments containing high levels of O_2, oxidative conversion of HNO/$^3NO^-$ to NO should be easily accessible. Therefore, it can be assumed that NO and HNO can be interconverted in biological media, depending on the redox state of the environment. From a general redox perspective, oxygen-rich and oxidizing environments will promote $NO^•$ conversion to nitrite and nitrate, peroxynitrite, NO_2, and N_2O_3. On the other hand, hypoxic and reducing environments are expected to lead to HNO. For example, NO· has been shown to have significant effects on mitochondrial respiration, mainly through the inhibition of cytochrome *c* oxidase.[35] Inhibition of cytochrome *c* oxidase could lead to accumulation of the reduced quinone pool (hydroquinone-like compounds), creating an ideal situation for NO· to HNO conversion, especially if oxygen tension is low. Also, myoglobin and other heme proteins in the absence of oxygen can reduce the abundant nitrate and nitrite ions to $NO^•$, producing it in a hypoxic environment, where its conversion to HNO could proceed using the available pool of circulating antioxidants, such as vitamin C.

References

1. Doctorovich F, Bikiel D, Pellegrino J, Suárez SA, Larsenc A, Martí MA. Nitroxyl (azanone) trapping by metalloporphyrins. *Coord Chem Rev* 2011;**255**(23–24):2764–84.
2. Doctorovich F, Bikiel DE, Pellegrino J, Suárez SA, Martí MA. Azanone (HNO) interaction with hemeproteins and metalloporphyrins. In: Chemistry, R. van E. and I.I.-B. B.T.-A. in I, editors. *Inorganic/bioinorganic reaction mechanisms*. Academic Press; 2012. p. 97–139, Chapter 4.
3. Shafirovich V, Lymar SV. Nitroxyl and its anion in aqueous solutions: spin states, protic equilibria, and reactivities toward oxygen and nitric oxide. *Proc Natl Acad Sci USA* 2002;**99**(11):7340–5.
4. Buchholz JR, Powell RE. The decomposition of hyponitrous acid. I. The non-chain reaction. *J Am Chem Soc* 1963;**85**(5):509–11.
5. Lymar SV, Shafirovich V. Photoinduced release of nitroxyl and nitric oxide from diazeniumdiolates. *J Phys Chem B* 2007;**111**(24):6861–7.
6. Smulik R, Dębski D, Zielonka J, Michałowski B, Adamus J, Marcinek A, et al. Nitroxyl (HNO) reacts with molecular oxygen and forms peroxynitrite at physiological pH: biological implications. *J Biol Chem* 2014;**289**(51):35570–81.
7. Goldstein S, Czapski G. The reaction of $NO^•$with $O_2^•$– and HO_2: a pulse radiolysis study. *Free Radic Biol Med* 1995;**19**(4):505–10.
8. Huie RE, Padmaja S. The reaction of NO with superoxide. *Free Radic Res Commun* 1993;**18**(4):195–9.
9. Goldstein S, Czapski G. Mechanism of the nitrosation of thiols and amines by oxygenated NO solutions: the nature of the nitrosating intermediates. *J Am Chem Soc* 1996;**118**(14):3419–25.

10. Lymar SV, Shafirovich V, Poskrebyshev GA. One-electron reduction of aqueous nitric oxide: a mechanistic revision. *Inorg Chem* 2005;**44**(15):5212–21.
11. Bartberger MD, Liu W, Ford E, Miranda KM, Switzer Ch, Fukuto JM, et al. The reduction potential of nitric oxide (NO) and its importance to NO biochemistry. *Proc Natl Acad Sci USA* 2002;**99**(17):10958–63.
12. Orgel LE. The electronic structures and spectra of some molecules related to oxygen. *J Chem Soc* 1953;1276.
13. Walsh AD. The electronic orbitals, shapes, and spectra of polyatomic molecules. Part III. HAB and HAAH molecules. *J Chem Soc* 1953;2288.
14. Szmytkowski C, Maciag K. Total cross section for electron impact on nitrogen monoxide. *J Phys B At Mol Phys* 1991;4273.
15. Tennyson J, Noble C. Low-energy electron scattering by the NO molecule. *J Phys B At Mol Phys* 1986;**19**:4025.
16. Kohout FC, Lampe FW. On the role of the nitroxyl molecule in the reaction of hydrogen atoms with nitric oxide. *J Am Chem Soc* 1965;**87**:5795.
17. Smith PAS, Hein GE. The alleged role of nitroxyl in certain reactions of aldehydes and alkyl halides. *J Am Chem Soc* 1960;**82**(21):5731–40.
18. Buchholz JR, Powell RE. The decomposition of hyponitrous acid. II. The chain reaction. *J Am Chem Soc* 1965;**87**(11):2350–3.
19. Miranda KM. The chemistry of nitroxyl (HNO) and implications in biology. *Coord Chem Rev* 2005;**249**(3–4):433–55.
20. Smith AL, Johnston HL. The magnetic susceptibility of liquid nitric oxide and the heat of dissociation of $(NO)_2$. *J Amer Chem Soc* 1952;12–14.
21. Bryukov MG, Kachanov AA, Timonnen R, Seetula J, Vandoren J, Sarkisov OM. Kinetics of HNO reactions with O_2 and HNO. *Chem Phys Lett* 1993;**208**(5–6):392–8.
22. Miranda KM, Espey MG, Yamada K, Krishna M, Ludwick N, Kim SM, et al. Unique oxidative mechanisms for the reactive nitrogen oxide species, nitroxyl anion. *J Biol Chem* 2001;**276**(3):1720–7.
23. Kirsch M, de Groot H. Formation of peroxynitrite from reaction of nitroxyl anion with molecular oxygen. *J Biol Chem* 2002;**277**(16):13379–88.
24. Miranda KM, Yamada K, Espey MG, Thomas DD, DeGraff W, Mitchell JB, et al. Further evidence for distinct reactive intermediates from nitroxyl and peroxynitrite: effects of buffer composition on the chemistry of Angeli's salt and synthetic peroxynitrite. *Arch Biochem Biophys* 2002;**401**(2):134–44.
25. Doyle MP, Mahapatro SN, Broene RD, Guy JK. Oxidation and reduction of hemoproteins by trioxodinitrate (II). The role of nitrosyl hydride and nitrite. *J Am Chem Soc* 1988;**110**(2):593–9.
26. Bartberger MD, Fukuto JM, Houk KN. On the acidity and reactivity of HNO in aqueous solution and biological systems. *Proc Natl Acad Sci USA* 2001;**98**(5):2194–8.
27. Wong PS-Y, Hyun J, Fukuto JM, Shirota FN, DeMaster EG, Shoeman DW, et al. Reaction between S-nitrosothiols and thiols: generation of nitroxyl (HNO) and subsequent chemistry. *Biochemistry* 1998;**37**(16):5362–71.
28. Shoeman DW, Shirota FN, DeMaster EG, Nagasawa HT. Reaction of nitroxyl, an aldehyde dehydrogenase inhibitor, with N-acetyl-L-cysteine. *Alcohol* 2000;**20**(1):55–9.
29. Fukuto JM, Carrington SJ, Tantillo DJ, Harrison JG, Ignarro LJ, Freeman BA, et al. Small molecule signaling agents: the integrated chemistry and biochemistry of nitrogen oxides, oxides of carbon, dioxygen, hydrogen sulfide, and their derived species. *Chem Res Toxicol* 2012;**25**(4):769–93.

30. Lind J, Merényi G. Kinetic and thermodynamic properties of the aminoxyl (NH_2O•) radical. *J Phys Chem A* 2006;**110**(1):192–7.
31. Benderskii VA, Krivenko AG, Ponomarev EA. Homogenous and electrode reactions of NO_2^{2-} ions. *Sov Electrochem Engl Tr* 1989;**25**:154–61.
32. Liochev SI, Fridovich I. The mode of decomposition of Angeli's salt ($Na_2N_2O_3$) and the effects thereon of oxygen, nitrite, superoxide dismutase, and glutathione. *Free Radic Biol Med* 2003;**34**(11):1399–404.
33. Suarez SA, Neuman NI, Muñoz M, Álvarez L, Bikiel DE, Brondino CD, et al. Nitric oxide is reduced to HNO by proton-coupled nucleophilic attack by ascorbate, tyrosine, and other alcohols. A new route to HNO in biological media? *J Am Chem Soc* 2015;**137**(14):4720–7.
34. Hamer M, Suarez SA, Neuman NI, Álvarez L, Muñoz M, Marti MA, et al. Discussing endogenous NO•/HNO interconversion aided by phenolic drugs and vitamins. *Inorg Chem Forum* 2015;**54**:9342–50.
35. Schumacker PT. Nitric oxide quenches the fire in heart mitochondria. *Nat Med* 2013;**19**(6):666–7.

HNO Donors: Angeli's Salt and Related Diazeniumdiolates

2

D. Basudhar[1], G. Bharadwaj[2], D.J. Salmon[3] and K.M. Miranda[4]
[1]National Cancer Institute, Frederick, MD, United States, [2]University of California, Davis School of Medicine, Sacramento, CA, United States, [3]University of Minnesota, Minneapolis, MN, United States, [4]University of Arizona, Tucson, AZ, United States

Chapter Outline

2.1 Introduction

As a class, small, inorganic nitrogen-containing compounds are characterized by extremely diverse chemical properties. For example, while molecular nitrogen is often categorized with the noble gases in terms of inertness, other species such as ammonium nitrate and hydrazine can be explosive. This variety has significant impacts in the biosphere as well as industrially and has led to an extensive literature. Nonetheless, the discovery in the 1980s of enzymatic biosynthesis of nitric oxide (NO)[1,2] has placed a central focus on this congener. Within the context of

The Chemistry and Biology of Nitroxyl (HNO). DOI: http://dx.doi.org/10.1016/B978-0-12-800934-5.00002-5

physiology and pathology, it is common to consider the impact of conversion of NO into other nitrogen oxides as well as to compare the effects of NO production to that of other related molecules.[3]

Interestingly, there is an accumulating literature that suggests that nitroxyl (HNO) may also be biosynthesized (see Ref. 4). However at present, only the pharmacological effects of HNO are definitive. Spontaneous interchange between HNO and NO is impeded by both thermodynamic and kinetic constraints,[5,6] suggesting a related but orthogonal response paradigm.[4,7–9] It is therefore often interesting to compare the effects of HNO donors to those of NO.

It is noteworthy that HNO has a far broader reactivity than its free radical redox cousin NO.[4,8] The higher reactivity of HNO limits its lifetime and spatial influence but broadens the target profile compared to NO. Additionally, unlike NO, HNO rapidly dimerizes. Although dimers can exhibit unique chemistry from the constituent monomers, this process perhaps would not be inherently significant. However, the HNO dimer hyponitrous acid is susceptible to rapid dehydration[10,11] ($8 \times 10^6\ M^{-1}\ s^{-1}$)[12] (Eq. 2.1).

$$2HNO \rightarrow [HONNOH] \rightarrow N_2O + H_2O \tag{2.1}$$

This irreversible consumption pathway has several important consequences, including severely complicating direct detection of HNO and necessitating in situ generation.

Although NO can be stored and also can be produced enzymatically, donor compounds[13–16] are often used as a matter of convenience as well as to reduce experimental artifacts. The ability to release NO with controlled rates in simulation of NO biosynthesis by NO synthase is also an advantage. The availability of NO donors has in fact been vital to the elucidation of the chemical biology of NO. Consequently, the need to produce HNO in situ (due to self-quenching by Eq. 2.1) is not inherently an experimental limitation.

In similarity to NO, the most common methods to generate HNO involve thermal or photochemical decomposition of donor compounds. The growing interest in HNO biology and pharmacology since the turn of the century[4,13,17–21] has led to expansion of existing donor classes and development of new donor species.[22,23] To date, a majority of studies on HNO have utilized Angeli's salt ($Na_2N_2O_3$, sodium trioxodinitrate), the synthesis of which was first published in 1896.[24] Angeli's salt has historically been and continues to be the reagent of choice for generating HNO in aqueous solution.[22]

Use of Angeli's salt offers a number of advantages, three of which are commercial availability, well-characterized decomposition, and ready comparison to related NO donors. In addition, Angeli's salt is highly water-soluble, is stable as a solid and in alkaline solution, and has a characteristic absorption spectrum for convenient determination of concentration.

As an inorganic salt, Angeli's salt is not amenable to alteration to tune properties such as decomposition half-life. Angeli's salt is of the diazeniumdiolate (NONOate) class, in which an NO dimer is attached to a nucleophile. Diazeniumdiolates include secondary amine derivatives, which are widely used as NO donors.[14,15,25] Primary amine diazeniumdiolates have been demonstrated to release HNO,[26,27] in similarity to Angeli's salt and offer organic and thus modifiable alternatives. This chapter describes the synthesis, structure, and mechanism of decomposition of Angeli's salt as well as recent developments of the diazeniumdiolate class to provide flexibility and specificity in HNO donation.

2.2 Angeli's salt

Angeli's salt is a spontaneous donor of HNO with a half-life of ~3 min under physiological conditions.[24,28,29] With the recent surge in interest in the biological activity of HNO, Angeli's salt has emerged as one of the most prevalent donors used for understanding HNO chemistry and pharmacology.

2.2.1 *Synthesis and use of Angeli's salt*

Translations and modifications of the original synthesis have been described in reports dating to at least the early 1950s (e.g., Refs. 29–32). At present, the synthesis is convenient and inexpensive, involving generation of hydroxylamine (NH_2OH) from the commercially available hydrochloride, followed by condensation with organic nitrates (Eqs. 2.2 and 2.3).[33]

$$NH_2OH \cdot HCl\ NH_2OH \underset{MeOH}{\overset{ex\ NaOH}{\Longrightarrow}} \tag{2.2}$$

$$NH_2OH + RONO_2 \xrightarrow[MeOH]{2NaOH} ROH + Na_2N_2O_3 \tag{2.3}$$

Bonner et al. reported the synthesis of ^{15}N-labeled Angeli's salt by use of $^{15}NH_2OH$, which was important to elucidation of the decomposition mechanism.[32,34]

One issue with diazeniumdiolates is that decomposition in water limits the ability to rigorously purify the salts. Dissolution in a 50 mM NaOH solution and filtration through celite or in minimum amount of water followed by recrystallization in methanol or ethanol have both been described, but result in significant loss of product.[29,32] Typically, organic species are removed by washing with ether, acetone, or ethanol. The relative purity of such samples is determined by comparing the extinction coefficient at 250 nm to the published value of 8000 M^{-1} cm^{-1}.[28–30,32] The concentrations of stock solutions in 10 mM NaOH are similarly determined.

Solid samples and solutions (> 10 mM in 10 mM NaOH) are stored at −20°C. For use, stock solutions should be diluted into buffered solutions (1:100 dilution in phosphate-buffered saline (PBS) allows maintenance of the pH by this relatively weak buffer system) and stored on ice. Typically, a metal chelator is added to reaction solutions in order to sequester contaminating metals that may oxidize HNO to

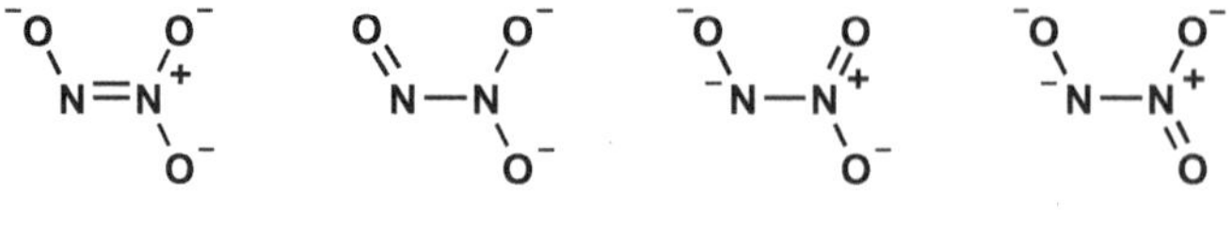

1 2 3 4

Scheme 2.1 Resonance structures of Angeli's dianion.

NO. Addition of diethylenetriaminepentaacetic acid (DTPA, 50 μM) reduces the concentration of NO from Angeli's salt to less than $<0.1\%$.[35]

2.2.2 Structural analysis

The characteristic UV absorbance indicates a nitrogen–nitrogen double bond. Photoelectron spectroscopy suggested two chemically nonequivalent nitrogen atoms.[36] The structure was further verified by X-ray crystallography, which showed a planar configuration for the anion.[37] Among the four possible resonance structures (Scheme 2.1), structure ***1*** is considered dominant, although significant contributions from the other three structures are indicated by thermochemical evidence.[38]

2.2.3 Mechanism of decomposition

Since Angeli's initial examination,[24] a number of studies have together led to elucidation of the decomposition mechanism.[27,28,32,39–42] The decomposition rate is readily monitored by loss of the maximum at 250 nm for the oxyhyponitrite dianion, $N_2O_3^{2-}$, or 237 nm (ε of 5500 M^{-1} cm^{-1} [32] or 6100 M^{-1} cm^{-1} [29]) for the monobasic species, $HN_2O_3^-$. Key data include that the thermal decomposition of Angeli's salt is first-order[43] and pH-dependent.[32,44] The decay profile with respect to pH has been instrumental in theoretical analysis. As alkalinity decreases, the decomposition rate increases, reaching a plateau from pH 8 to 4 (6.8×10^{-4} s^{-1} at 25°C and $4–5 \times 10^{-3}$ s^{-1} at 37°C[28,29]) and then increasing again below pH 4 (overall a 20-fold difference in rate constant at 25°C[28] but only 4.5-fold at 37°C[27]). This profile is reflective of the pK_as of 2.5 and 9.7 for oxyhyponitrous acid ($H_2N_2O_3$).[45]

The conventional mechanism of decomposition above pH 4 involves protonation of the dianion at the nitroso oxygen followed by tautomerization and heterolytic cleavage of the N–N bond to produce HNO and nitrite[28,32,39–42,46,47] (Eq. 2.4).

(2.4)

Decomposition of Angeli's salt is also suggested to be reversible, with excess nitrite modestly inhibiting the rate.[40,48] The mechanistic pathway is supported by

Raman and ^{15}N NMR spectroscopy.[32,34,49] ^{15}N-labeling demonstrated that the nitroso nitrogen is the source of HNO[32] while nitrite is derived from the nitro nitrogen.[47] The mechanism in Eq. (2.4) is also supported by a theoretical examination by Houk et al.[42] in which quantum mechanical calculations predicted that the most basic position of the dianion is the nitroso oxygen. The species protonated at this position is thermodynamically stable and kinetically inert to decomposition. Protonation instead at the nitroso nitrogen is suggested to produce a higher energy tautomer that is unstable due to accumulation of positive charge on adjacent nitrogen atoms.

At pH 3[32] (or at higher pH in the presence of excess nitrite[28]), Angeli's salt transitions to an NO donor.[43,50] A related mechanism to Eq. (2.4) in which decomposition requires tautomerization to a higher-energy species has been proposed (Eq. 2.5).[42] Calculations predict that the species diprotonated at a nitro oxygen spontaneously dissociates into water and an excited state of the NO dimer.

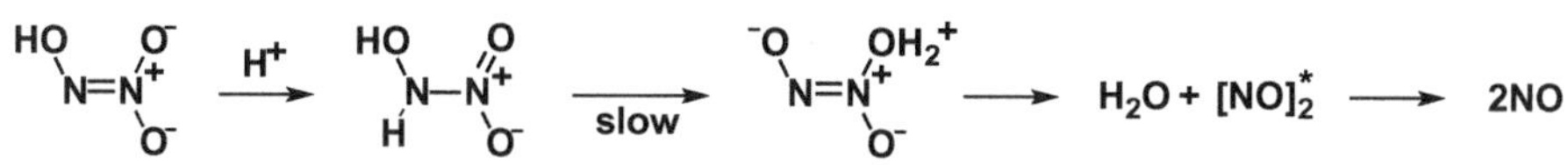

(2.5)

The slow decomposition rate for both pathways is understood to be a result of dissociation only of minor components of the equilibria.[42] Explanation of the insensitivity of rate between pH 4 and 8[28] and the sharp transition between HNO and NO production requires a more complete view of the equilibria established between stable and unstable isomers of varied protonation states (Scheme 2.2).[27]

The left side of Scheme 2.2 incorporates Eq. (2.4) with the caveat that the interchange process between monobasic species is not established and may involve tautomerization, complex acid–base equilibria, or intermolecular proton transfer. The resonance structure for the HNO-releasing species is also represented as predicted by Houk et al.[42] The right side of Scheme 2.2 in turn includes Eq. (2.5), again indicating that the pathway to the NO-producing tautomer is unclear. As the pH drops below 8, the stable dianion is no longer favored in the equilibrium (pK_a of 9.7[45]), which may simplify the process leading to HNO. As the pH approaches the next pK_a value (2.5[45]), a second protonation stabilizes against HNO formation. In fact,

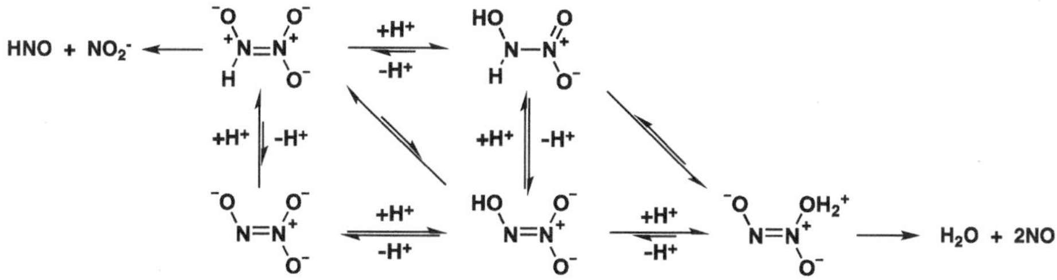

Scheme 2.2 Decomposition mechanisms available for Angeli's salt leading to release of HNO or NO.

protonation of the HNO-producing tautomer is calculated to provide the most stable diprotic species, protonated at both the nitroso oxygen and nitrogen, shuttling the system toward NO production.

2.2.4 Advantages and disadvantages

The advantages of Angeli's salt as an HNO donor include ease of synthesis or commercially availability, as preferred. Angeli's salt is straightforward to use, including concentration determination, and has a low toxicity (IC_{50} >1 mM). Stability in the solid state and in highly alkaline solution facilitates storage and use. The decomposition mechanism, though somewhat complicated, is well characterized and controllable via pH, and little NO is generated in aqueous solutions containing a suitable metal chelator. Such advantages have led Angeli's salt to be used extensively to understand the chemistry and pharmacology of HNO.

A disadvantage of ionic diazeniumdiolates is purification, given spontaneous decomposition in water. Decomposition of Angeli's salt also produces nitrite, which is considered to serve as a storage pool for NO.[51–54] Furthermore, the Angeli's anion can itself be oxidized, for instance, by metals or O_2[39,55,56] (Eq. 2.6).

$$N_2O_3^{2-} + M^{3+} \rightarrow NO + NO_2^- + M^{2+} \tag{2.6}$$

Thus, care must be taken to conduct experiments with appropriate controls.

Although less physiologically relevant, direct nucleophilic reactivity has also been demonstrated with various organic compounds including alkyl halides and aldehydes[31] (see Ref. 23 for further mechanistic discussion). Angeli's salt can also act as a dienophile to undergo Diels-Alder cycloaddition with conjugated dienes. The cycloadduct product formed from reaction with the diene 1,3-diphenylisobenzofuran is well characterized crystallographically.[57]

From a pharmacological perspective, Angeli's salt is an acute donor. As an inorganic salt, the structure cannot be modified to fine tune the half-life of HNO release. Other systems are therefore required for chronic exposure to HNO.

2.3 Diazeniumdiolates

In the early 1960s, Drago et al. reported the synthesis of diazeniumdiolates by reaction of NO with various amines.[58,59] Keefer and colleagues later developed secondary amine-based diazeniumdiolates as vehicles for the controlled delivery of NO in biological studies.[29,60–63] These donors have been extensively studied and utilized, and there are multiple reviews on the topic (e.g., Refs. 25,64–66).

Miranda and colleagues determined that primary amine-based diazeniumdiolates can serve as HNO donors.[26,27] The versatility of this class of donor to release NO or HNO depending upon the amine backbone and pH has been instrumental in the study of the chemical and pharmaceutical effects of both NO and HNO and their interplay.

2.3.1 Synthesis of primary amine diazeniumdiolates

While a large number of stable secondary amine-based diazeniumdiolates have been prepared, there are only a few stable primary amine derivatives, namely, the isopropylamine (IPA/NO), cyclopentylamine (CPA/NO), cyclohexylamine (CHA/NO), cycloheptylamine (CHPTA/NO), and cyclooctylamine (COA/NO) derivatives (Scheme 2.3; counterions omitted for clarity).[27,59,67]

Synthesis of primary amine-based diazeniumdiolates involves exposure of a solution of the primary amine in diethyl ether to 40 psi of NO for 24–48 h (Eq. 2.7).[27,29,67,68] The resulting ammonium salt is reacted with one equivalent of sodium methoxide (NaOMe) to give the corresponding sodium salt.

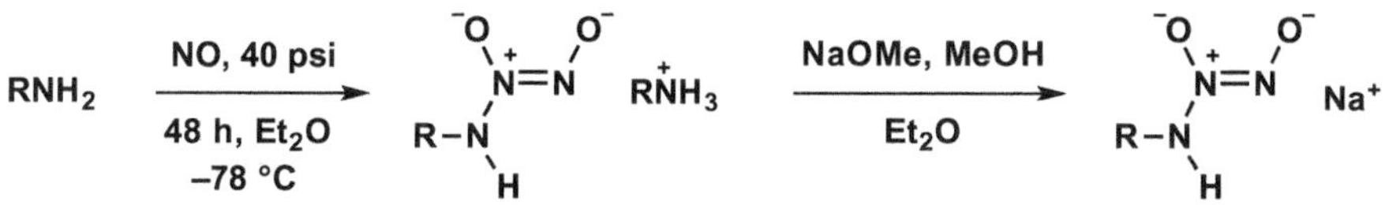

(2.7)

Using exactly one equivalent of NaOMe is crucial to avoid generation of a highly reactive diazoate ion and/or diazoalkane. Cold storage of solids and stock solutions is recommended for all diazeniumdiolates but is especially important for primary amine derivatives. Additionally, it is suggested that <250 mg is stored in a single container. *See Ref. 27 for full details on the recommended preparative method for IPA/NO, including a cautionary note.*

The compounds in Scheme 2.3 exhibit the characteristic diazeniumdiolate absorption spectra with maximum at 250 nm (ε of $8.2\text{–}10.0 \times 10^3\ \mathrm{M^{-1}\ cm^{-1}}$).[26,27,29,67] Rate constants and half-lives range from 2 to $5 \times 10^{-3}\ \mathrm{s^{-1}}$ and 2–6 min at pH 7.4 and 37°C. Storage and use are as described for Angeli's salt.

2.3.2 Structural analysis

Amine-based diazeniumdiolates can theoretically exist as either *E* or *Z* isomers[69,70] (Scheme 2.4). However, crystallographic studies have revealed that the *Z* configuration is preferred (see Refs. 64,71), due to a prohibitively high interconversion barrier (~40 kcal $\mathrm{mol^{-1}}$).[72]

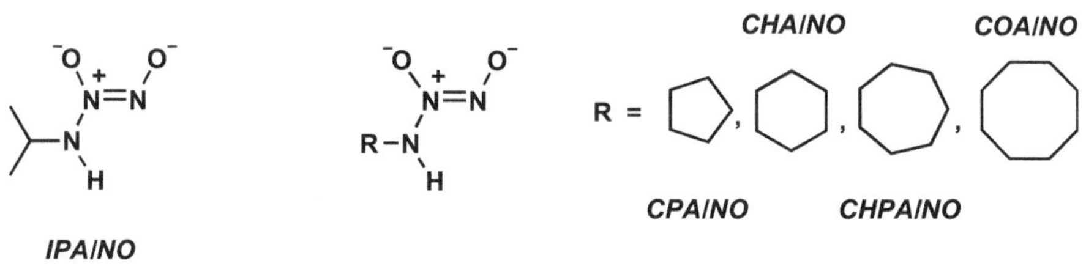

Scheme 2.3 HNO-releasing diazeniumdiolates.

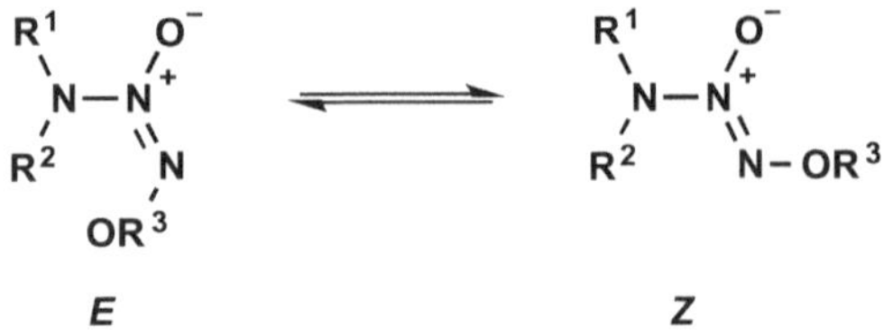

Scheme 2.4 *E* and *Z* configurations for amine-based diazeniumdiolates.

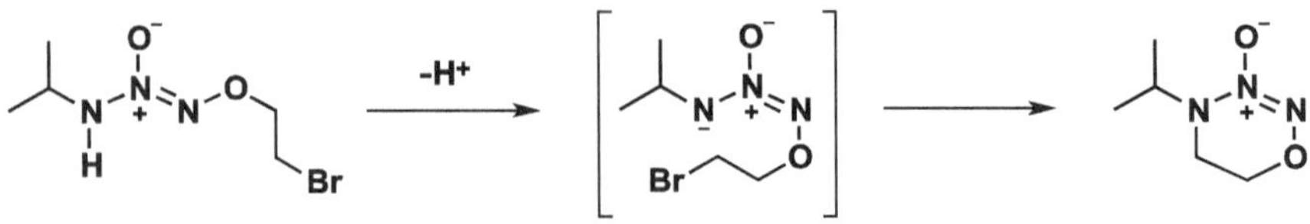

Scheme 2.5 Formation of a stable *E*-diazeniumdiolate.

In 2005, Wang et al. estimated that the interconversion barrier for MeHNN(O) = NOMe would be decreased by ~20 kcal mol^{-1} upon deprotonation of the NH proton, with the *E* form of the anion being favored.[72] This prediction was used to experimentally produce the first diazeniumdiolate in the *E* configuration by basifying iPrHNN(O)=$NOCH_2CH_2Br$ (Scheme 2.5).

2.3.3 Mechanism of decomposition

The decomposition mechanism of primary amine diazeniumdiolates has been studied both experimentally and theoretically, particularly for IPA/NO (sodium 1-(N-isopropylamino)diazen-1-ium-1,2-diolate).[26,27,67,73] As with Angeli's salt, the decay profile with respect to pH as well as analysis of product formation and pK_a are key data. Given that the two types of donors are related by substitution of a primary amine for the oxide in Angeli's salt, a number of similarities are observed. The following section is therefore structured to describe important differences between IPA/NO and Angeli's salt.

The observed transition from a higher decay rate in acidic medium to a slower, plateaued rate as pH increases is broader and shifted by several pH units for IPA/NO.[27] This is indicative of a weaker acid at the first observable deprotonation site (pK_a of 5.8 vs 2.5[27,45]). At pH 13, a slight increase in the decomposition rate occurs rather than the decrease observed for Angeli's salt. Furthermore, IPA/NO has a larger range in decomposition rate (250 vs 4.5-fold variance in rate constant at 37°C). The rate constants at pH 7.4 and 37°C for Angeli's salt and IPA/NO are comparable (2.7×10^{-3} s^{-1} and 1.7×10^{-3} s^{-1}, respectively[27,39]).

Although both Angeli's salt and primary amine diazeniumdiolates release HNO at alkaline pH and NO in acidic medium, the organic derivatives exhibit dual donor properties at intermediate pH.[26] That product crossover is not sharply delineated suggests a simple kinetic competition between two pathways, unlike the more

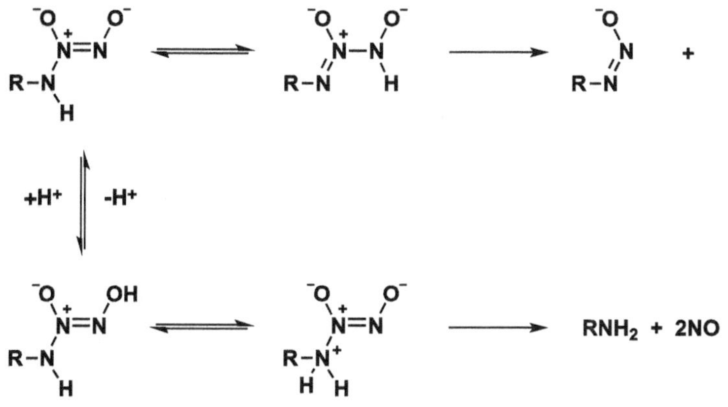

Scheme 2.6 Dual decomposition mechanisms available for primary amine diazeniumdiolates leading to release of HNO or NO.

complex equilibria for Angeli's salt. Again, quantum mechanical calculations by Houk and colleagues have clarified the mechanism of decomposition as a function of pH.[73]

Unlike Angeli's salt, which can be fully deprotonated to the dianion, the amine is predicted to not deprotonate in aqueous solution,[73] thus explaining observation of only a single transition in the decay rate profile. Tautomerization followed by heterolytic cleavage of the NO dimer bond provides access to an HNO-donating pathway independent of pH (Scheme 2.6). Note that although decomposition to HNO and a diazoate ion is relatively slow in alkaline solution, primary amine diazeniumdiolates are inherently less stable than those that do not contain a tautomerizable proton.

The NO-producing pathway is related to that for Angeli's salt in that the terminal oxygen is preferentially protonated, and decomposition requires tautomerization to the amine nitrogen to form an ammonium zwitterion.[73] Subsequent cleavage of the amine-NO dimer bond releases NO, presumably in the reverse of the synthetic pathway. Production of NO at a higher pH by IPA/NO is explained by the higher basicity of the nitroso oxygen ($pK_a \sim 6$)[27] compared to the nitroso nitrogen (pK_a of 2.5)[45] of Angeli's salt. Dual production of HNO from pH 4 to 8 by IPA/NO is a result of the kinetic competition between tautomerization and protonation.

Although the transition for IPA/NO occurs over four pH-units,[27] the ratio of HNO to NO is quite sensitive to pH near physiological pH ($\sim$60% and $\sim$35% HNO, respectively, at pH 7.4 and 7.0).[74,75] To ensure maximum HNO donation from IPA/NO, a solution of pH $\geq$ 8 is recommended; otherwise, donation of both HNO and NO should be considered. It may be possible to tune the point at which a primary amine diazeniumdiolate produces little NO by altering the basicity of the nitroso oxygen. The impact of amine identity on this parameter has yet to be examined. It is important to note that secondary amine diazeniumdiolates can be considered to exclusively donate NO as they only undergo decomposition through the protonation pathway in Scheme 2.6.[70]

2.3.4 Advantages/disadvantages

Many of the advantages of amine-based diazeniumdiolates mirror those of Angeli's salt. Although commercial availability has been sporadic, there is the potential for tuning rate and pH-sensitivity by altering the amine. For instance, available secondary amine analogs have half-lives ranging from 2 s to 20 h.[29] Progress in isolation of a variety of primary amine-based diazeniumdiolates has been slow. In part this may be due to inherent instability as a result of the tautomerization pathway (Scheme 2.6).

An interesting feature of the pH-independence of IPA/NO life-time in alkaline solution is the ability to synthesize peroxynitrite ($ONOO^-$) by a new, convenient route involving the nitroxyl anion (NO^-).[76] Decomposition of IPA/NO and subsequent deprotonation of HNO and autoxidation of NO^- (Eqs. 2.8 and 2.9) proved to provide a more reliable sample of peroxynitrite than the standard method of mixing acidified H_2O_2 with $NaNO_2$.

$$\text{IPA/NO} \rightarrow \text{HNO} \rightarrow \text{NO}^- \tag{2.8}$$

$$\text{NO}^- + \text{O}_2 \rightarrow \text{ONOO}^- \tag{2.9}$$

As with Angeli's salt, the impact of decomposition by-products must be considered. In the NO-producing pathway, the amine backbone is released. The products of the HNO pathway are more complex (Scheme 2.7). The diazoate ion is expected to rapidly protonate and lose water to form the diazonium ion,[77,78] which upon release of nitrogen is converted to a carbonium ion. This is turn can either deprotonate at the β-carbon to yield an olefin or react with water or other nucleophiles to generate stable adducts such as alcohols. In either case, decomposed solutions typically serve as reasonable controls.

While primary amine derivatives expand the class of spontaneous HNO donating diazeniumdiolates, isolated species are acute donors and are prone to decomposition on storage even at low temperatures. To circumvent such limitations, a variety of compounds capable of serving as caged HNO donors or prodrugs have been designed (e.g., Refs. 79–94). Here, we focus on derivatization at the O^2-position of the diazeniumdiolate functional group, which both enhances stability and facilitates purification.

2.4 Diazeniumdiolate-based HNO-releasing prodrugs

O^2-Derivatization of secondary amine-based diazeniumdiolates allows for design of enzymatic or photolytic activation (e.g., Refs. 95–99; see Ref. 64). The neutral products can also be rigorously purified, for instance by column chromatography, and exhibit increased stability in water.

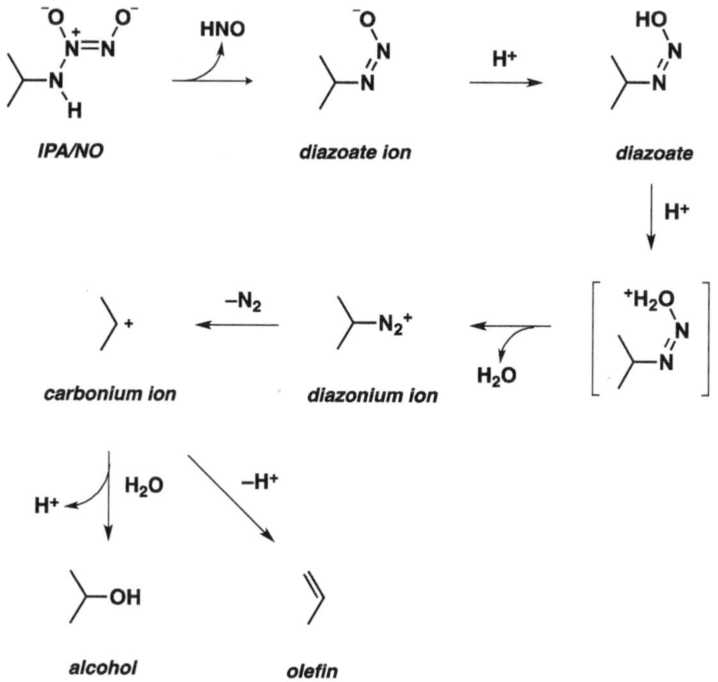

Scheme 2.7 Postulated mechanism of dissociation of IPA/NO to HNO and other products in aqueous media.

More recently, Saavedra et al. demonstrated that primary amine diazeniumdiolates were amenable to O^2-alkylation, esterification, and glycosylation.[71] Additional vinyl, ester, and sulfonylethyl O^2-derivatives have since been produced.[67,100–103] Such species have allowed harnessing of the HNO-releasing properties of primary amine diazeniumdiolates while incorporating controlled release either enzymatically or by slower hydrolysis than the parent donor.

2.4.1 Synthesis of primary amine diazeniumdiolate ester derivatives

Given the variety of O^2-derivatives, the reader is referred to the original publications for synthetic details. Synthesis of esterified derivatives is provided here as one example (Scheme 2.8). Reaction of IPA/NO (produced as in Eq. 2.7) with chloromethyl methyl sulfide in a slurry containing K_2CO_3 in DMF at room temperature for 3 h gives O^2-methylthiomethyl protected IPA/NO. The reactive O^2-chloromethyl derivative can then be generated by reaction with sulfuryl chloride under ice-cold conditions. Overnight reaction with a suitable carboxylic acid in an aprotic solvent such as DMSO in the presence of a base such as trimethylamine yields the desired esterified derivative.

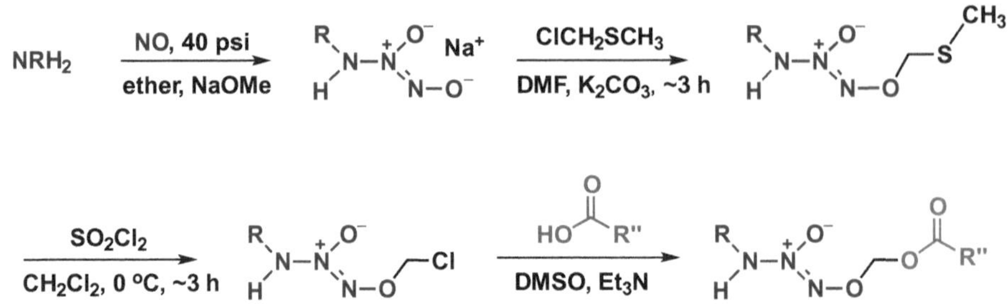

Scheme 2.8 Synthesis of esterified derivatives of primary-amine diazeniumdiolates.

2.4.2 Structural analysis

Detailed structural analysis has generally not been completed for primary amine diazeniumdiolates, with the exception of O^2-methylated CHA/NO.[71] The crystal structure revealed typical bonding modes for the nitrogen–nitrogen double bond (1.275(2) Å) in the diazeniumdiolate moiety and in the cyclohexyl ring (chair conformation). The oxygen atoms in the planar N_2O_2 group are oriented *cis* to each other, and the amine nitrogen is in an equatorial position on the cyclohexyl ring.

2.4.3 Mechanism of decomposition

Esterification is a common prodrug strategy, and secondary amine diazeniumdiolates have been esterified with a number of compounds.[64,97] The mechanism of spontaneous O^2-acetoxymethylated IPA/NO (AcOM-IPA/NO) decomposition was studied in detail.[100] As described above, the decomposition of Angeli's salt and IPA/NO is accelerated with decreasing pH. In contrast, esters undergo base hydrolysis, and the decomposition rate for AcOM-IPA/NO was observed to be first-order in hydroxide concentration from pH 6 to 12. The rate was extremely slow at lower pH, and as with IPA/NO, a small reversal in the rate trend was observed at pH 13. Under physiological conditions, the half-life of AcOM-IPA/NO (41 min) was 10-fold longer than for IPA/NO. Importantly, the decomposition rate of AcOM-IPA/NO exceeded that of IPA/NO at pH 8 and above, suggesting that the mechanism does not involve simple hydrolysis. Furthermore, AcOM-IPA/NO was found to be primarily an HNO donor, with little indication of NO production. These data, coupled with spectroscopic analysis, suggested the following unexpected mechanism (Scheme 2.9).

The first step involves deprotonation of the amine nitrogen (estimated pK_a of 12), which had previously been shown to be acidic compared to traditional amines.[71] The slight change in decomposition rate observed at pH 13 for both IPA/NO and its O^2-acetoxymethyl derivative may be a function of deprotonation at this position. Deprotonation produces a nucleophilic site that induces 1,4-acyl migration through a cyclic intermediate. Expulsion of formaldehyde and fragmentation forms an acyl nitroso intermediate and a diazoate ion. Acyl nitroso compounds are well-known

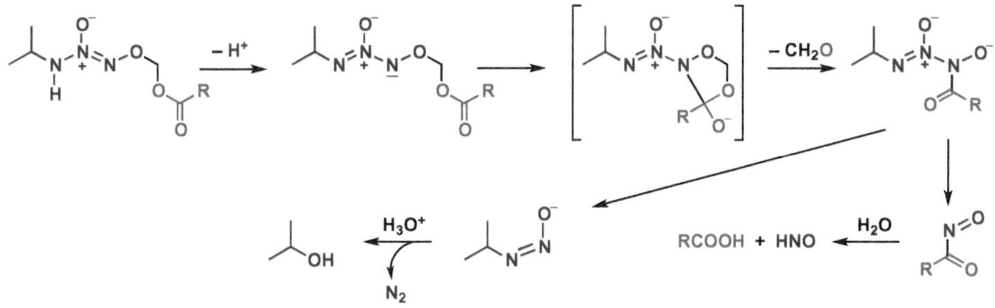

Scheme 2.9 Mechanism of spontaneous decomposition of O^2-acetoxymethylated IPA/NO.

donors of HNO and are intermediate in various HNO releasing prodrugs.[85,104] Diazoate ion reactivity is as in Scheme 2.7.

O^2-Acetoxymethylated CPA/NO was synthesized using a similar protocol and exhibited a shorter half-life (21 min) but a similarly high yield of HNO (96% HNO) at pH 7.4 and 37°C.[67] Thus, O^2-acetoxymethylation of primary amine diazeniumdiolates can be used to enhance both stability and HNO production compared to the parent donor. This class of compound therefore represents an important step forward in the ability to investigate the chemical and pharmaceutical properties of HNO in the absence of NO or nitrite. Furthermore, the slower rate of hydrolysis reduces consumption by dimerization such that the trapping efficiency for HNO is increased relative to the parent donor. As a consequence, lower doses of the O^2-acetoxymethylated derivative have been shown to be effective, for instance in sarcomere shortening.[100]

Esterase-mediated decomposition of AcOM-IPA/NO did lead to observable production of IPA/NO as well as acetate and formaldehyde, indicating simple hydrolysis at the O^2-position (Scheme 2.10). Esterase[100] did not impact the stability of IPA/NO or scavenging of HNO.

A galactosylated derivative of IPA/NO (Gal-IPA/NO) proved quite stable to spontaneous hydrolysis ($t_{½}$ >2 weeks under physiological conditions).[105] In the presence of β-D-galactosidase under saturating conditions, cleavage was first order in enzyme, and thus the half-life of decomposition was tunable via enzyme concentration. An impact of HNO or NO production on enzyme activity, efficiency, or structure was not observed. Based on the organic products, decomposition was assumed to involve hydrolysis at the O^2-position of IPA/NO, in similarity to AcOM-IPA/NO (Scheme 2.10). The Gal-IPA/NO/β-D-galactosidase system was presented as a platform for controlled generation of IPA/NO and thus its nitrogen oxide products, also without significant scavenging by the enzyme.

Hybrids of NO donors and nonsteroidal antiinflammatory drugs (NO-NSAIDs) have received significant attention as a means to harness the antiinflammatory properties of NSAIDs and the beneficial cardiovascular and gastrointestinal effects of NO[106–109] in order to reduce the side effects of traditional NSAIDs.[110,111] Such conjugates retain the antiinflammatory and analgesic properties of the NSAID while reducing gastrointestinal, cardiovascular, and renal side effects in animal models

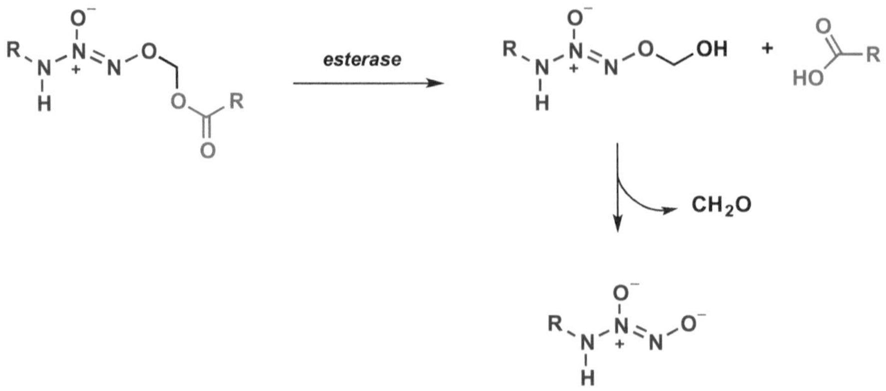

Scheme 2.10 Esterase-mediated cleavage of the protecting group to form a primary-amine diazeniumdiolate.

(see Ref. 112). A significantly lower degree of gastric injury has also been apparent in clinical trials. Early analysis also indicates efficacy in inhibition of cancer growth.[113–115] Interestingly, esterification of NSAIDs has been suggested to reduce toxicity by increasing stability as well as the absorption rate of the resulting prodrug in the gastrointestinal tract.[116] In turn, esterification of the donor can augment localized delivery of NO,[103,117] which is a critical feature in NO donor-based treatment of conditions other than hypertension.

Recently, detailed comparisons of adducts of primary and secondary amine diazeniumdiolates and aspirin were presented.[103,118] Readers are referred to the original articles for full discussion, as only the decomposition mechanism of the HNO-donating adduct (IPA/NO-aspirin) is presented here. Aspirin itself is an ester, which introduces a second potential hydrolysis site. Product analysis indicated that spontaneous hydrolysis of IPA/NO-aspirin ($t_{½}$ of 7.5 h) followed a related mechanism to AcOM-IPA/NO (Scheme 2.11 top), but that the acetyl group of aspirin was more susceptible to cleavage than the ester linker.[103] Thus, free salicylic acid rather than aspirin is produced. In the presence of guinea pig serum, the half-life was reduced to 18 min, and again, the acetyl group of aspirin was cleaved more rapidly than the linker (Scheme 2.11 bottom).

Interestingly, IPA/NO-aspirin was not cleaved by the esterases present in fetal bovine serum,[103] a commonly used additive in cell culture media. Such specificity provides controlled delivery in culture. In fact, derivatization was found to facilitate cellular uptake compared to IPA/NO, which primarily decomposes extracellularly. The requirement for diffusion of HNO across the cell membrane and to a specific target reduces effectiveness due to indiscriminant scavenging, e.g, by glutathione.

IPA/NO-aspirin was found to have significantly elevated cytotoxicity against various breast and lung cancer cells[103,118] compared to IPA/NO.[26] Interestingly, this effect was not apparent in nontumorigenic cells.[103,118] The observed specificity toward cancer cells suggests efficacy as a cancer therapeutic.

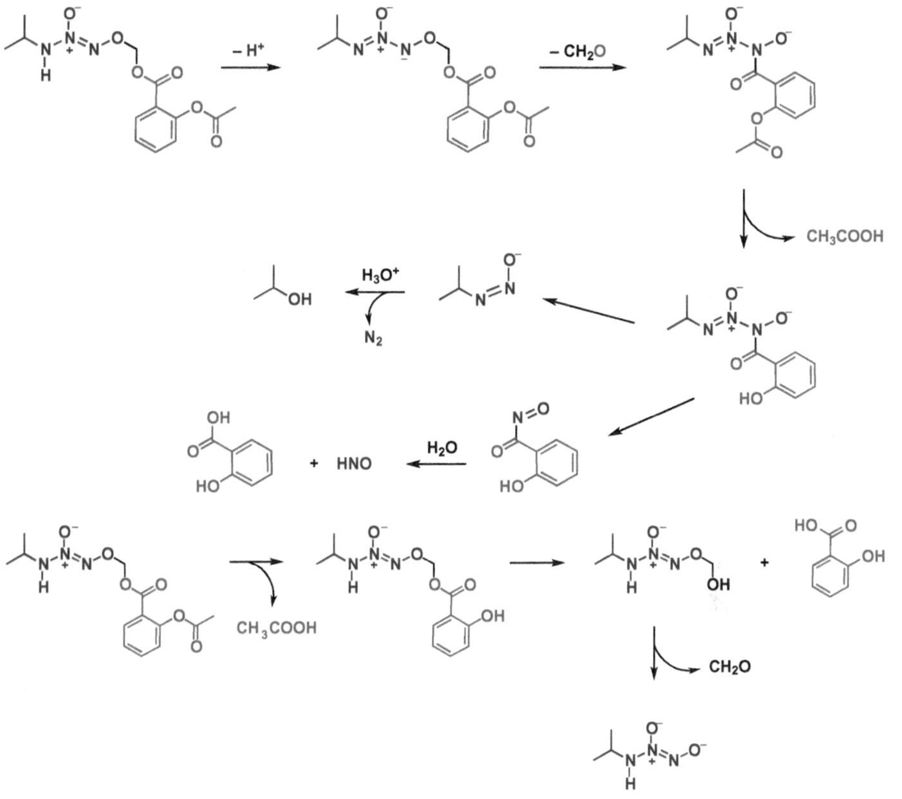

Scheme 2.11 Mechanism of decomposition of IPA/NO-aspirin: (*top*) spontaneous hydrolysis and (*bottom*) in the presence of serum.

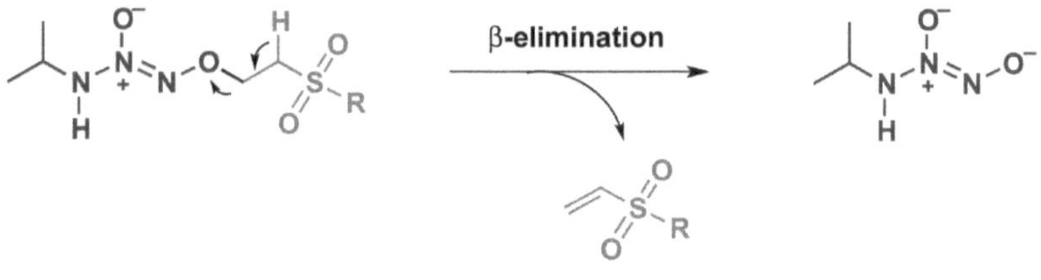

Scheme 2.12 Mechanism of decomposition of O^2-sulfonylethyl protected IPA/NO.

Recently Knaus et al. reported O^2-sulfonylethyl (-$CH_2CH_2SO_2R$ where R = Me, OMe, NHOMe, NHOBn) derivatives of IPA/NO, which can undergo β-elimination to release IPA/NO[119] (Scheme 2.12). Under physiological conditions, these compounds exhibited extended half-lives of 6.6 – 17.1 h depending on the R substituent.

2.4.4 Advantages/disadvantages

Advantages over the ionic parent donors include the ability to rigorously purify and the added stability induced by O^2-derivatization. The stability of the linker to spontaneous cleavage can be significant. Furthermore, spontaneous decomposition can occur through a pathway that limits production of NO. These characteristics in addition to specific sensitivity to enzymatic cleavage render O^2-derivatized primary amine diazeniumdiolates an attractive platform for systematically generating reliable, controlled fluxes of HNO at programmable rates.

While O^2-derivatization enhances the ability to purify the products, it also reduces water solubility. Furthermore, the products are often oils rather than solids, which can be less convenient to handle. The linker can introduce new by-products such as formaldehyde, and decomposition can produce a more complex mixture of organic species compared to the ionic precursors.

2.5 Acyl nitroso compounds as HNO donors

Decomposition of O^2-esterified IPA/NO produces an acyl nitroso intermediate as the HNO donating species (Schemes 2.9 and 2.11). Acyl nitroso compounds hydrolyze to HNO via a pathway that is suggested to involve nucleophilic acyl substitution reaction (Eq. 2.10).[33,87,120] Generation of nitrous oxide (Eq. 2.1) in these reactions has been used to provide support for HNO production.

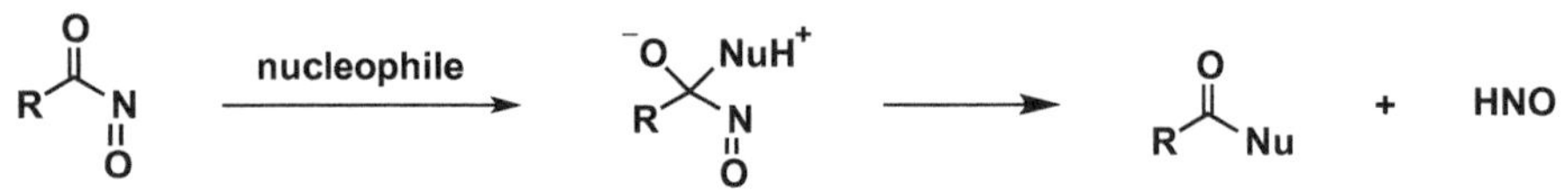

(2.10)

In addition to releasing HNO, acyl nitroso intermediates are used as an important intermediate in organic synthesis for production of critical functionalized molecules. For instance, they are used as N-O heterodienophiles to synthesize oxazines, pyrrolidines, amino alcohols, and aza sugars.[121–124]

Synthesis of acyl nitroso compounds is generally accomplished via oxidation of hydroxylamine derivatives using oxidants such as periodate salts, the Dess-Martin periodinane,[125] metal complexes,[126–128] using the Swern–Moffatt protocol,[129] and others.[130–132] Other routes to prepare acyl nitroso species include photochemical cleavage of 1,2,4-oxadiazole-4-oxides,[133] oxidation of nitrile oxides,[134] the retro-Diels-Alder reaction,[135,136] and rearrangement of nitrocarbenes generated from diazo compounds[137] (Scheme 2.13).

As acyl nitroso compounds are unstable, they are generated in situ and trapped as Diels-Alder adducts.[136,138] Hydroxyurea, used in treatment of sickle cell anemia and a number of cancers, has been shown to undergo horseradish peroxidase- and catalase-mediated oxidation in the presence of hydrogen peroxide to form an acyl nitroso compound that was successfully trapped in this fashion.[119]

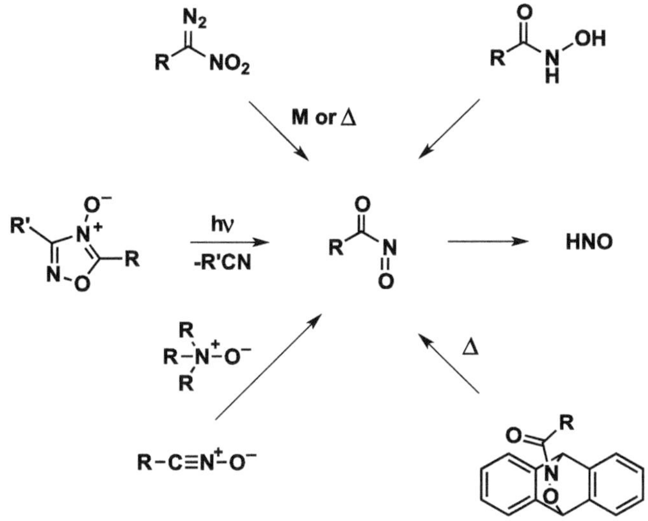

Scheme 2.13 Synthesis of acyl nitroso compounds.

Scheme 2.14 s-*cis* and s-*trans* isomers.

Schwarz and coworkers reported spectroscopic evidence for nitrosoformaldehyde in the gas phase using charge-reversal and neutralization-reionization mass spectrometry.[139] Toscano and colleagues recently reported the first direct detection of an acyl nitroso compound in organic solution using time-resolved infrared spectroscopy and estimated a lifetime of 1 Ms.[104] These compounds can exist in s-*cis* or s-*trans* conformations (Scheme 2.14) without a general preference for either conformer. The energy difference can vary from 0 to 15 kcal/mol.[140–142]

2.6 Conclusions

Diazeniumdiolates offer a versatile, tunable means to produce HNO (and/or NO) conveniently and with reliable fluxes. Detailed understanding of the decomposition mechanisms facilitates analysis, especially in biological systems, allows design of suitable control experiments and may aid in production of clinically viable agents. In this regard, enhanced water-solubility may be an attractive characteristic for new diazeniumdiolate derivatives.

Acknowledgments

This work is supported in part by grants from the National Institutes of Health grant (R01-GM076247 to KMM) and the National Science Foundation (CHE-0645818 to KMM).

Abbreviations

AcOM-IPA/NO	O^2-acetoxymethylated IPA/NO
Angeli's salt	sodium trioxodinitrate
CHA/NO	cyclohexylamine diazeniumdiolate
CHPTA/NO	cycloheptylamine diazeniumdiolate
COA/NO	cyclooctylamine diazeniumdiolate
CPA/NO	cyclopentylamine diazeniumdiolate
DTPA	diethylenetriaminepentaacetic acid
Gal-IPA/NO	O^2-galactosylated IPA/NO
HNO	nitroxyl
$H_2N_2O_3$	oxyhyponitrous acid
IPA/NO	isopropylamine diazeniumdiolate
IPA/NO-aspirin	hybrid of IPA/NO and aspirin
NaOMe	sodium methoxide
NH_2OH	hydroxylamine
NO	nitric oxide
NO^-	nitroxyl anion
NO-NSAIDs	hybrids of NO donors and nonsteroidal antiinflammatory drugs
NONOate	diazeniumdiolate
$ONOO^-$	peroxynitrite
PBS	phosphate-buffered saline

References

1. Palmer RMJ, Ferrige AG, Moncada S. Nitric oxide release accounts for the biological activity of endothelium-derived relaxing factor. *Nature* 1987;**327**:524–6.
2. Ignarro LJ, Buga GM, Wood KS, Byrns RE, Chaudhuri G. Endothelium-derived relaxing factor produced and released from artery and vein is nitric oxide. *Proc Natl Acad Sci USA* 1987;**84**:9265–9.
3. Wink DA, Mitchell JB. Chemical biology of nitric oxide: insights into regulatory, cytotoxic, and cytoprotective mechanisms of nitric oxide. *Free Radic Biol Med* 1998;**25**:434–56.
4. Miranda KM. The chemistry of nitroxyl (HNO) and implications in biology. *Coord Chem Rev* 2005;**249**:433–55.
5. Bartberger MD, Liu W, Ford E, Miranda KM, Switzer C, Fukuto JM, et al. The reduction potential of nitric oxide (NO) and its importance to NO biochemistry. *Proc Natl Acad Sci USA* 2002;**99**:10958–63.
6. Flores-Santana W, Salmon DJ, Donzelli S, Switzer CH, Basudhar D, Ridnour L, et al. The specificity of nitroxyl chemistry is unique among nitrogen oxides in biological systems. *Antioxid Redox Signal* 2011;**14**:1659–74.

7. Wink DA, Miranda KM, Katori T, Mancardi D, Thomas DD, Ridnour LA, et al. Orthogonal properties of the redox siblings nitroxyl and nitric oxide in the cardiovascular system: a novel redox paradigm. *Am J Physiol Heart Circul Physiol* 2003;**285**: H2264–76.
8. Miranda KM, Paolocci N, Katori T, Thomas DD, Ford E, Bartberger MD, et al. A biochemical rationale for the discrete behavior of nitroxyl and nitric oxide in the cardiovascular system. *Proc Natl Acad Sci USA* 2003;**100**:9196–201.
9. Pagliaro P. Differential biological effects of products of nitric oxide (NO) synthase: it is not enough to say NO. *Life Sci* 2003;**73**:2137–49.
10. Kohout FC, Lampe FW. On the role of the nitroxyl molecule in the reaction of hydrogen atoms with nitric oxide. *J Am Chem Soc* 1965;**87**:5795–6.
11. Lin MC, He Y, Melius CF. Theoretical interpretation of the kinetics and mechanisms of the HNO + HNO and HNO + 2NO reactions with a unified model. *Int J Chem Kinet* 1992;**24**:489–516.
12. Shafirovich V, Lymar SV. Nitroxyl and its anion in aqueous solutions: spin states, protic equilibria, and reactivities toward oxygen and nitric oxide. *Proc Natl Acad Sci USA* 2002;**99**:7340–5.
13. Feelisch M, Stamler JS. Donors of Nitrogen Oxides. In: Feelisch M, Stamler JS, editors. *Methods in nitric oxide research.* New York: John Wiley & Sons; 1996. p. 71–115.
14. Thomas DD, Miranda KM, Espey MG, Citrin D, Jourd'heuil D, Paolocci N, et al. Guide for the use of nitric oxide (NO) donors as probes of the chemistry of NO and related redox species in biological systems. *Methods Enzymol* 2002;**359**:84–105.
15. Keefer LK, Nims RW, Davies KM, Wink DA. NONOates (1-substituted diazen-1-ium-1,2-diolates) as nitric oxide donors: convenient nitric oxide dosage forms. *Methods Enzymol* 1996;**268**:281–93.
16. Wang PG, Xian M, Tang X, Wu X, Wen Z, Cai T, et al. Nitric oxide donors: chemical activities and biological applications. *Chem Rev* 2002;**102**:1091–134.
17. Bonner FT, Stedman GR. The chemistry of nitric oxide and redox-related species. In: Feelisch M, Stamler JS, editors. *Methods in nitric oxide research.* New York: John Wiley & Sons; 1996. p. 3–18.
18. Wink DA, Feelisch M. Formation and detection of nitroxyl and nitrous oxide. In: Feelisch M, Stamler JS, editors. *Methods in nitric oxide research.* New York: John Wiley & Sons; 1996. p. 403–12.
19. Bonner FT, Hughes MN. The aqueous solution chemistry of nitrogen in low positive oxidation states. *Comments Inorg Chem* 1988;**7**:215–34.
20. Hughes MN. Relationships between nitric oxide, nitroxyl ion, nitrosonium cation and peroxynitrite. *Biochim Biophys Acta-Bioenerg* 1999;**1411**:263–72.
21. Farmer PJ, Sulc F. Coordination chemistry of the HNO ligand with hemes and synthetic coordination complexes. *J Inorg Biochem* 2005;**99**:166–84.
22. Miranda KM, Nagasawa HT, Toscano JP. Donors of HNO. *Curr Top Med Chem* 2005;**5**:649–64.
23. DuMond JF, King SB. The chemistry of nitroxyl-releasing compounds. *Antioxid Redox Signal* 2011;**14**:1637–48.
24. Angeli A. Sopra la nitroidrossilammina. *Gazz Chim Ital* 1896;**26**:17–25.
25. Keefer LK. Fifty years of diazeniumdiolate research. From laboratory curiosity to broad-spectrum biomedical advances. *ACS Chem Biol* 2011;**6**:1147–55.
26. Miranda KM, Katori T, Torres de Holding CL, Thomas L, Ridnour LA, McLendon WJ, et al. Comparison of the NO and HNO donating properties of diazeniumdiolates: primary amine adducts release HNO in vivo. *J Med Chem* 2005;**48**:8220–8.

27. Salmon DJ, Torres de Holding CL, Thomas L, Peterson KV, Goodman GP, Saavedra JE, et al. HNO and NO release from a primary amine-based diazeniumdiolate as a function of pH. *Inorg Chem* 2011;**50**:3262–70.
28. Hughes MN, Wimbledon PE. The chemisry of trioxodinitrates. Part I. Decompostion of sodium trioxodinitrate (Angeli's salt) in aqueous solution. *J Chem Soc Dalton Trans* 1976;703–7.
29. Maragos CM, Morley D, Wink DA, Dunams TM, Saavedra JE, Hoffman A, et al. Complexes of NO with nucleophiles as agents for the controlled biological release of nitric oxide – vasorelaxant effects. *J Med Chem* 1991;**34**:3242–7.
30. Addison CC, Gamlen GA, Thompson R. The ultraviolet absorption spectra of sodium hyponitrite and sodium alpha-oxyhyponitrite – the analysis of mixtures with sodium nitrite and nitrate. *J Chem Soc* 1952;338–45.
31. Smith PAS, Hein GE. The alleged role of nitroxyl in certain reactions of aldehydes and alkyl halides. *J Am Chem Soc* 1960;**82**:5731–40.
32. Bonner FT, Ravid B. Thermal decomposition of oxyhyponitrite (sodium trioxodinitrate (II)) in aqueous solution. *Inorg Chem* 1975;**14**:558–63.
33. King SB, Nagasawa HT. Chemical approaches toward generation of nitroxyl. *Methods Enzymol* 1999;**301**:211–20.
34. Bonner FT, Degani H, Akhtar MJ. Nitrogen-15 magnetic resonance spectroscopy of trioxodinitrate: N-protonation of an oxoanion. *J Am Chem Soc* 1981;**103**: 3739–42.
35. Lopez BE, Shinyashiki M, Han TH, Fukuto JM. Antioxidant actions of nitroxyl (HNO). *Free Radic Biol Med* 2007;**42**:482–91.
36. Hendrickson DN, Hollander JM, Jolly WL. Nitrogen 1 s electron binding energies. Correlations with molecular orbital calculated nitrogen charges. *Inorg Chem* 1969;**8**:2642–6.
37. Hope H, Sequeira MR. Angeli's salt – crystal structure of sodium trioxodinitrate(II) monohydrate, $Na_2N_2O_3$ H_2O. *Inorg Chem* 1973;**12**:286–8.
38. Hunt HR, Cox JR, Ray JD. Heat of formation of crystalline sodium alpha-oxyhyponitrite – structure of aqueous alpha-oxyhyponitrite ion. *Inorg Chem* 1962;**1**:938–41.
39. Miranda KM, Dutton AS, Ridnour LA, Foreman CA, Ford E, Paolocci N, et al. Mechanism of aerobic decomposition of Angeli's salt (sodium trioxodinitrate) at physiological pH. *J Am Chem Soc* 2005;**127**:722–31.
40. Hughes MN, Wimbledon PE. The chemistry of trioxodinitrates. Part 2. The effect of added nitrite on the stability of sodium trioxodinitrate in aqueous solution. *J Chem Soc Dalton Trans* 1977;1650–3.
41. Donald CE, Hughes MN, Thompson JM, Bonner FT. Photolysis of the nitrogen-nitrogen double bond in trioxodinitrate: reaction between triplet oxonitrate(1-) and molecular oxygen to form peroxonitrite. *Inorg Chem* 1986;**25**:2676–7.
42. Dutton AS, Fukuto JM, Houk KN. Mechanisms of HNO and NO production from Angeli's salt: density functional and CBS-QB3 theory predictions. *J Am Chem Soc* 2004;**126**:3795–800.
43. Veprek-Siska J, Pliska V, Smirous F, Visely F. Anorganische stickstoffverbindungen. 1. Zersetzungsmechanismus wassriger losungen des natrium-salzes der nitrohydroxylaminsaure. *Collect Czech Chem Commun* 1959;**24**:687–93.
44. Hughes MN, Wimbledon PE. Decomposition of sodium hyponitrate (Angeli's salt) in aqueous solution. *Chem Ind* 1975;742–3.
45. Sturrock PE, Ray JD, McDowell J, Hunt HR. The dissociation constant of α-oxyhyponitrous acid. *Inorg Chem* 1963;**2**:649–50.

46. Angeli A, Angelico F. Nuove ricerche sopra l'acido nitroidrossilamminico. *Gazz Chim Ital* 1903;**33**:245–52.
47. Hendrickson DN, Jolly WL. The structure of the oxyhyponitrite ion. *Inorg Chem* 1969;**8**:693–4.
48. Akhtar MJ, Lutz CA, Bonner FT. Decomposition of sodium trioxodinitrate ($Na_2N_2O_3$) in the presence of added nitrite in aqueous solution. *Inorg Chem* 1979;**18**:2369–75.
49. Bonner FT, Akhtar MJ, King TV, Chen LH, Ishida T. Studies in the vibrational spectroscopy of sodium trioxodinitrate. *J Phys Chem* 1981;**85**:4051–6.
50. Cambi L. Eber das nitrosyl von A. Angeli. *Ber Dtsch Chem Ges* 1936;**B69**:2027–33.
51. Dejam A, Hunter CJ, Schechter AN, Gladwin MT. Emerging role of nitrite in human biology. *Blood Cells Mol Dis* 2004;**32**:423–9.
52. Rassaf T, Feelisch M, Kelm M. Circulating NO pool: assessment of nitrite and nitroso species in blood and tissues. *Free Radic Biol Med* 2004;**36**:413–22.
53. Gladwin MT. Nitrite as an intrinsic signaling molecule. *Nat Chem Biol* 2005;**1**:245–6.
54. Lundberg JO, Gladwin MT, Ahluwalia A, Benjamin N, Bryan NS, Butler A, et al. Nitrate and nitrite in biology, nutrition and therapeutics. *Nat Chem Biol* 2009;**5**: 865–9.
55. Akhtar MJ, Bonner FT, Hughes MN, Humphreys EJ, Lu CS. Kinetic and isotopic studies on the reaction between trioxodinitrate and the hexaammineruthenium(III) cation. *Inorg Chem* 1986;**25**:4635–9.
56. Raspi G, Cinquantini A. Oxidative determination of α-hyponitrate with alkaline hexacyanoferrate(III). *Anal Chim Acta* 1974;**72**:200–4.
57. Torun L, Mohammad T, Morrison H. Cycloaddition reactions of sodium dinitroxytrioxide (Angeli's salt). *Tetrahedron Lett* 1999;**40**:5279–82.
58. Drago RS, Paulik FE. The reaction of nitrogen(II) oxide with diethylamine. *J Am Chem Soc* 1960;**82**:96–8.
59. Drago RS, Karstetter BR. The reaction of nitrogen(II) oxide with various primary and secondary amines. *J Am Chem Soc* 1961;**83**:1819–22.
60. Hansen TJ, Croisy AF, Keefer LK. N-nitrosation of secondary amines by nitric oxide via the 'Drago complex'. *IARC Sci Publ* 1982;**41**:21–9.
61. Keefer LK, Flippen-Anderson JL, George C, Shanklin AP, Dunams TM, Christodoulou D, et al. Chemistry of the diazeniumdiolates. 1. Structural and spectral characteristics of the $[N(O)NO]^-$ functional group. *Nitric Oxide* 2001;**5**:377–94.
62. Davies KM, Wink DA, Saavedra JE, Keefer LK. Chemistry of the diazeniumdiolates. 2. Kinetics and mechanism of dissociation to nitric oxide in aqueous solution. *J Am Chem Soc* 2001;**123**:5473–81.
63. Hrabie JA, Klose JR, Wink DA, Keefer LK. New nitric oxide-releasing zwitterions derived from polyamines. *J Org Chem* 1993;**58**:1472–6.
64. Hrabie JA, Keefer LK. Chemistry of the nitric oxide-releasing diazeniumdiolate ("nitrosohydroxylamine") functional group and its oxygen-substituted derivatives. *Chem Rev* 2002;**102**:1135–54.
65. Keefer LK. Progress toward clinical application of the nitric oxide-releasing diazeniumdiolates. *Annu Rev Pharmacol Toxicol* 2003;**43**:585–607.
66. Keefer LK. Nitric oxide (NO)- and nitroxyl (HNO)-generating diazeniumdiolates (NONOates): emerging commercial opportunities. *Curr Top Med Chem* 2005;**5**:625–36.
67. Bharadwaj G, Benini PG, Basudhar D, Ramos-Colon CN, Johnson GM, Larriva MM, et al. Analysis of the HNO and NO donating properties of alicyclic amine diazeniumdiolates. *Nitric Oxide* 2014;**42**:70–8.

68. Wink DA, Feelisch M, Fukuto J, Christodoulou D, Jourd'heuil D, Grisham MB, et al. The cytotoxicity of nitroxyl: possible implications for the pathophysiological role of NO. *Arch Biochem Biophys* 1998;**351**:66–74.
69. Taylor DK, Bytheway I, Barton DHR, Bayse CA, Hall MB. Toward the generation of NO in biological systems theoretical studies of the N_2O_2 grouping. *J Org Chem* 1995;**60**:435–44.
70. Dutton AS, Fukuto JM, Houk KN. The mechanism of NO formation from the decomposition of dialkylamino diazeniumdiolates: density functional theory and CBS-QB3 predictions. *Inorg Chem* 2004;**43**:1039–45.
71. Saavedra JE, Bohle DS, Smith KN, George C, Deschamps JR, Parrish D, et al. Chemistry of the diazeniumdiolates. O- versus N-alkylation of the $RNH[N(O)NO]^-$ ion. *J Am Chem Soc* 2004;**126**:12880–7.
72. Wang YN, Bohle DS, Bonifant CL, Chmurny GN, Collins JR, Davies KM, et al. Chemistry of the diazeniumdiolates: Z $\rightleftharpoons$ E isomerism. *J Am Chem Soc* 2005;**127**:5388–95.
73. Dutton AS, Miranda KM, Wink DA, Fukuto JM, Houk KN. Mechanism of pH-dependent decomposition of monoalkylamine diazeniumdiolates to form HNO and NO, deduced from the model compound methylamine diazeniumdiolate, density functional theory, and CBS-QB3 calculations. *Inorg Chem* 2006;**45**:2448–56.
74. Reisz JA, Zink CN, King SB. Rapid and selective nitroxyl (HNO) trapping by phosphines: kinetics and new aqueous ligations for HNO detection and quantitation. *J Am Chem Soc* 2011;**133**:11675–85.
75. Johnson GM, Chozinski TJ, Salmon DJ, Moghaddam AD, Chen HC, Miranda KM. Quantitative detection of nitroxyl upon trapping with glutathione and labeling with a specific fluorogenic reagent. *Free Radic Biol Med* 2013;**63**:476–844.
76. Jorolan JH, Buttitta LA, Cheah C, Miranda KM. Comparison of the chemical reactivity of synthetic peroxynitrite with that of the autoxidation products of nitroxyl or its anion. *Nitric Oxide* 2015;**44**:39–46.
77. Archer S. Measurement of nitric oxide in biological models. *FASEB J* 1993;**7**:349–60.
78. Wink DA, Kasprzak KS, Maragos CM, Elespuru RK, Misra M, Dunams TM, et al. DNA deaminating ability and genotoxicity of nitric oxide and its progenitors. *Science* 1991;**254**:1001–3.
79. Lee MJC, Nagasawa HT, Elberling JA, DeMaster EG. Prodrugs of nitroxyl as inhibitors of aldehyde dehydrogenase. *J Med Chem* 1992;**35**:3648–52.
80. Fukuto JM, Hszieh R, Gulati P, Chiang KT, Nagasawa HT. N,O-Diacylated-N-hydroxyarylsulfonamides: nitroxyl precursors with potent smooth muscle relaxant properties. *Biochem Biophys Res Commun* 1992;**187**:1367–73.
81. Nagasawa HT, Yost Y, Elberling JA, Shirota FN, DeMaster EG. Nitroxyl analogs as inhibitors of aldehyde dehydrogenase. C-nitroso compounds. *Biochem Pharmacol* 1993;**45**:2129–34.
82. Nagasawa HT, Kawle SP, Elberling JA, Demaster EG, Fukuto JM. Prodrugs of nitroxyl as potential aldehyde dehydrogenase inhibitors vis-a-vis vascular smooth muscle relaxants. *J Med Chem* 1995;**38**:1865–71.
83. Lee MJC, Shoeman DW, Goon DJW, Nagasawa HT. N-hydroxybenzenecarboximidic acid derivatives: a new class of nitroxyl-generating prodrugs. *Nitric Oxide* 2001;**5**:278–87.
84. Reisz JA, Bechtold E, King SB. Oxidative heme protein-mediated nitroxyl (HNO) generation. *Dalton Trans* 2010;**39**:5203–12.

85. Kovacic P, Edwards CL. Hydroxamic acids (therapeutics and mechanism): chemistry, acyl nitroso, nitroxyl, reactive oxygen species, and cell signaling. *J Recept Signal Transduct Res* 2011;**31**:10–19.
86. Ware RW, King SB. P-Nitrosophosphate compounds: new N-O heterodienophiles and nitroxyl delivery agents. *J Org Chem* 2000;**65**:8725–9.
87. Zeng BB, Huang JM, Wright MW, King SB. Nitroxyl (HNO) release from new functionalized N-hydroxyurea-derived acyl nitroso-9,10-dimethylanthracene cycloadducts. *Bioorg Med Chem Lett* 2004;**14**:5565–8.
88. King SB. N-hydroxyurea and acyl nitroso compounds as nitroxyl (HNO) and nitric oxide (NO) donors. *Curr Top Med Chem* 2005;**5**:665–73.
89. Huang J, Kim-Shapiro DB, King SB. Catalase-mediated nitric oxide formation from hydroxyurea. *J Med Chem* 2004;**47**:3495–501.
90. Pennington RL, Sha X, King SB. N-Hydroxy sulfonimidamides as new nitroxyl (HNO) donors. *Bioorg Med Chem Lett* 2005;**15**:2331–4.
91. Sha X, Isbell TS, Patel RP, Day CS, King SB. Hydrolysis of acyloxy nitroso compounds yields nitroxyl (HNO). *J Am Chem Soc* 2006;**128**:9687–92.
92. Kirsch M, Büscher AM, Aker S, Schulz R, de Groot H. New insights into the S-nitrosothiol-ascorbate reaction. The formation of nitroxyl. *Org Biomol Chem* 2009;**7**:1954–62.
93. Keceli G, Moore CD, Labonte JW, Toscano JP. NMR detection and study of hydrolysis of HNO-derived sulfinamides. *Biochemistry* 2013;**52**:7387–96.
94. Guthrie DA, Kim NY, Siegler MA, Moore CD, Toscano JP. Development of N-substituted hydroxylamines as efficient nitroxyl (HNO) donors. *J Am Chem Soc* 2012;**134**:1962–5.
95. Saavedra JE, Dunams TM, Flippen-Anderson JL, Keefer LK. Secondary amine/nitric oxide complex ions, $R_2N[N(O)NO]^-$. O-Functionalization chemistry. *J Org Chem* 1992;**57**:6134–8.
96. Saavedra JE, Billiar TR, Williams DL, Kim YM, Watkins SC, Keefer LK. Targeting nitric oxide (NO) delivery in vivo. Design of a liver-selective NO donor prodrug that blocks tumor necrosis factor-alpha-induced apoptosis and toxicity in the liver. *J Med Chem* 1997;**40**:1947–54.
97. Saavedra JE, Shami PJ, Wang LY, Davies KM, Booth MN, Citro ML, et al. Esterase-sensitive nitric oxide donors of the diazeniumdiolate family: in vitro antileukemic activity. *J Med Chem* 2000;**43**:261–9.
98. Makings LR, Tsien RY. Caged nitric oxide. Stable organic molecules from which nitric oxide can be photoreleased. *J Biol Chem* 1994;**269**:6282–5.
99. Saavedra JE, Srinivasan A, Bonifant CL, Chu J, Shanklin AP, Flippen-Anderson JL, et al. The secondary amine/nitric oxide complex ion $R_2N[N(O)NO]^-$ as nucleophile and leaving group in S_NAr reactions. *J Org Chem* 2001;**66**:3090–8.
100. Andrei D, Salmon DJ, Donzelli S, Wahab A, Klose JR, Citro ML, et al. Dual mechanisms of HNO generation by a nitroxyl prodrug of the diazeniumdiolate (NONOate) class. *J Am Chem Soc* 2010;**132**:16526–32.
101. Nandurdikar RS, Keefer LK, Saavedra JE. Novel protection-deprotection strategies in diazeniumdiolate chemistry: synthesis of V-IPA/NO. *Chem Commun* 2011;**47**:6710–12.
102. Nandurdikar RS, Maciag AE, Cao Z, Keefer LK, Saavedra JE. Diazeniumdiolated carbamates: a novel class of nitric oxide donors. *Bioorg Med Chem* 2012;**20**:2025–9.
103. Basudhar D, Bharadwaj G, Cheng RY, Jain S, Shi S, Ridnour LA, et al. Synthesis and comparison of nitroxyl and nitric oxide releasing diazeniumdiolate-based aspirin derivatives. *J Med Chem* 2013;**56**:7804–20.

104. Cohen AD, Zeng BB, King SB, Toscano JP. Direct observation of an acyl nitroso species in solution by time-resolved IR spectroscopy. *J Am Chem Soc* 2003; **125**:1444–5.
105. Holland RJ, Paulisch R, Cao Z, Keefer LK, Saavedra JE, Donzelli S. Enzymatic generation of the NO/HNO-releasing IPA/NO anion at controlled rates in physiological media using β-galactosidase. *Nitric Oxide* 2013;**35**:131–1316.
106. Keeble JE, Moore PK. Pharmacology and potential therapeutic applications of nitric oxide-releasing non-steroidal anti-inflammatory and related nitric oxide-donating drugs. *Br J Pharmacol* 2002;**137**:295–310.
107. Bolla M, Almirante N, Benedini F. Therapeutic potential of nitrate esters of commonly used drug. *Curr Top Med Chem* 2005;**5**:707–20.
108. Velázquez C, Rao PNP, Knaus EE. Novel nonsteroidal antiinflammatory drugs possessing a nitric oxide donor diazen-1-ium-1,2-diolate moiety: design, synthesis, biological evaluation, and nitric oxide release studies. *J Med Chem* 2005;**48**:4061–7.
109. Velázquez CA, Rao PNP, Citro ML, Keefer LK, Knaus EE. O^2-Acetoxymethyl-protected diazeniumdiolate-based NSAIDs (NONO-NSAIDs): synthesis, nitric oxide release, and biological evaluation studies. *Bioorg Med Chem* 2007;**15**:4767–74.
110. Allison MC, Howatson AG, Torrance CJ, Lee FD, Russell RI. Gastrointestinal damage associated with the use of nonsteroidal antiinflammatory drugs. *N Engl J Med* 1992;**327**:749–54.
111. Wallace JL. Nonsteroidal anti-inflammatory drugs and gastroenteropathy: the second hundred years. *Gastroenterol* 1997;**112**:1000–16.
112. Wallace JL, Del Soldato P. The therapeutic potential of NO-NSAIDs. *Fundam Clin Pharmacol* 2003;**17**:11–20.
113. Rigas B, Kashfi K. Nitric-oxide-donating NSAIDs as agents for cancer prevention. *Trends Mol Med* 2004;**10**:324–30.
114. Harris RE, Beebe-Donk J, Doss H, Burr Doss D. Aspirin, ibuprofen, and other non-steroidal anti-inflammatory drugs in cancer prevention: a critical review of non-selective COX-2 blockade. *Oncol Rep* 2005;**13**:559–83.
115. Rigas B. Novel agents for cancer prevention based on nitric oxide. *Biochem Soc Trans* 2007;**35**:1364–8.
116. Jain S, Tran S, El Gendy MA, Kashfi K, Jurasz P, Velázquez-Martínez CA. Nitric oxide release is not required to decrease the ulcerogenic profile of nonsteroidal anti-inflammatory drugs. *J Med Chem* 2012;**55**:688–96.
117. Keefer LK. Nitric oxide-releasing compounds: from basic research to promising drugs. *Chem Tech* 1998;**28**:0–35.
118. Basudhar D, Cheng RC, Bharadwaj G, Ridnour LA, Wink DA, Miranda KM. Chemotherapeutic potential of diazeniumdiolate-based aspirin prodrugs in breast cancer. *Free Radic Biol Med* 2015;**83**:101–14.
119. Huang Z, Kaur J, Bhardwaj A, Alsaleh N, Reisz JA, DuMond JF, et al. O^2-Sulfonylethyl protected isopropylamine diazen-1-ium-1,2-diolates as nitroxyl (HNO) donors: synthesis, β-elimination fragmentation, HNO release, positive inotropic properties, and blood pressure lowering studies. *J Med Chem* 2012;**55**:10262–71.
120. Atkinson RN, Storey BM, King SB. Reactions of acyl nitroso compounds with amines: production of nitroxyl (HNO) with the preparation of amides. *Tetrahedron Lett* 1996;**37**:9287–90.
121. Vogt PF, Miller MJ. Development and applications of amino acid derived chiral acylnitroso hetero Diels-Alder reactions. *Tetrahedron* 1998;**54**:1317–48.

122. Streith J, Defoin A. Hetero Diels-Alder reactions with nitroso dienophiles: application to the synthesis of natural product derivatives. *Synthesis (Mass)* 1994;1107–17.
123. Ritter AR, Miller MJA. Amino acid-derived chiral acyl nitroso compounds: diastereoselectivity in intermolecular hetero Diels-Alder reactions. *J Org Chem* 1994;**59**:4602–11.
124. Surman MD, Miller MJ. Synthesis of a conformationally restricted substrate analogue of siderophore biosynthetases. *Org Lett* 2001;**3**:519–21.
125. Jenkins NE, Ware Jr. RW, Atkinson RN, King SB. Generation of acyl nitroso compounds by the oxidation of N-acyl hydroxylamines with the Dess-Martin periodinane. *Synth Commun* 2000;**30**:947–53.
126. Howard JA, Ilyashenko G, Sparkes HA, Whiting A. Development of new transition metal catalysts for the oxidation of a hydroxamic acid with in situ Diels-Alder trapping of the acyl nitroso derivative. *Dalton Trans* 2007;2108–11.
127. Flower KR, Lightfoot AP, Wan H, Whiting A. The development and application of ruthenium catalysed oxidations of a hydroxamic acid and in situ Diels–Alder trapping of the acyl nitroso derivative. *J Chem Soc Perkin Trans* 2002;**1**:2058–64.
128. Chaiyaveij D, Cleary L, Batsanov AS, Marder TB, Shea KJ, Whiting A. Copper(II)-catalyzed room temperature aerobic oxidation of hydroxamic acids and hydrazides to acyl-nitroso and azo intermediates, and their Diels-Alder trapping. *Org Lett* 2011;**13**:3442–5.
129. Martin SF, Hartmann M, Josey JA. Diastereoselective [4 + 2] cycloadditions of acyl nitroso compounds. *Tetrahedron Lett* 1992;**33**:3583–6.
130. Iwasa S, Tajima K, Tsushima S, Nishiyama H. A mild oxidation method of hydroxamic acids: efficient trapping of acyl nitroso intermediates. *Tetrahedron Lett* 2001;**42**:5897–9.
131. Palmer LI, Frazier CP, de Alaniz JR. Developments in nitrosocarbonyl chemistry: mild oxidation of N-substituted hydroxylamines leads to new discoveries. *Synthesis (Mass)* 2014;**46**:269–80.
132. Davey MH, Lee VY, Miller RD, Marks TJ. Synthesis of aryl nitroso derivatives by tert-butyl hypochlorite oxidation in homogeneous media. Intermediates for the preparation of high-hyperpolarizability chromophore seletons. *J Org Chem* 1999;**64**: 4976–9.
133. Quadrelli P, Mella M, Caramella P. A photochemical generation of nitrosocarbonyl intermediates. *Tetrahedron Lett* 1999;**40**:797–800.
134. Quadrelli P, Invernizzi AG, Caramella P. A mild oxidation of nitrile oxides: a new synthetic route to nitroso carbonyl intermediates. *Tetrahedron Lett* 1996;**37**:1909–12.
135. Xu YP, Alavanja MM, Johnson VL, Yasaki G, King SB. Production of nitroxyl (HNO) at biologically relevant temperatures from the retro-Diels-Alder reaction of N-hydroxyurea-derived acyl nitroso-9,10-dimethylanthracene cycloadducts. *Tetrahedron Lett* 2000;**41**:4265–9.
136. Corrie JET, Kirby GW, Mackinnon JWM. Reactions of transient C-nitrosocarbonyl compounds with dienes, monoolefins, and nucleophiles. *J Chem Soc Perkin Trans* 1985;**1**:883–6.
137. O'Bannon PE, Dailey WP. Rearrangement of nitrocarbenes to acyl nitroso-compounds. *Tetrahedron Lett* 1988;**29**:5719–22.
138. Kirby GW, Sweeny JG. Formation and dienophilic reactions of transient C-nitrosocarbonyl compounds. *J Chem Soc Perkin Trans* 1981;**1**:3250–4.
139. O'Bannon PE, Sulzle D, Dailey WP, Schwarz H. Spectroscopic detection of two neutral [$CHNO_2$] isomers – nitrocarbene and nitrosoformaldehyde. *J Am Chem Soc* 1992;**114**:344–5.

140. Bodnar BS, Miller MJ. The nitrosocarbonyl hetero-Diels-Alder reaction as a useful tool for organic syntheses. *Angew Chem Int Ed Engl* 2011;**50**:5630–47.
141. Miller A, Procter G. On the origin of the diastereoselectivity observed in cycloadditions of chiral acyl-nitroso compounds. *Tetrahedron Lett* 1990;**31**:1041–2.
142. Gouverneur V, McCarthy SJ, Mineur C, Belotti D, Dive G, Ghosez L. New acylnitroso compounds for the asymmetric oxyamination of dienes. *Tetrahedron* 1998;**54**: 10537–54.

Hydroxylamines With Organic-Based Leaving Groups as HNO Donors

3

D.A. Guthrie, S. Nourian and J.P. Toscano
Johns Hopkins University, Baltimore, MD, United States

Chapter Outline

3.1 Introduction

Hydroxylamine derivatives with good leaving groups substituted on nitrogen (HOHN−X) have the propensity to generate HNO (azanone, nitroxyl) under physiologically relevant conditions through a base-catalyzed deprotonation mechanism where the rate and amount of HNO production is dependent on pH, temperature, and the nature of the X group (Scheme 3.1).[1−3] Since our goal is to release HNO under mild conditions (i.e., pH 7.4, 37°C), we require relatively unstable *N*-substituted hydroxylamines. Unfortunately, due to this metastability, competing, non-HNO producing mechanisms are quite common for this general class of compounds. Therefore, much of the research done in this field is focused on determining appropriate leaving groups (i.e., X groups) that heavily favor HNO generation over the non-HNO producing decomposition mechanisms, such as oxidations, rearrangements, and hydrolysis-type mechanisms (*vide infra*). Another aim of this field is to develop specific HNO donor classes with tunable HNO release profiles, which is predicted to have a broad impact on optimizing HNO release rate for the treatment of any number of potential diseases, including congestive heart failure, alcoholism, vascular dysfunction, and cancer. Fortunately, the use of organic-based leaving groups has garnered dozens of HNO donors with half-lives ranging from minutes to hours under physiological conditions.[4−6]

The Chemistry and Biology of Nitroxyl (HNO). DOI: http://dx.doi.org/10.1016/B978-0-12-800934-5.00003-7

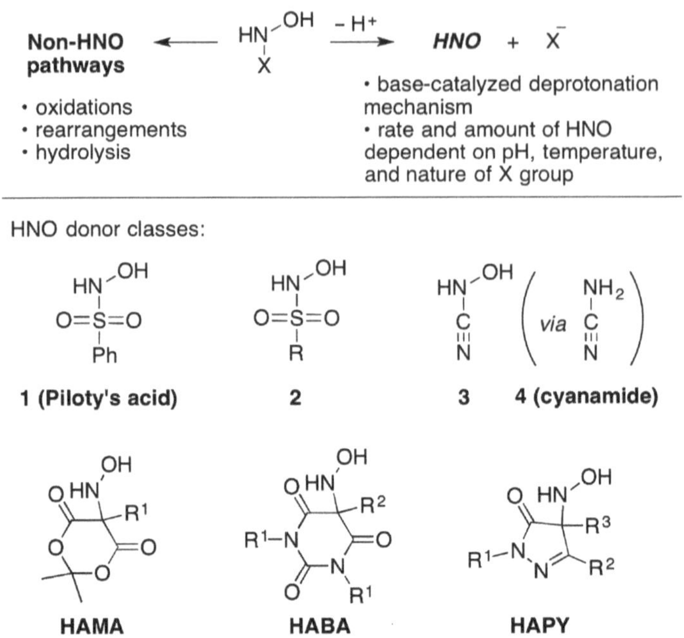

Scheme 3.1 General strategy for HNO release and related donor classes.

The most biologically relevant case of alternative, non-HNO producing decomposition mechanisms within the general class of *N*-substituted hydroxylamines occurs with the initial product of the reaction of HNO and a thiol (i.e., *N*-hydroxysulfenamide, HOHN−SR). This unstable intermediate does not reform HNO but rather rearranges to a sulfinamide ($R-S(O)NH_2$) or is oxidized to a disulfide (R−SS−R) and hydroxylamine depending on the concentration of thiol (see also chapter: Reactions of HNO With Nonheme Proteins).[7–14] On the other hand, these non-HNO producing mechanisms are not observed in the decomposition of Piloty's acid **1** and its derivatives **2** with sulfinate ($R-SO_2^-$) leaving groups (Scheme 3.1). Rather, Piloty's acid **1** and its derivatives **2** are susceptible to oxidation, producing NO along with the sulfinate leaving group (see also chapter: Azanone (HNO) Reactions With Metal Porphyrins and Corroles).[15] However, the HNO producing mechanism (Scheme 3.1) can prevail, particularly under basic, anaerobic conditions.[16]

Beyond Piloty's acid, Nagasawa has shown that cyanamide **4**, an alcohol deterrent used clinically in Europe, Canada, and Japan, generates HNO following metabolic activation to an unstable *N*-hydroxycyanamide **3** intermediate (Scheme 3.1).[17–19] Although *N*-hydroxycyanamide **3** is unstable and has not yet been isolated to confirm its reactivity, Nagasawa has demonstrated that bioactivated cyanamide **4** can hydrolyze to five end products: HNO, HCN, HNO_2, CO_2, and NH_3 (not measured).[18] As discussed later in this chapter, direct dissociation of *N*-hydroxycyanamide **3** to HNO and HCN, although possible, is one of several possible HNO and non-HNO decomposition pathways.

As organic chemists studying HNO, an *inorganic* compound, we strive to have more examples beyond cyanide where carbon is bound directly to the nitrogen of hydroxylamine. We have developed new *N*-substituted hydroxylamine derivatives with good carbon-bound leaving groups based on (hydroxylamino) Meldrum's acid **HAMA**, (hydroxylamino) barbituric acid **HABA**, and (hydroxylamino) pyrazolone **HAPY** scaffolds such that HNO is released along with a stable carbanion at neutral pH (Scheme 3.1).[3–5] Following a discussion concerning general synthetic strategies, this chapter will focus on the development of these new hydroxylamines with organic-based leaving groups as physiologically useful HNO donors.

3.2 Synthesis

Following the classical procedure traditionally used to synthesize sulfonamides, Piloty's acid **1** and its derivatives **2** are prepared through the condensation of hydroxylamine with a sulfonyl chloride in basic media (Scheme 3.2). Given the large variety of synthetic and commercially available sulfonyl chlorides, many Piloty's acid derivatives have been synthesized and examined for HNO production.[2,20–23] Recently, a modified procedure reports the use of MgO as a base for the efficient, chemoselective and mild coupling of hydroxylamine with different alkyl, aryl, and heteroaryl sulfonyl chlorides in moderate to excellent yields.[21–22] Catalytic dimethylaminopyridine (DMAP) in pyridine also provides suitable conditions for the coupling of hydroxylamine with sulfonyl chlorides.[23] Most Piloty's acid derivatives **2** are isolated as stable solids, but as a cautionary note, some derivatives can spontaneously decompose in the course of a few hours even in the solid form.[22]

The bromination–displacement–deprotection synthetic strategy for the synthesis of the **HABA** donor **6** is shown in Scheme 3.3.[3] Barbiturate **5**, in the presence of triethylamine (TEA), readily brominates on the carbon to be equipped with hydroxylamine. The *N*-selective nucleophilic displacement of the bromide by *N,O-bis(tert*-butoxycarbonyl) hydroxylamine under basic conditions overnight gives the desired *bis*-Boc-protected product as a gummy solid. Following acid deprotection and recrystallization, HABA donor **6** is isolated as a stable white solid that can be

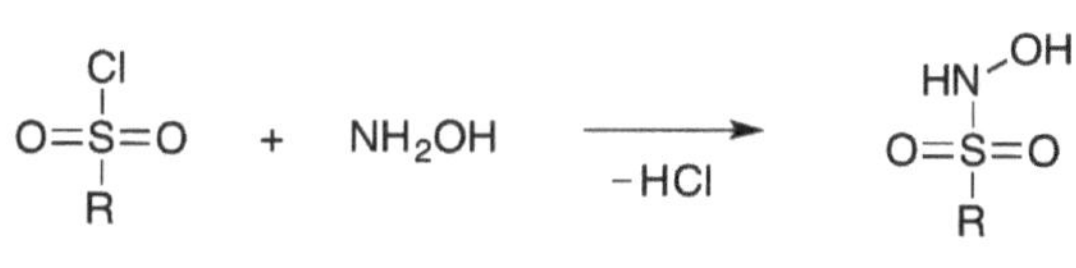

2

Conditions:
(1) $NH_2OH{\cdot}HCl$, $NaHCO_3$, THF, H_2O[2]
(2) aq NH_2OH, THF[20]
(3) $NH_2OH{\cdot}HCl$, MgO, MeOH, THF, H_2O[21, 22]
(4) $NH_2OH{\cdot}HCl$, DMAP, pyridine[23]

Scheme 3.2 Synthesis of Piloty's acid derivatives **2**.

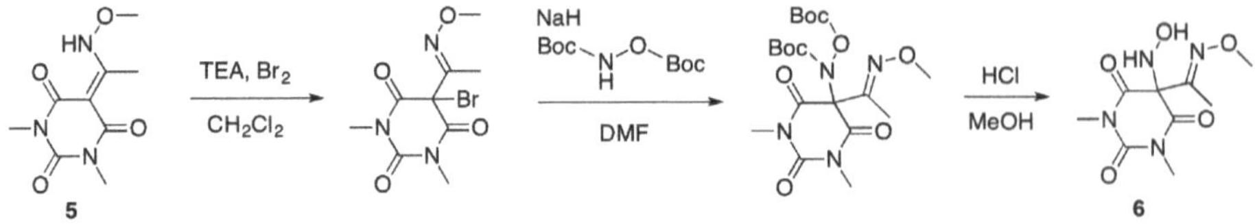

Scheme 3.3 Bromination–displacement–deprotection synthetic strategy.

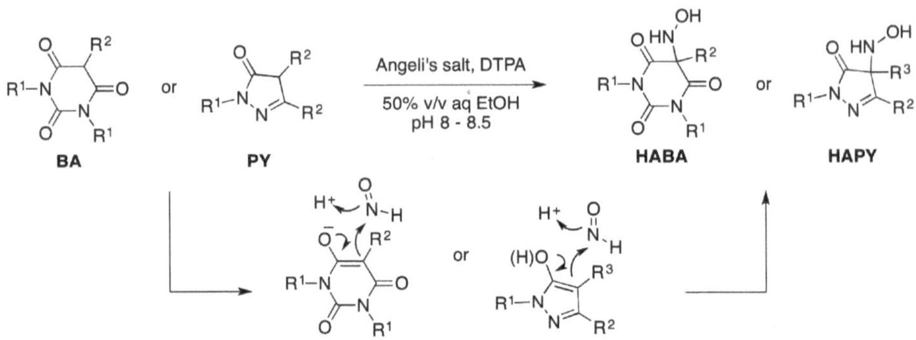

Scheme 3.4 HNO-aldol synthetic strategy.

stored under air at 4°C for, in some instances, greater than 5 years, which is significant considering that HABA donor **6** has a very short half-life at pH 7.4, 37°C of 0.7 min.[3] This synthetic strategy has also proven effective for other **HAMA**, **HABA**, and **HAPY** donors.[3]

The inorganic salt, Angeli's salt ($Na_2N_2O_3$), so named after the Italian chemist Angelo Angeli, was first prepared in 1896, and to this day remains the most widely used HNO donor for chemical and biological activity studies.[24–31] Angeli's salt has most recently been used as a reagent in the synthesis of many new **HABA** and **HAPY** derivatives (Scheme 3.4).[4,5] Angeli's salt-derived HNO reacts with barbiturates **BA** or pyrazolones **PY** via an HNO-aldol reaction to give the corresponding **HABA** and **HAPY** derivatives, respectively, typically in moderate to excellent yields. Diethylenetriaminepentaacetic acid (DTPA) is used primarily to maintain an optimal buffered solution (pH 8.0–8.5) for this reaction. The HNO-aldol reaction has been effective in broadening the scope of **HABA** and **HAPY** donors; however, in some instances, the traditional synthetic strategy (e.g., **HABA** donor **6**, Scheme 3.3) is advantageous.

3.3 Piloty's acid and its derivatives

In 1896, the German chemist, Oskar Piloty, published on benzenesulfohydroxamic acid,[32] known today as Piloty's acid **1**, which beyond its HNO donating capabilities, has found use throughout the 20th century in the Angeli-Rimini

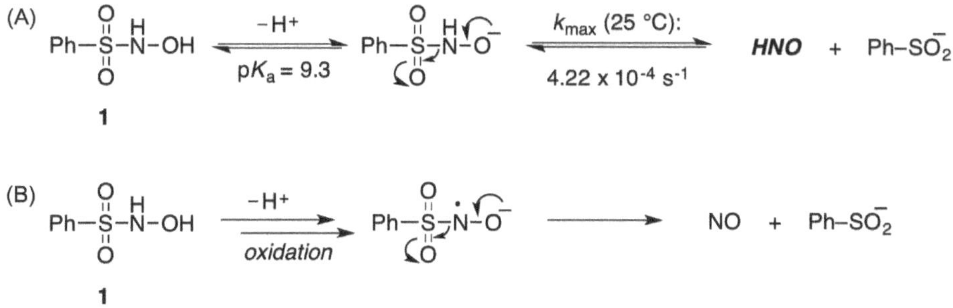

Scheme 3.5 HNO and NO-producing pathways of Piloty's acid **1**.

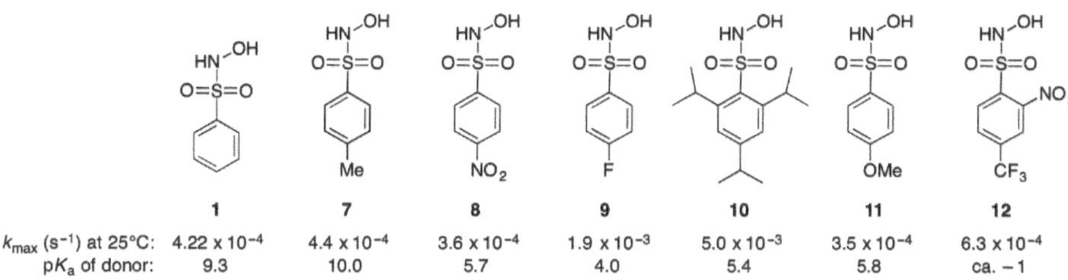

Figure 3.1 Steric and electronic effects on the pK_a and HNO release rate for Piloty's acid **1** and its derivatives **7–12**.[22]

colorimetric assay for aldehydes,[33] as well as in the synthesis of hydroxamic acids,[34] *N*-hydroxylactams,[35] and dimethylacetals.[36] Piloty's acid is an *N*-substituted hydroxylamine derivative equipped with a sulfinate leaving group and releases HNO under alkaline, anaerobic conditions (Scheme 3.5A).[16] However, under physiological conditions it is oxidized, presumably to the corresponding nitroxide, and becomes an NO donor (Scheme 3.5B).[15] We and others have added electron withdrawing groups or otherwise bulky groups to the phenyl ring in order to favor the HNO formation pathway.[6,20–23]

Recently, Doctorovich and coworkers synthesized and examined a number of Piloty's acid derivatives (**7–12**) that display a wide range of pK_a values (-1 to 10) (Fig. 3.1).[22] Regarding the acidity of Piloty's acid **1** itself, both hydroxylamine protons are acidic depending on the conditions, as **1** is considered an *N*-acid in the solid-state[37] or in dioxane,[38] yet in aqueous media, the dissociable hydrogen in **1** is bound to the oxygen rather than to nitrogen.[16] Nevertheless, HNO release is at its maximum rate in the range of 10^{-3} to 10^{-4}/s (25°C) at pH values where Piloty's acid **1** and its derivatives **2** are fully deprotonated (Fig. 3.1). This suggests a modest barrier to HNO release from the oxyanion ($^{-}OHN-SO_2R$), which could be a reflection of the reversible generation of HNO, as sodium benzenesulfinate is known to slow the decomposition rate of Piloty's acid **1**.[16] In addition, crystallographic studies show that Piloty's acid derivatives substituted with electron withdrawing

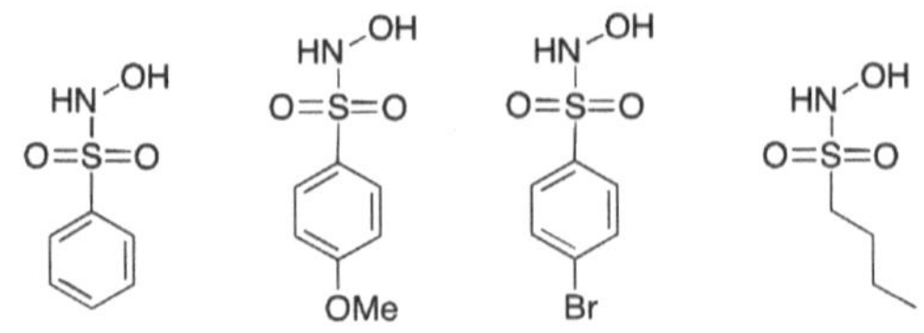

	1	11	13	14
k (M^{-1} s^{-1}): pH 7.0, 25°C $[Fe(bipy)_2(CN)_2]^+$	610	740	1100	230

Figure 3.2 Steric and electronic effects on NO release rate for the oxidative decomposition of Piloty's acid **1** and its derivatives **11**, **13**, and **14**.[39]

groups tend to have a shorter N–O bond distance that corresponds to a more facile HNO generation.[22]

The vasodilator activity of Piloty's acid **1** is attributed to its oxidation and subsequent release of NO, rather that HNO, at physiological conditions, and quite weak oxidants, such as dioxygen and hydrogen peroxide, are sufficient to promote this reaction.[15] The oxidative degradation of Piloty's acid **1** in the presence of the one-electron oxidant, hexacyanoferrate(III), is more facile at pH 8 than pH 7.4, indicating a strong pH dependence.[15] However, a series of Piloty's acid derivatives, **11**, **13**, and **14**, were synthesized and shown to have similar reactivity with a variety of metal oxidants (Fig. 3.2),[39] highlighting the difficulty in developing strategies that disfavor the oxidative, NO-producing pathway to favor the base-catalyzed, HNO producing pathway.

As a result, the most effective HNO donors in the Piloty's acid class have been restricted to relatively fast HNO release profiles with half-lives on the order of a few minutes. As shown in Fig. 3.3, Piloty's acid derivative **15** was developed as one of the first alternative HNO donors to Angeli's salt that had comparable HNO yield and release rate under physiological conditions ($t_{1/2}$ c.2 min at pH 7.4, 37°C).[6] This led to Piloty's acid derivative **16** (CXL-1020), which was used to provide the first-in-man evidence for HNO's potential usefulness in heart failure (Fig. 3.3).[40]

3.4 Cyanamide

Nagasawa and coworkers were among the first to indicate the pharmacological utility and thiophilic nature of HNO through their work with the alcohol aversion therapeutic, cyanamide **4**.[17–19] HNO has been shown to be a potent inhibitor of the enzyme aldehyde dehydrogenase (AlDH) through a reaction with the active site cysteine residue (Cys-302).[41] AlDH catalyzes the conversion of the acetaldehyde generated in the oxidative metabolism of ethanol to acetate, and AlDH inhibition in combination with consumption of alcohol results in acetaldehydemia, a condition

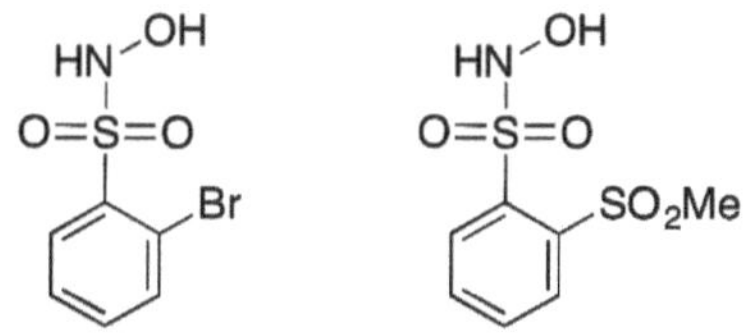

15 16 (CXL-1020)

Figure 3.3 Piloty's acid derivatives **15** and **16**.

with general "hangover"-like symptoms such as facial flushing, light-headedness, palpitations, and nausea.

The alcohol-deterrent activity of cyanamide **4** has a rapid onset and relatively short duration, as AlDH inhibition is maximal 1–2 h after drug administration, and AlDH activity is restored to 80% of control activity within 24 h, with complete recovery occurring after 6 days.[42] This indicates both reversible and irreversible AlDH inhibition. The inhibition of AlDH by cyanamide **4** is dependent on a catalase-mediated oxidation of cyanamide **4** to an active metabolite, *N*-hydroxycyanamide **3**, which can dissociate to HNO and HCN (Scheme 3.6).[17] However, the major metabolic pathway for cyanamide **4** in rat, rabbit, dog, and man involves acetylation catalyzed by *N*-acetyltransferase to give the acetylated derivative **17** (up to 94%), which, like cyanamide **4**, is devoid of AlDH inhibitory activity in vitro and is rapidly excreted in the urine.[43] As a result, in Canada and other countries where cyanamide **4** is approved for safety and efficacy, cyanamide **4** has not been widely used because of its short duration of alcohol-deterrent activity.[42]

Cyanamide **4** has been shown to generate HNO and HCN following metabolic activation. However, *N*-hydroxycyanamide **3** is unstable and has not yet been isolated to confirm its reactivity. In addition, evidence exists that it can be further oxidatively metabolized to a nitrosyl cyanide **18** that can also generate HNO following hydrolysis of the nitrile group to form an acyl nitroso compound (Scheme 3.6).[18] Moreover, both bioactivated cyanamide **4** and synthetic nitrosyl cyanide **18** can hydrolyze to five end products: HNO, HCN, HNO_2, CO_2, and NH_3 (not measured). However, the hydrolysis of nitrosyl cyanide **18** to form HNO is relatively slow, and it has been suggested that **18** itself is the inhibitory species of AlDH following cyanamide activation by *S*-nitrosylating the active site cysteine directly. Due to its instability, protected derivatives of *N*-hydroxycyanamide **3** have been synthesized and examined. An *N*,*O*-dibenzoyl derivative of *N*-hydroxycyanamide **3** has been reported, but this precursor releases HNO not via *N*-hydroxycyanamide **3**, but presumably via an acyl nitroso intermediate, and only following treatment with esterase or base.[44]

3.5 Hydroxylamines with carbon-bound leaving groups

Hydrogen cyanide is a weak carbon-based acid (pK_a of 9.2). With *N*-hydroxycyanamide **3** as a starting point, we recognized that the development of hydroxylamines

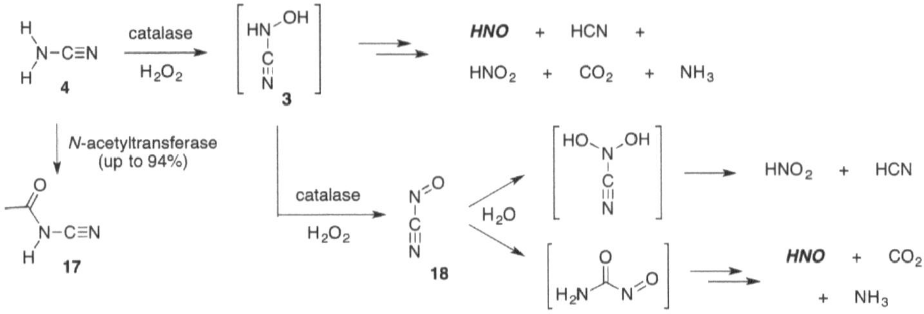

Scheme 3.6 Oxidative metabolism of cyanamide.

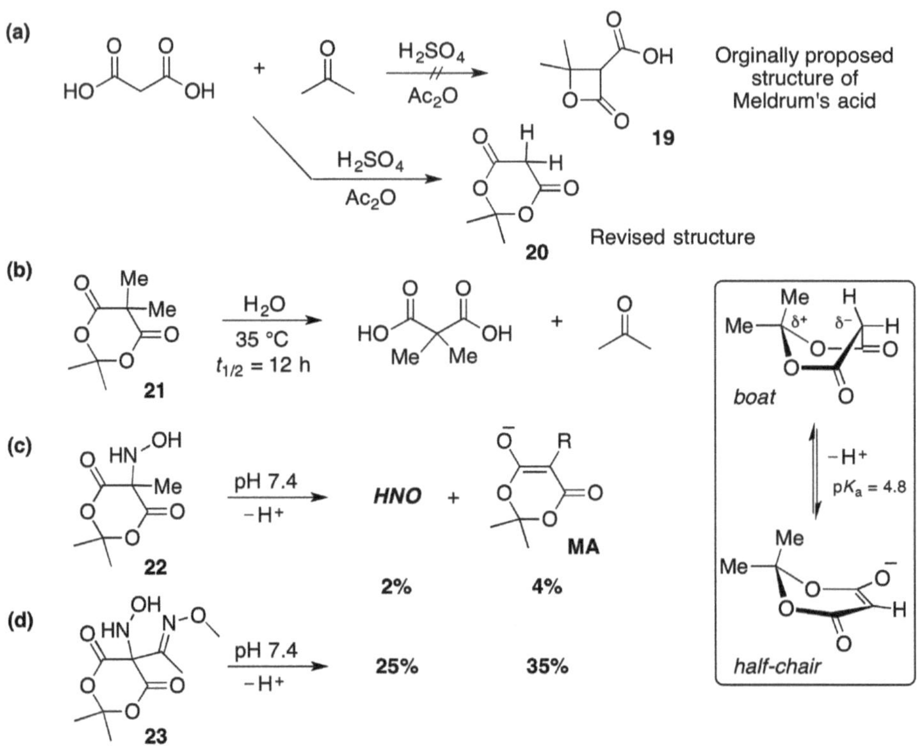

Scheme 3.7 HNO and non-HNO release decomposition pathways of **HAMA** derivatives.

with new carbon-bound leaving groups would require carbon-based acids of equal or greater strength to that of hydrogen cyanide. The acidity of Meldrum's acid **20** in aqueous solution (pK_a = 4.8) is comparable to that of acetic acid (pK_a = 4.75).[45] So understandably, in 1908 Andrew Meldrum wrongly assigned its structure as the β-lactone of β-hydroxyisopropymalonic acid **19** (Scheme 3.7A), especially in light of the fact that carbon dioxide is lost upon heating.[46] Forty years elapsed before the correct structure was deduced, as the properties of Meldrum's acid are comparable to other 1,3-cyclic diones, such as dimedone and barbituric acid.[47] Indeed, the acid-catalyzed condensation product of malonic acid and acetone is the *bis*-lactone of 2,2-dimethyl-1,3-dioxane-4,6-dione **20**, which exists primarily in the boat conformation due to the electrostatic interaction between the C-2 (δ^+) and C-5 (δ^-) carbons.[48,49] Upon deprotonation, there is substantial resonance stabilization with the adjacent carbonyl groups, and the resultant carbanion adopts a half-chair conformation.[49]

However, the boat conformation makes Meldrum's acid and its derivatives susceptible to nucleophilic attack at the carbonyl carbons, particularly in 5,5-disubstituted examples. For example, the dehydration process in the preparation of Meldrum's acid is reversible under hydrolytic conditions, such as in Scheme 3.7B, where 5,5-dimethyl Meldrum's acid **21** has a hydrolysis rate of 12 h at neutral pH, 35°C.[50] Nevertheless, we reasoned that HNO release would be competitive with hydrolysis in the **HAMA** scaffold (Scheme 3.7C,D).[3] To our surprise, the hydrolysis pathway is rapid in the **HAMA** donors **22** and **23**, as the major product observed by ^{1}H NMR spectroscopy is acetone. Evidently, a more robust scaffold that is stable to hydrolysis is necessary to have a good HNO donor.

Barbituric acid is another carbon-based acid that features a more robust ring to hydrolysis when compared to Meldrum's acid. Although **HABA** donor **6** has proven to be an efficient HNO donor with a half-life at pH 7.4, 37°C of 0.7 min, examination of **HABA** donor **24** reveals yet another non-HNO producing decomposition mechanism for *N*-substituted hydroxylamines.[3,5] Rather than forming HNO and the corresponding **BA** derivative, **HABA** donor **24** rearranges to hydantoin **25**, thus potentially limiting the scope of HNO release rates in the **HABA** class.

Recently, the factors that favor the rearrangement pathway have been examined and two independent strategies that protect against rearrangement to favor HNO generation have been developed (Scheme 3.8).[5] 5-Benzyl **HABA** donor **26** and 1,3-unsubstituted derivatives, such as **HABA** donor **28**, are inherently protected from nucleophilic attack by the hydroxylamine nitrogen, as the rate of rearrangement is substantially reduced when compared to **HABA** donor **24**. The rearrangement rate is also much slower in barbiturate **27**, the *O*-methylated derivative of **24**, suggesting H-bond activation is important for rearrangement, as is also likely for HNO release. As a result of increasing the barrier to rearrangement, **HABA** donors **26** and **28** have extended the HNO release profile of the **HABA** class to examples with half-lives on the order of an hour (*vide infra*).

The **HAPY** class of HNO donors offers several distinct advantages: (1) the pyrazolone ring structure is stable to non-HNO producing hydrolysis and rearrangement mechanisms; (2) multiple synthetic handles are available to investigate structure–activity relationships; and (3) the organic byproducts **PY** are related to edaravone, a potent antioxidant already in clinical use for the treatment of stroke and cardiovascular disease.[51,52] At pH 7.4, **HAPY** derivatives can be quite stable, as they have the ability to undergo reversible generation of HNO (Scheme 3.9).[4] Of note, this reversibility is also observed in **HABA**[5] and **HAMA**[53] derivatives.

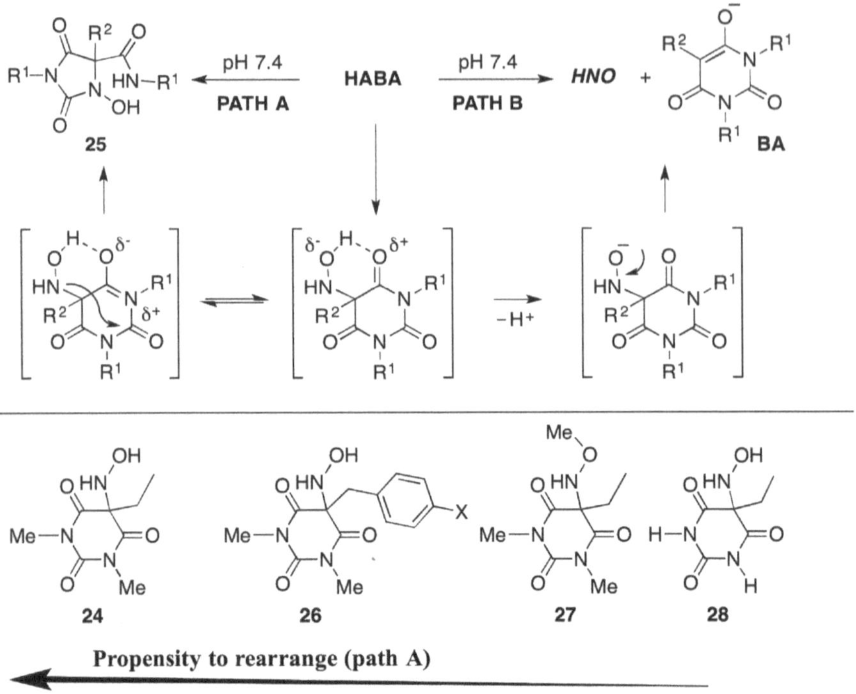

Scheme 3.8 HNO and non-HNO release decomposition pathways of **HABA** derivatives.

Although bimolecular rate constants for the reaction of HNO with **PY** at pH 7.4, 37°C can reach 8×10^5/M/s, HNO formation is rendered irreversible in the presence of slight excess of an irreversible trap for HNO, such as the thiol, glutathione (GSH), or the phosphine, *tris*(4,6-dimethyl-3-sulfanatophenyl)phosphine trisodium salt (TXPTS).[4] The HNO release rate is highly tunable, spanning three orders of magnitude, by varying the groups at the R^1, R^2, and R^3 positions on the pyrazolone ring; the relative effect of the R-position on half-life is $R^3 > R^2 > R^1$.[4]

Like Piloty's acid **1** and its derivatives **2** (Scheme 3.5A), HNO release is pH dependent and expected to occur through the oxyanion ($^-$OHN−X) of each **HAMA**, **HABA**, and **HAPY** derivative based on experimental and computational results (Fig. 3.4A).[3–5] Hence, the pK_a value of the donor itself is important to HNO release. However, unlike Piloty's acid **1** and its derivatives **2**, the hydroxylamine group on these carbon-bound derivatives remains largely neutral at physiologically pH, but the relatively high pK_a value range is compensated by the high rate of HNO release ($k_{max} > 0.15$/s) from the corresponding oxyanion ($^-$OHN−X), which indicates a very low barrier to HNO formation once the oxyanion is formed. **HAMA**, **HABA**, and **HAPY** derivatives have been examined for HNO release at pH 7.4, 37°C utilizing UV−vis and ^{1}H NMR spectroscopy in the presence of a good trap for HNO, such as GSH or TXPTS, respectively (Fig. 3.4B).[3–5] Based on

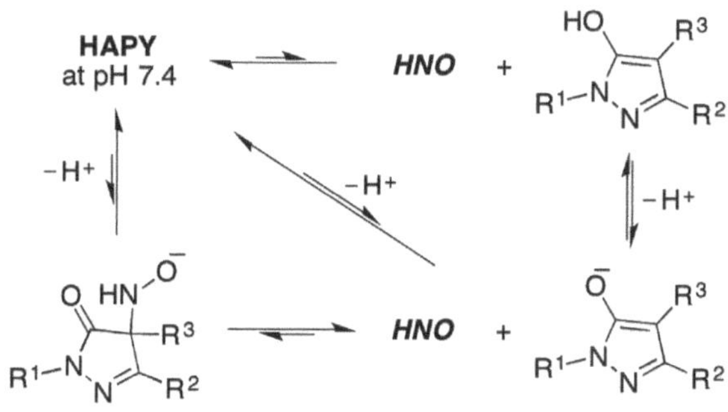

Scheme 3.9 Reversible generation of HNO for **HAPY** derivatives.

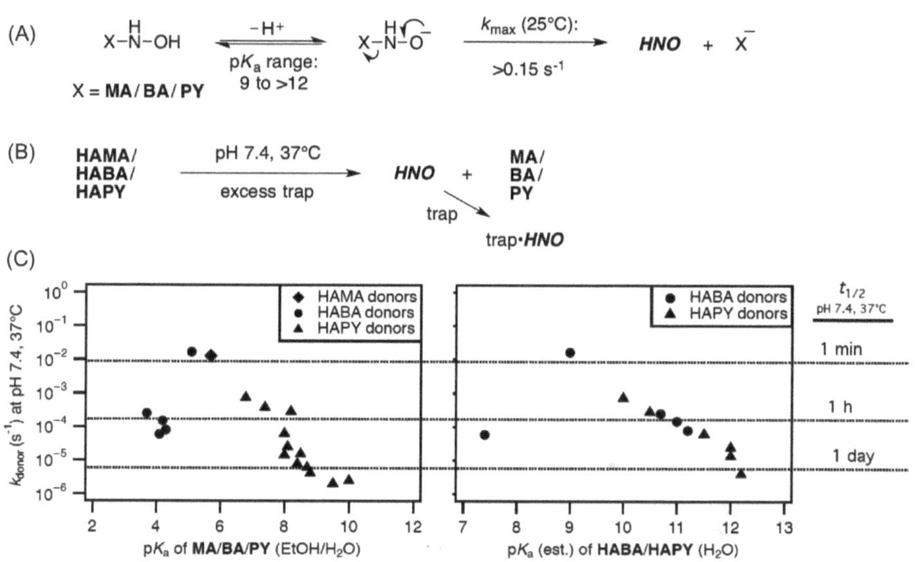

Figure 3.4 Decomposition of HNO donors **HAMA**, **HABA**, and **HAPY** (A) under alkaline conditions, (B) under neutral conditions in the presence of a chemical trap for HNO, and (C) rates of decomposition at pH 7.4, 37°C plotted as a function of byproduct **MA/BA/PY** and donor **HABA/HAPY** pK_a.[3–5]

the kinetic data from these experiments, the correlation between HNO release rate and pK_a is shown in Fig. 3.4C. Although there is a correlation between HNO release and the corresponding byproduct pK_a, there are a few unexplained exceptions. However, the correlation is strong between HNO release and the corresponding donor pK_a (estimated). The one **HABA** donor outlier is the 1,3-unsubstituted derivative **28** (Scheme 3.8), where the ring nitrogen proton is more acidic (pK_a = 7.4) than the hydroxylamine group.

3.6 Summary

Despite the historical significance in regards to HNO donor development, Piloty's acid **1** and cyanamide **4** are not ideal candidates for therapeutic treatments responsive to HNO. Under physiological conditions, Piloty's acid **1** is readily oxidized and becomes an NO donor. Fortunately, the organic framework of **1** allows for a wide range of Piloty's acid derivatives **2** to be examined, and a number of examples readily form HNO rather than NO under physiological conditions. To avoid the release of toxic cyanide in the oxidative metabolism of cyanamide, alternative *N*-substituted hydroxylamines with carbon-bound leaving groups suitable for HNO generation with a broad range of release rates at neutral pH without enzymatic activation have been developed.

Acknowledgment

J.P.T. gratefully acknowledges the National Science Foundation (CHE-1213438) and Cardioxyl Pharmaceuticals for generous support of this research.

References

1. Miranda KM, Nagasawa HT, Toscano JP. Donors of HNO. *Curr Top Med Chem* 2005;**5**:649–64.
2. King B, Nagasawa HT. Chemical approaches toward generation of nitroxyl. *Methods enzymol* 1999;**301**:211–20.
3. Guthrie DA, Kim NY, Siegler MA, Moore CD, Toscano JP. Development of N-substituted hydroxylamines as efficient nitroxyl (HNO) donors. *J Am Chem Soc* 2012;**134**:1962–5.
4. Guthrie DA, Ho A, Takahashi CG, Collins A, Morris M, Toscano JP. "Catch-and-Release" of HNO with pyrazolones. *J Org Chem* 2015;**80**:1338–48.
5. Guthrie DA, Nourian S, Takahashi CG, Toscano JP. Curtailing the hydroxylaminobarbituric acid–hydantoin rearrangement to favor HNO generation. *J Org Chem* 2015;**80**:1349–56.
6. Toscano, J.P.; Brookfield, F.A.; Cohen, A.D.; Courtney, S.M.; Frost, L.M.; Kalish, V.J. *Preparation of N-hydroxylsulfonamide derivatives as nitroxyl (HNO) donors. US 20110306614.*
7. Doyle MP, Mahapatro SN, Broene RD, Guy JK. Oxidation and reduction of hemoproteins by trioxodinitrate-(II). The role of nitrosyl hydride and nitrite. *J Am Chem Soc* 1988;**110**:593–9.
8. Wong PSY, Hyun J, Fukuto JM, Shirota FN, DeMaster EG, Shoeman DW, et al. Reaction between S-nitrosothiols and thiols: generation of nitroxyl (HNO) and subsequent chemistry. *Biochemistry* 1998;**37**:5362–71.
9. Keceli G, Toscano JP. Reactivity of C-terminal cysteines with HNO. *Biochemistry* 2014;**53**:3689–98.

10. Flores-Santana W, Salmon DJ, Donzelli S, Switzer CH, Basudhar D, Ridnour L, et al. The specificity of nitroxyl chemistry is unique among nitrogen oxides in biological systems. *Antioxid Redox Signal* 2011;**14**:1659–74.
11. Fukuto JM, Cisneros CJ, Kinkade RL. A comparison of the chemistry associated with the biological signaling and actions of nitroxyl (HNO) and nitric oxide (NO). *J Inorg Biochem* 2013;**118**:201–8.
12. Keceli G, Moore CD, Labonte JW, Toscano JP. NMR detection and study of hydrolysis of HNO-derived sulfinamides. *Biochemistry* 2013;**52**:7387–96.
13. Johnson GM, Chozinski TJ, Salmon DJ, Moghaddam AD, Chen HC, Miranda KM. Quantitative detection of nitroxyl upon trapping with glutathione and labeling with a specific fluorogenic reagent. *Free Rad Bio Med* 2013;**63**:476–84.
14. Donzelli S, Espey MG, Thomas DD, Mancardi D, Tocchetti CG, Ridnour LA, et al. Discriminating formation of HNO from other reactive nitrogen oxide species. *Free Rad Bio Med* 2006;**40**:1056–66.
15. Zamora R, Grzesiok A, Weber H, Feelisch M. Oxidative release of nitric oxide accounts for guanylyl cyclase stimulating, vasodilator and anti-platelet activity of Piloty's acid: a comparison with Angeli's salt. *Biochem J* 1995;**312**:333–9.
16. Bonner F, Ko Y. Kinetic, isotopic, and nitrogen-15 NMR study of N-hydroxybenzenesulfonamide decomposition: an nitrosyl hydride (HNO) source reaction. *Inorg Chem* 1992;**31**:2514–19.
17. Nagasawa HT, DeMaster EG, Redfern B, Shirota FN, Goon DJW. Evidence for nitroxyl in the catalase-mediated bioactivation of the alcohol deterrent agent cyanamide. *J Med Chem* 1990;**33**:3120–2.
18. Shirota FN, Goon DJW, DeMaster EG, Nagasawa HT. Nitrosyl cyanide, a putative metabolic oxidation product of the alcohol-deterrent agent cyanamide. *Biochem Pharmacol* 1996;**52**:141–7.
19. DeMaster EG, Redfern B, Nagasawa HT. Mechanisms of inhibition of aldehyde dehydrogenase by nitroxyl, the active metabolite of the alcohol deterrent agent cyanamide. *Biochem Pharmacol* 1998;**55**:2007–15.
20. Toscano, J.P.; Brookfield, F.A.; Cohen, A.D.; Courtney, S.M.; Frost, L.M.; Kalish, V.J. *N-Hydroxylsulfonamide derivatives as new physiologically useful nitroxyl donors. US 0080330356.*
21. Porcheddu A, De Luca L, Giacomelli G. A straightforward route to Piloty's acid derivatives: a class of potential nitroxyl-generating prodrugs. *Synlett* 2009;**13**:2149–53.
22. Sirsalmath K, Suárez SA, Bikiel DE, Doctorovich F. The pH of HNO donation is modulated by ring substituents in Piloty's Acid derivatives: azanone donors at biological pH. *J Inorg Biochem* 2013;**118**:134–9.
23. Aizawa K, Nakagawa H, Matsuo K, Kawai K, Ieda N, Suzuki T, et al. Piloty's acid derivative with improved nitroxyl-releasing characteristics. *Bioorganic Med Chem Lett* 2013;**23**:2340–3.
24. Angeli A. Nitrohydroxylamine. *Gazz Chim Ital* 1896;**26**:17–25.
25. Paolocci N, Saavedra WF, Miranda KM, Martignani C, Isoda T, Hare JM, et al. Nitroxyl anion exerts redox-sensitive positive cardiac inotropy in vivo by calcitonin gene-related peptide signaling. *Proc Natl Acad Sci USA* 2001;**98**:10463–8.
26. Paolocci N, Katori T, Champion HC, John St. ME, Miranda KM, Fukuto JM, et al. Positive inotropic and lusitropic effects of HNO/NO–in failing hearts: independence from β-adrenergic signaling. *Proc Natl Acad Sci USA* 2003;**100**:5537–42.

27. Paolocci N, Jackson MI, Lopez BE, Miranda K, Tocchetti CG, Wink DA, et al. The pharmacology of nitroxyl (HNO) and its therapeutic potential: not just the Janus face of NO. *Pharmacol Ther* 2007;**113**:442–58.
28. Tocchetti CG, Wang W, Froehlich JP, Huke S, Aon MA, Wilson GM, et al. Nitroxyl improves cellular heart function by directly enhancing cardiac sarcoplasmic reticulum Ca^{2+} Cycling. *Circ Res* 2007;**100**:96–104.
29. Froehlich JP, Mahaney JE, Keceli G, Pavlos CM, Goldstein R, Redwood AJ, et al. Phospholamban thiols play a central role in activation of the cardiac muscle sarcoplasmic reticulum calcium pump by nitroxyl. *Biochemistry* 2008;**47**:13150–2.
30. DeMaster EG, Redfern B, Nagasawa HT. Mechanisms of inhibition of aldehyde dehydrogenase by nitroxyl, the active metabolite of the alcohol deterrent agent cyanamide. *Biochem Pharmacol* 1998;**55**:2007–15.
31. Espey MG, Miranda KM, Thomas DD, Wink DA. Ingress and reactive chemistry of nitroxyl-derived species within human cells. *Free Rad Bio Med* 2002;**33**:827–34.
32. Piloty O. Ueber eine oxydation des hydroxylamins durch benzolsulfochlorid. *Ber Dtsch Chem Ges* 1896;**29**:1559–67.
33. Rimini E. About a new reaction of the aldehydes. *Gazz Chim Ital* 1901;**31**:84–93.
34. Banerjee R, King SB. Synthesis of cyclic hydroxamic acids through –NOH insertion of ketones. *Org Lett* 2009;**11**:4580–3.
35. Panizzi L, Di Maio G, Tardella PA, d'Abbiero L. Action of nitroxyl on ketonic compounds. I. cyclic ketones. *Ricerca Sci* 1961;**1**:312–18.
36. Hassner A, Wiederkehr E, Kascheres AJ. Reaction of aldehydes with N-hydroxybenzenesulfonamide. Acetal formation catalyzed by nucleophiles. *J Org Chem* 1970;**35**(6):1962–4.
37. Exner O, Juška T. X-ray photoelectron spectroscopic study of sulphonamides: charge distribution and tautomerism. *Collect Czech Chem Commun* 1984;**49**:51–7.
38. Exner O. Acyl derivatives of hydroxylamine. IX. A spectroscopic study of tautomerism of sulphohydroxamic acids. *Collect Czech Chem Commun* 1964;**29**:1337–43.
39. Wilkins PC, Jacobs HK, Johnson MD, Gopalan AS. Mechanistic variations in the oxidation of Piloty's acid by metal complexes. *Inorg Chem* 2004;**43**:7877–81.
40. Sabbah HN, Tocchetti CG, Wang M, Daya S, Gupta RC, Tunin RS, et al. Nitroxyl (HNO): a novel approach for the acute treatment of heart failure. *Circ Heart Fail* 2013;**6**:1250–8.
41. Shoeman DW, Shirota FN, DeMaster EG, Nagasawa HT. Reaction of nitroxyl, an aldehyde dehydrogenase inhibitor, with N-acetyl-L-cysteine. *Alcohol* 2000;**20**:55–9.
42. Peachey JE, Brien JF, Roach CA, Loomis CW. A comparative review of the pharmacological and toxicological properties of disulfiram and calcium carbimide. *J Clin Psychopharmacol* 1981;**1**:21–6.
43. Shirota FN, Nagasawa HT, DeMaster EG. N-acetylcyanamide, the major urinary metabolite of cyanamide in rat, rabbit, dog and man. *Drug Metab Dispos* 1984;**12**:337–44.
44. Nagasawa HT, Lee MJC, Kwon CH, Shirota FN, DeMaster EG. An N-hydroxylated derivative of cyanamide that inhibits yeast aldehyde dehydrogenase. *Alcohol* 1992;**9**:349–53.
45. Pihlaja K, Seilo M. The acidity and general base-catalyzed hydrolysis of Meldrum's acid and its methyl derivatives. *Acta Chem Scand* 1969;**23**:3003–10.
46. Meldrum AN. A β-Lactonic acid from acetone and malonic acid. *J Chem Soc, Trans* 1908;**93**:598–601.
47. Davidson D, Bernhard SA. The structure of Meldrum's supposed β-lactonic acid. *J Am Chem Soc* 1948;**70**:3426–8.

48. Pfluger CE, Boyle PD. Conformation and intermolecular interactions of Meldrum's acid: an X-ray structural investigation of 2,2-dimethyl-1,3-dioxane-4,6-dione. *J Chem Soc Perkin Trans* 1985;**2**:1547–649.
49. Lee I, Han IS, Kim CK, Lee HW. Theoretical studies on the structure and acidity of Meldrum's acid and related compounds. *Bull Korean Chem Soc* 2003;**24**:1141–9.
50. Pihlaja K, Seilo M. The kinetics and mechanisms of the uncatalyzed and acid-catalyzed decomposition reactions of Meldrum's acid and its methyl derivatives. *Acta Chem Scand* 1968;**22**:3053–62.
51. Higashi Y, Jitsuiki D, Chayama K, Yoshizumi M. Edaravone (3-methyl-1-phenyl-2-pyrazolin-5-one), a novel free radical scavenger, for treatment of cardiovascular diseases. *Recent Pat Cardiovasc Drug Discov* 2006;**1**:85–93.
52. Watanabe T, Tahara M, Todo S. The novel antioxidant edaravone: from bench to bedside. *Cardiovasc Ther* 2008;**26**:101–14.
53. Guthrie DA. *Development of new, physiologically useful nitroxyl (HNO) donors, Ph.D. Dissertation*. Baltimore, MD: Johns Hopkins University; 2014.

Mechanistic Aspects of HNO Production from Hydroxylamine and Derivatives

4

S.A. Suarez, M.A. Morales Vásquez, M. Hamer and F. Doctorovich
Universidad de Buenos Aires and CONICET, Buenos Aires, Argentina

Chapter Outline

In the present chapter, we will discuss mechanistic aspects of reactions involving hydroxylamine and derivatives which produce HNO as an intermediate, but are normally not used as HNO donors. Inorganic as well as biologically relevant reactions will be addressed. Nitroxyl can be generated from hydroxylamine-derived compounds such as Piloty's acids and Meldrum's acids, which are HNO donors described in detail in Chapter 3.

4.1 Hydroxylamine in chemical reactions

Hydroxylamine and its derivatives are undoubtedly the most extended source of HNO. Apart from the hydroxylamine-derived HNO donors described in Chapter 3, hydroxylamine itself disproportionates in alkaline media to produce variable amounts of nitrogen, nitrous oxide (via the formation of hyponitrite), and ammonia as final products.[1] Surprisingly enough, the decomposition of hydroxylamine itself has been a matter of controversy which has not been completely solved until today. Nast et al. suggested that the reaction proceeded via HNO,[2] since evidence for the presence of nitroxyl in the reaction mixture was found when tricyanonitrosonickelate(II) (a bent NO complex, or in other words, a nitroxyl complex) appeared by addition of tetracyanonickelate. The primary process proposed is shown below (Eq. 4.1) and the ratio of gaseous products is modified depending on the pH. For example, the reduction of HNO to N_2 predominates in alkaline solution but dimerization to N_2O predominates at acidic pH.

The Chemistry and Biology of Nitroxyl (HNO). DOI: http://dx.doi.org/10.1016/B978-0-12-800934-5.00004-9

$$2NH_2OH \rightarrow NH_3 + HNO + H_2O \quad (4.1)$$

However, Luňák and Vepřek-Šiška pointed out that a putative intermediate in the formation of tricyanonitroso nickelate is not free nitroxyl, but a complex $[Ni(CN)_3NH_2OH]^-$, produced by replacement of a cyanide ligand in tetracyano nickelate by a molecule of hydroxylamine.[1] The mechanism shown in Eqs. (4.2)–(4.4) was proposed for the formation of tricyanonitroso nickelate(II):

$$[Ni(CN)_4]^{2-} + NH_2OH \rightarrow [Ni(CN)_3NH_2OH]^- + CN^- \quad (4.2)$$

$$[Ni(CN)_3NH_2OH]^- + NH_2OH + OH^- \rightarrow [Ni(CN)_3NO]^{2-} + NH_3 + H_2O \quad (4.3)$$

$$2[Ni(CN)_3NH_2OH]^- + OH^- \rightarrow [Ni(CN)_3NO]^{2-} + [Ni(CN)_3NH_3]^- + H_2O \quad (4.4)$$

Moreover, in another somewhat neglected work of 1974, Luňák and Vepřek-Šiška have shown that the decomposition of hydroxylamine is catalyzed by metal ions such as Fe^{II}, Ni^{II}, and Cu^{II}.[1] Anderson had studied the Cu^{II} catalyzed oxidation of hydroxylamine in the presence of air in the mid-1960s,[3–4] finding N_2O as a product, and proposing the intermediacy of nitroxyl but not hyponitrite. A fact that is a contradicting idea, since HNO dimerizes rapidly to produce hyponitrite which decomposes to N_2O. As established by Luňák and Vepřek-Šiška,[1] the proportion of the gaseous products N_2 and N_2O in these metal-catalyzed reactions depends not only on pH and concentration of hydroxylamine, but also on the concentration of the catalyzing cations. N_2:N_2O has a value of 0.6 at pH = 7, while the addition of Fe^{II}, Co^{II}, and Cu^{II} consistently reduced the ratio, and the use of ultrapure KOH as a base raised it up to 7. The rate equation was found to be third order in hydroxide and first order in hydroxylamine, with $k_{(20°C)} = 7 \times 10^{-6}\ M^{-3}s^{-1}$. The intermediacy of free nitroxyl was objected based on the above described results, the main one being the finding of the complex $[Ni(CN)_3NH_2OH]^-$ as a precursor of $[Ni(CN)_3NO]^{2-}$ (instead of the last complex being produced by reaction of HNO with $[Ni(CN)_4]^{2-}$). With regard to the observed metal ion catalysis, the authors proposed that the decomposition intermediates were hydroxylamine complexes with the metal ions. These complexes would decompose to a nitrosyl complex that either reacts with hydroxylamine to produce dinitrogen, or two nitrosyl ligands would produce hyponitrite and N_2O. In any case, although they claim that free HNO is not produced, the complexes proposed by Luňák and Vepřek-Šiška are NO^- complexes, id est, the anion derived from HNO is found in a coordinated fashion.

Many years later the reaction of a hydroxylamine complex coordinated to a relatively robust $[Fe^{II}(CN)_5]^{3-}$ fragment was studied by Olabe et al.[5] Spectroscopic and kinetic evidence for the coordination of NH_2OH was found, and N_2, N_2O, NH_3, and $[Fe^{II}(CN)_5(NO)]^{2-}$ (nitroprusside) were found as reaction products. HNO was proposed as an intermediate, produced by reactions (4.5)–(4.7), in a radical-chain process:

$$[Fe^{II}(CN)_5O(H)NH_2]^{2-} + H^+ \rightarrow [Fe^{III}(CN)_5H_2O]^{2-} + {}^{\bullet}NH_2 \tag{4.5}$$

$${}^{\bullet}NH_2 + NH_2OH \rightarrow NH_3 + {}^{\bullet}NHOH \tag{4.6}$$

$${}^{\bullet}NHOH + NH_2OH \rightarrow HNO + H_2O + {}^{\bullet}NH_2 \tag{4.7}$$

More recently, it has been found that carbon in different allotropic forms (Mogul L, graphite, Black Pearls L) also catalyzes the decomposition of hydroxylamine;[6] N_2, NO, N_2O, and NH_3 were identified as products. Hydrogen atom transfer from one molecule of hydroxylamine to another to produce HNO, NH_3, and H_2O was proposed as a plausible reaction mechanism.

Another reaction involving hydroxylamine for which nitroxyl has been proposed as an intermediate is the reduction of NO. Cooper et al. reported equimolar quantities of N_2 and N_2O as products in basic solution.[7] A tracer experiment indicated the appearance of one N atom of NO origin in each of the two product molecules. The authors proposed initial H atom abstraction from hydroxylamine by NO, followed by formation of *N*-nitrosohydroxylamine which would decompose to N_2O (Eqs. 4.8–4.9). The formation of N_2 was rationalized as an attack of nitroxyl anion to NH_2OH (Eq. 4.10). N_2O is also formed by HNO dimerization (Eq. 4.11). However, given the relative N–H bond strengths in HNO versus hydroxylamine, Eq. (4.8) may seem unlikely.

$${}^{\bullet}NO + NH_2OH \rightarrow HNO(or\ NOH) + {}^{\bullet}NHOH(or\ H_2NO) \tag{4.8}$$

$${}^{\bullet}NHOH + {}^{\bullet}NO \rightarrow ON\text{-}NHOH \rightarrow N_2O + H_2O \tag{4.9}$$

$$NO^- + NH_2OH \rightarrow N_2 + H_2O + OH^- \tag{4.10}$$

$$2\,HNO \rightarrow N_2O + H_2O \tag{4.11}$$

A few years later, this mechanism was reanalyzed and confirmed by Bonner et al.,[8] by using a combination of tracer ($^{15}N^{18}O$) and kinetic experiments, and a detailed study of the stoichiometry of the final products. In the same work Bonner found that the nitroxyl intermediate encountered in trioxodinitrate ($[ON\text{-}NO_2]^{2-}$, Angeli's salt) decomposition had properties significantly different from those of the intermediate of the NO-NH_2OH reaction. Whereas the first nitroxyl was found to be much more reactive toward NH_2OH than toward itself at pH = 8, the second one was consumed almost exclusively by self-reaction at the same pH, indicating that the first form is fully deprotonated but the second one is not. Bonner rationalized this behavior as the presence of a different tautomer of azanone in each case: HNO for Angeli's salt decomposition and NOH in the NO-NH_2OH reaction.[9]

Bonner also studied the reactions of *N*- and *O*-substituted hydroxylamines with NO. He proposed that in the case of CH_3NHOH, NH_2OCH_3, and CH_3NHOCH_3 the reaction proceeds by N-bound H atom abstraction to form HNO. In the case of the dialkyl compounds $(CH_3)_2NOH$ and $(C_2H_5)_2NOH$, there is no N-bound H atom, and abstraction is postulated to occur at the α-carbon.[10] However, these works involving NO and azanone should be reinvestigated in view of the fact that both HNO and NO^- react with NO at fast rates,[11] producing N_2O and nitrite as final products.

4.2 Hydroxylamine derivatives and related compounds

Formation of HNO and NO from hydroxamate derivatives (RC(O)NHOH,HXs) involves the loss of two and three electrons, respectively.[12–13] One-electron oxidation of HXs to their respective transient nitroxide radicals in aqueous solutions has been shown by EPR spectroscopy.[14–16] The fate of the nitroxide radical is not clear, and several routes have been suggested for its decomposition,[17] as described in Scheme 4.1, including:

1. Dismutation of the nitroxide radicals (reaction a), forming the parent HX and RC(O) N = O, which hydrolyzes to RC(O)OH and HNO (reaction b).
2. Unimolecular decomposition of the nitroxide radical to the respective aldehyde and NO (reaction c).
3. Unimolecular decomposition of the nitroxide radical via homolysis along the C–N bond, yielding the respective acyl radical $RC(O)^{\bullet}$ and HNO (reaction d).

Regarding hydroxylamine derivatives, apart from the HNO donors already mentioned in Chapter 3, hydroxylamine-*N*-sulfonate was postulated to decompose in

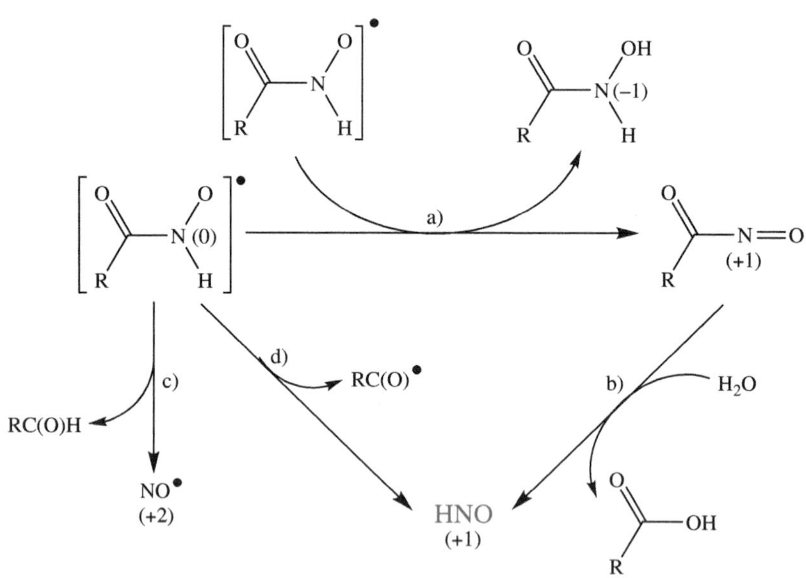

Scheme 4.1 Possible routes for the decomposition of the nitroxide radical formed via one-electron oxidation of RC(O)NHOH. In brackets: redox state of all nitrogen species.

alkaline media,[18] producing HNO or NOH, depending on the involved tautomers of the deprotonated reagent, as it is shown in Eq. (4.12):

$$HONH\text{-}SO_3^- + OH^- \rightarrow HNO(\text{or } HON) + SO_3^{2-} + H^+ \quad (4.12)$$

Related with the above-mentioned reaction, the reaction of sulfamic acid with nitric acid, with produces N_2O (Eq. 4.13), was postulated to proceed via an intermediate nitramide (H_2NNO_2) that could decompose to HNO by cleavage of the N–N bond.[8]

$$HNO_3 + NH_2SO_3H \rightarrow N_2O + H_2O + H_2SO_4 \quad (4.13)$$

Finally, nitroxyl has been postulated as an intermediate in NH_3-doped low-pressure $H_2/N_2O/Ar$ flames by molecular beam mass spectroscopy.[19,20,21] The $\Delta H°_f$ (298K) was estimated to be 25.4 kcal/mol by calculations. The reactions proposed for its formation involve hydroxylamine-related N-centered free radicals, as shown in Eqs. (4.14) and (4.15).

$$NH_2^\bullet + O^\bullet \rightarrow HNO + H^\bullet \quad (4.14)$$

$$NH + OH^\bullet \rightarrow HNO + H^\bullet \quad (4.15)$$

Condensed phase thermal decomposition of ammonium perchlorate at temperatures of 95°C and above produces HNO, as observed by high-resolution mass spectrometry.[21] As in the case of the low-pressure flames, HNO is postulated as a product of the reaction of NH_2 free radicals with oxygen, this time in its molecular form (Eq. 4.16):

$$NH_2^\bullet + O_2 \rightarrow HNO + OH^\bullet \quad (4.16)$$

The $NH_2^\bullet$ radicals, also detected by mass spectrometry, were suggested to be produced from ammonia (formed by decomposition of NH_4ClO_4), by reactions with $Cl^\bullet$ and $ClO_3^\bullet$ radicals formed in situ.[22]

4.2.1 Hydroxylamine in biological systems

Organic hydroxylamines are endogenous compounds produced by reduction of nitrates or nitrites in normal mammalian cell metabolism. Hydroxylamine itself is an intermediate in the oxidation of ammonia. All these reactions are components of an intracellular detoxification system. Hydroxylamines are also formed in the body from drugs such as amphetamine and ephedrine.[23]

In biological systems, hydroxylamine was found to exert on DNA a mutagenic effect at high concentrations, and an inactivating effect at low concentrations.[24] The inactivating effect was ascribed to nitroxyl and hydrogen peroxide produced by reaction between NH_2OH and O_2. The combination of hydroxylamine and H_2O_2 produced a much more rapid inactivation than either of the two compounds separately or even the additive effect that would result if the two agents acted separately. Formation of HNO and H_2O_2 was suggested to occur by reaction of hydroxylamine with O_2 (Eq. 4.17).

$$O_2 + NH_2OH \rightarrow H_2O_2 + HNO \tag{4.17}$$

Erlenmeyer et al. found that the peroxidase-like reaction between H_2O_2 and NH_2OH is catalyzed both by Cu^{II} and by the Cu^{II}(2,2′-bipyridyl) complex.[25] The authors proposed coordination of both H_2O_2 and NH_2OH to Cu^{II} (Scheme 4.2), and H-atom transfer from hydroxylamine to produce HNO.

The catalytic reaction between H_2O_2 and NH_2OH (or *N*-hydroxy-L-arginine) was more recently studied by Donzelli et al. in the presence of heme proteins, in which the iron metal center acts as the catalytic site.[26] Formation of HNO was evaluated with a selective assay in which the release of HNO was indicated by formation of sulfinamide ($GS(O)NH_2$), detected by mass spectra, when glutathione (GSH) was added (Eq. 4.18).

$$HNO + GSH \rightarrow GSNHOH \rightarrow GS(O)NH_2 \tag{4.18}$$

Sulfinamide was observed upon oxidation of NH_2OH, whereas *N*-hydroxy-L-arginine (NOHA), the primary intermediate in the oxidation of L-arginine by NO synthase, was resistant to oxidation by the heme proteins utilized. The highest yields of $GS(O)NH_2$ (and therefore HNO) were observed with proteins in which the heme was coordinated to a histidine (e.g., horseradish peroxidase, myoglobin, and other proteins), in contrast to a tyrosine (e.g., catalase) or cysteine (e.g., cytochrome P450). It was proposed oxidation of the metal center by H_2O_2 to produce compound I, which in turn would oxidize hydroxylamine to HNO (Scheme 4.3).

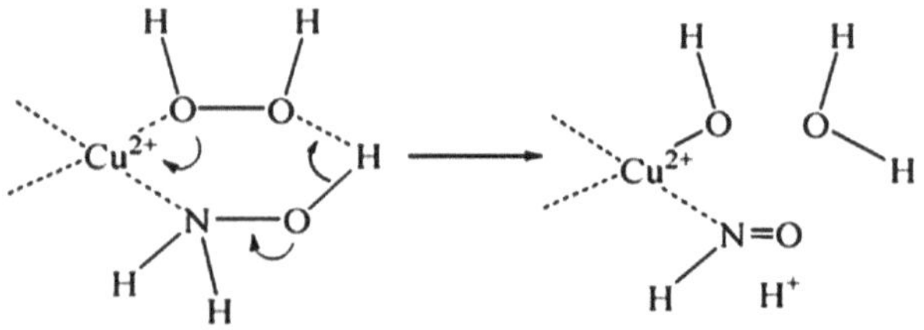

Scheme 4.2 Peroxidase-like reaction between H_2O_2 and NH_2OH catalyzed by Cu^{II}.

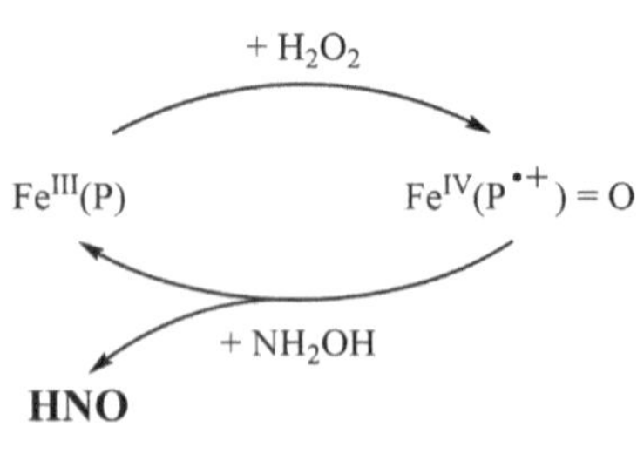

Scheme 4.3 Azanone production by catalytic reaction of heme proteins with hydroxylamine and H_2O_2.

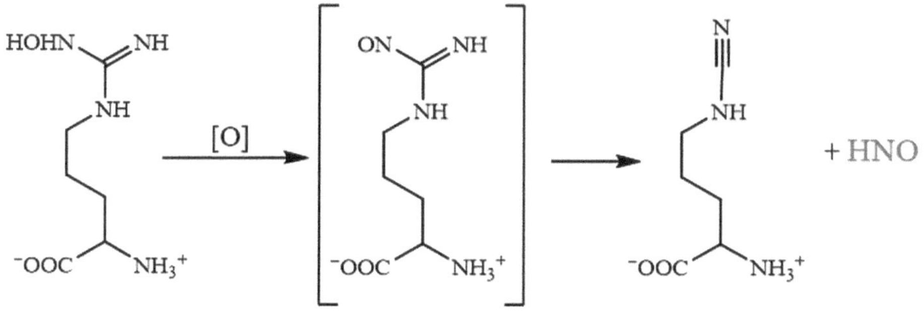

Scheme 4.4 Production of HNO by oxidation of NOHA.

Toscano and coworkers have recently shown NOHA is oxidized by excess hypochlorous acid (HOCl) to form HNO-derived N_2O.[27] Hypochlorous acid is generated in vivo from hydrogen peroxide and chloride by the heme enzyme myeloperoxidase, while NOHA is an intermediate in the enzymatic production of nitric oxide by NO synthases. Azanone is presumably produced via a nitroso intermediate which decomposes to the corresponding cyanamide derivative (Scheme 4.4). HNO was also observed from the HOCl oxidation of hydroxylamine, hydroxyurea, and acetohydroxamic acid.

NOHA has been shown to be chemically oxidized to HNO by other oxidants such as lead oxide and silver carbonate,[28] while *N*-hydroxyguanidine is oxidized by NO itself.[29] In the last case the proposed mechanism involves H-atom abstraction by NO to produce HNO and a free radical intermediate.

In another work, Adak et al. showed that nitroxyl could be produced by nitric oxide synthase (NOS) under certain conditions.[30] NOS is a rather complex enzyme that utilizes multiple redox-active cofactors and substrates to catalyze the five-electron oxidation of its substrate L-arginine to citrulline and NO. There are three known types of NOS: neuronal nitric oxide synthase (nNOS), endothelial nitric oxide synthase (eNOS), and inducible nitric oxide synthase (iNOS). Its functions include synaptic plasticity in the central nervous system (CNS), central regulation of blood pressure, smooth muscle relaxation, and vasodilatation via peripheral nitrergic nerves.[31] Two flavins, a cysteine-coordinated heme cofactor and a rare tetrahydrobiopterin cofactor, are used to deliver electrons from NADPH to molecular oxygen.[32] Neuronal NOS free of the cofactor tetrahydrobiopterin (BH_4) catalyzes arginine oxidation to NOHA and citrulline in both NADPH and H_2O_2-driven reactions.[33] It was suggested that a ferrous heme–NO complex (formally $Fe^{II}NO$ or $Fe^{III}NO^-$) built up after initiating catalysis in both NADPH and H_2O_2 reactions, consistent with formation of nitroxyl anion as a product. For the BH_4-replete enzyme, an NO-releasing $Fe^{III}NO$ heme complex was produced.[30] A model for heme-dependent oxygen activation and NO/NO^- synthesis were proposed (Scheme 4.5). A heme-peroxo intermediate forms and can react with NOHA in the presence or absence of BH_4. If the reductase domain provides the electron to the ferrous-dioxy species (path A), then the reaction generates NO^-. However, if BH_4 provides this electron (path B), then the BH_4 radical can accept an electron and NO is produced.[30]

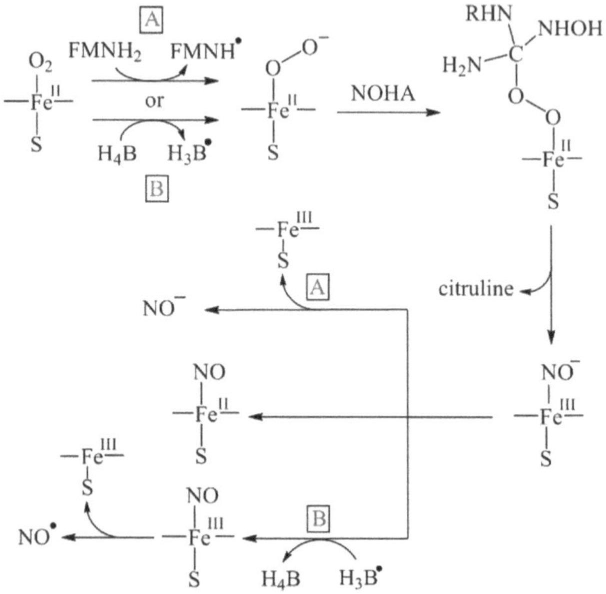

Scheme 4.5 Heme-dependent oxygen activation and NO/NO^- synthesis from NOS and NOHA.

The previous mechanism was revised 10 years later.[34,35] The peroxide-driven reaction with both Mn- and Fe-containing heme domain constructs of NOS was used to characterize the formation of HNO as the initial inorganic product formed when O_2 activation occurs without BH_4 radical formation. Since hydrogen peroxide can directly from the ferric-peroxo intermediate, generation of a pterin-centered radical does not occur under these conditions, and path B shown in Scheme 4.3 is not operative, leading to the formation of the Fe^{II}-NO complex that either releases HNO or is oxidized by peroxide to generate nitrite and nitrate. In the presence of the pterin-centered radical generated by preturnover of the iron-containing enzyme with L-arginine, NO was the observed product.

A modified mechanism was proposed for the NOS oxidation of NOHA in which an Fe^{III}-peroxo complex is formed initially (Scheme 4.6), instead of the Fe^{II}-peroxo as proposed in Scheme 4.5. Activation of O_2 occurs with one-electron derived from NADPH and one from BH_4, generating a pterin-centered radical $^{\bullet}BH_3$. Nucleophilic attack of NOHA by this peroxo intermediate leads to the formation of an addition complex, which decomposes to generate Fe^{III}-NO^- and citrulline. At this point it should be noted that Fe^{III}-NO^- and Fe^{II}-$NO^{\bullet}$ are resonance Lewis structures, and therefore the release of nitroxyl is probably aided by protonation, producing azanone by a competing reaction, or alternatively, HNO is released when BH_4 is not present. In the free form, nitroxyl is expected to be protonated at physiological pH (i.e., as HNO). Since the nitric oxide ligand in Fe^{II}-$NO^{\bullet}$ is not labile, one could expect that this species could live enough time as to be protonated. On the other hand, NO can

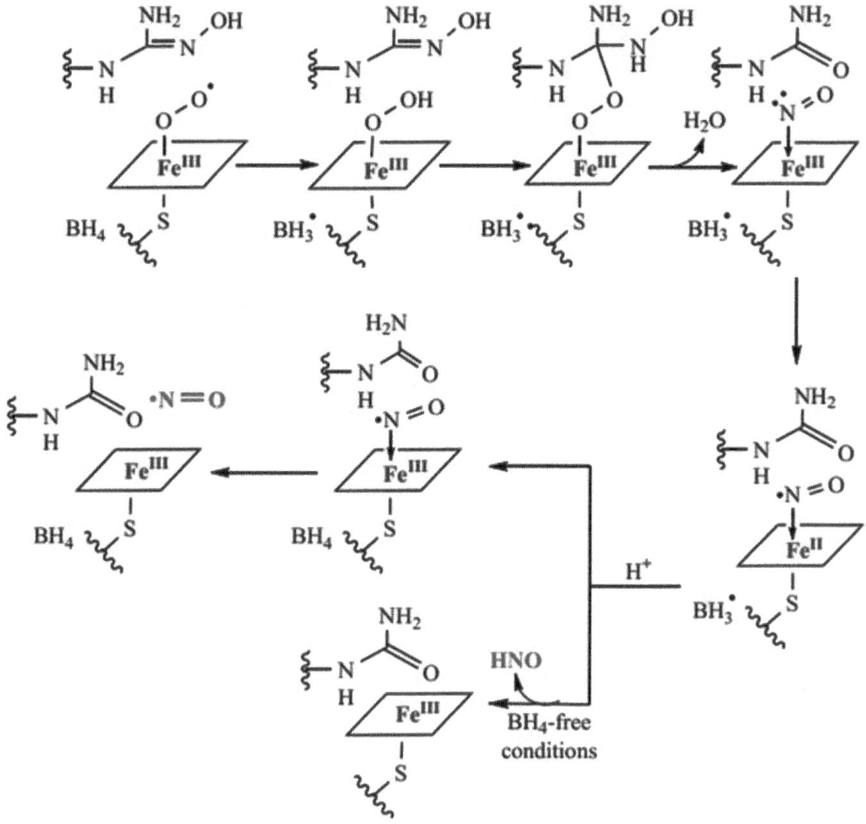

Scheme 4.6 Updated mechanism for the NOS-mediated oxidation of NOHA.

be produced if Fe^{II} is oxidized by $^{\bullet}BH_3$ to Fe^{III}-$NO^{\bullet}$ (Scheme 4.6), nitric oxide becomes labile and is rapidly released as the final product.

Cytochrome c nitrite reductase (ccNiR), which catalyzes the six-electron reduction of nitrite to ammonia in vivo,[36–40] was shown to oxidize hydroxylamine in the presence of large quantities of this substrate, yielding nitrite as the sole free nitrogenous product (Scheme 4.7).[41–42] This global process involves "recharging" of the catalytic heme with electrons obtained through transfer from the other hemes of the enzyme and of the heme environment with protons. The proposed mechanism suggests HNO formation through a sequence of one-electron intramolecular electron transfers, which leaves the heme pool reduced by two electrons and an HNO moiety bound at the active site (C_2, Scheme 4.7).[36,42] Ditto, aceto-HX demonstrates synergic pro-oxidative effects on cells exposed to H_2O_2 and $MbFe^{III}$, which may be attributed in the case of *Bacillus subtilis* to the formation of HNO and to further oxidation of HNO to NO in the case of *Escherichia coli*.[43]

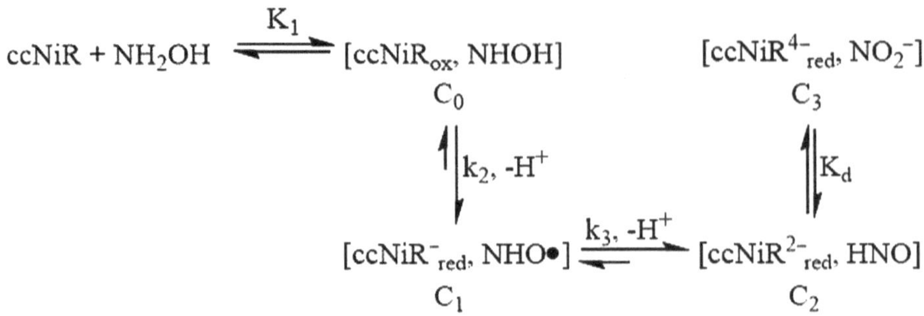

Scheme 4.7 Interaction of hydroxylamine with ccNiRox.

4

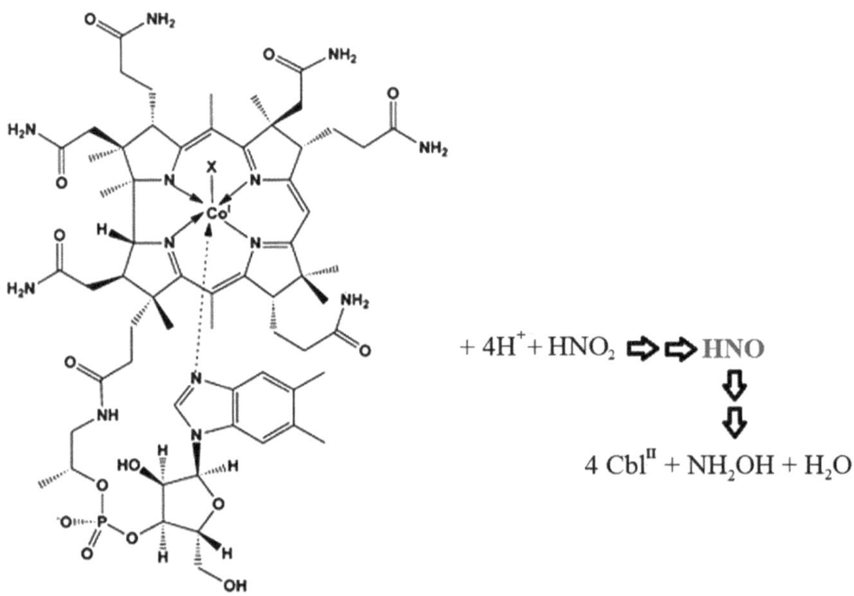

Scheme 4.8 Reaction between Cbl(I) and (H)NO_2 yield 4Cbl(II) and NH_2OH producing HNO as intermediate.

Finally, the cob(III)alamins are essential cofactors for two enzymes in mammals and multiple enzymes in bacteria.[44] Reduced cobalamins, cob(II)alamin and cob(I) alamin, are also important intracellular complexes. The reaction between Cbl(I) and (H)NO_2 at neutral pH conditions, which involves a 2e− rate-determining step, to form Cbl(III) and HNO was postulated by Brasch *et al.* (Scheme 4.8).[45,46] The latter species reacts further with Cbl(I), which ultimately results in the oxidation of 4Cbl(I) by HNO_2 to yield 4Cbl(II) and NH_2OH.

References

1. Luňák S, Vepřek-Šiška J. The catalytic effect of cations on the decomposition of alkaline solutions of hydroxylamine. *Collect Czechoslov Chem Commun* 1974;**39**:391–5.
2. Nast R, Foppl I. Uber die Bildung von hyponitritdurch disproportionierung des hydroxylamins. *Zeitschrift fur Anorg und Allg Chemie* 1950;**263**:310–15.
3. Anderson JH. The copper-catalysed oxidation of hydroxylamine. *Analyst* 1964;**89**: 357–62.
4. Anderson JH. The oxidation of hydroxylamine in sodium hydroxide in the presence of copper(II). *Analyst* 1966;**91**:532–5.
5. Alluisetti GE, Almaraz AE, Amorebieta VT, Doctorovich F, Olabe JA. Metal-catalyzed anaerobic disproportionation of hydroxylamine. Role of diazene and nitroxyl intermediates in the formation of N_2, N_2O, NO^+, and NH_3. *J Am Chem Soc* 2004;**126**:13432–42.
6. Larsen JW, Jandzinski J, Sidovar M, Stuart JL. Carbon-catalyzed decompositions of hydrazine and hydroxylamine. *Carbon NY* 2001;**39**:473–6.
7. Cooper JN, Chilton JE, Powell RE. Chemistry, I. Reaction of nitric oxide with alkaline hydroxylamine. *Inorg Chem* 1970;**9**:2303–4.
8. Bonner FT, Dzelzkalns LS, Bonucci JA. Properties of nitroxyl as intermediate in the nitric oxide-hydroxylamine reaction and in trioxodinitrate decomposition. *Inorg Chem* 1978;**17**:2487–94.
9. Bonner FT, Wang NY. Reduction of nitric oxide by hydroxylamine. 1. Kinetics and mechanism. *Inorg Chem* 1986;**25**:1858–62.
10. Wang NY, Bonner FT. Reduction of nitric oxide by hydroxylamine. 2. Reactions of the alkylated hydroxylamines. *Inorg Chem* 1986;**25**:1863–6.
11. Lymar SV, Shafirovich V, Poskrebyshev GA. One-electron reduction of aqueous nitric oxide: a mechanistic revision. *Inorg Chem* 2005;**44**:5212–21.
12. Samuni Y, Samuni U, Goldstein S. The mechanism underlying nitroxyl and nitric oxide formation from hydroxamic acids. *Biochimica et Biophysica Acta* 2012;**1820**(10):1560–6.
13. Shirota FN, DeMaster EG, Lee MJC, Nagasawa HT. Generation of nitric oxide and possibly nitroxyl by nitrosation of sulfohydroxamic acids and hydroxamic acids. *Nitric Oxide* 1999;**3**(6):445–53.
14. Ackermann MN, Powell RE. Alkaline hydrolysis of hydroxylamine-*N*-sulfonate. *Inorg Chem* 1966;**5**:1334–7.
15. Einsle O, Messerschmidt A, Huber R, Kroneck PMH, Neese F. Mechanism of the six-electron reduction of nitrite to ammonia by cytochrome c nitrite reductase. *J Am Chem Soc* 2002;**124**(39):11737–45.
16. Bykov D, Plog M, Neese F. Heme-bound nitroxyl, hydroxylamine, and ammonia ligands as intermediates in the reaction cycle of cytochrome c nitrite reductase: a theoretical study. *J Biol Inorg Chem* 2014;**19**(1):97–112.
17. Youngblut M, Pauly DJ, Stein N, Walters D, Conrad JA, Moran GR, et al. Shewanella oneidensis cytochrome c nitrite reductase (ccNiR) does not disproportionate hydroxylamine to ammonia and nitrite, despite a strongly favorable driving force. *Biochemistry* 2014;**53**(13):2136–44.
18. Shaw S, Lukoyanov D, Danyal K, Dean DR, Hoffman BM, Seefeldt LC. Nitrite and hydroxylamine as nitrogenase substrates: mechanistic implications for the pathway of N_2 reduction. *J Am Chem Soc* 2014;**136**(36):12776–83.
19. Brown HW, Pimentel GC. Photolysis of nitromethane and of methyl nitrite in an argon matrix; infrared detection of nitroxyl, HNO. *J Chem Phys* 1958;**29**:883.

20. Quee MJY, Thynne JCJ. Reactions of *iso*-propoxyl radicals. *Trans Faraday Soc* 1968; **64**:1296.
21. Sausa RC, Singh G, Lemire GW, Anderson WR. Flame structure studies of neat and NH3-doped low-pressure H2/N2O/Ar flames by molecular beam mass spectroscopy. *Twentieth Army Conference (Award winning papers)*; 1996; 67–71.
22. Hackman E, Hesser H, Beachell H. Detection of species resulting from condensed phase decomposition of ammonium perchlorate. *J Phys Chem* 1972;**76**:3545–54.
23. Gross Paul, Smith RP. Biologic activity of hydroxylamine. *CRC Crit Rev Toxicol* 1985; **14**(1):87–99.
24. Freese EBE, Rutherford E. The oxygen effect on deoxyribonucleic acid inactivation by hydroxylamines. *Biochemistry* 1965;**4**(11):2419–33.
25. Erlenmeyer H, Flierl C, Sigel H. On the kinetics and mechanism of the reactions of hydro en peroxide with hydrazine or hydroxylamine, catalyzed by Cu2 and Cu 2-212′-Bipyrieyl Complex. *J Am Chem Soc* 1969;**91**:1065.
26. Donzelli S, Graham M, Flores-Santana W, Switzer CH, Yeh GC, Huang J, et al. Generation of nitroxyl by heme protein-mediated peroxidation of hydroxylamine but not *N*-hydroxy-L-arginine. *Free Radic Biol Med* 2008;**45**:578–84.
27. Cline MR, Chavez TA, Toscano JP. Oxidation of *N*-hydroxy-L-arginine by hypochlorous acid to form nitroxyl (HNO). *J Inorg Biochem* 2013;**118**:148–54.
28. Fukuto JM, Wallace GC, Hszieh R, Chaudhuri G. Chemical oxidation of *N*-hydroxy-guanidine compounds: release of nitric oxide, nitroxyl and possible relationship to the mechanism of biological nitric oxide generation. *Biochem Pharmacol* 1992;**43**(3): 607–13.
29. Yoo J, Fukuto JM. Oxidation of *N*-hydroxyguanidine by nitric oxide and the possible generation of vasoactive species. *Biochem Pharmacol* 1995;**50**(12):1995–2000.
30. Adak S, Wang Q, Stuehr DJ. Arginine conversion to nitroxide by tetrahydrobiopterin-free neuronal nitric-oxide synthase. Implications for mechanism. *J Biol Chem* 2000;**275**: 33554–61.
31. Förstermann U, Sessa W. Nitric oxide synthases: regulation and function. *Eur Heart J* 2012;**33**(7):829–37.
32. Andrew P, Mayer B. Enzymatic function of nitric oxide synthases. *Cardiovasc Res* 1999;**43**:521–31.
33. Zhou Li, Zhu D. Neuronal nitric oxide synthase: structure, subcellular localization, regulation, and clinical implications. *Nitric Oxide* 2009;**20**:223–30.
34. Woodward JJ, Nejatyjahromy Y, Britt RD, Marletta MA. Pterin-centered radical as a mechanistic probe of the second step of nitric oxide synthase. *J Am Chem Soc* 2010;**132**:5105–13.
35. Sabat J, Egawa T, Lu C, Stuehr DJ, Gerfen GJ, Rousseau DL, et al. Catalytic intermediates of inducible nitric oxide synthase stabilized by the W188H mutation. *J Biol Chem* 2012;**288**:6095–106.
36. Bykov D, Neese FJ. Reductive activation of the heme iron-nitrosyl intermediate in the reaction mechanism of cytochrome c nitrite reductase: a theoretical study. *J Biol Inorg Chem* 2012;**17**:741–60.
37. Attia AA, Silaghi-Dumitrescu R. Computational investigation of the initial two-electron, two-proton steps in the reaction mechanism of hydroxylamine oxidoreductase. *J Phys Chem B* 2014;**118**(42):12140–5.
38. Bykov D, Neese F. Six-electron reduction of nitrite to ammonia by cytochrome c nitrite reductase: insights from density functional theory studies. *Inor Chem* 2015;**54**(19): 9303–16.

39. Chamulitrat W, Mason RP, Riendeau DJ. Nitroxide metabolites from alkylhydroxylamines and *N*-hydroxyurea derivatives resulting from reductive inhibition of soybean lipoxygenase. *Biol Chem* 1992;**267**:9574–9.
40. Gabricevic M, Besic E, Birus M, Zahl A, van Eldik RJ. Oxidation of hydroxyurea with oxovanadium(V) ions in acidic aqueous solution. *Inorg Biochem* 2006;**100**:1606–13.
41. Budimir A, Besic E, Birus M. Kinetics and mechanism of oxidation of hydroxyurea with hexacyanoferrate(III) ions in aqueous solution. *Croat Chem Acta* 2009;**82**:807–18.
42. Samuni A, Goldstein S. One-electron oxidation of acetohydroxamic acid: the intermediacy of nitroxyl and peroxynitrite. *J Phys Chem A* 2011;**115**(14):3022–8.
43. Yadav R, Goldstein S, Nasef M, Lee W, Samuni U. Synergistic activity of acetohydroxamic acid on prokaryotes under oxidative stress: the role of reactive nitrogen species. *Free Radical Biol Med* 2014;**77**:291–7.
44. Brown KL. Chemistry and enzymology of vitamin B12. *Chem Rev* 2005;**105**:2075–150.
45. Subedi H, Brasch NE. Studies on the reaction of reduced vitamin B12 derivatives with the nitrosyl-hydride (HNO) Donor Angeli's salt: HNO oxidizes the transition-metal center of cobalamin. *Eur J Inorg Chem* 2015;**2015**(23):3825–34.
46. Plymale NT, Dassanayake RS, Hassanin HA, Brasch NE. Kinetic and mechanistic studies on the reactions of the reduced vitamin B12 complex cobalamin with nitrite and nitrate. *Eur J Inorg Chem* 2012;**2012**(6):913–21.

HNO Generation From $NO^{\bullet}$, Nitrite, Inorganic or Organic Nitrosyls, and Crosstalk With H_2S

5

I. Ivanović-Burmazović
Friedrich-Alexander University Erlangen-Nuremberg, Erlangen, Germany

Chapter Outline

Nitroxyl (HNO) can be produced from various substrates through either oxidative or reductive pathways. In this chapter, we will summarize and discuss the nonenzymatic reductive pathways for the generation of nitroxyl starting from $NO^{\bullet}$, inorganic or organic nitrosyl species (metal-nitrosyls and *S*-nitrosothiols), and nitrite. These mechanisms can be mediated by the metal centers. In last years, a number of experimental evidences emerged that indicated a synergy between nitric oxide, nitroxyl, and hydrogen sulfide (H_2S) in their redox activity that induces a specific transduction of biological signals. In this context, the role of H_2S, as an important physiologically available reducing agent, in the HNO generation will be also discussed. Examples of $H_2S/NO^{\bullet}/HNO$ orchestrated reactions that are responsible for certain biological events will be tackled as well. The herein presented library of chemical reactions could offer a perspective for further understanding and developing of mechanisms for the HNO generation that are of physiological and therapeutic relevance, respectively.

The Chemistry and Biology of Nitroxyl (HNO). DOI: http://dx.doi.org/10.1016/B978-0-12-800934-5.00005-0

5.1 HNO generation from $NO^{•}$ and $\{MNO\}^{n}$

5.1.1 Basic considerations

As mentioned in Chapter 1, Introduction, pure electron transfer to $NO^{•}$ that generates $^{3}NO^{-}$ is thermodynamically quite demanding (see Eq. 1.3). But if proton-coupled electron transfer (PCET) is in play, $NO^{•}$ reduction to ^{1}HNO can be considered attainable. From thermodynamic point of view (see Fig. 1.1), it is quite feasible to obtain ^{1}HNO by reduction of $NO^{•}$ at lower pH, since ^{1}HNO is just for c. 14 kJ/mol less stable than $NO^{•}$ under standard conditions (pH = 0). At pH 7, based on the $NO^{•}/^{1}HNO$ redox potential (see Eq. 1.4), the $NO^{•}$ reduction becomes more difficult. But it might still be achievable if it is coupled to other processes that consume the initial products (HNO and an oxidized form of a reducing agent), which have a character of reactive intermediates and react/decay shifting the equilibrium toward a final thermodynamic sink.

However, kinetics could be more critical. The fact that protonation is associated with the spin change (vide supra) implies that the H^{+} transfer is significantly slower than the electron transfer, which disfavors an efficient PCET process (see Eq. 1.4). Therefore, a sort of catalyst would be quite useful. Consequently, we can assume that a direct reduction of $NO^{•}$ to ^{1}HNO according to an outer-sphere mechanism (i.e., an intermolecular reduction of free noncoordinated $NO^{•}$) is less viable, and that an inner-sphere reduction (i.e., an intramolecular reduction of $NO^{•}$ coordinated to a metal center) coupled to the proton transfer would be a preferable mechanism. In that respect, coordination of $NO^{•}$ to a transition metal center with an acidic ligand (M^{II}-LH in Fig. 5.1) or formation of a (cage-)adduct with a PCET reagent (HX in Fig. 5.1) could considerably facilitate formation of HNO. One can envision an inner-sphere complex, HL-M^{II}-$NO^{•}$, or (cage-)adduct, HX−$NO^{•}$, as an intermediate or transient species, respectively, within either stoichiometric or catalytic processes leading to HNO (Fig. 5.1). An alternative mechanism to PCET is a hydrogen atom transfer (HAT). In this context, we will at first in the next section address a metal-mediated production of HNO/NO^{-}.

5.1.2 Metal-mediated processes

5.1.2.1 Outer-sphere electron transfer

Direct outer-sphere one-electron reduction of $NO^{•}$ to form $^{3}NO^{-}$ requires metal centers with very negative redox potentials, lower than that given in Eq. 1.3. A reverse process, i.e., oxidation of nitroxyl to $NO^{•}$, was studied by applying two different 2,2′-bipyridinium compounds ($V1^{2+}$ and $V2^{2+}$ with $E_{1/2} = -0.547$ V and -0.690 V, respectively) or methyl viologen (MV^{2+}, $E_{1/2} = -0.44$ V), and facile reduction of $V1^{2+}$ and VM^{2+}, on the one hand, and slow reduction of $V2^{2+}$, on the other hand, were observed.[1] Based on these experiments, one would expect that a metal center of the redox potential ≤ -0.7 V would be required for direct $NO^{•}$ reduction to nitroxyl. Therefore, there are not many examples of such processes.

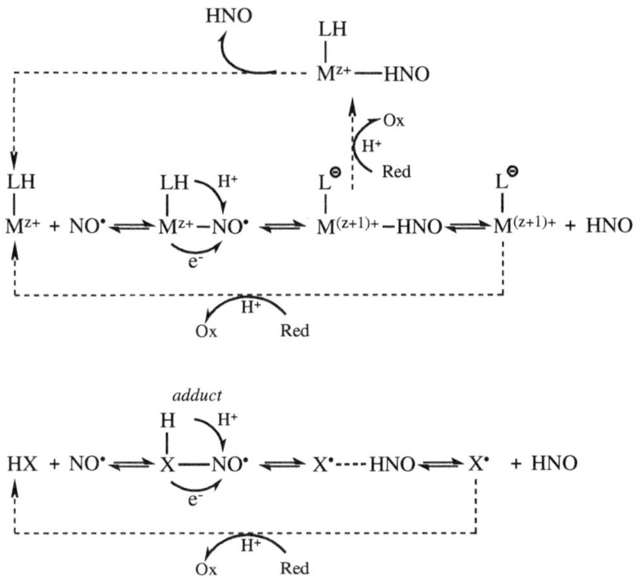

Figure 5.1 Suggested mechanisms for the stoichiometric and catalytic HNO generation according to an inner-sphere proton-coupled electron transfer mediated by coordination to the transition metal center (A) or formation of a cage adduct (B).

Interestingly, it has been proposed that the mechanism of the reaction between $[Ru(NH_3)_6]^{3+}$ and trioxodinitrate ($N_2O_3^{2-}$) involves an intermediate NO^- generation in an outer-sphere reaction between $[Ru(NH_3)_6]^{2+}$ and $NO^•$ (Eq. 5.1).[2a] Knowing that the redox potential of the $[Ru(NH_3)_6]^{3+}/[Ru(NH_3)_6]^{2+}$ couple is +0.05 V versus NHE,[2b] the suggested mechanism raises the question of putative nitroxyl formation. Furthermore, in a study of the electrochemical analytical procedure for sensing $NO^•$ based on its reduction, ferrocene (Fc) incorporated onto the electrode surface was used as a mediator molecule.[3] This study demonstrates a chemical one-electron oxidation of Fc by $NO^•$ at pH 2 (Eq. 5.2), which suggests that $NO^•$ reduction proceeds at a potential close to Fc/Fc^+ conversion, ranging from 295 ± 5 to 380 ± 5 mV versus NHE (depending on electrode configuration) under chosen experimental conditions. Authors argue in favor of the previously reported value for $E°(NO^•/^3NO^-)$ (+0.39 V)[4] rather than the revised one (-0.8 ± 0.2 V),[1] or an apparent contradiction with the revised value could be explained by the difference in the pH conditions.[3]

Recently, instead of obtaining experimental evidence for the N–N bond formation between free $NO^•$ and ferrous heme-nitroxyl species, Fe^{II}-NO^-, related to the catalytic cycles of NorBC (bacterial nitric oxide reductase) and P450nor (cytochrome P450 nitric oxide reductase), reduction of free $NO^•$ to NO^- by a porphyrin $[Fe^{II}(Por)NO]^-$ species was reported (Eq. 5.3).[5] This reaction can be in an agreement with an outer-sphere electron transfer mechanism, because these low-spin (ls) heme Fe^{II}-NO^- complexes are very strong reducing agents, with half wave potentials for the ferrous porphyrin nitrosyl reduction, $[Fe^{II}(Por)NO] + e^- \rightarrow [Fe^{II}(Por)NO]^-$, at −1.78 and −1.18 V versus Fc/Fc^+ for [Fe(3,5-Me-BAFP) NO]$^-$ and [Fe(T*o*-F_2PP)NO]$^-$ complexes, respectively.[5] However, such negative redox potentials do not operate in vivo. Taking into consideration that in the physiological milieu pH values as low as 4.5 can be found, e.g., in the organelles involved in the removal of cellular waste such as lysosomes,[6] the potential around −0.4 V versus SHE (based on the value at pH = 0 in Eq. 1.4) would be close to an upper limit for a reducing agent still capable of outer-sphere electron transfer to free $NO^•$ and production of HNO. But, metal centers of such redox properties and mechanistic outer-sphere electron transfer behavior toward nitric oxide have not been identified in biological systems yet.[2–5]

$$[Ru(NH_3)_6]^{2+} + NO^• \rightarrow [Ru(NH_3)_6]^{3+} + NO^- \quad (5.1)$$

$$Fc + NO^• \leftrightarrow Fc^+ + NO^- \quad (5.2)$$

$$[Fe^{II}(Por)NO]^- + NO^• \rightarrow [Fe^{II}(Por)NO] + NO^- \quad (5.3)$$

5.1.2.2 Coordination and inner(outer)-sphere electron transfer

Different from pure outer-sphere reduction of free $NO^•$ by metal centers/complexes, chemistry of electron transfer to metal-bound $NO^•$ within a complex coordination sphere according to an inner-sphere mechanism, or by an external electron(-donor), is much richer in examples. In principle, a redox catalysis and activation of small molecules, in general, and $NO^•$, in particular, by metal centers are based on the fact that upon coordination their redox potentials are significantly changed (Fig. 5.2)[7] due to the presence of adjacent positive charge of the metal cation and/or newly established electronic structure (formation of new molecular orbitals) of the

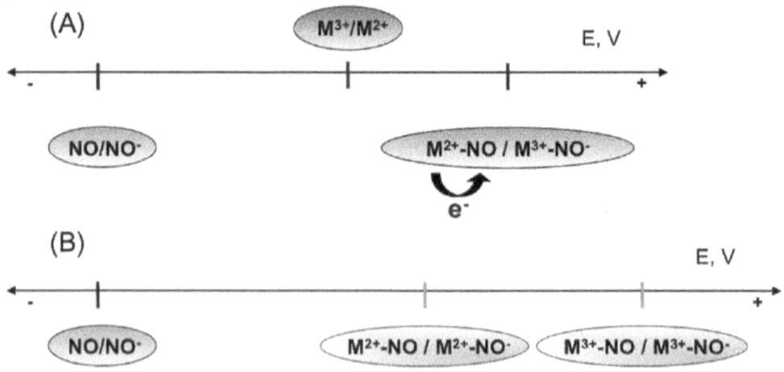

Figure 5.2 Effects of coordination of $NO^•$ to a metal center on its redox potential and reduction according to an inner- (A) or outer-sphere (B) electron transfer mechanism. (Note that in an inner-sphere electron transfer, indicated by the yellow arrow (black in print versions), there is no net charge change.)

complex. Consequently, in the case of NO$^•$ such changes support its reduction (Fig. 5.2). These processes strongly depend on the electronic structure of metal–NO$^•$ complexes (metal-nitrosyls), in the first place on the total number of the *d*-type electrons that distinguishes different class of $\{MNO\}^n$ nitrosyls (according to the Enemark-Feltham notation; n = number of electrons in metal d and π^*_{NO} orbitals),[8] and on the effects of other ligands present in the coordination sphere. Below are discussed some representatives of $\{MNO\}^n$ nitrosyl complexes that might offer a platform for approaching nitroxyl from nitric oxide.

5.1.2.2.1 $\{MNO\}^8$ complexes

In general, electron-rich $\{MNO\}^8$ complexes are good candidates for attaining metal-bound nitroxyl that might be released or transferred to other molecules, thus potentially considered either as sources of free HNO (HNO donors) or as transnitroxylation agents. There are different approaches to generate $\{MNO\}^8$ complexes, the "simplest" one is formal NO$^•$ coordination to the d^7 metal centers (Eq. 5.4).

$$\begin{matrix} Re^0 \\ Os^I \\ Fe^I \\ Co^{II} \end{matrix} + NO^\bullet \rightarrow \{MNO\}^8 \qquad (5.4)$$

But, the electronic nature of the NO-moiety within $\{MNO\}^8$ complexes, i.e., whether it has the NO$^•$, NO^+, or NO^- character, its basicity and reactivity, is determined by the overall ligand situation. Namely, the coordination of certain ligands is capable of triggering an electron transfer from the metal center toward NO-moiety, which increases its basicity and enables its protonation, thus assisting in an inner-sphere PCET and resulting in M-HNO species. A type of such mechanism is depicted in Fig. 5.1A, and demonstrates how powerful coordination chemistry tools are in the activation of small molecules.

These principles were behind the synthesis of the first characterized M-HNO species, starting from the nitrosyl osmium(0) $[OsCl(CO)(NO)(PPh_3)_2]$ complex.[9] In this reaction (Scheme 5.1), coordination of Cl^- induces a formal inner-sphere proton-coupled two-electron transfer, resulting in $[OsCl_2(CO)(HNO)(PPh_3)_2]$. One can also consider it as a hydrogen atom transfer (see Section 5.1.2.4.) from HCl to nitrosyl ligand, which based on the quite strong N–H bond of the Os-coordinated HNO[10] cannot be excluded. Using the analogues procedure (Eq. 5.5–5.6), starting from formally Re(-I)-nitrosonium complexes $[Re(CO)_2(NO)(PR_3)_2]$, upon 1,2-addition of HCl across Re–N bond the nitroxyl complex $[ReCl(CO)_2(HNO)(PR_3)_2]$ can be obtained.[11] This mechanistic principle for generation of nitroxyl complexes can further be extended to $\{MNO\}^{10}$ species as $[Ir(NO)(PPh_3)_3]$, where oxidative addition of two HCl equivalents induces a formal inner-sphere four-electron oxidation of the Ir center accompanied by two-electron reduction of NO^+ moiety (or a formal three-electron oxidation of Ir^0 and one-electron reduction of NO$^•$), resulting again in the $\{M\text{-}HNO\}^8$ complex, $[IrCl_2(H)(HNO)(PPh_3)_2]$ (Eq. 5.7).[12]

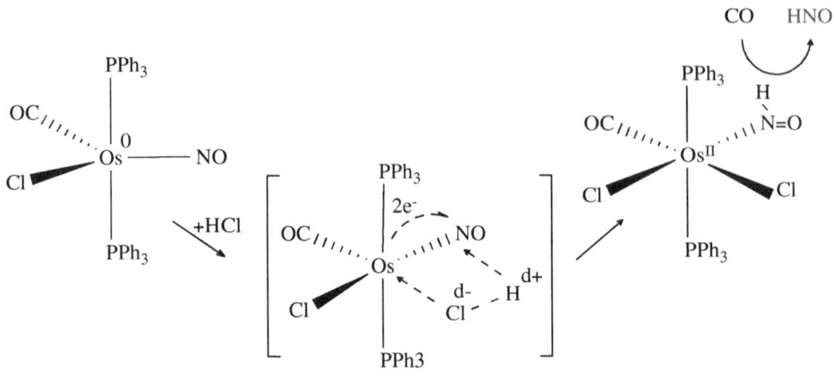

Scheme 5.1

$$Os^{0}\text{-}NO^{+}/Os^{I}\text{-}NO^{\bullet} \xrightarrow{HCl} Cl\text{-}Os^{II}\text{-}HNO \xrightarrow{CO} HNO \tag{5.5}$$

$$Re^{-I}\text{-}NO^{+}/Re^{0}\text{-}NO \xrightarrow{HCl} Cl\text{-}Re^{I}\text{-}HNO \tag{5.6}$$

$$Ir^{-I}\text{-}NO^{+}/Ir^{0}\text{-}NO^{\bullet} \xrightarrow{2HCl} Cl_2Ir^{III}H(HNO) \tag{5.7}$$

$$(PPh_3)_2(CO)_3Re^{I}\text{-}HNO \xrightarrow{Br^-,\ OTf^-} HNO \tag{5.8}$$

$$Co^{I}\text{-}NO^{+} \xrightarrow{Br^-} Br\text{-}Co^{III}\text{-}NO^{-} \xrightarrow{H^+} Br\text{-}Co^{III}\text{-}HNO \xrightarrow{MeOH} HNO \tag{5.9}$$

Although strong binding of HNO by 5*d* transition metal centers is expected,[13] substitution of Os-bound HNO by CO (Scheme 5.1 and Eq. 5.5) was reported,[9] suggesting that inner-sphere PCET within $\{MNO\}^8$ complexes followed by the HNO release or ligand exchange is a viable mechanism for the production of free HNO. Displacement of Rebound HNO by triflate or bromide was also observed in the case of $[Re(CO)_3(HNO)(PPh_3)_2]$ (Eq. 5.8), though this particular complex was prepared by oxidation of coordinated hydroxylamine.[14] Nevertheless, this example just confirms that $\{M(H)NO\}^8$ species indeed can serve as HNO donors.

In that respect, Co^{II}-$NO^{\bullet}$ adducts represent another class of potential HNO sources. Here again, the electronic structure of $\{CoNO\}^8$ complexes and its HNO-releasing ability are tuned by other electron-donating groups present in the coordination sphere. Whereas five-coordinate $[Co(NO)(das)_2]^{2+}$ is best formulated as Co^{I}-NO^{+}, the six-coordinate $[Co(NO)(NCS)(das)_2]^{+}$ has a character of

Co^{III}-NO^- (Eq. 5.9).[15] Although the presence of metal center in the +3 oxidation state stabilizes the nitroxyl moiety and certainly reduces its basicity, the analogues $[Co(NO)Br(das)_2]^+$ complex could be protonated resulting in the rare spectroscopically characterized Co-HNO species, $[Co(HNO)Br(das)_2]^{2+}$.[16] Even more, a rapid HNO displacement by the weak MeOH ligand was observed (Eq. 5.9).[16] It has recently been shown that NO• binding to $[Co^{II}(LN_4^{PhCl})]$ produces the five-coordinate complex $[Co(LN_4^{PhCl})(NO)]$ that has a nitroxyl character (Scheme 5.2), most probably promoted by the presence of the electron-rich doubly negatively charged tetradentate (LN_4^{PhCl}) ligand in the coordination sphere.[17] Although neither free nor coordinated HNO could be demonstrated upon proton addition, protons were able to induce a sort of *trans*-nitroxylation from $[Co(LN_4^{PhCl})(NO)]$ to Fe^{III}-porphyrin ([Fe(TPP)Cl]) and PPh_3 (Scheme 5.2). A role of protons might be to enhance the rate of NO^- transfer within an inner-sphere bridging intermediate.[17] Importantly, irreversible electrochemical oxidation of such Co^{III}-NO^- species with a strong nitroxyl character occurs at the significantly positive potential of +0.75 V versus Fc^+/Fc in the case of $[Co(LN_4^{PhCl})(NO)]$ (Scheme 5.2).[17] This confirms high stabilization of nitroxyl by its coordination to the Co^{III} center, in another words that NO• can be easily reduced if it is coordinated to Co^{III} or other metal +3 centers, in general.

These *trans*-nitroxylation reactions by five-coordinate Co^{III}-NO^- complexes (Scheme 5.2) and substitution of HNO from six-coordinate X-Co^{III}-HNO by other ligands (Eq. 5.9), as well as the easy to access them through the NO• coordination to Co^{II}, open a question of whether such species and reactions can be of biological relevance. The six-coordinate Co^{III}-NO^- species of physiological relevance is nitroxylcobalamin, $Cbl^{III}(NO^-)$,[18] which results from fast and thermodynamically favorable NO• scavenging by Co^{II} form of cobalamin (the reduced vitamin B_{12}) as its major intracellular form.[19,20] But whether nitroxylcobalamin may serve as a pull of HNO/NO^- remains unknown (Scheme 5.3).

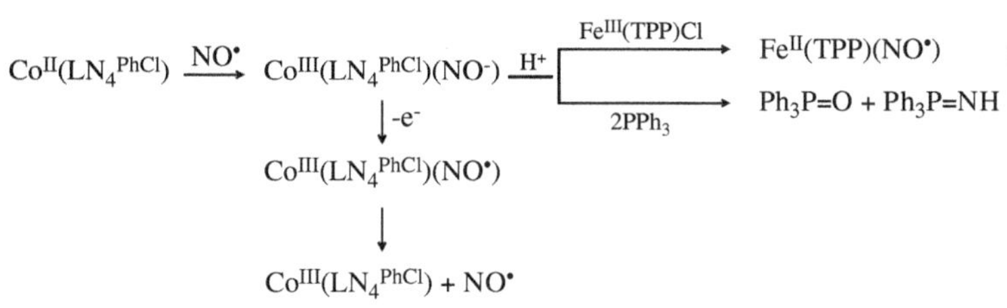

Scheme 5.2

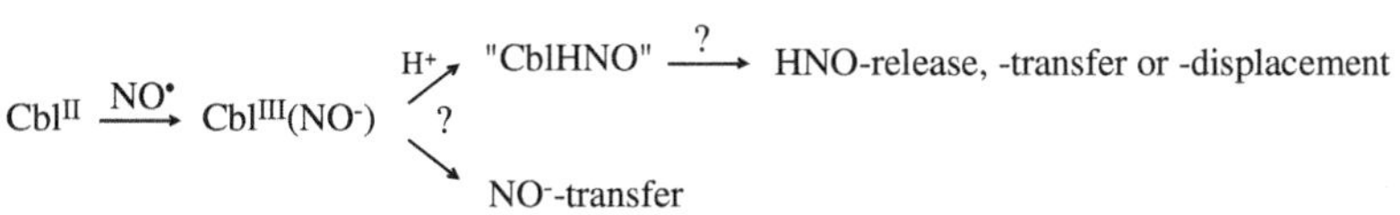

Scheme 5.3

Interestingly, although it is well established that Co^{III} does not have a strong affinity to react with $NO^•$, in general, and that at physiological conditions it does not react with aquacobalamin, $Cbl^{III}(H_2O)$, in particular, its reductive nitrosylation at low pH in the presence of HNO_2, is observed.[21] This reaction is a sort of stoichiometric $NO^•$ dismutation to NO^- and HNO_2/NO_2^- (Eq. 5.10) mediated by metal center. Since $Cbl^{III}(H_2O)$ does not react with $NO^•$, the reaction can occur only if: (1) the dimethylbenzimidazole group is dissociated from the Co^{III} center, upon protonation at low pH, leaving a vacant site for $NO^•$ coordination and (2) NO_2^- is present in the *trans*-position to it. This again represents an example of how spectator ligands control the redox behavior of metal centers toward $NO^•$.[21]

$$Cbl^{III}(NO_2^-) + 2NO^\bullet + H_2O \rightleftarrows Cbl^{III}(NO^-) + 2HNO_2 \quad (5.10)$$

Another class of $\{MNO\}^8$ species, the $\{FeNO\}^8$ complexes, will be discussed below in the context of reduction of $\{FeNO\}^7$ nitrosyls.

5.1.2.2.2 $\{MNO\}^7$ complexes and their reduction

Classical and widely studied examples of $\{MNO\}^7$ complexes are ferrous-NO adducts with the very thermodynamically stable Fe–NO bond due to its high degree of covalency.[5,22] Their electronic structure varies depending on the overall spin state and electronic nature of the ligands that are commonly divided into heme (porphyrin based) and nonheme systems.

5.1.2.2.2.1 Heme/porphyrin $\{MNO\}^7$ systems Ferrous heme-nitrosyls are exclusively ls complexes ($S = ½$) with the σ-bonding character of SOMO (singly occupied molecular orbital) that is distributed over Fe-NO moiety, but with the somewhat stronger NO localization.[5] Thus, they can be described as Fe^{II}-$NO^•$ species, which means that $NO^•$ binding to ferrous heme is not coupled to the electron transfer and does not result in nitroxyl species. Only upon addition of an external electron, via chemical or electrochemical reduction, the $\{FeNO\}^8$ complexes of nitroxyl-like character can be obtained. Their electronic nature is something between two ls resonance structures Fe^{II}-NO^- ↔ Fe^{I}-$NO^•$, but with predominant NO^- character. Consequently, the reduction of ls heme $\{FeNO\}^7$ is mostly NO-centered, though not entirely. An alternative way to prepare heme $\{FeNO\}^8$ complexes is the reaction of $NO^•$ with a Fe^{I}(Por) system, previously obtained by reduction of corresponding ferrous heme (Scheme 5.4).[5,22–24] However, both paths require application of very low redox potentials or very strong reducing agents (within electrochemical or chemical reduction). The reduction of heme $\{FeNO\}^7$ to $\{FeNO\}^8$ occurs usually at potentials between −0.6 and −1.3 V versus NHE,[5,22,24]

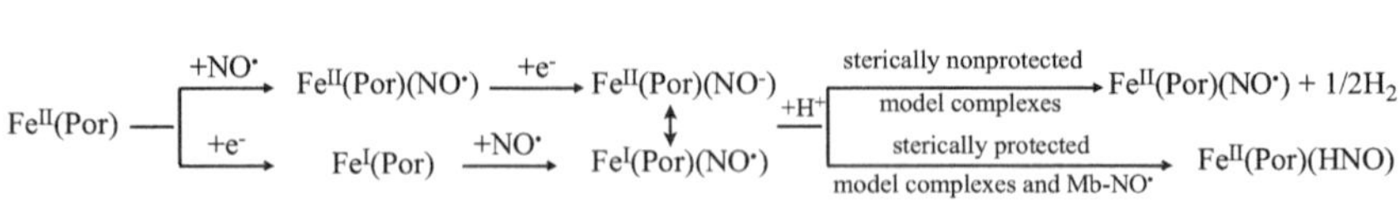

Scheme 5.4

that are far from biologically accessible conditions. By using highly electron-deficient porphyrin systems, reduction can be achieved at less negative potential such as in the case of [Fe(TFPPBr$_8$)(NO)], which could be converted into [Fe(TFPPBr$_8$)(NO^-)] at -0.65 V versus Fc^+/Fc.[25]

Once formed, heme $\{FeNO\}^8$ complexes cannot release NO^- due to the strong Fe-N bond and the bonding character of the corresponding HOMO (highest occupied molecular orbital).[5] Having very negative redox potentials, heme $\{FeNO\}^8$ complexes can reduce free NO•,[5] probably according to an outer-sphere electron transfer (Eq. 5.3, see Section 5.1.2.1.). Based on the electronic structure of heme $\{FeNO\}^8$, heme-bound nitroxyl is quite basic and can be easily protonated generating heme Fe^{II}-HNO species.[5,22] However, in the case of porphyrin model complexes that are usually sterically nonprotected, heme $\{FeHNO\}^8$ are not stable and disproportionate to starting $\{FeNO\}^7$ and H_2 (Scheme 5.4).[5,25,26] Recently, the first ferrous heme-HNO model complex [Fe(3,5-Me-BAFP)(HNO)] has been generated by utilizing the bis-picket fence porphyrin that sterically hinders the disproportionation reaction (Scheme 5.4).[5] A steric protection of the heme center within myoglobin (Mb) and hemoglobin (Hb) is probably also responsible for stable formation of nitroxyl ferrous heme forms. By way of comparison, a half-life of Mb-HNO is greater than weeks at pH between 6 and 10.[23,24] Mb-HNO can be generated from ferrous Mb-NO at −0.65 V versus NHE even at pH 10, which suggests high basicity of obtained Mb-NO^- and the corresponding pK_a value being higher than 10.[23,24] Thus, there are no evidences that Mb-HNO can be a source of free HNO, which even could not be substituted by CO (Eq. 5.11).[24] In contrary, it was reported that NO• could replace the weaker HNO ligand in the nitroxyl heme model complex [Fe(3,5-Me-BAFP)(HNO)] (Eq. 5.12).[5] A mechanism of this reaction is not known and can in principle be considered either as the ligand substitution or as a H-atom transfer from coordinated HNO to free NO• (see Section 5.1.2.4.).

$$\text{Mb-HNO} + \text{CO} \xrightarrow{\times} \tag{5.11}$$

$$\text{Fe}^{\text{II}}(3,5\text{-Me-BAFP})(\text{HNO}) \xrightarrow{+\text{NO}^\bullet} \text{HNO} + \text{Fe}^{\text{II}}(3,5\text{-Me-BAFP})(\text{NO}^\bullet) \tag{5.12}$$

Importantly, NO^- has a very strong *trans*-effect, stronger than that of NO• and CO.[5] Consequently, all known heme $\{FeNO\}^8$ complexes are five-coordinate. On the other hand, their protonated $\{FeHNO\}^8$ forms might be six-coordinate due to the significantly weaker *trans*-effect of HNO, but such heme model complexes are not characterized yet. An exception is the six-coordinate ruthenium porphyrin complex, [Ru(ttp)(HNO)(1-MeIm)], which can release HNO upon substitution by nitrosobenzene (Eq. 5.13).[27] Release of HNO from the $\{RuHNO\}^8$ complexes is not common, since 4*d* and 5*d* metals in general bind HNO more strongly than 3*d* metal centers.[13] It seems that the nature of the sixth ligand *trans* to HNO determines the reactivity of the $\{RuHNO\}^8$ complexes as well. While [Ru(Me$_3$[9]aneN$_3$)(bpy)(HNO)]$^{2+}$ is inert toward substitution,[28] generation of N_2O, an indirect marker of

HNO, was observed upon reduction (electrochemical and by Zn(Hg)) of six-coordinate *trans*-$[Ru(NO)(NH_3)_4(P(OEt)_3)]^{2+}$ at pH = 4.2, probably due to decomposition of transient/unstable $\{RuHNO\}^8$ species with the triethylphosphite ligand *trans* to nitroxyl (Eq. 5.14).[29]

$$Ru^{II}(ttp)(HNO)(1\text{-}MeIm) \rightarrow \begin{cases} PhNO \\ HNO \end{cases} \tag{5.13}$$

$$[Ru(NO)(NH_3)_4(P(OEt)_3)]^{2+} \xrightarrow[+H^+]{+e^-} HNO + [Ru(H_2O)(NH_3)_4(P(OEt)_3)]^{2+} \tag{5.14}$$

This may suggest that coordination of the sixth ligand *trans* to HNO in the heme $\{FeHNO\}^8$ complexes could also activate heme $\{FeHNO\}^8$ toward HNO release or the reaction with $NO^{\bullet}$ and the N–N bond formation, the latter being relevant for the mechanisms of $NO^{\bullet}$ reductases.[5,22] However, it should be stressed once more that generation of heme $\{FeNO\}^8/\{FeHNO\}^8$ species by one-electron reduction of corresponding heme $\{FeNO\}^7$ is difficult to achieve with the biologically available reducing agents and therefore this route is not very probable for the generation of the potential sources of free (H)NO. Considering a relatively high abundance of heme centers in the biological systems, this high thermodynamic barrier for the direct outer-sphere one-electron reduction of $\{FeNO\}^7$ might be seen as a protection of $NO^{\bullet}$ from its exceeded/uncontrolled transformation into (H)NO species that requires more subtle mechanistic regulation.

Interestingly, there are some experimental evidences[30–32] that heme $\{FeNO\}^7$ species itself, without external reduction, can serve as the potential, thermodynamically accessible source of free HNO. Similar to the processes described in Eq. 5.5–5.9 and Scheme 5.1, coordination of appropriate sixth ligand (L) can induce a sort of an inner-sphere electron transfer that stabilizes the L-Fe^{III}-NO^- resonance structures over L-Fe^{II}-$NO^{\bullet}$, which upon protonation releases HNO (Eq. 5.15).

$$Fe^{II}(por)(NO) \xrightarrow[L^- = NO_2^-, HS^-]{+L^-, +H^+} HNO + (L)Fe^{III}(por) \tag{5.15}$$

HNO generation was reported in the course of oxygen atom transfer reaction from nitrite to variety of substrates (phosphine, dimethyl sulfide, cysteine, and glutathione) mediated by the water-soluble ferriheme model Fe^{III}(TPPS).[30,31] Protonation of transient sixth-coordinate $Fe^{II}(TPPS)(NO_2^-)(NO)$ was proposed as a step that resulted in HNO.[31] We have recently demonstrated by using highly negatively charged iron porphyrin complex that HNO can be generated within the catalytic cycle of nitrite reduction by H_2S (Fig. 5.3, see Section 5.2).[32] Also here, one of the paths that lead to HNO includes protonation of the transient sixth-coordinate Fe^{II}(P(8-)) $(NO_2^-)(NO)$ or $Fe^{II}(P(8\text{-}))(HS^-)(NO)$ species (Eq. 5.15; Fig. 5.3, path A).

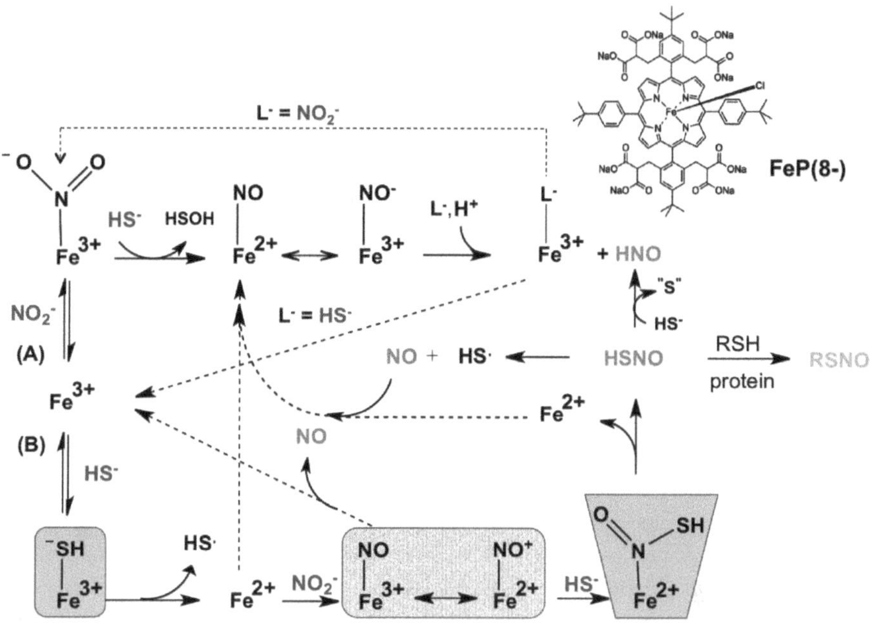

Figure 5.3 Cogeneration of HNO and HSNO by heme model complex FeP(8-).

Also in biological systems there is an evidence that heme $\{FeNO\}^7$ species can be a source of HNO in the presence of appropriate sixth ligand, which regulates the electronic nature of coordinated nitric oxide. For example, in the nitric oxide synthase (NOS) catalytic cycle, if there is no tetrahydrobiopterin radical available, the ferrous heme—NO$^•$ complex builds up, which is related to the production of nitroxyl instead of NO$^•$.[33] Importantly, the heme center of NOS has axially coordinated cysteine thiolate as the sixth ligand *trans* to the NO-binding site.

5.1.2.2.2.2 Low-spin nonheme $\{MNO\}^7$ systems Low-spin $\{FeNO\}^7$ complexes are also available with nonheme ligand systems, but there is no report about HNO generation neither from five- nor from six-coordinate ls nonheme $\{FeNO\}^7$. They can be reduced to corresponding ls nonheme $\{FeNO\}^8$ species at quite negative potentials, similar to those required for electrochemical reduction of heme $\{FeNO\}^7$. As an example, the ls six-coordinate $[Fe^{II}(NO)(cyclam\text{-}ac)]^{1+}$, with monoanionic pentadentate ligand, has a predominant Fe^{II}-NO$^•$ character and exhibits entirely NO-centered reduction at -1.39 V versus Fc/Fc$^+$ (-0.89 V vs Ag/AgCl) in acetonitrile.[34] Though this potential is quite low, it is still significantly more positive than that for the free NO$^•$/^{1}NO$^-$ couple (c. -2 V vs NHE, see chapter: Introduction), demonstrating that reduction of NO$^•$ becomes thermodynamically more favorable upon its coordination (Fig. 5.2B). The effect of the Fe(II) center on the outer-sphere electron transfer to NO$^•$ can be compared with the effect of protonation at c. pH 9, since at this pH the redox potential for NO$^•$/^{1}HNO (i.e., H^+-^{1}NO$^-$) resembles that of the Fe^{II}-NO$^•$/Fe^{II}-^{1}NO$^-$ couple. Almost identical redox potential of −1.38 V versus Fc/Fc^+, assigned to the reversible $\{FeNO\}^7/\{FeNO\}^8$ couple, has been reported for five-coordinate ls $[Fe(LN_4)(NO)]$, with electron-rich dianionic tetradentate ligand.[35] The reduction occurs rather over entire Fe-NO moiety than at the Fe-center or the NO unit, resulting in $\{FeNO\}^8$ with an electronic structure between ls Fe^{II}-NO^- ↔ ls Fe^I-NO·. Although neither free HNO nor corresponding $\{FeHNO\}^8$ was demonstrated, a sort of NO^- transfer to metMb at pH 7.2 was reported (Eq. 5.16).[35] An unusual case of the ls nonheme $\{FeNO\}^7$ species for which the Fe^{III}-NO^- electronic structure was proposed is the five-coordinate [Fe(TC-5,5)(NO)] tropocoronand complex that exhibits a reversible reduction at −1.81 V versus Fc/Fc^+.[36]

By way of comparison, for $\{RuNO\}^7$ that are exclusively ls complexes significantly more positive NO-centered reduction potentials were reported. For example, a reversible $[Ru(Me_3[9]aneN_3)(bpy)(NO)]^{2+/1+}$ conversion was observed at −0.44 V versus standard Ag/AgCl (3 M NaCl) in CH_3CN (c. −0.23 V vs NHE) and for the $[Ru(NO)(\text{'py}^{bu}S_4\text{'})]^{0/-1}$ redox couple at −0.32 V versus NHE in DMF, whereas *trans*-$[Ru(NO)(NH_3)_4(P(OEt)_3)]^{2+}$ was irreversibly reduced at −0.46 V versus NHE in aqueous solution at pH = 4.2 (which is proposed to be subsequently coupled with HNO release).[28,29,37] These potentials are much more positive than that for the free $NO^\bullet/^1NO^-$ and even $NO^\bullet/^3NO^-$ couple, demonstrating a dramatic effect of the metal coordination on the affinity of nitric oxide toward reduction (Fig. 5.2B). Interestingly, for the proton-coupled reduction of $[Ru(Me_3[9]aneN_3)(bpy)(NO)]^{2+}$ to yield the corresponding $\{RuHNO\}^8$ species, $[Ru(Me_3[9]aneN_3)(bpy)(HNO)]^{2+}$, redox potential of c. −0.44 V versus standard Ag/AgCl at pH = 7.4 was reported, which is almost identical to that for the pure electron transfer process.[28] On the other hand, the $\{RuHNO\}^8$ species, $[Ru(HNO)(\text{'py}^{bu}S_4\text{'})]^0$, exhibits an irreversible oxidation to $[Ru(NO)(\text{'py}^{bu}S_4\text{'})]^0$ at +0.24 V versus NHE in THF.[37] Though these redox potentials are biologically available, ruthenium compounds are not. However, they may have a potential pharmaceutical application if they are able to release either NO or HNO in biological systems, as it is the case with *trans*-$[Ru(NO)(NH_3)_4(P(OEt)_3)]^{2+}$ upon its reduction (Eq. 5.14).[38]

$$[Fe^{II}(LN_4)(NO^-)]^- \xrightarrow[\text{pH} = 7.2]{\text{metMb}^{III}} Mb^{II}NO \qquad (5.16)$$

5.1.2.2.2.3 High-spin nonheme $\{MNO\}^7$ systems Different from majority of nonheme ls $\{FeNO\}^7$, the high-spin $\{FeNO\}^7$ species are best described as Fe^{III}-NO^-, with hs Fe(III) antiferromagnetically coupled to $^3NO^-$.[22,39] The Fe–NO bond has a high degree of covalency and release of NO^-, as well as its protonation is not favorable.[22] For comparison, pK_a of the aqua ligand within hs $[Fe(H_2O)_6]^{3+}$ is c. 2, which means it is for ~12 pH units lower than that of free H_2O and $[Fe^{III}(H_2O)_5(OH^-)]^{2+}$ can be fully protonated only below pH 1. Since pK_a of free HNO is c. 11.4 (see chapter: Introduction), it is not expected to protonate

$[Fe^{III}(H_2O)_5(NO^-)]^{2+}$ even at pH 0. But electron-rich ligand systems can change the basicity of coordinated NO^- significantly. As an example, pK_a of aqua ligand in $[Fe^{III}(EDTA)(H_2O)]^-$ is c. 7 and one could envision that putative $[Fe^{III}(EDTA)(HNO)]^-$ may exist at pH <5. Though such species has not been identified, it was proposed as an intermediate responsible for the N_2O production (Eq. 5.17).[40,41] Thus, electron-rich ligands can decrease Fe−NO bond strength[42] and facilitate a stronger Fe^{III}-NO^- character as well as basicity of hs $\{FeNO\}^7$ species that upon protonation can result in HNO.

$$[Fe^{II}(EDTA)(NO)]^{2-} \xrightarrow{slow} [Fe^{III}(EDTA)(NO^-)]^{2-} \xrightarrow{+H^+} [Fe^{III}(EDTA)]^- + HNO \;\; (HNO \dashrightarrow N_2O) \quad (5.17)$$

The second way to labilize the Fe−NO bond of the hs $\{FeNO\}^7$ complexes implies its reduction to corresponding hs $\{FeNO\}^8$, which similar to the effect of electron-rich ligands reduces a degree of Fe−NO bond covalency and its strength. The reduction of hs $\{FeNO\}^7$ is a metal-centered process and occurs at lower potentials. However, these potentials can be tuned by ligand systems and in this way they can enter in the range of biologically feasible redox processes, as in case of the model complex $[Fe(BMPA\text{-}Pr)(NO)]^+$ of the Fe_B center of bacterial nitric oxide reductase (c. −0.3 V vs SHE).[42] If the ligand is a strong donor, the reduction of hs $\{FeNO\}^7$ occurs at quite negative potentials, similar to those for ls $\{FeNO\}^7$ systems. An example is the $[Fe(TMG_3tren)(NO)]^{2+}$ complex that is reduced at −1.34 V versus Fc/Fc^+.[43] Reduction not only weakens the Fe−NO bond, but it also increases the basicity of the NO moiety, both of which contribute to reactivity of hs $\{FeNO\}^8$ toward HNO release. However, until now the only stable hs $\{FeNO\}^8$ complex is the one obtained by reduction of $[Fe(TMG_3tren)(NO)](OTf)_2$,[43] and the HNO generation from hs $\{FeNO\}^8$ is still unrevealed research field. In conclusion, hs $\{FeNO\}^7$ species can be activated toward HNO generation either by reduction or by electron-rich ligands, which is a feasible mechanism under physiological conditions.

5.1.2.2.3 $\{MNO\}^6$ complexes

In the $\{MNO\}^6$ complexes, the NO ligand typically has the nitrosonium character, and the M−NO and N−O bonds in the linear M-NO^+ units have the multiple character.[44] Therefore, they cannot be considered as a direct source of (H)NO. However, upon one-electron reduction to $\{MNO\}^7$, two-electron reduction (reduction by H^-) to $\{M(H)NO\}^8$ (see Section 5.1.2.3), as well as upon interaction with H_2S (Fig. 5.3) resulting in intermediate HSNO formation (see Section 5.2 and Scheme 5.11), they can also contribute to the HNO generation.

Within $\{FeNO\}^6$ complexes the Fe^{II}-NO^+ electronic structure dominates, as in the case of Ru and Os species,[44] though $[Ru(NH_3)_5(NO)]^{3+}$ was firstly assigned as Ru^{IV}-NO^-.[45] Interestingly, $\{MnNO\}^6$ complexes can serve as a source of (H)NO. It has been reported that five-coordinate Mn(II) tropocoronand [Mn(THF)(TC-5,5)] can activate NO toward its disproportionation.[46] The reaction involves an intermediate trigonal-bipyramidal Mn-nitrosyl [Mn(NO)(TC-5,5)] that has characteristics of high-spin Mn^{III}-NO^-, which is different from the Mn(II) porphyrin systems and in general ls $\{MnNO\}^6$ complexes that have a Mn^I-NO^+ character.[44,47] The strongly electron-releasing nature of the tropocoronand ligand favors the nitroxyl character of [Mn(NO)(TC-5,5)]. [Mn(NO)(TC-5,5)] does not release (H) NO but reacts further with excess NO, resulting in N_2O and metal-bound NO_2 (Eq. 5.18).[46]

$$[(TC\text{-}5,5)Mn(THF)] \xrightarrow[\text{-THF}]{+NO} [(TC\text{-}5,5)Mn^{III}\text{-}NO^-] \xrightarrow{+2NO} [(TC\text{-}5,5)Mn\text{-}NO_2] + N_2O \quad (5.18)$$

Importantly, it was demonstrated that manganese superoxide dismutase enzyme (MnSOD) from *Escherichia coli* catalyzes NO disproportionation (dismutation) both under anaerobic[48] and aerobic[49] conditions (Scheme 5.5).

This has implied that natural manganese enzyme is not strictly selective toward superoxide and that such type of the $NO^•$ activation may be a more general feature of high-spin Mn species capable of catalytic superoxide dismutation. Indeed, we have shown that the Mn pentaazamacrocyclic SOD mimetic [Mn(pyane)]Cl_2, previously considered to be selective for superoxide,[7b] does react with $NO^•$, according to a catalytic mechanism in which one-electron oxidation and reduction of $NO^•$ are accompanied by cycling of the Mn center from Mn(II) to Mn(III) (Fig. 5.4).[50] A single-cell study confirmed that this Mn complex is capable of activating $NO^•$ in biological environment as well.[51]

Our preliminary new data indicate that there is a substantial increase of the HNO monitored (by a fluorescent HNO sensor) in the cells treated with [Mn(pyane)]Cl_2. In the presence of biologically relevant reducing agents, such as glutathione (GSH), the predominant reaction mechanism is the catalytic $NO^•$ reduction (Scheme 5.6) instead of $NO^•$ dismutation. Thus, the overall reaction represents the $NO^•$ reduction by GSH catalyzed by the Mn^{II} center. Involvement of HNO and the oxidized form of reducing agent (e.g., GSH, Scheme 5.6) in further chemical transformations (e.g., formation of N_2O, NH_2OH and GSNO, all detected in the product mixture) leads the overall process to a thermodynamic sink.

In the absence of reducing agents or scavengers of the nitrosonium cation (e.g., thiols), there is a decline in the complex catalytic activity due to the ligand *N*-nitrosation (Fig. 5.5). This is confirmed by the detection of the {[Mn(pyane)]-3H + 3NO} and {(pyane)-3H + 3NO} adducts of the complex and the free ligand, respectively, by the ESI Ms analysis.[50] Most probably, *N*-nitrosation of the three

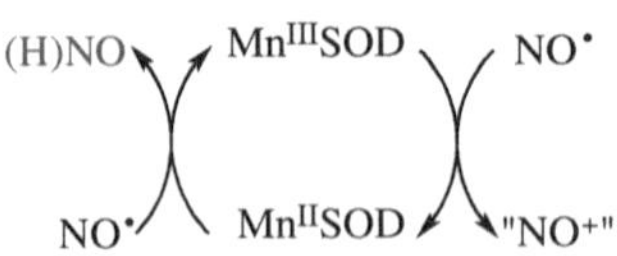

Scheme 5.5

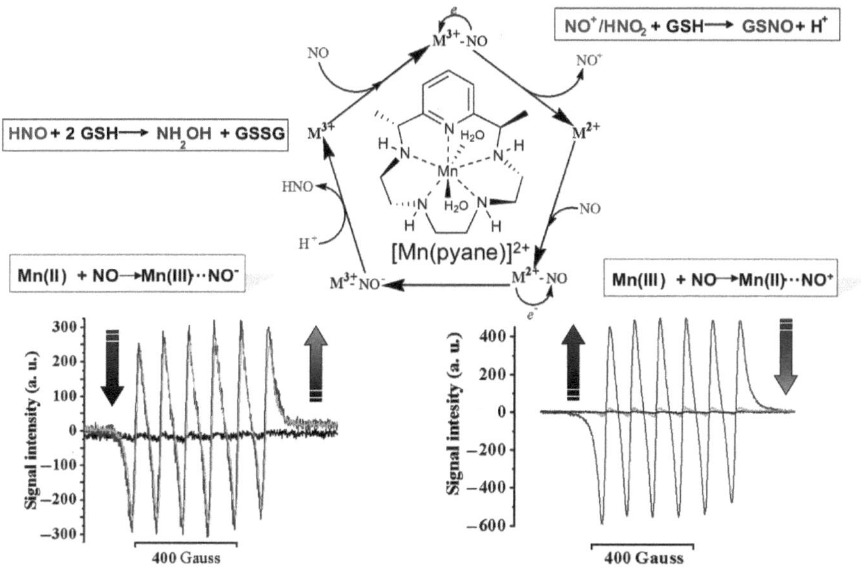

Figure 5.4 Generation of HNO within the catalytic dismutation of NO by the manganese SOD mimetic $[Mn(pyane)]Cl_2$ (in aqueous solutions present as $[Mn(pyane)(H_2O)_2]^{2+}$) that cycles between Mn(II) and Mn(III) state. Catalytic cycles monitored by EPR: signal of the starting Mn(II) form of the complex (red signal (dark gray in print versions); left) and upon addition of one (black signal; left) and two equivalents of NO• (green signal (light gray in print versions); left); signal of the starting Mn(III) form obtained by electrochemical oxidation (black signal; right) and upon addition of one (red signal (dark gray in print versions); right) and two equivalents of NO• (green signal (light gray in print versions); right).

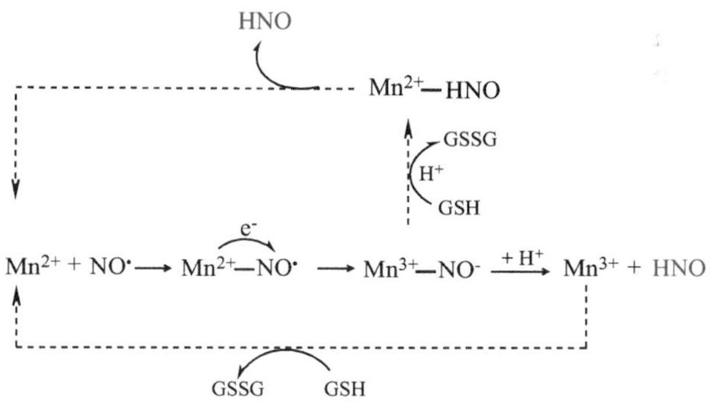

Scheme 5.6

secondary amine groups out of four leads to the Mn dechelation. This ligand modification together with the fact that the formation of the Mn(III) form of this type of complexes is accompanied by the ligand deprotonation ($\{Mn^{III}(L-2H^+)\}^+$, $\{Mn^{III}(L\text{-}H^+)(OH)\}^+$, and $\{Mn^{III}(L\text{-}H^+)(OH)(H_2O)\}^+/\{Mn^{III}(L\text{-}2H^+)(H_2O)_2\}^+/$

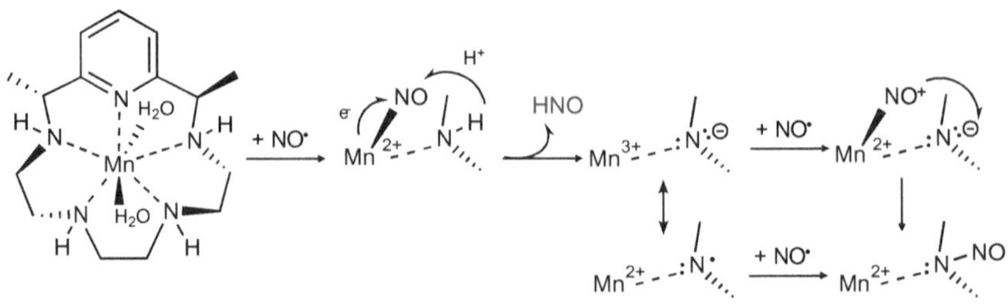

Figure 5.5 Schematic presentation of the proposed mechanisms of HNO generation and *N*-nitrosation of $[Mn(pyane)(H_2O)_2]^{2+}$ by NO$^•$ in the absence of external reducing agents, protons or scavengers of the nitrosonium cation, via an inner-sphere PCET resulting in the intermediate Mn(III) form with the deprotonated secondary amine group.

$\{Mn^{III}(L)(OH)_2\}^+$ were detected as the main species by ESI-Ms)[52] indicates the operation of an inner-sphere PCET mechanism (according to a general scheme depicted in Fig. 5.1A) in the production of HNO both in the presence (Scheme 5.6) and absence (Fig. 5.5) of reducing agents or nitrosonium scavengers, respectively. Since the redox potential of the corresponding Mn^{III}/Mn^{II} couple of these Mn pentaazamacrocyclic complexes is quite high (c. +0.8 V vs SHE),[53,54] a direct outer-sphere reduction of NO$^•$ by their Mn(II) forms is not feasible, which emphasizes the importance of the inner-sphere electron transfer and the proton-donating ability of the ligand. The latter assists in the protonation of the coordinated NO^-, an increase of the ligand negative charge, and consequently the electron density on the metal center that cooperatively stabilizes the Mn(III) form of the complex and contributes to the HNO release. Above all, an important factor that enables such reactivity of high-spin manganese species and easy HNO release from the Mn(III) coordination sphere is the hs d^4 configuration of Mn(III) that exhibits a strong Jahn–Teller effect, which is crucial for its kinetic lability and consequently its catalytic activity.[55]

5.1.2.3 Reduction by hydride

In general, generation of the ls $\{Fe(H)NO\}^8$ nitroxyl complexes by one-electron reduction of corresponding $\{FeNO\}^7$ is thermodynamically unfavorable (see Section 5.1.2.2.2) and can be overcome by utilizing hydride (H^-) for the reduction of ferric nitrosyl $\{FeNO\}^6$ species[56] in a formal proton-coupled two-electron transfer process (Eq. 5.19).

$$M^{III} \xrightarrow{+NO} M^{III}\text{-}NO^\bullet \leftrightarrow M^{II}\text{-}NO^+ \xrightarrow{+H^-} M^{II}\text{-}HMO \quad (M = \text{Fe, Ru}) \qquad (5.19)$$

This approach was used to obtain an HNO adduct of ferrous myoglobin (Mb^{II}-HNO) in a one-flask reaction of either met- or deoxymyoglobin with nitrite and borohydride (Eq. 5.20).[24,56] Similar reaction was observed with hemoglobin, as well.[56] Importantly, ferrous nitrosyl myoglobin (Mb^{II}-NO) does not react with borohydride, again demonstrating that electron reduction of $\{FeNO\}^7$ is not a feasible route for the synthesis of ferrous nitroxyl complexes. However, the mechanism of the formation of Mb^{II}-HNO and Hb^{II}-HNO in the nitrite/borohydride procedure is not necessarily a direct hydride transfer to $\{FeNO\}^6$. One possible mechanism is the generation of free HNO from nitrite and hydride, which is then trapped by Mb^{II}.[56] On the other hand, a direct hydride reduction of ferric nitrosyl to ferrous nitroxyl was demonstrated in the reaction between sodium nitroprusside (SNP, i.e., $Na_2[Fe(CN)_5(NO)]$) and borohydride (Eq. 5.20),[56] though only with c. 20% yield.[21] This synthetic path was also previously used for the preparation of the first ruthenium nitroxyl complex, $[Ru(HNO)(\text{'py}^{bu}S_4\text{'})]^0$, by addition of $NaBH_4$ to a cationic octahedral nitrosyl complex, $[Ru(NO)(\text{'py}^{bu}S_4\text{'})]^+$, (Eq. 5.20).[37]

$$\left.\begin{array}{r} Mb^{II/III} + NaNO_2 \\ [Fe(CN)_5(NO)]^{2-} \\ [Ru(NO)(\text{'py}^{bu}S_4\text{'})]^+ \end{array}\right] \xrightarrow{NaBH_4} \left[\begin{array}{l} Mb^{II}\text{-HNO} \\ [Fe(CN)_5(HNO)]^{3-} \\ [Ru(HNO)(\text{'py}^{bu}S_4\text{'})]^0 \end{array}\right. \quad (5.20)$$

As mentioned above (see Section 5.1.2.2.2), release of free HNO is usually not expected from Ru−HNO complexes[13] and is also not observed from Mb^{II}-HNO.[24] We have recently demonstrated release of free HNO from $[Fe(CN)_5(HNO)]^{3-}$ that was generated in the reaction of sodium nitroprusside with H_2S, through the intermediate formation of $[Fe(CN)_5(HSNO)]^{3-}$ (Fig. 5.6, see Section 5.2).[57]

Based on the reactions of the heme Fe^{III}-NO• species (of a predominant Fe^{II}-NO^+ character) with hydride donors, it can be concluded that under physiological conditions NAD(P)H can be used to directly convert them into Fe^{II}-HNO and in that way avoid unfavorable one-electron reduction of stable Fe^{II}-NO adducts. However, this process of a direct delivery of hydride in biological systems requires a binding of NADH in a close proximity to the heme Fe^{III}-NO• moiety, which can be seen as a mechanism for a control of the production of Fe^{II}-HNO adducts. This process operates within the catalytic cycle of nitric oxide reductase cytochrome P450nor, where NADH binds close to heme Fe^{III}-NO• and directly transforms it into an intermediate, most probably of a Fe^{II}-HNO character, which further reacts with additional NO• resulting in N_2O and recovery of ferric heme.[58,59]

5.1.2.4 Hydrogen atom transfer

Since HNO was originally synthesized in the reaction between NO• and hydrogen atoms, H•, (Eq. 1.1)[60] it is to be expected that free or coordinated HNO can result from the HAT process, where free or coordinated NO· abstracts H• from appropriate H• donor, H-X (Eq. 5.21).

$$\left.\begin{array}{r} NO^\bullet \\ M\text{-NO} \end{array}\right] \xrightarrow[\text{[HAT]}]{H\text{-}X} \left[\begin{array}{l} HNO \\ M\text{-HNO} \end{array}\right. \quad (5.21)$$

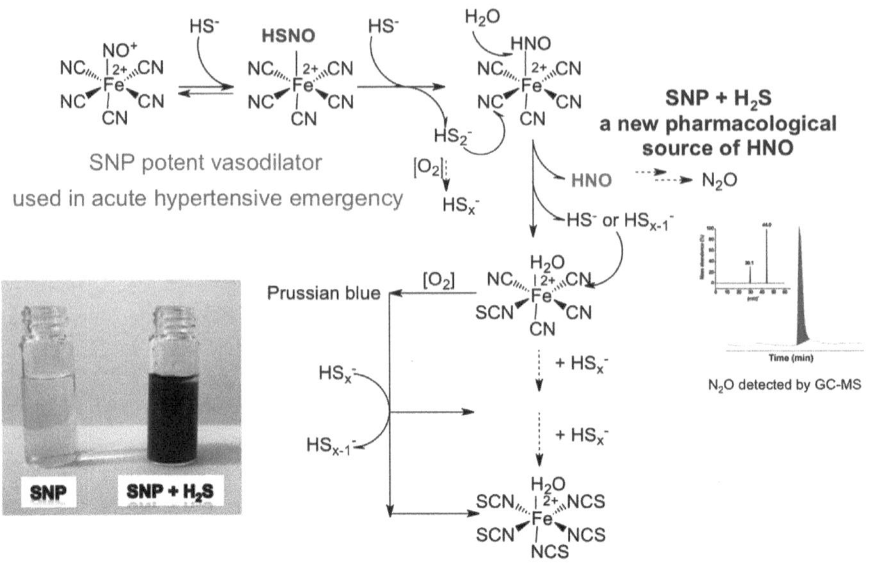

Figure 5.6 Mechanism of the reaction of sodium nitroprusside (SNP, i.e., $Na_2[Fe(CN)_5(NO)]$) with H_2S that generates free HNO (through an intermediate formation of Fe-HSNO species) and converts toxic cyanide ions into benign thiocyanate, both of which results in beneficial therapeutic effects for the treatment or prevention of cardiovascular diseases.

In order to be able to abstract $H^•$, a metal-bound $NO^•$ species needs to have a strong NO-centered radical character, $M\text{-}NO^•$, and a sort of interconversion between these species ($NO^•$, HNO, $M\text{-}NO^•$, and M-HNO; Eq. 5.22) is reported in the case of $[M(H)Cl(NO)(CO)L_2]/[M(H)Cl(HNO)(CO)L_2]$ complexes (M = Ru, Os; L = P^iPr_3).[61] In these studies of the reaction between the 16 valence electron $[Ru(H)Cl(CO)(P^iPr_3)_2]$ complex and $NO^•$, it was observed that a metal-hydride functionality, M-H, can serve as a hydrogen atom donor for the formation of free and coordinated HNO (Scheme 5.7).[61] Though metal-hydrides do not exist under physiological conditions, some other $H^•$ donors are available in biological systems such as ascorbate and that route for the HNO formation will be discussed in Section 1.3.

$$\text{M-HNO} + NO^• \rightleftharpoons \text{M-NO}^• + \text{HNO}$$

$$\text{M-HNO} + NO^• \text{ (in excess)} \rightleftharpoons \text{M-NO}^• + \underset{\substack{\downarrow \\ \text{decay/consumption}}}{\text{HNO}} \qquad (5.22)$$

It has been reported, based on DFT calculations, that the actual H–NO bond dissociation energy of free HNO is lower by c. 9–16 kcal/mol than that of HNO coordinated to the ls d^6 metal centers.[11,61] This would imply that free $NO^•$ is a

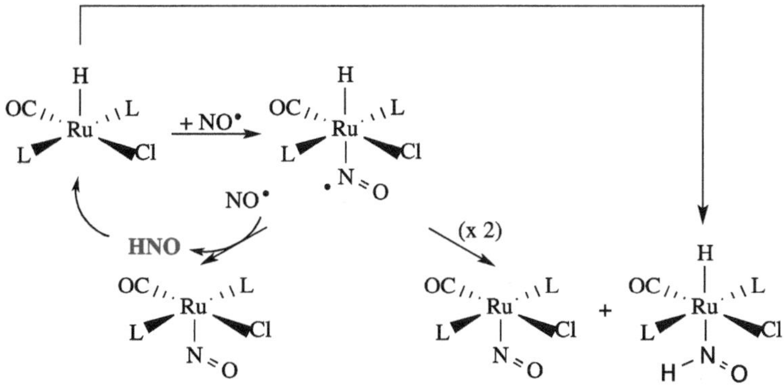

Scheme 5.7

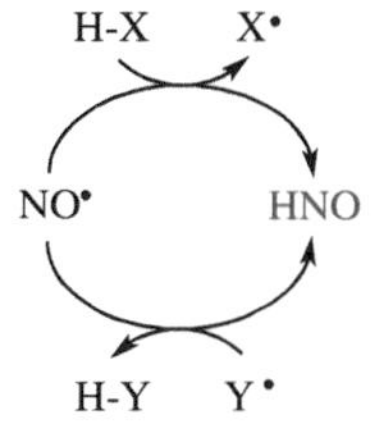

Scheme 5.8

somewhat weaker hydrogen atom abstractor than the coordinated one, though the difference is not so big and supports the equilibrium presented in Eq. 5.22. Under conditions of excess NO$^•$, this equilibrium can be pushed toward free HNO, whose consumption (e.g., $HNO + 2NO \rightarrow N_2O + HNO_2$) further facilitates an equilibrium shift in the direction of its intermediate formation (Eq. 5.22). Such HAT processes between M-HNO and NO$^•$ are responsible for the mechanisms of the reported [M(H)Cl(NO)(CO)L_2]/[M(H)Cl(HNO)(CO)L_2]/[MCl(NO)(CO)L_2] conversions (Scheme 5.7) and might explain the observed transformation of Fe^{II}(Por)HNO into Fe^{II}(Por)NO$^•$ (Eq. 5.12).[5] They also imply that relatively stable M-HNO species can act as a source of free HNO, not via release or substitution of HNO but via HAT to NO$^•$.

Having such chemistry in mind, one can consider NO$^•$ as an acceptor of relatively weakly bound H· atoms (in H-X; Scheme 5.8) that in the form of HNO, as a small and mobile H· atom "shuttle"/"carrier,"[61b] can easily be transferred to form a stronger H—Y bond (Scheme 5.8). In another word, NO$^•$ may catalyze HAT reactions ($H\text{-}X + Y^• \rightarrow H\text{-}Y + X^•$), similar to its observed catalytic activity in the case of recombination of H$^•$ atoms to form H_2 (Eqs. 1.1 and 1.2).[62] This is in line with a previous assumption that HNO complexes may be generally considered as H$^•$ atom sources.[24] As possible illustration of similar type of reactivity of coordinate HNO, we can find in the literature reports on a disproportionation of Fe^{II}(Por)(HNO) complexes (Por = TPP, TFPPBr$_8$, and To-F$_2$PP)[5] into Fe^{II}(Por)(NO) and H_2 that also can be seen as recombination of H$^•$ atoms, carried by the M-HNO moiety. However, whether this type of reactions, where HNO ligand is involved in the HAT process, represents a more general feature of nitroxyl complexes remains to be investigated in the future.

5.1.3 Metal-free processes

In the previous section (see Section 5.1.2), we have described a number of reactions and possible mechanism for the reduction of $NO^{\bullet}$ in a stoichiometric (Eqs. 5.5, 5.8, 5.9, 5.12–5.15, 5.17, 5.22; and Schemes 5.1 and 5.7) or catalytic manner (Scheme 5.5, 5.6; and Fig. 5.4) mediated by metal centers. The question is whether there are physiologically available nonmetal-based reducing agents that are capable of reducing nitric oxide notwithstanding its low redox potential (see Section 5.1.1 and chapter: Introduction). In this section, we will present our findings that biologically important reductants, ascorbate ($AscH^-$) and hydrogen sulfide (H_2S/HS^-), do generate HNO by forming (cage-)adducts with $NO^{\bullet}$, within which an inner-sphere electron transfer occurs coupled to a proton transfer (Fig. 5.1B). In that way, the activation barrier for the reduction of $NO^{\bullet}$ to singlet $(H)NO^-$ significantly drops. Thermodynamically, further consumption/decay of HNO and of the reactive, primarily formed, oxidation products of $AscH^-$ and H_2S, such as ascorbyl radical (AscH·) and $HS\cdot/H_2S_2$, shifts the reaction equilibrium enabling the intermediate formation of HNO.

5.1.3.1 Reduction of $NO^{\bullet}$ to HNO by vitamin C

In a recent study coauthored by us, it has been demonstrated that the reaction of $NO^{\bullet}$ with aromatic or "pseudoaromatic" alcohols, tyrosine (Y), hydroquinone (HQ), phenol (PhOH). and ascorbic acid (vitamin C) provides a possible physiologically relevant route for the generation of HNO.[63] The rate and yield of the reaction decreases in the order $AscH^- > HQ > PhOH > Y$ and shows a certain correlation with the redox potentials of the studied alcohols, although all of them are much more positive than $E°(NO^{\bullet}, H^+/^1HNO)$ (Eq. 1.4), which would imply that the reaction is thermodynamically not feasible. However, as mentioned above, generation of HNO is only the first and rate-limiting step, followed by the subsequent reactions of RO· radicals and HNO. Kinetically, the reaction becomes more favorable if we consider a mechanism that involves an intermediate formation of a (cage-)adduct $RO(H)-NO^{\bullet}$ (Scheme 5.9). The inner-sphere electron transfer that occurs within this adduct that is coupled to a proton transfer formally represents a sort of HAT. Based on the bond dissociation energies for the H-NO and corresponding RO–H bond of ascorbate that is for c. 25 kcal/mol bigger in the case of $AscH^-$, $NO^{\bullet}$ would not be able to abstract the hydrogen atom from ascorbate. However, as in the case of the HAT between

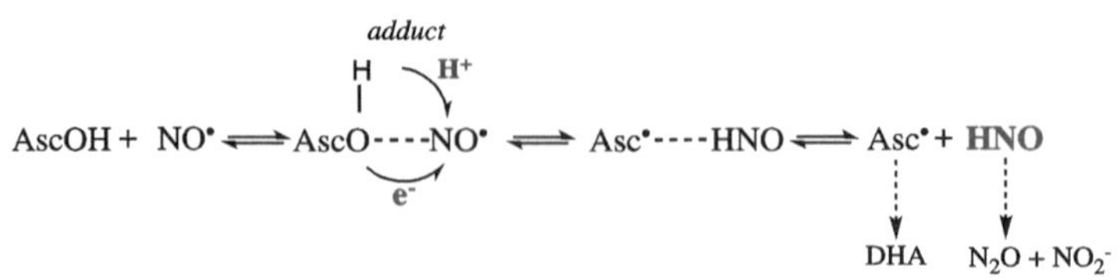

Scheme 5.9

M-HNO and NO$^•$ (Eq. 5.22), consumption of the reactive intermediates (RO· radicals and HNO) leads to a final thermodynamic sink of the overall reaction.

Evidences for the existence of such adduct we have provided by monitoring the reaction between NO$^•$ and ascorbate by the real time cryo-spray ionization mass spectrometry.[63] These experiments have revealed the intermediate formation of the adduct between ascorbic acid and NO$^•$ ([HAscONOH] in Fig. 5.7), its decay to HNO and ascorbyl radical (AscH· detected by EPR), as well as the subsequent formation of the adduct between AscH· and NO$^•$ (*O*-nitrosoascorbate, [HAscONO] in Fig. 5.7) that also releases HNO and results in dehydroascorbate (DHA in Fig. 5.7), confirming the mechanism depicted in Scheme 5.9. The cell studies provided a support for the $AscH^-$-NO$^•$ reaction as a source of HNO under physiological conditions.[63] In bovine arterial endothelial cells pretreated with 1 mM $AscH^-$, an increased intracellular level of HNO was observed, visualized by the HNO fluorescence sensor (CuBOT1) loaded in cells after treatment with $AscH^-$. Also, an extracellular production of HNO was detected by HNO electrode immersed in the medium containing millimolar ascorbate and macrophages cells stimulated to produce NO$^•$.

5.1.3.2 Reduction of NO$^•$ to HNO by H_2S

Like NO$^•$, in biological systems hydrogen sulfide is recognized as a gasotransmitter involved in redox signaling processes.[64] Different from NO$^•$ that can undergo both oxidations and reductions, H_2S can only be oxidized and acts as a good reducing agent. Despite its reactivity, its physiologically available steady-state concentrations

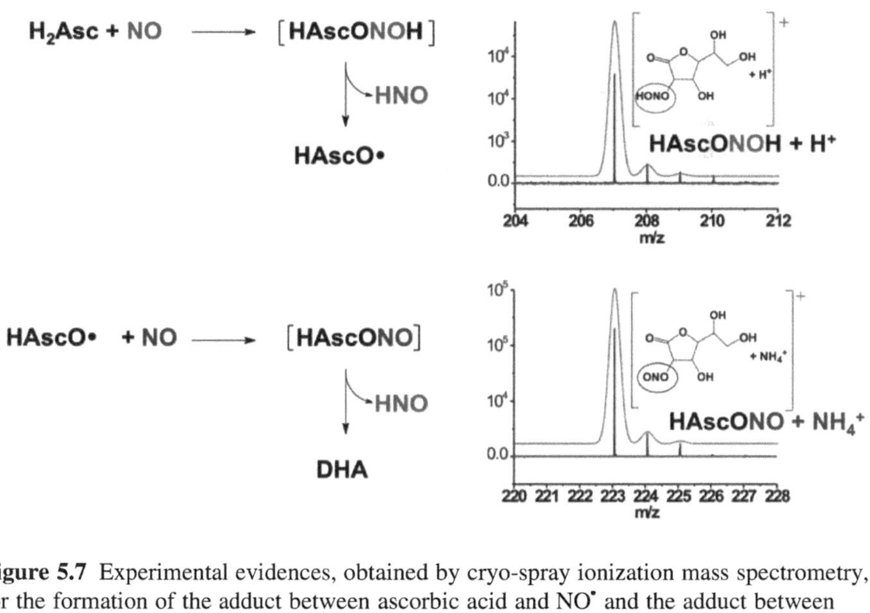

Figure 5.7 Experimental evidences, obtained by cryo-spray ionization mass spectrometry, for the formation of the adduct between ascorbic acid and NO$^•$ and the adduct between ascorbyl radical and NO$^•$, both of which release HNO.

are usually higher than that of $NO^{\bullet}$, which corresponds to its huge generating flux that is almost as that of glutathione.[65] In addition, it is small and can easily diffuse through the membranes,[66] thus, being a very good candidate as a reducing agent for $NO^{\bullet}$.

And indeed, in extensive in vitro and in vivo studies[67] we have shown that H_2S and $NO^{\bullet}$ react directly generating HNO. We have demonstrated that endogenous HNO production is strongly dependent on both H_2S and $NO^{\bullet}$ by showing that intracellular levels of HNO dramatically drop upon inhibition or substrate depletion of NOS and/or CBS (cystathionine beta synthase), the enzymes that generate $NO^{\bullet}$ and H_2S, respectively (Fig. 5.8). The reaction kinetics was amperometrically followed by $NO^{\bullet}$-, H_2S-, and HNO-sensitive electrodes (Fig. 5.9), and the measurements have revealed very fast kinetics of the HNO generation. Also, we have shown that the rate-determining step is first order in both $NO^{\bullet}$ and H_2S (Fig. 5.9E). The HNO production in the H_2S-$NO^{\bullet}$ reaction is the fastest ever observed. It is more rapid than in the case of the ascorbate-$NO^{\bullet}$ reaction and is even faster than HNO release from Angeli's salt. The rate of the HNO generation from 2 mM $NO^{\bullet}$ and H_2S corresponds to that observed in the case of 1 mM Angeli's salt![67]

Fast interaction between $NO^{\bullet}$ and H_2S is of crucial importance for efficient, high-yield HNO production in a physiological environment, since there is a variety of competing interactions with small molecules and macromolecules in which $NO^{\bullet}$ and H_2S are involved in biological systems. Consumption of $NO^{\bullet}$ and H_2S under

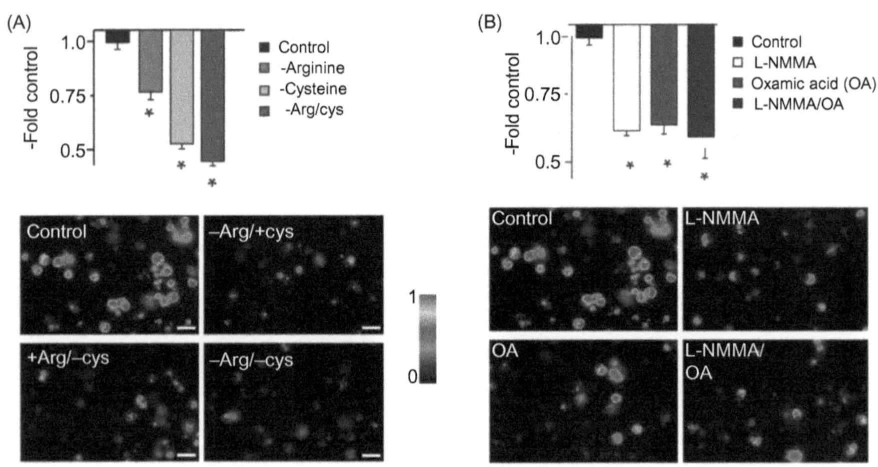

Figure 5.8 Endogenous HNO formation is dependent on both $NO^{\bullet}$ and H_2S: (A) Basal fluorescence of the HNO sensor in sensory neurons was reduced by arginine and cysteine depletion, substrates for endogenous $NO^{\bullet}$ and H_2S production (ANOVA HSD post hoc test; $^{*}P < 0.001$, treated versus control respectively; $n = 150$; all error bars represent SEM, scale bar, 50 μm); (B) Basal fluorescence of the HNO sensor CuBOT1 in sensory neurons was reduced by inhibition of NO-synthase by L-NMMA (1 mM, 2 h of pretreatment) and cystathionine beta synthase (CBS) by oxamic acid (1 mM, 2 h of pretreatment, ANOVA HSD post hoc test; $P < 0.001$, treated versus control respectively; $n > 150$ per group; error bars represent SEM, scale bar = 25 μm).

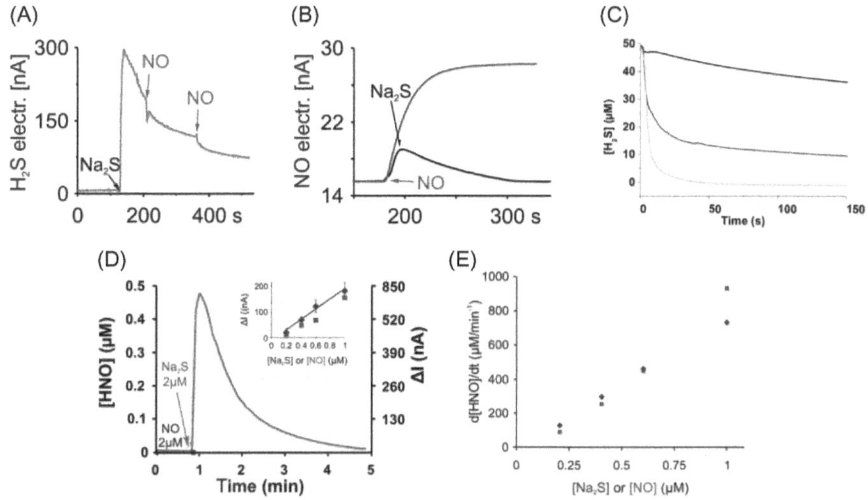

Figure 5.9 Kinetics of the reaction between NO• and H_2S: (A, B) H_2S and NO• react in vitro as observed by the drop of H_2S electrode response upon subsequent additions of NO• solution (A) and by the drop of NO• electrode response when H_2S was injected into solution containing DEA NONOate (B). (C) Kinetics of H_2S (50 mM) decay upon addition of NO solution (100 mM) in the presence (red (dark gray in print versions)) and absence (green (light gray in print versions)) of 2 mM glutathione in 300 mM potassium phosphate buffer (pH 7.4). Black line represents the spontaneous decay/removal of H_2S in the same time scale. The obtained kinetic traces (red (dark gray in print versions) and green (light gray in print versions)) suggest the same initial reaction rate in the presence and absence of glutathione. Addition of glutathione only results in a shift of the baseline. (D) Amperometric signal of the HNO-selective electrode after addition of H_2S (2 μM) to a solution of 2 μM NO• (left axis: (HNO) after calibration, right axis: measured current). Inset: Signal peak versus H_2S (blue (gray in print versions)) and NO• (red (dark gray in print versions)) concentration, while the other reactant concentration is maintained constant and in excess. (E) Rate of HNO formation (determined from the slope of the HNO electrode signal vs time plots) versus H_2S (blue (gray in print versions)) and NO• (red (dark gray in print versions)) concentration, while the other reactant concentration is maintained constant and in large excess. The points represent the mean of at least three measurements.

physiological conditions is strongly oxygen dependent,[65] and the cross-reactions of different oxidized forms of NO• and H_2S result in a number of by-products[68] making the production of HNO through such pathways biologically less significant. Nature has found a solution to overcome these side reactions by colocalization of the NOS and CBS enzymes, which facilitates direct interaction between H_2S and NO• and thus reduces a number of reaction steps that result in the efficient HNO production.[67] Even more, for efficient HNO signaling a colocalization of its protein target with the production sites of its precursors, NO• and H_2S, is desired. Such signaling strategy was elucidated by our studies that demonstrated strong coexpression of CBS and the HNO target, the calcium channel TRPA1 (transient receptor potential channel A1), in sensory neurons and fibers (Fig. 5.10).[67] This together with previously reported colocalization of neuronal NOS with TRPA1 stresses the importance of the production of H_2S and $NO^•$ in a close proximity to the HNO target for the operation of the neuroendocrine signaling pathway (Fig. 5.11) that is essential for the control of vascular tone throughout the cardiovascular system. HNO

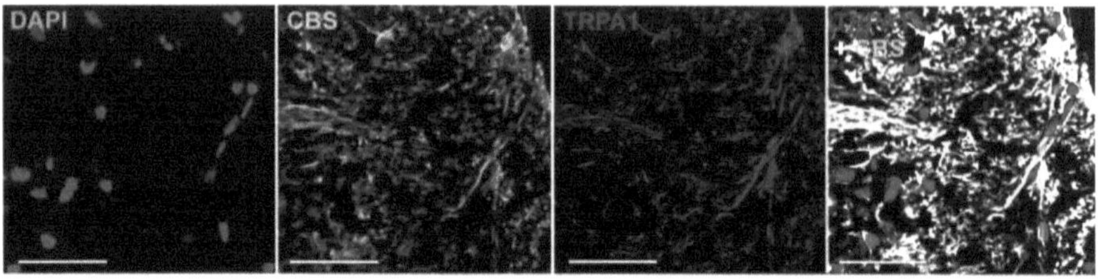

Figure 5.10 Colocalization of TRPA1 and CBS: Confocal images of a cross section through the rat spinal trigeminal nucleus caudalis immunohistochemically double-stained with antibodies against TRPA1 (Cy3, red (dark gray in print versions)) and CBS (FITC, green (light gray in print versions)) combined with nuclear DAPI staining (blue (gray in print versions)). Bundles of immunostained afferent nerve fibers run through the trigeminal tract (right side of images) and into the superficial laminae of the trigeminal nucleus (lamina 1 seen on left side of images). Most of the nerve fibers show the signal for TRPA1 and CBS, producing the yellow color (white in print versions) in the combined image. Neuronal cell bodies are not TRPA1 or CBS immunopositive (scale bar = 50 μm).

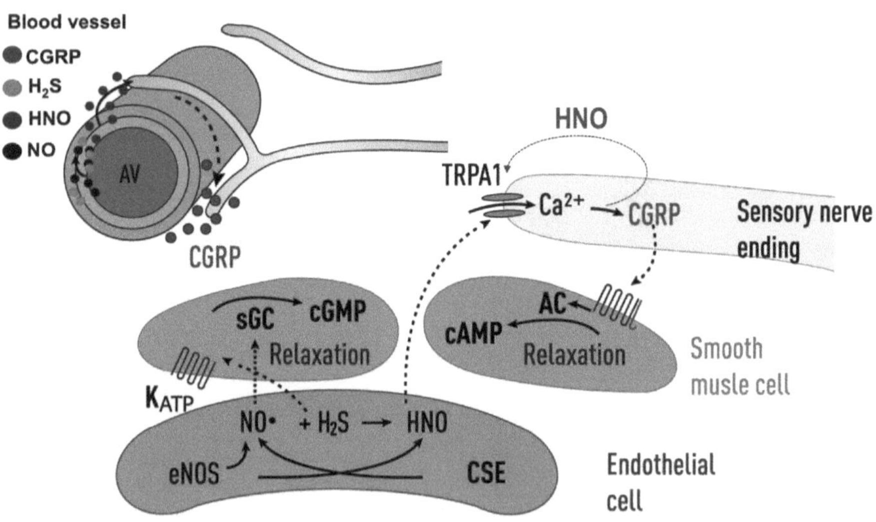

Figure 5.11 $NO^•$ and $H_2S \rightarrow HNO \rightarrow TRPA1 \rightarrow CGRP$ signaling path: TRPA1/CGRP expressing nerve endings in the periphery communicate with the smooth muscle cells surrounding the endothelium of blood vessels. Endothelial cells are known to produce H_2S and $NO^•$, both of which freely diffuse and activate guanidyl cyclase and K_{ATP} channels, respectively, to induce vasodilatation. However, H_2S and $NO^•$ also react with each other to give HNO, which could reach paravascular TRPA1-expressing sensory nerve fibers, inducing Ca^{2+} influx and CGRP release.

activates TRPA1 via formation of amino-terminal disulfide bonds, which results in sustained calcium influx. As a consequence, CGRP (calcitonin gene-related peptide) is released (Fig. 5.11), which induces local and systemic vasodilation.

From chemical point of few, the mechanism of the production of HNO in the H_2S-NO• reaction is still under investigation. As in the reaction between $AscH^-$ and NO• (Scheme 5.10), we propose formation of the (cage)-adduct $HSNO^{\bullet -}$ that either through an inner-sphere PCET or through an outer-sphere electron transfer to NO• produces (H)NO (Scheme 5.11). The latter process would resemble reduction of free NO• to NO^- by $[Fe^{II}(Por)NO]^-$ as observed in the literature[5] (Eq. 5.3, see Section 5.1.2.1). And indeed, in our preliminary experiments we could identify the existence of the intermediate $HSNO\cdot^-$ and it seems that its redox potential for the HSNO/ $HSNO^{\bullet -}$ couple is similar to that of *S*-nitrosothiols, c. −1.3 V versus SCE,[69] which would allow an outer-sphere electron transfer to NO•. It is known that clustering of an anion (A^-) with a neutral molecule (HX) is a stabilizing process that results in a complex structure $A^{-} \bullet HX$ with a significantly higher potential to be oxidized in comparison to the free A^- anion.[70] It is due to the relocation of the hydrogen atom within the cluster, i.e., inner-sphere PCET, which results in a stable structure upon electron release. This supports the existence of the $HS^{-} \bullet NO^{\bullet}$ adduct with very negative redox potential that upon oxidation can result in

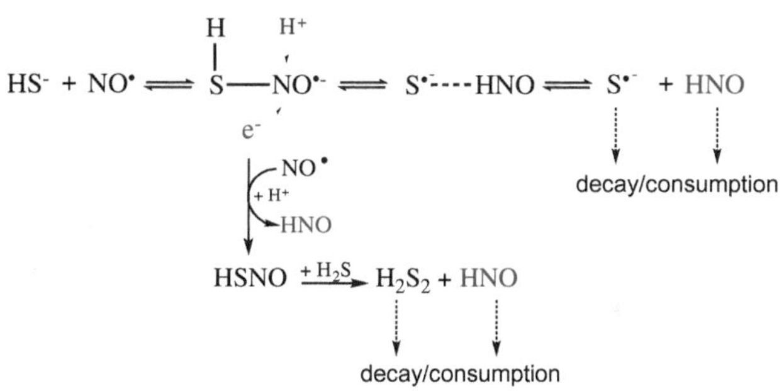

Scheme 5.10

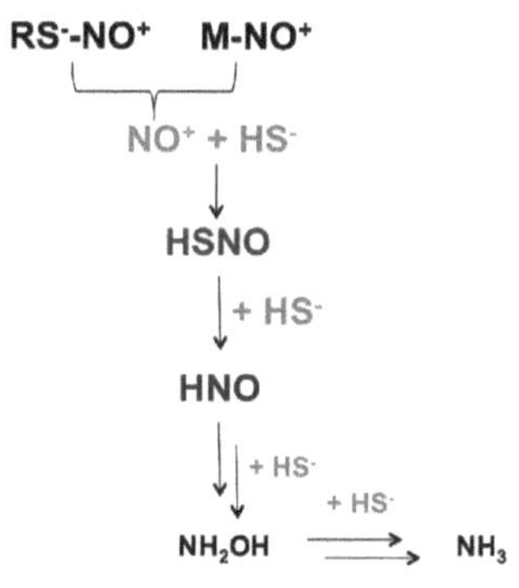

Scheme 5.11

stable HSNO. Alternatively, due to the mentioned relocation of electrons and protons the $S^{\bullet-}$ • HNO resonance structure of the adduct can be envisioned, which upon oxidation will result in HNO and S^0. These two pathways are in line with the proposed mechanistic Scheme 5.10, but there are other variations such as the subsequent reaction of $S^{\bullet-}$ with H_2S to form $H_2S_2^{\bullet-}$. Further mechanistic investigations are in progress.

5.2 HNO generation from nitrite and nitroso, NO^+, species through the interplay with H_2S and HSNO formation

Generation of HNO from nitrite or nitrosonium species, such as metal-NO^+ or *S*-nitroso thiols, formally requires a two-electron reduction. Application of hydride as reducing agent represents one possible route to provide two electrons in one-step reaction (Eq. 5.20), as discussed in Section 5.1.2.3. We have recently demonstrated that reaction with H_2S represents the biologically/pharmacologically relevant alternative mechanism that goes through intermediate formation of HSNO (Scheme 5.11).[32, 57b, 71] Formation of HSNO, the smallest *S*-nitrosothiol, also enables HNO generation in a one-step reaction with H_2S (Eq. 5.23).[71,72] In this section, we will further describe these processes.

$$HSNO + H_2S \rightarrow HNO + H_2S_2 \tag{5.23}$$

5.2.1 *HNO generation through HSNO in the reaction of S-nitrosothiols or nitrite with H_2S*

Transnitrosation between *S*-nitrosothiols (RSNO) and H_2S leads to formation of HSNO as demonstrated in the case of *S*-nitrosoglutathione (GSNO) (Eq. 5.24). In an analogy with the preparation of RSNO in the reaction between corresponding thiol and acidified nitrite solutions,[73] we also generated HSNO from the reaction of acidified nitrite and H_2S (Eq. 5.25). Even more, by applying pulse radiolysis experiments we obtained HSNO in a radical reaction between HS· and $NO^{\bullet}$ (Eq. 5.26).

$$GSNO + H_2S \rightarrow HSNO + GSH \tag{5.24}$$

$$[H_2NO_2]^+ + H_2S \rightarrow HSNO + H_3O^+ \tag{5.25}$$

$$HS^{\bullet} + NO^{\bullet} \rightarrow HSNO \tag{5.26}$$

Generation of HSNO according to these three different synthetic pathways enabled us to chemically characterize it and its reactivity, as well as its bioactivity by variety of experimental techniques, as well as computational methods.[71] HSNO generation by pulse radiolysis was monitored by time-resolved absorbance spectroscopy that revealed its absorbance maximum at c. 340 nm. The transnitrosation reaction (Eq. 5.24) was followed by time-resolved FTIR spectroscopy in solution that resulted in the ν_{NO} vibration of HSNO at $\sim 1568\ cm^{-1}$, which shifted to $\sim 1549\ cm^{-1}$ (calculated at $1540\ cm^{-1}$) in the case of $HS^{15}NO$. ^{15}N NMR measurements demonstrated appearance of a previously unidentified ^{15}N resonance at 322 ppm, which we assign to HSNO. HSNO was also identified by high-resolution ESI-Ms measurements performed both in the case of the reaction of H_2S with GSNO and acidified nitrite, respectively (Eqs. 5.24 and 5.25). A recent attempt to reproduce HSNO peak in the ESI-Ms experiments failed.[68] In the 63–66 m/z range, authors reported disturbing peaks predominantly attributed to the presence of DMSO in the multiuser core instrument and its application for the preparation of some RSNO samples.[68] These aspects apply neither for our instrument nor for the sample preparation, confirmed by the absence of DMSO-related peaks in the control measurements of the buffer solutions and the solutions of RSNO in the absence of any added sulfide, as well as by the clear 1 m/z shift to the higher masses of the HSNO peak in the case of its generation from $GS^{15}NO$,[71] in accord with the corresponding ν_{NO} vibration and NMR shift of $HS^{15}NO$. One should also stress that if the DMSO-related cations such as CH_3SO^+ were observed in the ESI-Ms experiments as a consequence of the breaking of S–C bond that requires c. 51 kcal/mol,[74] this implies that the chosen ionization conditions were so harsh and inappropriate to detect compounds such as HSNO with the S–N bond dissociation energy (BDE) of 29.4 kcal/mol.[72]

Reactivity studies have shown that HSNO is a source for NO$^{\bullet}$, HNO, and NO^+-moiety, which explains its bioactivity. The NO$^{\bullet}$ release is a consequence of the homolysis of the S–N bond (BDE 29.4 kcal/mol)[72] and is slightly less stable than that in *S*-nitrosothiols. Upon reaction with H_2S, it generates HNO (Eq. 5.23). Considering that the physiological concentrations of H_2S are in excess of what was proposed for RSNOs, cellular HSNO formation (Eq. 5.24) can be followed by HNO production according to Eq. 5.23. This was confirmed by HNO trapping by hemoglobin and detection of N_2O. Importantly, the cells loaded with the nitroxyl-responsive dyes, CuBOT1,[75] provided a clear proof of intracellular HNO formation only upon incubation with both GSNO and H_2S.

Although a direct reaction of RSNO and H_2S (Eq. 5.27) could be considered as an alternative route for the HNO production, we have not detected neither corresponding RSSH nor N_2O in the reaction with a 1:1 molar ration of GSNO and H_2S. This route was excluded based on computational studies and the estimated unfavorable energetics.[71]

$$GSNO + H_2S \rightarrow GSSH + HNO \qquad (5.27)$$

Furthermore, in a series of in vitro and cell studies it was demonstrated that HSNO can serve as a shuttle for NO^+ equivalents according to the *trans-S*-nitrosation reaction (Scheme 5.12). Interestingly, it was shown that HSNO-enabled exchange of NO^+ equivalents between proteins separated by a membrane, most

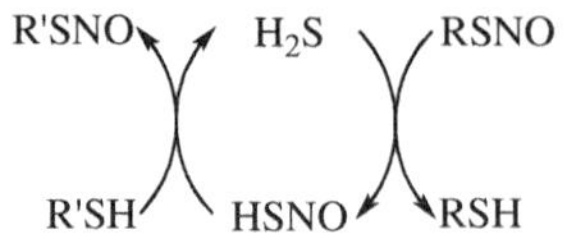

Scheme 5.12

probably due to its small size and neutral nature that facilitates its diffusion across membranes.[71] In that way it might link plasma *S*-nitrosothiols to intracellular *S*-nitrosohemoglobins, among other potential biological functions.

These different type of reactivity of HSNO could be explained by probable existence of its different isomeric forms: *trans*- and *cis*-HSNO, *trans*- and *cis*-SNOH and SN(H)O.[72] Whereas the *trans*- and *cis*-HSNO structures with the terminal NO fragment might predominantly be involved in the release of $NO^{\bullet}$ or NO^+, the so-called Y-isomer, SN(H)O, is predicted to be responsible for the reaction with H_2S leading to the generation of HNO.[72]

Biological effects of HSNO could alternatively be substituted by those of $SSNO^-$ species.[68] To evaluate this possibility, we prepared $SSNO^-$ by two different approaches, which lead to two distinct species: perthionitrite ($SSNO^-$) and dithionitric acid (HON(S)S/HSN(O)S). Their properties were studied by ^{15}N NMR, IR, EPR, and high-resolution ESI-TOF mass spectrometry, as well as by X-ray structure analysis, cyclic voltammetry, and computational methods.[76] The obtained results pointed toward the inherent instability of $SSNO^-$ in water solutions. X-ray structure analysis and computation studies revealed much higher instability of the corresponding S−N bond in (H)SSNO, with bond dissociation energy of only 16 kcal/mol in comparison to 29.4 kcal/mol for HSNO. The bond lengths in $SSNO^-$ are indicative for an overall delocalization of the negative charge and even imply a slightly polarized $S_2{-}NO^-$-like structure. A calculated charge distribution reveals a prominent negative charge at the NO moiety. Together with other spectroscopic/spectrometric characteristics of $SSNO^-$, all this explains the experimentally observed generation of HNO and sulfane sulfur-containing compounds from $SSNO^-$ salts in water solutions. Using pure crystals in nonaqueous solutions as a well-defined source of $SSNO^-$, we studied its potential biological effects.[67] The observed effects did not differ from those of polysulfides, and neither *S*-nitrosation nor *S*-sulfhydration of proteins could be observed. Overall, the structural and electronic features, as well as the observed reactivity of $SSNO^-$, do not support its direct NO^+- or NO•-donating ability, in contrary to effects of HSNO. A difference with the reported observations can be explained by the fact that in those literature studies[68] the samples that presumably contained $SSNO^-$ were mixtures of NO-donors or RSNOs with sulfide, which in some experiments additionally contained oxygen. Importantly, in these reaction mixtures claimed to contain stable $SSNO^-$ in aqueous solutions, neither ^{15}N NMR signal nor N-O stretching vibration other than that for nitrite could be observed.[71] This excludes the presence of $SSNO^-$, which as a nitrogen atom-containing species would have displayed a corresponding ^{15}N peak and N-O vibration band (^{15}N NMR shift at 354 ppm and IR vibrations at

1335, 1304, 1297, and 1271 cm^{-1} were observed in the acetone solutions of $SS^{15}NO^-$ and $SS^{14}NO^-$ salts, respectively).[76] Therefore, application of solely $SSNO^-$ salt[76] versus application of undefined complex reaction mixtures[68] distinguishes these two experimental approaches.

5.2.2 Metal-mediated HNO generation through HSNO in the reaction of nitrite or nitrosonium complexes with H_2S

5.2.2.1 Heme-catalyzed reaction between nitrite and H_2S generates HSNO, HNO and $NO^•$

In clinical praxis, nitrite is used as an antidote for hydrogen sulfide poisoning, being even superior to oxygen.[77] An answer to the decades-old question about the mechanism by which nitrite acts in such severe cases has been recently offered by us.[32] We have discussed how H_2S reacts with acidified nitrite (Eq. 5.25) and generates HNO (Eq. 5.23) through intermediate HSNO formation (Eq. 5.25).[71] However, under physiological pH pure nitrite and sulfide solutions do not react. On the other hand, we have demonstrated that cells consume H_2S when nitrite is present in the medium and in a concentration-dependent manner. As a consequence of this reaction, we have detected increased levels of intracellular $NO^•$ and HNO by using selective fluorescence sensors. Even more, we have demonstrated that production of HNO is localized in mitochondria and that it does not occur in the cells depleted of mitochondria. Also, only in cells incubated with sulfide and nitrite a strong EPR signal of the heme Fe^{II}-$NO^•$ moiety was observed. Since mitochondria is the organelle that is richest in heme-iron centers, these findings pointed toward the conclusion that the NO and HNO formation induced by nitrite-/sulfide-interaction is a heme-iron-catalyzed process that takes place within mitochondria. They also suggest that H_2S may be the elusive thiol responsible for the reduction of nitrite in biological milieu.

This heme-iron-catalyzed mechanism was proved by using a water-soluble porphyrin $FeP(8)^-$ as a heme model system and studying the reaction between nitrite and sulfide in its presence. The mechanism is depicted in Fig. 5.3. At excess of nitrite, it proceeds via pathway A (Fig. 5.3) that is similar to that described for thiol (cysteine and gluthathione) stimulated nitrite reductase activity of Fe^{III}-porphyrins through an oxygen atom transfer (OAT) mechanism that results in corresponding sulfenic acid and Fe^{II}-$NO^•$.[30,31] As discussed in Section 5.1.2.2.2, upon coordination of the sixth ligand and protonation of the L-Fe^{II}-$NO^• \leftrightarrow$ L-Fe^{III}-NO^- intermediate, HNO gets released (Eq. 5.15) and the Fe^{III} species can start the catalytic cycle again.[31,32] However, under the physiological conditions an excess of H_2S over nitrite is a more common situation, and under these conditions the reaction proceeds according to pathway B (Fig. 5.3). There in the first step the Fe^{III}(Por) complex gets reduced through formation of the ls $Fe^{III}(Por)(HS^-)$ that was characterized[32] as a very rare case of sulfide bound to synthetic ferric porphyrin,[78] which has not yet been observed in aqueous solution. HS· was also the product of this reaction and was characterized by EPR. Since it is formed within catalytic cycle together with

$NO^{\bullet}$, two of them could also produce HSNO according to Eq. 5.26. In the next step, Fe^{II}(Por) in the reaction with nitrite produces Fe^{III}-$NO^{\bullet} \leftrightarrow Fe^{II}$-$NO^{+}$, the $\{FeNO\}^6$ species (Eq. 5.28).

$$Fe^{II}(por) + NO_2^- + 2H^+ \rightarrow Fe^{III}\text{-}NO^{\bullet} \leftrightarrow Fe^{II}\text{-}NO^+ + H_3O \quad (5.28)$$

As described in Section 5.1.2.3, such species upon two-electron reduction coupled by the protonation can be transferred into Fe^{II}-HNO (Eq. 5.19). But here, the Fe^{II}-NO^+ species that is prone to react with nucleophiles is quickly scavenged by the excess HS^- resulting in Fe^{II}(Por)(HSNO) (Scheme 5.13). Finally, either free or coordinated HSNO in the reaction with H_2S (Eq. 5.23) generates HNO (Scheme 5.13, Fig. 5.3).

These findings elucidate some general pathways that under physiological conditions can result in HNO and that comprise both Fe^{II}- and Fe^{III}-nitrosyls (Eq. 5.15 and Scheme 5.13) that among other biochemically relevant reactions can also result from the heme mediated interplay between nitrite and sulfide species. Taken together, these data give an insight into the possible mechanism for intracellular nitrite reduction and its subsequent physiological effects. They also offer a possible explanation for the antidote effect of nitrite in H_2S poisoning: generation of $NO^{\bullet}$, HNO, and HSNO could lead to an increase in the blood flow, and finally, better oxygen delivery to the tissue.

5.2.2.2 *Sodium nitroprusside with H_2S generates HNO accompanied by conversion of toxic cyanide into thiocyanate*

Since the above-discussed studies have revealed a general potential of H_2S to generate HNO from the nitrogen species in the oxidation state +3 through the formation of HSNO (Scheme 5.11), with this knowledge in hand we wanted to prove such mechanism in the case of other, e.g., nonheme metal-nitrosyls. A well-known nonheme $\{FeNO\}^6$ complex with the Fe^{II}-NO^+ functionality is sodium nitroprusside (SNP), $Na_2[Fe(CN)_5(NO)]$.

Despite the long interest in the reactions of H_2S with SNP (since 1850),[79] the reaction mechanism has eluded researchers. It was studied previously under anaerobic and alkaline conditions,[80] but nothing was definitively known about the products of this reaction under physiological conditions.[57a] This prompted us to study

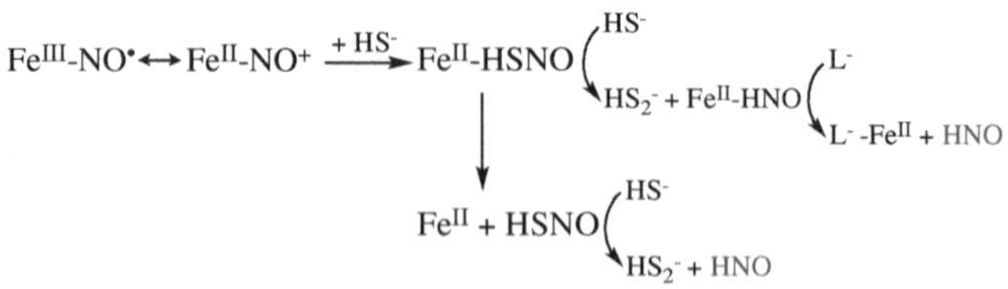

Scheme 5.13

in detail the reaction of H_2S and SNP under physiologically relevant conditions (physiological pH and aerobic conditions) using both chemical and physiological/pharmacological tools.[57b,c] Based on these results, the operation of a new mechanism was revealed (Fig. 5.6). The key feature of this mechanism is the rapid formation of an $[Fe(CN)_5(HSNO)]^{3-}$ intermediate (or its deprotonated form, $[Fe(CN)_5(SNO)]^{4-}$), with coordinated (H)SNO and its reduction to HNO, i.e., $[Fe(CN)_5(HNO)]^{3-}$, which confirms the general reactivity profile of the NO^+ + HS^- system in an excess of sulfide (Scheme 5.11). During the generation of HNO, H_2S gets oxidized initially to H_2S_2 that under aerobic conditions further generates polysulfides. In the reaction course some free CN^- can be obtained in the solution that reacts with H_2S_2 and polysulfides, in a way any other S-S containing compound does with cyanide,[81] i.e., CN^- gets converted into SCN^-. Through dynamic solution processes, e.g., ligand substitutions, coordinated thiocyanate species can be generated. Such conversions may labilize HNO that gets released and the remaining aqua complexes $[Fe(SCN)_x(CN)_{5-x}(H_2O)]^{3-}$ continue transformation with polysulfides to final $[Fe(SCN)_5(H_2O)]^{3-}$, and partially become oxidized to the corresponding Fe^{III} complex that can form the mixed-valent bridged complexes known as Prussian blue. They are short-lived and further react with polysulfides to give the final thiocyanate products as well.

This mechanism has physiological/therapeutic relevance and we have also demonstrated the intracellular HNO formation and the subsequent release of calcitonin gene-related peptide (CGRP) from a mouse heart induced by the SNP/H_2S reaction mixture (Fig. 5.12). One of the main biological markers for HNO formation is the increased release of CGRP that is a potent vasodilator.[82] Unlike with other thiols, H_2S reacts with SNP in the same way as rhodanese, i.e., the cyanide transforms into a thiocyanate.

SNP is still used in acute hypertensive crises to regulate blood pressure; however, it must be combined with thiosulfate to minimize the toxic effects of cyanide

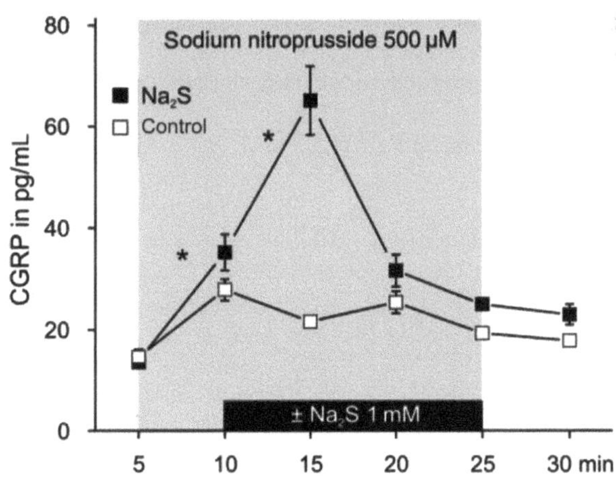

Figure 5.12 HNO-induced CGRP released from isolated mouse heart upon stimulation with reaction mixture containing 500 M SNP and 1 mM Na_2S.

through natural rhodanese activity.[83] Our results offer a new perspective on the clinical application of SNP. The combination of SNP with H_2S, which causes the HNO-induced release of CGRP and transforms toxic cyanide into thiocyanate, could be a new, more effective, and less toxic therapeutic alternative to the combining of SNP with thiosulfate.[57c]

It should be pointed out that the reaction between SNP and H_2S or sulfides is a complex process with a number of transient, intermediate and product species, cross-reactions of which are also possible. These cross-interactions are very much dependent on reaction conditions such as concentration, pH, temperature, presence of impurities, presence/absence of oxygen and its concentration, observation time. In that line, recently it has been reported that one of intermediate species is blue $[Fe(CN)_5N(O)SS]_4^-$ with an absorption band at c. 575 nm, which can be obtained in the reaction between SNP and Na_2S_2 in strong alkaline solutions.[84] These studies are predominantly based on NMR measurements. Structural investigations would certainly be helpful to further confirm/reveal a chemical identity of that and other reaction intermediates/products. Thus, this so-called Gmelin reaction continues to be a challenging subject of scientific interest.

5.3 Conclusion

Considering all reactions and mechanisms that we have discussed above, here are summarized possible reductive pathways for the HNO generation in biological systems using $NO^•$, nitrite, and inorganic or organic nitrosyl species (metal-nitrosyls and *S*-nitrosothiols, including the smallest one HSNO) as precursors, as following:

- direct reaction between $NO^•$ and H_2S without the interferences of oxidized species, boosted by the colocalized $NO^•$ and H_2S production (Scheme 5.10, Figs. 5.8–5.11);
- reaction between $NO^•$ and ascorbate (Scheme 5.9 and Fig. 5.7);
- reaction of *S*-nitrosothiols and H_2S that generates HSNO and subsequent reaction of HSNO with H_2S (Eq. 5.23, Scheme 5.11);
- heme-catalyzed reaction between nitrite and H_2S predominantly localized in mitochondria (Fig. 5.3);
- from the heme and even nonheme $Fe^{III}\text{-}NO^• \leftrightarrow Fe^{II}\text{-}NO^+$, $\{FeNO\}^6$, species upon hydride (NAD(P)H) reduction and formation of Fe^{II}-HNO, $\{FeHNO\}^8$, (Eq. 5.19), that either might undergo HAT to free $NO^•$ (Eq. 5.12) or HNO substitution (in an analogy to Eq. 5.13), or upon coordination of the sixth ligand *trans* to HNO might be labilized/activated toward HNO release;
- reaction of heme and nonheme $Fe^{III}\text{-}NO^• \leftrightarrow Fe^{II}\text{-}NO^+$ species with H_2S through intermediate formation of Fe^{II}-HSNO (Scheme 5.11 and 5.13)
- from heme and hs nonheme $Fe^{II}\text{-}NO^•$, $\{FeNO\}^7$ upon coordination of an appropriate sixth ligand that stabilizes the $Fe^{III}\text{-}NO^-$ resonance form, which upon protonation might release HNO (Eq. 5.15);
- from nitroxylcobalamin upon substitution of the dimethylbenzimidazole group by appropriate sixth ligand that might induce *trans*-labilization of HNO (Scheme 5.3);
- reduction of NO· catalyzed by hs Mn centers of, e.g., MnSOD or pharmacologically applied MnSOD mimetics.

In future investigations, we should harness a potential of all this mechanisms for the HNO generation in order to learn how to manipulate/control the redox signaling induced by them, for the benefit of our health.

References

1. Bartberger MD, Liu W, Ford E, Miranda KM, Switzer C, Fukuto JM, et al. The reduction potential of nitric oxide (NO) and its importance to NO biochemistry. *PNAS* 2002;**99**:10958–63.
2. (a) Akhtar MJ, Bonner FT, Hughes MN, Humphreys EJ, Lu C-S. Kinetics and isotopic studies on the reaction between trioxodinitrate and the hexaammineruthenium(III) cation. *Inorg Chem* 1986;**25**:4635–9.
   (b) Stanbury DM, Haas O, Taube H. Reduction of oxygen by ruthenium(II) ammines. *Inorg Chem* 1980;**19**:518–24.
3. Vergheese TM, Berchmans S. Selective NO reduction using blue ferrocenyl cation. *Electrochim Acta* 2006;**52**:567–74.
4. Stanbury DM. Reduction potentials involving inorganic free radicals in aqueous solution. In: Sykes AG, editor. *Advances in Inorganic Chemistry*, vol. 33. Amsterdam: Elsevier; 1989. p. 69–138.
5. Goodrich LE, Roy S, Alp EE, Zhao J, Hu MY, Lehnert N. Electronic structure and biologically relevant reactivity of low-spin $\{FeNO\}^8$ porphyrin model complexes: new insight from a bis-picket fence porphyrin. *Inorg Chem* 2013;**52**:7766–80.
6. Mindell JA. Lysosomal acidification mechanisms. *Annu Rev Physiol* 2012;**74**:69–86.
7. (a) Ivanović-Burmazović I, van Eldik R. Metal complex-assisted activation of small molecules. From NO to superoxide and peroxides. *Dalton Trans* 2008;5259–75.
   (b) Ivanović-Burmazović I, Filipović MR. Reactivity of manganese superoxide dismutase mimetics toward superoxide and nitric oxide: selectivity vs. cross-reactivity. *Adv Inorg Chem* 2012;**64**:53–95.
8. Enemark JH, Feltham RD. Principles of structure, bonding, and reactivity for metal nitrosyl complexes. *Coord Chem Rev* 1974;**13**:339–406.
9. Roper WR, Grundy KR, Reed CA. Hydrogen transfer from HCl to the nitrosyl ligand. Examples of co-ordinated HNO, NHOH-, and NH_2OH. *J Chem Soc Chem Commun* 1970;1501–2.
10. Wilson RD, Ibers JA. Coordinated nitrosyl hydride: structural and spectroscopic study of dichlorocarbonyl(nitrosyl hydride)bis(triphenylphosphine)osmium(II). *Inorg Chem* 1979;**18**:336–43.
11. Southern JS, Green MT, Hillhouse GL, Guzei IA, Rheingold AL. Chemistry of coordinated nitroxyl. reagent-specific protonations of trans-$Re(CO)_2(NO)(PR_3)_2$ (R = Ph, Cy) that give the neutral nitroxyl complexes *cis, trans*-$ReCl(CO)_2(NH=O)(PR_3)_2$ or the cationic hydride complex [*trans,trans*-$ReH(CO)_2(NO)(PPh_3)_2{}^{+}$][$SO_3CF_3{}^{-}$]. *Inorg Chem* 2001;**40**:6039–46.
12. Melenkivitz R, Hillhouse GL. Synthesis, structure, and reactions of a nitroxyl complex of iridium(III), cis,trans-$IrHCl_2(NH=O)(PPh_3)_2$. *Chem Commun* 2002;660–1.
13. Zhang Y. Computational investigations of HNO in biology. *J Inorg Biochem* 2013;**118**:191–200.
14. Southern JS, Hillhouse GL, Rheingold AL. Preparation of a nitroxyl (HN = O) complex of rhenium by selective oxidation of coordinated hydroxylamine. *J Am Chem Soc* 1997;**119**:12406–7.

15. Enemark JH, Feltham RD. Stereochemical control of valence and its application to the reduction of coordinated NO and N_2. *PNAS, USA* 1972;**69**:3534–6.
16. Enemark JH, Feltham RD, Riker-Nappier J, Bizot KF. Stereochemical control of valence. IV. Comparison of the structures and chemical reactivities of five- and six-coordinate diarsine complexes of the cobalt nitrosyls $\{CoNO\}^8$ group. *Inorg Chem* 1975;**14**:624–32.
17. Rhine MA, Rodrigues AV, Bieber Urbauer RJ, Urbauer JR, Stemmler TL, Harrop TC. Proton-induced reactivity of NO^- from a $\{CoNO\}^8$ complex. *J Am Chem Soc* 2014; **136**:12560–3.
18. Hassanin HA, El-Shahat MF, DeBeer S, Smith CA, Brasch NE. Redetermination of the X-ray structure of nitroxylcobalamin: base-on nitroxylcobalamin exhibits a remarkably long Co–*N*(dimethylbenzimidazole) bond distance. *Dalton Trans* 2010;**39**: 10626–30.
19. Zheng D, Birke RL. Spectroscopic evidence for nitric oxide binding with Cob(II)alamin. *J Am Chem Soc* 2001;**123**:4637–8.
20. Wolak M, Zahl A, Schneppensieper T, Stochel G, van Eldik R. Kinetics and mechanism of the reversible binding of nitric oxide to reduced cobalamin B_{12r} (Cob(II)alamin). *J Am Chem Soc* 2001;**123**:9780–91.
21. Franke A, Oszajca M, Brindell M, Stochel G, van Eldik R. Metal-assisted activation of nitric oxide—mechanistic aspects of complex nitrosylation processes. In: van Eldik R, Olabe J, editors. *Advances in Inorganic Chemistry*, vol. 67. Amsterdam: Elsevier; 2015; p. 171–236.
22. Speelman AL, Lehnert N. Heme versus non-heme iron-nitroxyl $\{FeN(H)O\}^8$ complexes: electronic structure and biologically relevant reactivity. *Acc Chem Res* 2014;**47**:1106–16.
23. Farmer PJ, Kumar MR, Almaraz E. The coordination Chemistry of HNO: from Warren Roper to Hemoglobin. *Comments Inorg Chem* 2010;**31**:130–43.
24. Farmer PJ, Sulc F. Coordination chemistry of the HNO ligand with hemes and synthetic coordination complexes. *J Inorg Biochem* 2005;**99**:166–84.
25. Pellegrino J, Bari SE, Bikiel DE, Doctorovich F. Successful stabilization of the elusive species $\{FeNO\}^8$ in a heme model. *J Am Chem Soc* 2010;**132**:989–95.
26. Wei Z, Ryan MD. Infrared spectroelectrochemical reduction of iron porphyrin complexes. *Inorg Chem* 2010;**49**:6948–54.
27. Lee J, Richter-Addo GB. A nitrosyl hydride complex of a heme model [Ru(ttp)(HNO)(1-MeIm)] (ttp = tetratolylporphyrinato dianion). *J Inorg Biochem* 2004;**98**(7):1247–50.
28. Codesido NO, Weyhermüller T, Olabe JA, Slep LD. Nitrosyl-centered redox and acid–base interconversions in $[Ru(Me_3[9]aneN_3)(bpy)(NO)]^{3,2,1+}$. The p$K_a$ of HNO for its nitroxyl derivative in aqueous solution. *Inorg Chem* 2014;**53**:981–97.
29. Metzker G, Stefaneli EV, Pereira JCM, Lima FCA, da Silva SC, Franco DW. Nitric oxide and nitroxyl formation in the reduction of trans-tetraamminenitrosyltriethylphosphiteruthenium (II) ion. *Inorg Chim Acta* 2013;**394**: 765–9.
30. Khin C, Heinecke JL, Ford PC. Oxygen atom transfer from nitrite mediated by Fe(III) porphyrins in aqueous solution. *J Am Chem Soc* 2008;**130**:13830–1.
31. Heinecke JL, Khin C, Pereira JCM, Suárez SA, Iretskii AV, Doctorovich F, et al. Nitrite reduction mediated by heme models. Routes to NO and HNO? *J Am Chem Soc* 2013;**135**:4007–17.

32. Miljković J Lj, Kenkel I, Ivanović-Burmazović I, Filipovic MR. H_2S generates HNO and HSNO from nitrite by a heme iron-catalyzed metabolism. *Angew Chem Int Ed* 2013;**52**(46):12061–4.
33. Adak S, Wang Q, Stuehr DJ. Arginine conversion to nitroxide by tetrahydrobiopterin-free neuronal nitric-oxide synthase: implications for mechanism. *J Biol Chem* 2000;**275**:33554–61.
34. Serres RG, Grapperhaus CA, Bothe E, Bill E, Weyhermüller T, Neese F, et al. Structural, spectroscopic, and computational study of an octahedral, non-heme $\{Fe-NO\}^{6-8}$ series: $[Fe(NO)(cyclam\text{-}ac)]^{2+/+/0}$. *J Am Chem Soc* 2004;**126**:5138–53.
35. Patra AK, Dube KS, Sanders BC, Papaefthymiou GC, Conradie J, Ghosh A, et al. A thermally stable $\{FeNO\}^8$ complex: properties and biological reactivity of reduced MNO systems. *Chem Sci* 2012;**3**:364–9.
36. Franz KJ, Lippard SJ. NO disproportionation reactivity of Fe tropocoronand complexes. *J Am Chem Soc* 1999;**121**:10504–12.
37. Sellmann D, Gottschalk-Gaudig T, Häußinger D, Heinemann FW, Hess BA. [Ru(HNO)('$py^{bu}S_4$')], the first HNO complex resulting from hydride addition to a NO complex ('$py^{bu}S_4$'$^{2-}$ = 2,6-bis(2-mercapto- 3,5-di-tert-butylphenylthio)dimethylpyridine(2−)). *Chem Eur J* 2001;**7**:2099–103.
38. de Barros BF, Toledo Jr. JC, Franco DW, Tfouni E, Krieger MH. A new inorganic vasodilator, *trans*-$[Ru(NO)(NH_3)_4(POEt)_3](PF_6)_3$: hypotensive effect of endothelium-dependent and -independent vasodilators in different hypertensive animals models. *Nitric Oxide* 2002;**7**:50–6.
39. Wanat A, Schneppensieper T, Stochel G, van Eldik R, Bill E, Wieghardt K. Kinetics, mechanism, and spectroscopy of the reversible binding of nitric oxide to aquated iron (II). An undergraduate text book reaction revisited. *Inorg Chem* 2002;**41**:4–10.
40. Schneppensieper T, Finkler S, Czap A, van Eldik R, Heus M, Nieuwenhuizen P, et al. Tuning the reversible binding of NO to iron(II) aminocarboxylate and related complexes in aqueous solution. *Eur J Inorg Chem* 2001;**2**:491–501.
41. van der Maas P, van de Sandt T, Klapwijk B, Lens P. Biological reduction of nitric oxide in aqueous Fe(II)EDTA solutions. *Biotechnol Prog* 2003;**19**:1323–8.
42. Berto TC, Hoffman MB, Murata Y, Landenberger KB, Alp EE, Zhao J, et al. Structural and electronic characterization of non-heme Fe(II)–nitrosyls as biomimetic models of the Fe_B center of bacterial nitric oxide reductase. *J Am Chem Soc* 2011;**133**:16714–17.
43. Speelman AL, Lehnert N. Characterization of a high-spin non-heme $\{FeNO\}^8$ complex: implications for the reactivity of iron nitroxyl species in biology. *Angew Chem In Ed* 2013;**52**:12283–7.
44. Bari SE, Olabe JA, Slep LD. Three redox states of metallonitrosyls in aqueous solution. In: van Eldik R, Olabe JA, editors. *Advances in Inorganic Chemistry*, vol. 67. Amsterdam: Elsevier; 2015. p. 87–144.
45. Armor JN, Scheidegger HA, Taube H. A bimolecular mechanism for substitution. *J Am Chem Soc* 1968;**90**:5928–9.
46. Franz KJ, Lippard SJ. Disproportionation of nitric oxide promoted by a Mn tropocoronand. *J Am Chem Soc* 1998;**120**:9034–40.
47. Awasabisah D, Richter-Addo GB. NOx linkage isomerization in metal complexes. In: van Eldik R, Olabe JA, editors. *Advances in Inorganic Chemistry*, vol. 67. Amsterdam: Elsevier; 2015. p. 1–86.
48. Niketić V, Stojanović S, Nikolić A, Spasić M, Michelson AM. Exposure of Mn and FeSODs, but not Cu/ZnSOD, to NO leads to nitrosonium and nitroxyl ions generation

which cause enzyme modification and inactivation: an in vitro study. *Free Rad Biol Med* 1999;**27**:992–6.
49. Filipovic MR, Stanic D, Raicevic S, Spasic M, Niketic V. Consequences of MnSOD interactions with nitric oxide: nitric oxide dismutation and the generation of peroxynitrite and hydrogen peroxide. *Free Rad Res* 2007;**41**:62–72.
50. Filipović MR, Duerr K, Mojović M, Simeunović V, Zimmermann R, Niketić V, et al. NO dismutase activity of seven-coordinate manganese(II) pentaazamacrocyclic complexes. *Angew Chem Int Ed* 2008;**47**:8735–9.
51. Filipovic MR, Koh ACW, Arbault S, Niketic V, Debus A, Schleicher U, et al. Striking inflammation from both sides: manganese(II) pentaazamacrocyclic SOD mimics act also as nitric oxide dismutases: a single-cell study. *Angew Chem Int Ed* 2010;**49**:4228–32.
52. Lieb D. *Mechanistic and synthetic approaches for activation of water- and oxygen-species by biomimetic systems (Containing manganese and cobalt). Ph.D. Thesis*. Friedrich-Alexander-Universität Erlangen-Nürnberg; 2013.
53. Muscoli C, Cuzzocrea S, Riley DP, Zweier JL, Thiemermann C, Wang ZQ, et al. On the selectivity of superoxide dismutase mimetics and its importance in pharmacological studies. *Br J Pharmacol* 2003;**140**:445–60.
54. Salvemini D, Mazzon E, Dugo L, Riley DP, Serraino I, Caputi AP, et al. Protective effects of a new stable, highly active SOD mimetic, M40401 in splanchnic artery occlusion and reperfusion. *Br J Pharmacol* 2001;**132**:19–29.
55. Lieb D, Zahl A, Shubina TE, Ivanović-Burmazović I. Water exchange on manganese (III) porphyrins. Mechanistic insights relevant for oxygen evolving complex and superoxide dismutation catalysis. *J Am Chem Soc* 2010;**132**:7282–4.
56. Kumar MR, Fukuto JM, Miranda KM, Farmer PJ. Reactions of HNO with heme proteins: new routes to HNO-heme complexes and insight into physiological effects. *Inorg Chem* 2010;**49**:6283–92.
57. (a) Filipović MR, Ivanović-Burmazović I. The kinetics and character of the intermediates formed in the reaction between sodium nitroprusside and hydrogen sulfide need further clarification. *Chem Eur J* 2012;**18**:13538–40.
(b) Filipovic MR, Eberhardt M, Prokopovic V, Mijuskovic A, Orescanin-Dusic Z, Reeh P, et al. Beyond H_2S and NO interplay: hydrogen sulfide and nitroprusside react directly to give nitroxyl (HNO). A new pharmacological source of HNO. *J Med Chem* 2013;**56**:1499–508.
(c) Ivanović-Burmazović I, Filipović M. Combination of nitroprusside and a sulfide salt as an HNO-releasing therapeutic for the treatment or prevention of cardiovascular diseases. *PCT Int Appl* 2014. *WO 2014122074 A1 20140814*
58. Shoun H, Fushinobu S, Jiang L, Kim S-W, Wakagi T. Fungal denitrification and nitric oxide reductase cytochrome $P450_{nor}$. *Phil Trans R Soc B* 2012;**367**:1186–94.
59. (a) Shiro Y, Fujii M, Iizuka T, Adachi S, Tsukamoto K, Nakahara K, et al. Spectroscopic and kinetic studies on reaction of cytochrome P450nor with nitric oxide: implication for its nitric oxide reduction mechanism. *J Biol Chem* 1995; **270**:1617–23.
(b) Oshima R, Fushinobu S, Su F, Zhang L, Takaya N, Shoun H. Structural evidence for direct hydride transfer from NADH to cytochrome P450nor. *J Mol Biol* 2004;**342**:207–17.
60. Harteck P. Die darstellung von HNO bzw. $[HNO]_n$ (the preparation of HNO or $[HNO]_n$). *Berichte der Deutschen Chemischen Gesellschaft* 1933;**66B**:423–6.

61. (a) Marchenko AV, Vedernikov AN, Dye DF, Pink M, Zaleski JM, Caulton KG. An electron-excessive nitrosyl complex: reactivity of a ligand-centered radical leading to coordinated HNO. *Inorg Chem* 2002;**41**:4087–9.
    (b) Marchenko AV, Vedernikov AN, Dye DF, Pink M, Zaleski JM, Caulton KG. Reactivity of the hydrido/nitrosyl radical $MHCl(NO)(CO)(P^iPr_3)_2$, M = Ru, Os. *Inorg Chem* 2004;**43**:351–60.
62. Smallwood HM. The rate of recombination of atomic hydrogen. *J Am Chem Soc* 1929;**51**:1985–99.
63. Suarez SA, Neuman NI, Munoz M, Alvarez L, Bikiel DE, Brondino C, et al. Nitric oxide is reduced to HNO by proton-coupled nucleophilic attack (PCNA) by ascorbate, tyrosine, and other alcohols. A new route to HNO in biological media? *J Am Chem Soc* 2015;**137**:4720–7.
64. Filipovic MR. Persulfidation (*S*-sulfhydration) and H_2S. In: Moore PK, Whiteman M, editors. *Chemistry, biochemistry and pharmacology of hydrogen sulfide, Handbook of experimental pharmacology*, 230. 2015. p. 29–59.
65. Vitvitsky V, Kabil O, Banerjee R. High turnover rates for hydrogen sulfide allow for rapid regulation of its tissue concentrations. *Antioxid Redox Signal* 2012;**17**:22–31.
66. Cuevasanta E, Denicola A, Alvarez B, Moller MN. Solubility and permeation of hydrogen sulfide in lipid membranes. *PLoS One* 2012;**7**:e34562.
67. Eberhardt M, Dux M, Namer B, Miljkovic J, Cordasic N, Will C, et al. H_2S and NO cooperatively regulate vascular tone by activating a neuroendocrine HNO-TRPA1-CGRP signaling pathway. *Nat Commun* 2014;**5**:4381. Available from: http://dx.doi.org/10.1038/ncomms5381
68. Cortese-Krott M, Kuhnle G, Fernandez BO, Grman M, DuMond J, Barrow M, et al. Key bioactive reaction products of the NO/H2S interaction are S/N-hybrid species, polysulfides, and nitroxyl. *Proc Natl Acad Sci USA* 2015;1–10. Available from: http://dx.doi.org/10.1073/pnas.1509277112
69. Arulsamy N, Bohle DS, Butt JA, Irvine GJ, Jordan PA, Sagan E. Interrelationships between conformational dynamics and the redox chemistry of *S*-nitrosothiols. *J Am Chem Soc* 1999;**121**:7115–23.
70. Wang X-B, Kass SR. Anion A^- • HX clusters with reduced electron binding energies: proton vs hydrogen atom relocation upon electron detachment. *J Am Chem Soc* 2014;**136**:17332–6.
71. Filipović MR, Miljković J, Nauser T, Royzen M, Klos K, Shubina TE, et al. Chemical characterization of the smallest *S*-nitrosothiol, HSNO; cellular cross-talk of H_2S and *S*-nitrosothiols. *J Am Chem Soc* 2012;**134**:12016–27.
72. Ivanova LV, Anton BJ, Timerghazin QK. On the possible biological relevance of HSNO isomers: a computational investigation. *Phys Chem Chem Phys* 2014;**16**:8476–86.
73. Stamler JS, Feelisch M. Preparation and detection of *S*-nitrosothiols. In: Feelisch M, Stamler JS, editors. *Methods in nitric oxide research*. Chichester: John Wiley & Sons; 1996. p. 521–39.
74. Blank DA, Northa SW, Strangesb D, Suits AG, Yuan TL. Unraveling the dissociation of dimethyl sulfoxide following absorption at 193 nm. *J Chem Phys* 1996;**106**:539–50.
75. Rosenthal J, Lippard S. Direct detection of nitroxyl in aqueous solution using a tripodal copper(II) BODIPY complex. *J Am Chem Soc* 2010;**132**:5536–7.
76. Wedmann R, Zahl A, Shubina TE, Duerr M, Heinemann FW, Bugenhagen BEC, et al. Does perthionitrite ($SSNO^-$) account for sustained bioactivity of NO? A (Bio)chemical characterization. *Inorg Chem* 2015. http://dx.doi.org/10.1021/acs.inorgchem.5b00831, Article ASAP

77. Butler AR, Feelisch M. Therapeutic uses of inorganic nitrite and nitrate from the past to the future. *Circulation* 2008;**117**:2151–9.
78. (a) Pavlik JW, Noll BC, Oliver AG, Schulz CE, Scheidt WR. Hydrosulfide (HS^-) coordination in iron porphyrinates. *Inorg Chem* 2010;**49**:1017–26.
    (b) Pietri P, Lewis A, León RG, Casabona G, Kiger L, Yeh SR, et al. Factors controlling the reactivity of hydrogen sulfide with hemeproteins. *Biochemistry* 2009;**48**:4881–94.
79. Sidgwick NV. *The chemical elements and their compounds II*. London: Oxford Univeristy Press; 1950. p. 1345.
80. (a) Butler AR, Calsy-Harrison AM, Glidewell C, Sorenses PE. The pentacyanonitrosylferrate ion-V. The course of the reactions of nitroprusside with a range of thiols. *Polyhedron* 1988;**7**:1197–202.
    (b) Rock PA, Swinehart JH. The kinetics of the aqueoushydrogen sulfide-nitroprusside system. *Inorg Chem* 1965;**5**:1078–9.
    (c) Quiroga SL, Almaraz AE, Amorebieta VT, Perissinotti LL, Olabe JA. Addition and redox reactivity of hydrogen sulfides (H_2S/HS^-) with nitroprusside: new chemistry of nitrososulfide ligands. *Chem Eur J* 2011;**17**:4145–56.
81. Luthy RG, Bruce Jr. SG. Kinetics of reaction of cyanide and reduced sulfur species in aqueous solution. *Environ Sci Technol* 1979;**13**:1481–7.
82. Paolocci N, Saavedra WF, Miranda KM, Martignani C, Isoda T, Hare JM, et al. Nitroxyl anion exerts redox-sensitive positive cardiac inotropy in vivo by calcitonin gene-related peptide signaling. *Proc Natl Acad Sci USA* 2001;**98**:10463–8.
83. (a) Bisset WIK, Butler AR, Glidewell C, Reglinski J. Sodium nitroprusside and cyanide release: reasons for re-appraisal. *Br J Anaesth* 1981;**53**:1015–18.
    (b) Pasch T, Schulz V, Hoppelshaüser G. Nitroprusside-induced formation of cyanide and its detoxication with thiosulfate during deliberate hypotension. *J Cardiovasc Pharmacol* 1983;**5**:77–86.
84. Gao Y, Toubaei A, Kong X, Wu G. Solving the 170-year-old mystery about red-violet and blue transient intermediates in the gmelin reaction. *Chem Eur J* 2015;**21**:17172–7.

HNO–Thiol Relationship

6

Milos Filipovic
Université de Bordeaux, IBGC, UMR 5095, Bordeaux, France;
CNRS, IBGC, UMR 5095, Bordeaux, France

Chapter Outline

6.1 Reaction of HNO with thiols: biochemical foundations

As an electrophile HNO reacts with nucleophiles, but it appears that it has preference for soft nucleophiles, one of which is thiol-containing biomolecule.[1–3] Unlike NO, which reacts with thiols very slowly, the reaction of HNO with thiols is very fast and it probably depends on the pK_a of the reacting thiolate. For example, the rate constant for the reaction of glutathione (GSH) with HNO is reported to be $2{-}8 \times 10^6\ M^{-1}\ s^{-1}$.[2,3] However, the reaction with a thiolate in papain is one order of magnitude faster.[4]

The first described "thiol-dependent" mechanism of biological actions of HNO is the one explaining the mechanism of cyanamide.[5,6] HNO generated from cyanamide modifies cysteines in aldehyde dehydrogenase (ALDH). However, the observed lack of toxicity suggested minor changes in overall GSH/GSSG levels as confirmed on hepatocytes treated with cyanamide.[7] Furthermore, yeast treated with Angeli's salt (AS) were found to have significant inhibition of GAPDH,[8,9] an effect that is also observed during the treatment of cancer cells with HNO donors.[10]

Mechanistically the reaction between HNO and a thiolate starts when a nucleophilic thiol attacks on the electrophilic nitrogen of HNO to initially form the *N*-hydroxysulfenamide (Eq. 6.1).[11–14] In the presence of vicinal thiol, the reaction

The Chemistry and Biology of Nitroxyl (HNO). DOI: http://dx.doi.org/10.1016/B978-0-12-800934-5.00006-2

proceeds further to generate the corresponding disulfide and hydroxylamine (Eq. 6.2). *N*-hydroxysulfenamide also serves as intermediate that is prone to isomerization, consequently leading to formation of sulfinamide (Eq. 6.3).[2,15,16] Formation of sulfinamide and intramolecular disulfides in the proteins reacting with AS in vitro was also demonstrated by electrospray ionization mass spectrometry.[15] Sulfinamides are not stable and could be further hydrolyzed to sulfinic acids.[17] However, under certain conditions, like in the specific environments of the protein active sites, sulfinamides could be stabilized. Additionally, *N*-hydroxysulfenamide could also react with the vicinal lysine residue to generate an internal sulfinamide.[15] This internal cyclic sulfinamide could be an intermediate toward the formation of sulfinic acid.

$$\mathrm{RSH + HNO \rightarrow RSNHOH} \tag{6.1}$$

$$\mathrm{RSNHOH + R'SH \rightarrow RSSR' + NH_2OH} \tag{6.2}$$

$$\mathrm{RSNHOH \rightarrow [RS^+NH + HO^-] \rightarrow RS^+(OH)NH \rightarrow RS(O)NH_2} \tag{6.3}$$

The fate of the initially formed *N*-hydroxysulfenamide would definitely be determined by the microenvironment of the cysteine thiol as shown by Shen and English.[15] In addition, recent study addressing the effects of HNO on platelets proteome identified several proteins with sulfinic acid modifications, presumably generated from the corresponding sulfinamide.[17]

6.2 Specific thiolates as targets for HNO-induced physiological/pharmacological effects

As mentioned above, the growing body of evidence has been suggesting that HNO interacts only with certain proteins (or at least more selectively) without significantly changing the redox status of GSH in the cell. This ability of HNO to modify thiols on specific proteins without any apparent effect on GSH is rather puzzling. Flores-Santana and colleagues reported recently an interesting calculation that based on the low concentration of GAPDH in the yeast (2 μM), the high concentration of intracellular GSH (1−10 mM), and relatively high rate for the reaction of HNO with GSH (2−8 10^6 M^{-1} s^{-1}), it seems that the reaction of HNO with GAPDH in the cells proceeds in a diffusion-controlled rate.[4] The most probable explanation would be the significant difference in the pK_a of the protein thiols, when compared to the GSH. Thiols with lower pK_a will exist predominantly as thiolates and consequently be better nucleophiles. Several proteins have been suggested as sole targets responsible for biological effects of HNO.

6.2.1 Soluble guanlyate cyclase (sGC)

First biological effects of Angeli's salt were related to its ability to induce vasodilation of blood vessels. AS caused vasodilation of large conduit and small resistance arteries, as well as of coronary and pulmonary blood vessels.[18–24] These effects were related to the activation of soluble guanylate cyclase.

Indeed, activation of sGC and the increase of cGMP have been demonstrated to facilitate most of the above-mentioned effects, although the actual mechanism is not clear. Most recently, Zhu et al., demonstrated that HNO-induced systemic vasodilation is dependent on soluble guanylate cyclase.[25] They used mice lacking the sGC-β subunit, and in these mice vasodilatory effects of a new HNO donor, CXL-1020, were absent. The authors proposed that it is HNO-induced thiol modification of sGC that leads to its activation, contrary to the previous studies showing that HNO cannot activate sGC and that oxidation of thiols by HNO causes a loss of enzyme activity when purified sGC was used.[26] The reason for such discrepancies is unclear and it remains to be elucidated.

6.2.2 Transient receptor potential ion channel type A1 (TRPA1): the role of HNO-TRPA1-CGRP signaling cascade

One of the early discovered biomarkers of HNO-induced physiological effects that distinguishes it from the NO-induced effects is the release of calcitonin gene-related peptide release (CGRP). CGRP is a nonadrenergic/noncholinergic type peptide composed of 37 amino acids,[27] stored in dense-core vesicles that are located in terminal endings of sensory nerve fibers.[28] Release of CGRP from terminal vesicles is triggered by Ca^{2+} influx as a consequence of neuronal depolarization. Released CGRP binds to its corresponding CGRP receptor, a 461 amino acid long transmembrane heterodimeric macromolecule that is composed of "calcitonin receptor-like receptor" (CLR) protein, one of the three peptide isoforms of "receptor activity modifying protein" (RAMPs: RAMP1, RAMP2, RAMP3) and receptor component protein (RCP).[29,30]

CGRP is the strongest known vasodilator produced in our body. When it binds to CGRP receptor it may initiate several different signaling pathways (Fig. 6.1). The release of CGRP was also demonstrated by using capsaicin, which is the active principle of chili peppers responsible for its burning effect, sensitization of nerve fibers and local inflammation. This led to further studies, which confirmed that CGRP is localized in sensory nerve endings. More importantly, the discovery of "capsaicin receptor," vanilloid transient receptor potential (TRPV1) channels,[31–33] suggested the existence of physiological receptor/agonist complex that could control CGRP release.

Activation of another member of TRP ion channels: transient receptor potential ankyrin (TRPA1) also provokes the CGRP release.[34] Originally discovered as cold temperature sensors,[35] TRPA1 channels are polymodal voltage-dependent Ca^{2+}-permeable channels that can be activated by different gating modes. TRPA1 is expressed in ubiquitous peripheral sensory neurons and activated by numerous noxious chemicals like pungent natural and environmental irritants.[36,37] The actual mechanism for the chemo sensing of electrophilic agonists by TRPA1 is the covalent modification of particular cysteine residues of ankyrin-repeat domain of

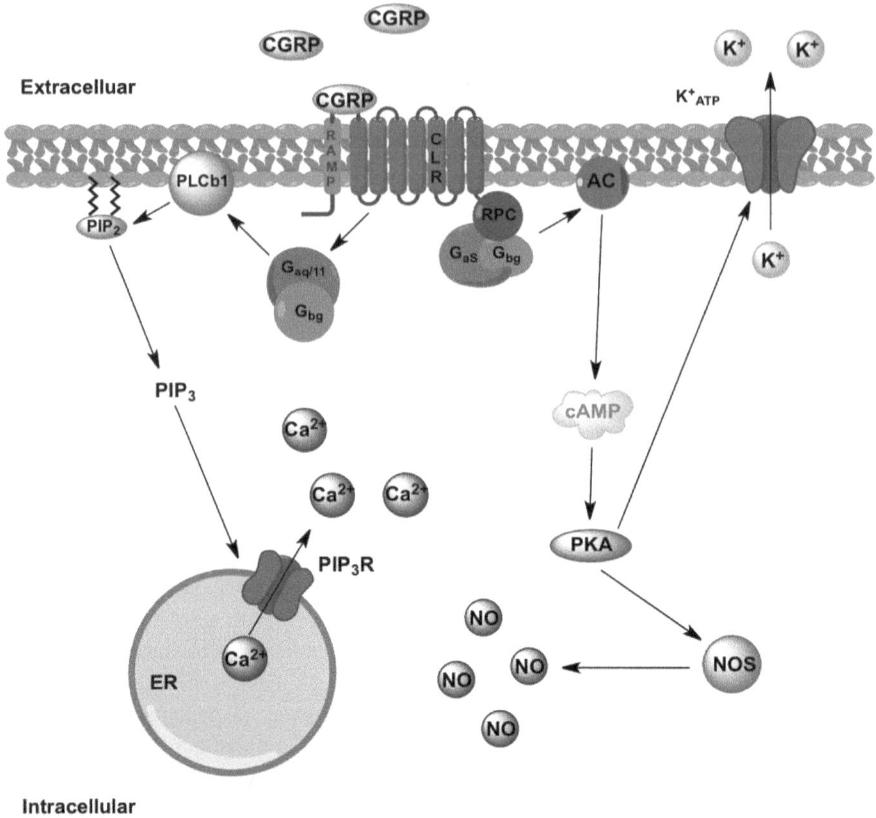

Figure 6.1 Intracellular signaling by CGRP. Stimulation of CGRP receptor induces activation of adenylate cyclase (AC), an enzyme that generates cAMP. Generated cAMP subsequently activates protein kinase A (PKA), an enzyme that phosphorylates multiple intracellular molecules. Increased PKA activity may activate NOS by phosphorylation, and consequently increases the levels of intracellular NO. Other target molecules that undergo PKA dependent phosphorylation, induced by CGRP activation are: ATP-dependent K^+ channels involved in vasodilatation, cAMP response element-binding protein (CREB), and extracellular signal-related kinases (ERKs). Additionally, phospholipase C-β1 (PLC-β1) could cleave phosphatidylinositol 4,5-bisphosphate (PIP_2) ultimately leading to an increase of the intracellular Ca^{2+}, and activation of the DAG-mediated upregulation of protein kinase Cε (PKCε).

N-terminal part of the channel.[38,39] The intracellular N-terminus of human TRPA1 (hTRPA1) is rich in cysteine residues, of which C621, C641, and C665 are responsible for activation.[38] Recent structural studies of mouse TPRA1 channel suggested that all cysteines (critical and noncritical ones) are in close proximity to each other, which could also favor disulfide formation.[40,41]

Serving as a chemosensor for electrophilic compounds TRPA1 could be a target for HNO. Indeed, Eberhardt et al. demonstrated for the first time that HNO-induced

CGRP release goes *via* activation of TRPA1 (Fig. 6.2A).[42] The authors showed that TRPA1 is responsible for the AS-induced Ca^{2+} influx into neurons from dorsal root ganglion (Fig. 6.2B), an effect that is completely absent in the cells prepared from TRPA1 knockout mice. In addition, the authors showed that the intramolecular disulfide formation between critical Cys 621 and the neighboring Cys 633, as well as between Cys 651 and the critical Cys 665, is responsible for the channel activation. This opens the channel to allow Ca^{2+} influx, an effect that is long lasting and could only be reversed by the reducing agent DTT (Fig. 6.2C).[42]

CGRP release from mice dura or sciatic nerve preparations was completely abolished in $TRPA1^{-/-}$ mice treated with AS, undoubtedly showing that CGRP is released only by the reaction between HNO and TRPA1 cysteine residues (Fig. 6.2A). This study finally answered a decade-long dilemma on how HNO actually stimulates CGRP release. Surprisingly, despite several other, redox sensitive Ca^{2+} channels, only TRPA1 was responsible for HNO induced CGRP release. Consequently, the HNO induced blood pressure drop in mice was significantly

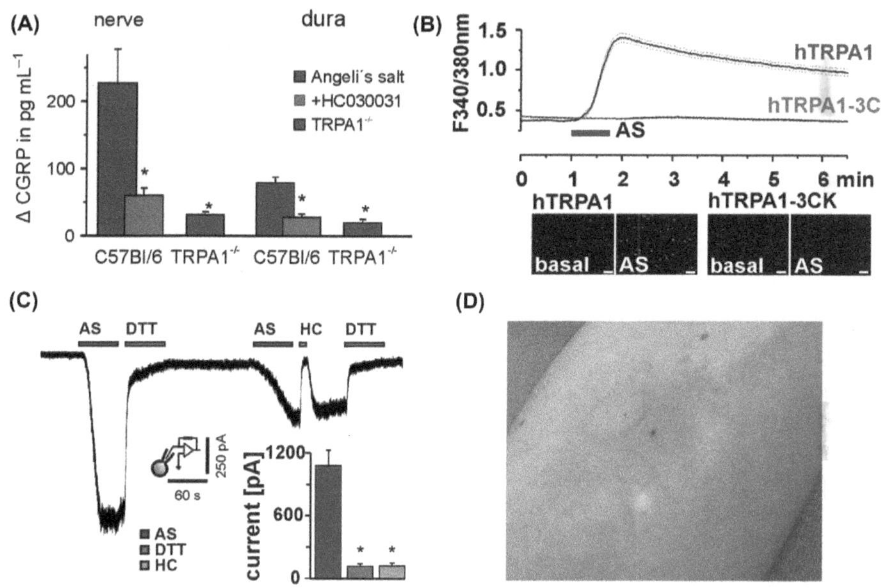

Figure 6.2 Experimental dissection of HNO-TRPA1-CGRP pathway. (A) Maximal CGRP released from C57Bl/6 and TRPA1 −/− mouse sciatic nerve and dura mater upon stimulation with AS. AS effects were inhibited by HC030031, a selective TRPA1 inhibitor, and absent in TRPA1 −/− mice. (B) AS increases intracellular Ca^{2+} in hTRPA1 transfected HEK cells but not in cells expressing hTRPA1-3C mutant; Representative pseudocolor images (scale bar = 100 μm). (C) AS-evoked inward currents can be reversed by application of DTT, or temporarily blocked by HC030031. (D) Photograph of a subject's volar forearm with noticeable axon reflex erythema upon intradermal injection of AS. Adapted from Eberhardt M, Dux M, Namer B, Miljkovic J, Cordasic N, Will C, et al. H2S and NO cooperatively regulate vascular tone by activating a neuroendocrine HNO-TRPA1-CGRP signalling pathway. *Nat Commun* 2014;15:4381.

reduced in TRPA1 knockout mice. The meningeal blood flow effects of AS were diminished by the application of TRPA1 antagonist or CGRP receptor antagonists, suggesting that this pathways plays an important role in the control of meningeal blood flow. The control of the peripheral blood flow, in the sites, which are innervated with TRPA1 containing fibers, could represent the way by which HNO-TRPA1-CGRP signaling cascade contributes to the control of systemic blood pressure.

More interesting observation came from the psychophysics study. Subcutaneous injection of Angeli's salt lead to a rather painful (4/10) sensation followed by the axon-flare erythrema, characteristic for the CGRP release (Fig. 6.2D). The effect was histamine-independent. This suggested that in certain pathologies HNO could be a pro-nociceptive agent.

Characterization of HNO-TRPA1-CGRP pathway opened up a new perspective on the potential role of HNO in both central and peripheral nervous system. Being a nociceptive transduction channel in the first place, TRPA1 contributes to pain/itch sensations.[34] Eberhardt et al. showed the strong and persistent HNO-induced activation of TRPA1 due to the disulfide formation, which led to a strong pain sensation when injected subcutaneously.[42] Furthermore, the role of CGRP release has never been considered outside the cardiovascular system, although CGRP is long known as important mediator of migraine attacks. In humans suffering from migraine, increased level of CGRP is found in both the external and internal jugular blood during migraine attacks. The most recent development of the drugs to treat migraine goes in two direction: (i) development of CGRP receptor antagonists and (ii) development of anti-CGRP antibodies, the latter showing some promise and have moved further in clinical trial.[43,44] All this shed a different light on the potential application of HNO donors and further studies are needed in order to clarify whether the drawback of HNO therapy could be strong pro-nociceptive stimulation.[45]

6.2.3 Protein targets in cardiomyocites

The most promising of the beneficial pharmacological effects of HNO donors is certainly their effect on heart. In the seminal work by Paolocci and colleagues, the authors showed previously unknown effect of HNO: an increase of myocardial contractility, known as positive inotropy, in combination with accelerated relaxation, known as positive lusitropy.[46] These effects could be of particular importance when treating congestive heart failure (CHF), a serious and life-threatening disease. Paolocci and colleagues also studied the pressure–volume relationship in a dog model of CHF.[47] They showed that Angeli's salt still has positive inotropic and lusitropic effects like it does in the control healthy dogs. Furthermore, coadministration of beta-agonist mimetics (beta-agonist are the most used therapeutic approach to treat CHF), had an additive effect. It is worth mentioning that these effects of HNO are completely opposite to what has been reported for NO. This sparkled the research in the field of new HNO donor development for the treatment of heart failure. One such donor, named CXL-1020, a product of a company Cardioxyl, has been recently showed to have the beneficial effects in acute decompensated heart failure in humans, being the first reported example of the HNO donors tested in humans.[48]

The mechanism by which HNO causes these effects is still a matter of intense research. The first reports indicated the role of cardiac ryanodine receptors (RyR).[49,50] RyRs are known to be redox sensors, suggesting that they could be easily modified by the reactive oxygen and nitrogen species.[51] Therefore, it came as no surprise that HNO-mediated Ca^{2+} release from the sarcoplasmic reticulum is caused by the RyR activation, which could be reversed by DTT.[50] Later studies broadened the view by introducing the direct interaction with the sarcoplasmic reticulum Ca^{2+} pump (SERCA). The authors suggested that HNO increases the opening probability of RyR and accelerates Ca^{2+} reuptake into sarcoplasmic reticulum by stimulating SERCA.[50,52] Contraction therefore improves with no net rise in diastolic Ca^{2+}. The mechanism by which HNO activates SERCA has been proposed to be the oxidation of cysteine residues in the SERCA's regulatory protein phoshpolamban (PLN) (Fig. 6.3). Mutations of critical cysteines in PLN abolished the HNO effects, and the DTT treatment reversed them, confirming essential role of disulfide formation. Recently Paolocci's group observed that HNO enhances SERCA activity by promoting redox-dependent PLN oligomerization. The final effect is therefore the enhanced Ca^{2+} cycling which can explain both the HNO-induced increase of contractile force and myocardial relaxation.[50]

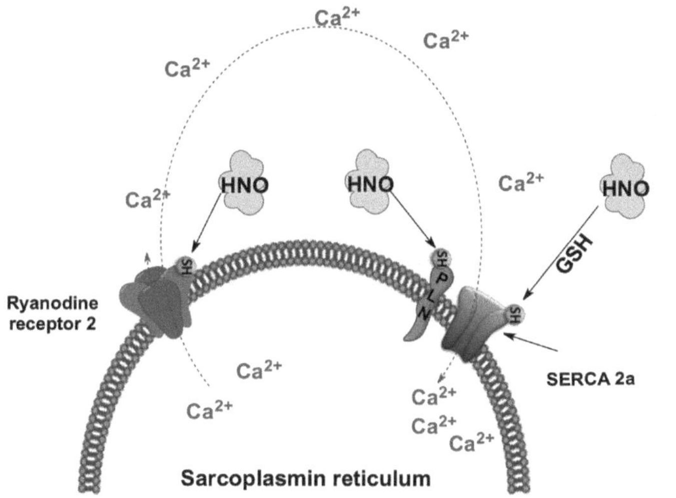

Figure 6.3 Proposed mechanism(s) for the HNO-induced cardiovascular effects. HNO causes oxidation of cysteine residues on ryanodine receptor 2, opening the channel to pump the calcium (Ca^{2+}) out of SR. This makes more Ca^{2+} ions available for myofilaments. HNO also activates SERCA2a either directly, initially forming *N*-hydroxysulfenamide which further reacts with glutathione (GSH) to cause glutathionylation of C674, or by modifying cysteines of the SERCA's regulatory protein phoshpolamban (PLN). This leads to the pumping of the calcium (Ca^{2+}) back to the sarcoplasmic reticulum (SR).

Conversely, in their attempt to elucidate the mechanism by which HNO exerts inotropic effects, Lancel et al. found that HNO causes oxidation of a particular cysteine residue on SERCA.[53] In particular, they observed *S*-glutathionylation of the channel at the cysteine 674 (Fig. 6.3). Overexpression of glutaredoxin-1 prevented HNO-stimulated oxidative modification and therefore activation of SERCA as it did the cysteine 674 replacements with serine. Although the actual mechanism has not been solved, the authors proposed that HNO causes *N*-hydroxysulfenamide intermediate, which then reacts with GSH.

In the most recent, extensive study, Gao et al. have identified 12 sites of HNO-induced disulfides in the key myofilament proteins, such as actin, tropomyosin, myosin heavy chain, myosin light chain (MLC1), alpha-actinin, myosin binding protein c, and troponin C. By modifying the myofilament proteins, HNO increase both maximum force and Ca^{2+} sensitivity of force in intact muscles.[54] For example, Cys257 in the actin subdomain 4 formed a disulfide with Cys190 in the tropomyosin, while Cys81 in the myosin light chain formed a disulfide with Cys37 in the myosin heavy chain's head region. Interestingly, HNO also increased Ca^{2+} sensitivity but not maximum force in skeletal muscle. Skeletal muscle cell lack Cys81 in MLC1. This suggests that this particular residue is critical redox switch responsible for HNO-stimulated increase in cardiac force production. The study also compared two different HNO donors, AS with 1-nitrosocyclohexyl acetate (which release HNO slower and upon cleavage of the ester bond). A fewer modified proteins were observed in AS-treated cells suggesting that the rate of HNO release and its location could change the overall outcome of HNO donating drugs.[54]

Shiva et al. tried to unravel the effects of HNO on isolated mitochondria.[55] The authors demonstrated that mitochondria are capable of re-oxidizing HNO to NO, but that HNO exhibits inhibition of mitochondrial respiration through the inhibition of complexes I and II, most probably via modification of specific cysteine residues in the proteins.[55] The proteomic approach identified extensive modifications of mitochondrial protein thiols. The protective effects of HNO in ischemia/reperfusion injury are reminiscent of those seen with diazoxide, a mitochondrial K_{ATP} channel opener, which the authors later proved to indeed be the case: HNO indeed caused the modifications of critical cysteines in mitochondrial K_{ATP} channels, leading to the channel opening, which is considered to be an important event for ischemic-preconditioning-induced cardioprotection.[56]

6.3 Thiol-based HNO generation

Beside the well-established reactivity of HNO with thiols, thiols could in fact play a role in endogenous HNO generation as well. Although not studied in detail and with many unsolved questions, studies published to date suggest that the reaction of thiols with *S*-nitrosothiols, but also the direct reaction of thiols with NO could lead to HNO generation. Recent emergence of hydrogen sulfide as a new gasotransmitter introduced another thiol-like player, which could be involved in the same type of reactions.

6.3.1 Reaction of S-nitrosothiols with thiols

Until recently there was no evidence for the endogenous HNO formation. This was mainly due to the short half-life that HNO is expected to have in the cells, which provides a methodological limitation for its intracellular detection. Several in vitro observations have suggested a thiol-assisted decomposition of *S*-nitrosothiols (RSNO) as a possible endogenous sources of HNO.[14,57]

Protein *S*-nitrosation is considered to be an important posttranslational modification of proteins. The number of proteins found to be controlled by this modification is constantly increasing. The role of *S*-nitrosation in controlling the protein function has been covered in several extensive reviews.[58–60]

One of the possible ways for the intracellular HNO generation is the reaction of *S*-nitrosothiols with some other thiol:[14,57]

$$RSNO + RSH \rightarrow RSSR + HNO \qquad (6.4)$$

This reaction is, however, considered to occur only in special protein environments as the trans-nitrosation reaction is thermodynamically more favourable:[61]

$$RSNO + R^1SH \rightarrow RSH + R^1SNO \qquad (6.5)$$

Charged and polar residues can strongly affect the electronic structure of the –SNO group in the protein environment. This can be explained using the concept of the antagonism between two of the three –SNO group resonance structures: zwitterionic, $RS^+ = N - O^-$, vs ion-pair $RS^- \cdots NO^+$. Charged residue interactions with opposite formal charge in the antagonistic structures promote one structure and reduce the contribution of another. This results in significant modification of the –SNO reactivity: either shortening and strengthening of the S–N bond and increasing electrophilicity of the S atom (making it prone for the nucleophilic attack by thiolate and elimination of HNO, Eq. 6.4) or lengthening and weakening of the S–N bond and increasing electrophilicity of the N atom (making it prone for the nucleophilic attack by thiolate and transnitrosation, Eq. 6.5). It is also worth mentioning that, although thermodynamically hindered, the reaction shown in Eq. (6.4) could still occur if the following reaction (such as generation of N_2O from HNO dimerization) provides the necessary thermodynamic push.

Although not really a thiol in strict chemical clasification, hydrogen sulfide (H_2S) could be considered as close-to-the-smallest thiol found in the cells, and ever since the first report of its potential physiological role,[62] there has been a growing literature on the subject of H_2S signaling. Very fast, H_2S joined the other two gases, nitric oxide (NO) and carbon monoxide (CO) as the third gasotransmitter.[63,64] The incredible similarity between H_2S and NO signaling effects lead to the several important studies addressing the direct cross-talk between H_2S and NO signaling pathways and the possible regulation of NO redox status by H_2S.

As already mentioned, the reaction of thiols with RSNOs has been suggested to be one of the mechanism by which HNO could be produced in the cells

(Eq. 6.4). Being the smallest thiol-like molecule, which diffuses without any constraints,[65,66] H_2S could be a player in such process.

We studied the reaction of *S*-nitrosothiols with H_2S to observe the generation of the smallest *S*-nitrosothiol, thionitrous acid (HSNO) (Fig. 6.4A).[67] The stability of the RSNOs usually decreases as the "size" of the R substituent decreases, so in the case of HSNO it was not surprising that HSNO was less stable than its cysteine analogue. Despite its intrinsic instability due to relatively low S–N bond energy, HSNO lived long enough to cause further trans-nitrosation, i.e. transfer of NO^+ moiety, from one protein to another but more importantly, when HSNO reacted further with H_2S, nitroxyl was formed. The formation of HNO was observed by using methemoglobin as a scavenger and by monitoring the hydroxylamine, a final product of the reaction of nitroxyl with thiols.

Furthermore we were also able, using recently developed fluorescence sensor for nitroxyl (CuBOT1),[68] to detect HNO generation in the cells. An obvious increase of the CuBOT1 fluorescence was observed when the cells were treated with the combination of *S*-nitrosoglutathione and H_2S, but even stronger intracellular HNO generation was observed in cells pretreated with *S*-nitrosoglutathione (in order to increase the intracellular content of *S*-nitrosothiols) and then washed and subsequently treated with H_2S. This was the first direct evidence for the intracellular generation of HNO, albeit not endogenous but rather pharmacological. The fact that HSNO could freely diffuse from one cell to another (or between cellular compartments within one cell) suggests that HSNO could be a "carrier" of HNO. Aside its role in *trans*-nitrosation, if it would meet another H_2S molecule, HSNO would form HNO.

The similarity between metal-nitrosyls and *S*-nitrosothiols was addressed in a follow-up study, which examined the reaction of sodium nitroprusside with H_2S.[69] Sodium nitroprusside (SNP) is a vasodilator still used in acute hypertensive crises. In SNP, $Na_2[Fe(CN)_5(NO)]$, NO coordinated to iron has an NO^+ character (similar like in RSNO) and undergoes chemistry that is best described as direct nitrosation biochemistry. SNP's effects are fast and efficient but the problem of cyanide toxicity exists, and this limits its application only to situations monitored by medical doctors. In a study performed at physiological pH and aerobic conditions, combining the chemical tools with pharmacological/physiological experiments we suggested the reaction mechanism depicted in Fig. 6.4B. The reaction steps are reminiscent of the ones found in the reaction of *S*-nitrosothiols with H_2S (Fig. 6.4A).

Recently, Gao et al. suggested that the blue product in this reaction is $SSNO^-$ is coordinated to the iron.[70] The authors used nitrogen and oxygen NMR to suggest such structure, although the actual chemical shifts for such species could only be speculated (and computed values are significantly off). Most importantly, it remains unclear how that second sulfur athom got oxidized by two electrons when metal center remained in the same oxidation state as before the reaction. Further structural studies (such as X-ray crystallography) will prove the structure of the blue product.

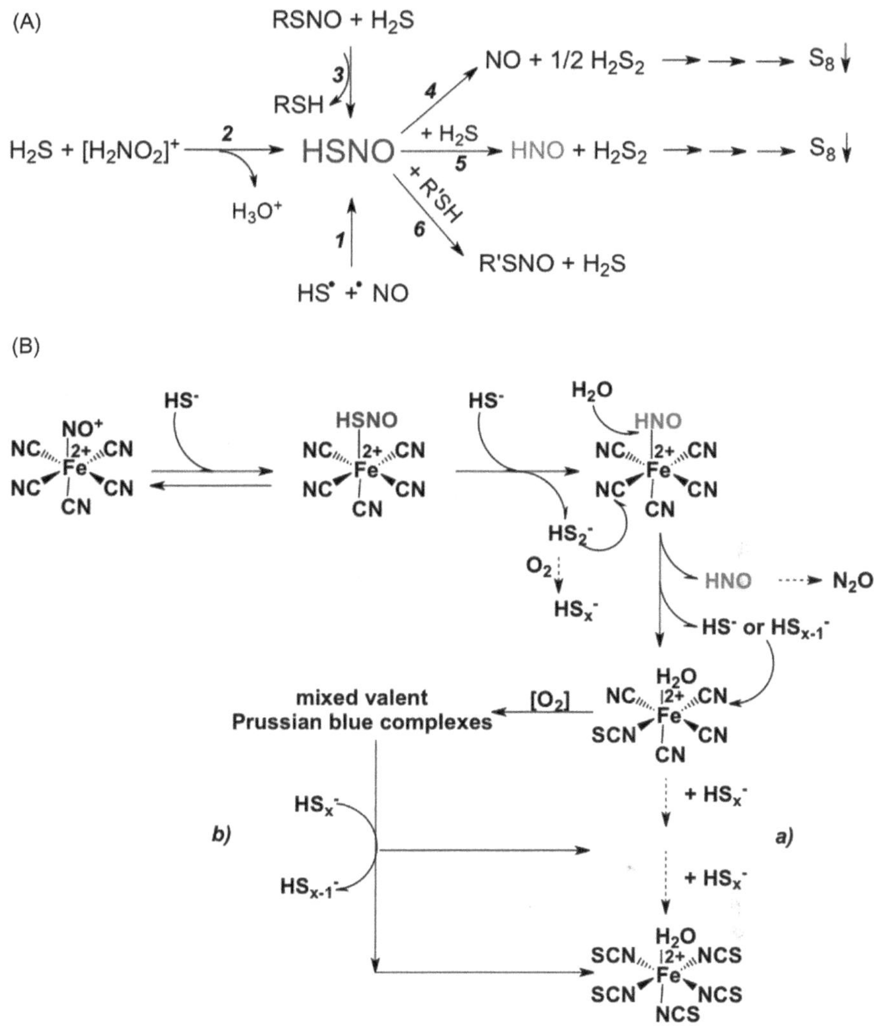

Figure 6.4 HSNO-mediated HNO generation. (A) Reaction paths for HSNO generation (1–3) and subsequent biologically relevant reactions (4–6). HNO is generated in the reaction 5. (B) Mechanistic depiction of the reaction steps behind sodium nitroprusside and H_2S reaction at pH 7.4 and aerobic conditions. SNP reacts very fast with H_2S to form intermediate $[(CN)_5FeN(O)SH]^{3-}$. In the second reaction step, sulfide reduces coordinated HSNO/SNO^- and gets initially oxidized to disulfide, which can be further oxidized to polysulfides along the reaction progress. As a result of this step, the $[(CN)_5Fe(HNO)]^{3-}$ intermediate is formed which is unreactive toward H_2S but does react with polysulfides forming thiocyanate adduct with elimination of HNO. HNO can dimerize to give N_2O, while the other cooridinated cyanides are consecutively transformed to thiocyanate (main reaction path, *a*). Oxygen can induce oxidation of the iron(II) center that would lead to mixed-valent bridged complexes, known as Prussian blue, which are short-lived and react further with polysulfides to give the final thiocyanate product as well (minor reaction route, *b*).

6.3.2 Direct reaction of thiols with NO

Direct reaction of thiols with NO (under anaerobic conditions) is considered to be very slow and of no biological relevance. Early studies[71–73] showed that NO slowly react with thiols leading to formation of disulfides, N_2O and eventually N_2. Interestingly, Pryor et al. showed that the reaction gets much faster as pH gets closer to pK_a of the used thiol (t = 6 h at pH 7 and t = 15 min at pH 9 for cysteine).[73] This was confirmed by Aravindakumar et al. who determined that the rate constant for NO and cysteine at 25°C is $0.37 \times 10^3\ M^{-1}\ s^{-1}$.[74] Knowing that the pK_a of cysteine residues in proteins varies, and that those responsible for signaling are often much more acidic (therefore present as thiolates), despite its apparent slow kinetics for the reaction of cysteine or glutathione with NO, direct reaction of NO with protein thiolates is quite plausible.

A direct product of this reaction is proposed to be:

$$RS^- + NO \rightarrow RSNO^{\bullet -} \tag{6.6}$$

Pryor and colleagues proposed that two of those radical species would react with each other (Eq. 6.7) to form disulifde with elimination of hyponitrous acid (Eq. 6.8) that finally ends up as N_2O (Eq. 6.9).

$$2\ RSNO^{\bullet -} \rightarrow RSN(OH) - N(OH)SR \tag{6.7}$$

$$RSN(OH) - N(OH)SR \rightarrow RSSR + HON = NOH \tag{6.8}$$

$$HON = NOH \rightarrow N_2O + H_2O \tag{6.9}$$

Thionitroxide radical anion is also suggested to be a product of direct reaction of NO with thiolates in hemoglobin. In fact, computational studies by Zhao and Houk[75] showed that the experimentally observed bond lengths for the *S*-nitroso hemoglobin[76] correspond better to thionitroxide-like species. $RSNO\cdot^-$ represent a one electron reduced *S*-nitrosothiol and have redox potential ~ -1.2 V (vs SCE).[77] This makes this species a very strong reducing agent, being able to transfer the electron to molecules such as oxygen or NO. In fact, Zhao and Houk propose that thionitroxides would react with NO to give HNO.[75] Due to the widely accepted opinion that the reaction between NO and thiols is too slow to have any physiological meaning, the reaction between thiols and NO to form HNO has not been studied in detail. Our preliminary studies suggest that cysteine readily reacts with NO to form HNO and *S*-nitrosocysteine under strict anaerobic conditions. Considering that NO has tendency to dimerize and form $(NO)_2$ in aromatic hosts,[78] a condition which can be achieved in specific protein environments and lipid bilayers, formation of HNO seems even more energetically favorable as confirmed by Houk's group in both computational and experimental studies.[78]

Drawing an analogy with protein thiols we studied the possibility that direct reaction between NO and H_2S could be a source of HNO. Several groups have considered that H_2S effects could be linked to NO, including the first study that showed physiological effects of H_2S.[62] Additionally, Yong et al. proposed that H_2S and NO together could give HNO [79] and recently demonstrated positive inotropic effects of these combined gasotransmitters on the heart,[80] which is one of the hallmarks of HNO physiology.[46,47]

Reaction of NO with thiols is considered to be thermodynamically unfavorable and very slow to proceed at any detectable rate under physiological conditions.[81] The same could be extended to the reaction of NO with H_2S, if outer sphere electron transfer is to be the mechanism and the −0.81 V is a redox potential for the reduction of NO.

However, in the study by Eberhardt and colleagues, the authors show the first chemical evidence for the HNO formation from a direct reaction of NO and H_2S.[42] Using a recently developed HNO electrode,[82] the authors demonstrated that a 2 μM combination of NO and H_2S yields a peak HNO concentration of c.0.5 μM, similar to the effects of 1 mM AS concentration (Fig. 6.5A). The reaction proceeds even in the absence of oxygen and metal ions, therefore suggesting that it could be a major source of HNO in vivo. In fact, the intracellular levels of HNO, detected with a fluorescence sensor, were diminished when cells were treated with NOS and CSE inhibitors. Furthermore, the combination, just like Angeli's salt, activated HNO-TRPA1-CGRP pathway (Fig. 6.5B).[42]

Unpublished data from our laboratory further support this mechanism. We managed to identify the reaction intermediate and study its decomposition to give HNO. The following reaction steps leading to HNO formation have been identified:

$$HS^- + NO \rightarrow HSNO^{\cdot -} \tag{6.10}$$

$$HSNO^{\cdot -} + H^+ \rightarrow [H_2SNO^{\cdot}] \rightarrow HNO + HS^{\cdot}/\text{''S''} \tag{6.11}$$

$$HSNO^{\cdot -} + HS^- \rightarrow HNO + H_2S_2/\text{''S''} \tag{6.12}$$

$$HSNO^{\cdot -} + NO + H^+ \rightarrow HSNO + HNO \tag{6.13}$$

$$HSNO + H_2S \rightarrow HNO + H_2S_2/\text{''S''} \tag{6.14}$$

A significant portion of H_2S-induced vasodilation was inhibited in TRPA1 −/− and CGRP −/− mice or mice treated with NOS inhibitors. The possibility that H_2S/NO reaction is the main endogenous source of HNO is further strengthen by the observation that TRPA1 and CBS coexpress in small to medium-sized sensory neurons and axons, which together with the recent demonstration of the coexpression of TRPA1 and nNOS [82] suggests the existence of an functional unit for the HNO generation and subsequent activation of TRPA1-dependent CGRP release. This functional unit can be of importance in the regulation of peripheral blood flow (as demonstrated in dura mater) and even of systemic blood pressure.[42]

All three, NO, H_2S, and HNO, are freely diffusible, so several possibilities could be envisioned for TRPA1 activation and CGRP release: (i) NO and H_2S produced

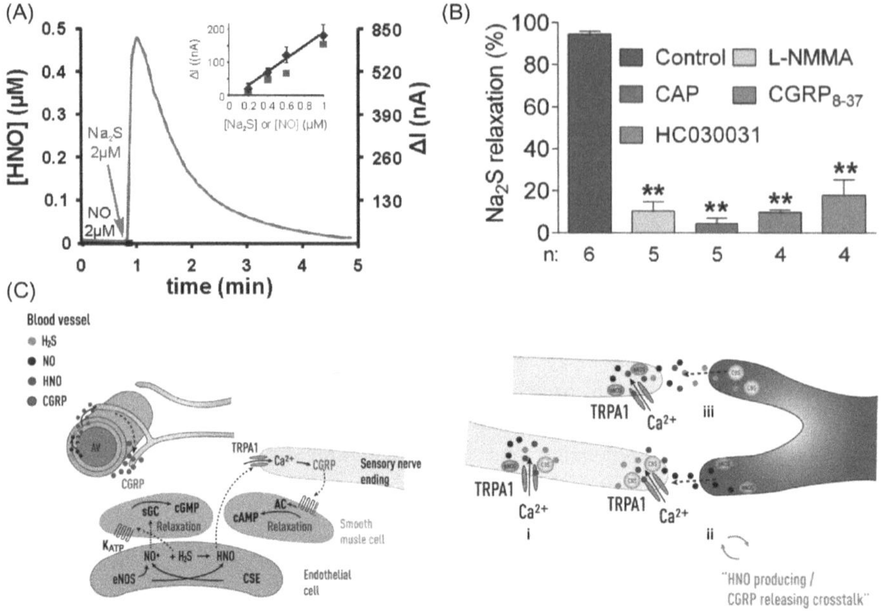

Figure 6.5 H_2S and NO react directly to endogenously produce HNO and stimulate HNO-TRPA1-CGRP pathway. (A) Amperometric signal of the HNO-selective electrode after addition of H_2S (2 μM) to a solution of 2 μM NO (left axis: [HNO] after calibration, right axis: measured current). Inset: Signal peak vs H_2S (blue) and NO (red) concentration, while the other reactant concentration is maintained constant and in excess. (B) Isometric tension recordings of phenylephrine- preconstricted ring segments of rat second order mesenteric artery branches: Na_2S (10 μM)-induced vasodilation was reversed by $CGRP_{8\text{-}37}$ and abolished by $CGRP_{8\text{-}37}$. NOS inhibitor, L-NMMA, and TRPA1 antagonist, HC030031, pretreatment inhibited vasodilation induced by Na_2S. CGRP depletion by capsaicin (CAP) also abolished the Na_2S-induced blood vessel relaxation. (C) TRPA1/CGRP expressing nerve endings in the periphery communicate with the smooth muscle cells surrounding the endothelium of blood vessels. Endothelial cells are known to produce NO and H_2S, both of which freely diffuse and activate guanylyl cyclase and K_{ATP} channels, respectively, to induce vasodilatation. However, H_2S and NO also react with each other to give HNO, which could reach paravascular TRPA1-expressing sensory nerve fibers, inducing Ca^{2+} influx and CGRP release. Other potential sites of NO-H_2S interaction is in neurons: (i) TRPA1 channels are coexpressed with nNOS and CBS in primary afferents forming a functional signaling complex that leads to confined HNO generation and TRPA1 gating upon activation of the gasotransmitter-generating enzymes. Additionally, NO (ii) or H_2S (iii) could originate from either side of a synaptic cleft (or from nearby axons of passage) and freely diffuse into adjacent neurons (or nerve fibers). There, they react with their counterpart producing HNO in vicinity of TRPA1.
Adapted from Eberhardt M, Dux M, Namer B, Miljkovic J, Cordasic N, Will C, et al. H2S and NO cooperatively regulate vascular tone by activating a neuroendocrine HNO-TRPA1-CGRP signalling pathway. *Nat Commun* 2014;15:4381.

in endothelium react to give HNO that could diffuse and activate nerve endings expressing TRPA1 which would consequently stimulate CGRP release and smooth muscle relaxation (unpublished data); (ii) production of H_2S and NO from colocalized CBS and neuronal NOS leads to intracellular HNO formation and TRPA1 activation as shown by Eberhardt et al. [42] (iii) taking into account that constitutive levels of NO in neurons are very low,[83] it is also plausible that, for instance, in CNS, astrocyte-derived NO, as a paracrine signal, reaches the network of CGRP- and TRPA1-expressing nerve fibers that surround small blood vessels, reacts with H_2S produced in vicinity of TRPA1 to give HNO, which then induces CGRP release and vasodilatation in the periphery (Fig. 6.5C). Recent work showed that the NO vasodilatory effect on aorta rings is partially blocked by inhibiting H_2S production whereas H_2S effects were diminished by inhibiting NO production, further strengthening the link between these gasotransmitters.[84]

HNO-TRPA1-CGRP activation may have important roles in the gastrointestinal tract, in particular in chronic inflammation of the bowel, a site with the highest local H_2S generation provided by commensal bacteria.[85] On the physiological side, it is tempting to speculate that the HNO-TRPA1-CGRP pathway could contribute to local vasodilatation in the gut and enhanced nutrient resorption, which would be achieved by needs-based channel activation. However, in inflammation, characterized by high NO production, the pathway appears Janus-faced; it could contribute to the completion and resolution of the process by the protective effects of CGRP, or it can perpetuate the disease by TRPA1-controlled corelease of substance P that turns the immune system auto-aggressive.[86,87]

Similar to Angeli's salt, combination of H_2S and NO, when injected subcutaneously, induced pain and itch, with an obvious axon-flare erythrema, suggestive of CGRP signaling. When co-infused (but in close proximity) *via* separate microdialysis tubes, H_2S and NO also induced local vasodilation, but more importantly NH_2OH was detected in the effluate as a final product (unpublished data).

These recent data have several implications on the understanding of HNO signaling: (i) they offer an endogenous source for HNO formation and complete the cycle spanning from HNO production to CGRP release; (ii) they offer mechanistic explanation of H_2S-induced vasodilation and positive inotropic effect; (iii) they suggest a potential use of H_2S/NO combination as a pharmacological tool for HNO generation and treatment of cardiac failure. Interestingly, the simultaneous pharmacological NO/H_2S donors have been reported and they show anticancer affect, just like other HNO donors.[88,89]

It is worth mentioning that recently Cortese-Krott and colleagues suggested that as a product of the reaction between RSNO and H_2S, as well as between NO and H_2S, both aerobically and anaerobically, $SSNO^-$ is formed.[90] Based on authors' claims, $SSNO^-$ can be obtained in high yields in aqueous solutions at pH 7.4 (even in the presence of oxygen) and it is stable for hours. It decomposes slowly in time to give HNO, NO, and polysulfides, among other products.[90] However, everything known about the chemical/spectroscopic properties of SSNO contradicts authors' conclusions. This paper suffers from serious methodological errors; the authors failed to prove HSNO using contaminated demo Ms instrument with background noise at the detection limit of that particular instrument, the same spectra were used to explain the existence of isotopic distribution of two different species, etc. The results also fail the test of basic chemical scrutiny.

Namely, a history of the identification of S–S bond-containing compounds in solutions was rich in contradicting conclusions,[91] and "some of those who dared to tackle this challenging task were victims of delusions because such species that were optically (or even by other methods) observed in nonaqueous solutions could not easily be established as defined substances."[92] This is a translated quotation from Seel and Wagner, who were the first to synthesize $SSNO^-$ salt under exclusion of water and dioxygen.[93] However, although citing work of Seel and Wagner, Cortese-Krott and colleagues do not mention that (i) $SSNO^-$ solutions are sensitive to oxygen and water[92] and (ii) even in very alkaline aqueous solutions only 10 % of $SSNO^-$ was obtained from NO and S_2^-, although even that was questionable, as $SSNO^-$ could not be confirmed by ^{15}N-NMR in aqueous solutions.[94] This contradicts that $SSNO^-$ is stable for hours at pH 7.4, especially in high concentrations in "$SSNO^-$-enriched mixtures" ("$SSNO^-$ mix").[90]

In agreement with the original work of Seel and Wagner, Wedmann et al. demonstrated $SSNO^-$ inherent instability by preparing pure crystals of PNP^+SSNO^- and characterizing its properties by ^{15}N-NMR, IR, EPR, Ms, X-ray analysis, electrochemical, and computational methods.[95] For example, when ca. 10% water was added to an acetone solution of a pure $SSNO^-$ salt, it decomposed within ca. 100s.[95] Cortese-Krott et al. report that $SSNO^-$ does not react with thiols, H_2S and cyanide.[90] However, solutions of a pure SSNO- salt, which Cortese-Krott and colleagues never used, quickly decompose in the presence of thiols, H_2S, and cyanide.[95] Furthermore, $SSNO^-$ is reduced at a physiological potential of −0.2 V vs. NHE.[95] Being unstable at pH 7.4, in the presence of thiols and biological reducing agents, $SSNO^-$ cannot exist under physiological conditions in any relevant concentration.

It is claimed that HSSNO is more stable than HSNO, because HSSNO supposedly has increased electron density on the proximal sulfur (although no experimental support for such claim was provided) and therefore does not easily react with HS^- and positive metal centers (which is contradictio in adjecto).[90] The facts are quite different: (i) the proximal-S has a +0.24 charge, (ii) the S–N bonds in HSSNO and SSNO− (calculated BDE 16.0 and 22.1 kcal/mol, respectively; B3LYP/aug-cc-pv5z, in the presence of solvent/water) are weaker than those in HSNO and SNO − (BDE 27.74 and 36.21 kcal/mol, respectively), which makes (H)SSNO more prone to homolysis than (H)SNO, and (iii) SSNO− reacts with metal centers (as evidenced by the reaction with [Fe3 + (TPP)]).[95] Furthermore, Cortese-Krott and colleagues have previously observed that (H)SNO forms at room temperature from a 1:1 mixture of RSNO and sulfide in water (pH 7.4) at even higher yield than their "$SSNO^-$."[96]

It remains unclear why the authors did not use any other structural and analytical method to prove $SSNO^-$, since it was so stable as they claim, or why they did not prepare pure $SSNO^-$ to compare the reactivity of their undefined mixture with the reactivity of the pure compound.

6.4 Conclusions

Relationship between HNO and thiols (and H_2S) is dual, thiols (and H_2S) could be involved in HNO formation but they are also the main targets for HNO bioactivity. The role of H_2S (and other thiols as well), in endogenous formation of HNO warrants further investigation. Despite the widespread opinion, a proton-coupled nucleophilic attack under physiological conditions may seem like a rather feasible way. Due to the high reactivity of HNO, it is expected that its steady state concentration in the cells should be several orders of magnitude lower than that of NO. Therefore, the "physiological" redox potential for NO, H^+/HNO couple is probably even more positive than that of O_2/O_2^-, i.e. easily accessible. With the expanding arsenal of analytical tools for HNO detection in both solution and cells there is no doubt that we are getting closer to answering that question. Modifications of thiol residues on several proteins are accounted as responsible for HNO's cardiovascular effects, but many more have certainly passed unnoticed. This is particularly true for those proteins that contain only one reactive cysteine and therefore the product of the reaction with HNO is sulfinamide. Due to the lack of analytical tools to study sulfonamides their formation in the cells is completely unexplored field. Development of a new method for a wide proteomic approach could lead to identification of endogenous targets for HNO formation, as well as for potential pharmacological applications.

References

1. Bartberger MD, Fukuto JM, Houk KN. On the acidity and reactivity of HNO in aqueous solution and biological systems. *Proc Natl Acad Sci USA* 2001;**98**:2194–8.
2. Jackson MI, Han TH, Dutton A, Ford E, Miranda KM, Houk KN, et al. Kinetic feasibility of nitroxyl (HNO) reduction by physiological reductants and biological implications. *Free Radic Biol Med.* 2009;**47**:1130–9.
3. Miranda KM, Paolocci N, Katori T, Thomas DD, Ford E, Bartberger MD, et al. A biochemical rationale for the discrete behavior of nitroxyl and nitric oxide in the cardiovascular system. *Proc Natl Acad Sci USA* 2003;**100**:9196–201.
4. Flores-Santana W, Salmon DJ, Donzelli S, Switzer CH, Basudhar D, Ridnour L, et al. The specificity of nitroxyl chemistry is unique among nitrogen oxides in biological systems. *Antioxid Redox Signal.* 2011;**14**(9):1659–74.
5. DeMaster EG, Kaplan E, Shirota FN, Nagasawa HT. Metabolic activation of cyanamide by liver mitochondria, a requirement for the inhibition of aldehyde dehydrogenase enzymes. *Biochem Biophys Res Commun.* 1982;**107**:1333–9.
6. DeMaster EG, Redfern B, Nagasawa HT. Mechanisms of inhibition of aldehyde dehydrogenase by nitroxyl, the active metabolite of the alcohol deterrent agent cyanamide. *Biochem Pharmacol.* 1998;**55**:2007–15.
7. Hammond AH, Fry JR. Effect of cyanamide on toxicity and glutathione depletion in rat hepatocyte cultures: differences between two dichloropropanol isomers. *Chem Biol Interact.* 1999;**122**:107–15.
8. Lopez BE, Rodriguez CE, Pribadi M, Cook NM, Shinyashiki M, Fukuto JM. Inhibition of yeast glycolysis by nitroxyl (HNO): mechanism of HNO toxicity and implications to HNO biology. *Arch Biochem Biophys.* 2005;**442**:140–8.

9. Lopez BE, Wink DA, Fukuto JM. The inhibition of glyceraldehyde-3-phosphate dehydrogenase by nitroxyl (HNO). *Arch Biochem Biophys.* 2007;**465**:430–6.
10. Norris AJ, Sartippour MR, Lu M, Park T, Rao JY, Jackson MI, et al. Nitroxyl inhibits breast tumor growth and angiogenesis. *Int. J. Cancer.* 2008;**122**(8):1905–10.
11. Doyle MP, Mahapatro SN, Broene RD, Guy JK. Oxidation and reduction of hemoproteins by trioxodinitrate(II). The role of nitrosyl hydride and nitrite. *J Am Chem Soc.* 1988;**110**:593–9.
12. Miranda KM. The chemistry of nitroxyl (HNO) and implications in biology. *Coord Chem Rev.* 2005;**249**:433–55.
13. Fukuto JM, Carrington SJ. HNO signalling mechanisms. *Antioxid. Redox. Signal.* 2011;**14**:1649–57.
14. Wong PSY, Hyun J, Fukuto JM, Shirota FN, DeMaster EG, Shoeman DW, et al. Reaction between S-Nitrosothiols and Thiols: Generation of Nitroxyl (HNO) and Subsequent Chemistry. *Biochemistry.* 1998;**37**:5362–71.
15. Shen B, English AM. Mass spectrometric analysis of nitroxyl-mediated protein modification: comparison of products formed with free and protein-based cysteines. *Biochemistry.* 2005;**44**:14030–44.
16. Sherman MP, Grither WR, McCulla RD. Computational investigation of the reaction mechanisms of nitroxyl and thiols. *J Org Chem.* 2010;**75**:4014–24.
17. Hoffman MD, Walsh GM, Rogalski JC, Kast J. Identification of nitroxyl-induced modifications in human platelet proteins using a novel mass spectrometric detection method. *Mol Cell Proteomics.* 2009;**8**:887–903.
18. Andrews KL, Irvine JC, Tare M, Apostolopoulos J, Favaloro JL, Triggle CR, et al. A role for nitroxyl (HNO) as an endothelium- derived relaxing and hyperpolarizing factor in resistance arteries. *Br J Pharmacol.* 2009;**157**:540–50.
19. Ellis A, Li CG, Rand MJ. Differential actions of L-cysteine on responses to nitric oxide, nitroxyl anions and EDRF in the rat aorta. *Br J Pharmacol.* 2000;**129**:315–22.
20. Favaloro JL, Kemp–Harper BK. The nitroxyl anion (HNO) is a potent dilator of rat coronary vasculature. *Cardiovasc Res.* 2007;**73**:587–96.
21. Fukuto JM, Chiang K, Hszieh R, Wong P, Chaudhuri G. The pharmacological activity of nitroxyl: A potent vasodilator with activity similar to nitric oxide and/or endothelium-derived relaxing factor. *J Pharmacol Exp Ther.* 1992;**263**:546–51.
22. Li CG, Karagiannis J, Rand MJ. Comparison of the redox forms of nitrogen monoxide with the nitrergic transmitter in the rat anococcygeus muscle. *Br J Pharmacol.* 1999;**127**:826–34.
23. Wanstall JC, Jeffery TK, Gambino A, Lovren F, Triggle CR. Vascular smooth muscle relaxation mediated by nitric oxide donors: a comparison with acetylcholine, nitric oxide and nitroxyl ion. *Br J Pharmacol.* 2001;**134**:463–72.
24. Irvine JC, Favaloro JL, Widdop RE, Kemp–Harper BK. Nitroxyl anion donor, Angeli's salt, does not develop tolerance in rat isolated aortae. *Hypertension.* 2007; **49**:885–92.
25. Zhu G, Groneberg D, Sikka G, Hori D, Ranek MJ, Nakamura T, et al. Soluble guanylate cyclase is required for systemic vasodilation but not positive inotropy induced by nitroxyl in the mouse. *Hypertension.* 2015;**65**(2):385–92.
26. Miller TW, Cherney MM, Lee AJ, Francoleon NE, Farmer PJ, King SB, et al. The effects of nitroxyl (HNO) on soluble guanylate cyclase activity interactions at ferrous heme and cysteine thiols. *J Biol Chem.* 2009;**284**:21788–96.
27. Morris HR, Panico M, Etienne T, Tippins J, Girgis SI, MacIntyre I. Isolation and characterization of human calcitonin gene-related peptide. *Nature.* 1984;**308**:746–8.

28. Matteoli M, Haimann C, Torri-Tarelli F, Polak JM, Ceccarelli B, De Camilli P. Differential effect of alphalatrotoxin on exocytosis from small synaptic vesicles and from large dense-core vesicles containing calcitonin gene-related peptide at the frog neuromuscular junction. *Proc Natl Acad Sci USA* 1989;**85**:7366–70.
29. Russell FA, King R, Smillie SJ, Kodji X, Brain SD. Calcitonin gene-related peptide: physiology and pathophysiology. *Physiol Rev.* 2014;**94**(4):1099–142.
30. Chang CP, Pearse Ii RV, O'Connell S, Rosenfeld MG. Identification of a seven trans-membrane helix receptor for corticotropin-releasing factor and sauvagine in mammalian brain. *Neuron.* 1993;**11**:1187–95.
31. Caterina MJ, Julius D. The vanilloid receptor: a molecular gateway to the pain pathway. *Annu Rev Neurosci.* 2001;**24**:487–517.
32. Caterina MJ, et al. The capsaicin receptor: a heat-activated ion channel in the pain pathway. *Nature.* 1997;**389**:816–24.
33. Ramsey IS, Delling M, Clapham DE. An introduction to TRP channels. *Annu. Rev. Physiol.* 2006;**68**:619–47.
34. Zygmunt PM, Högestätt ED. TRPA1. *Handb Exp Pharmacol.* 2014;583–630.
35. Story GM, Peier AM, Reeve AJ, Eid SR, Mosbacher J, Hricik TR, et al. ANKTM1, a TRP-like Channel Expressed in Nociceptive Neurons, Is Activated by Cold Temperatures. *Cell.* 2003;**112**:819–29.
36. Bautista DM, Jordt SE, Nikai T, Tsuruda PR, Read AJ, Poblete J, et al. TRPA1 mediates the inflammatory actions of environmental irritants and proalgesic agents. *Cell.* 2006;**124**:1269–82.
37. Bautista DM, Movahed P, Hinman A, Axelsson HE, Sterner O, Högestätt ED, et al. Pungent products from garlic activate the sensory ion channel TRPA1. *Proc Natl Acad Sci USA* 2005;**102**:12248–52.
38. Hinman A, Chuang HH, Bautista DM, Julius D. TRP channel activation by reversible covalent modification. *Proc. Natl. Acad. Sci. USA* 2006;**103**:19564–8.
39. Macpherson LJ, Dubin AE, Evans MJ, Marr F, Schultz PG, Cravatt BF, et al. Noxious compounds activate TRPA1 ion channels through covalent modification of cysteines. *Nature.* 2007;**445**:541–5.
40. Wang L, Cvetkov TL, Chance MR, Moiseenkova-Bell VY. Identification of in vivo Disulfide Conformation of the TRPA1 Ion Channel. *J Biol Chem.* 2012;**287**: 6169–76.
41. Cvetkov TL, Huynh KW, Cohen MR, Moiseenkova-Bell VY. Molecular Architecture and Subunit Organization of TRPA1 Channel Revealed by Electron Microscopy. *J. Biol. Chem.* 2011;**286**:38168–76.
42. Eberhardt M, Dux M, Namer B, Miljkovic J, Cordasic N, Will C, et al. H2S and NO cooperatively regulate vascular tone by activating a neuroendocrine HNO-TRPA1-CGRP signalling pathway. *Nat Commun.* 2014;**15**:4381.
43. Edvinsson L. Calcitonin gene-related peptide (CGRP) and the pathophysiology of headache: therapeutic implications. *CNS drugs* 2001;**15**:745–53.
44. Elvidge S. Anti-CGRP antibodies for migraine turn industry heads. *Nat Biotechnol.* 2014;**32**(8):707.
45. Dux M, Will C, Vogler B, Filipovic MR, Messlinger K. Meningeal blood flow is controlled by H2S-NO crosstalk activating a HNO-TRPA1-CGRP signalling pathway. *Br J Pharmacol.* 2015. Available from: http://dx.doi.org/10.1111/bph.13164
46. Paolocci N, Saavedra WF, Miranda KM, Martignani C, Isoda T, Hare JM, et al. Nitroxyl anion exerts redox-sensitive positive cardiac inotropy in vivo by calcitonin gene-related peptide signaling. *Proc. Natl. Acad. Sci. USA* 2001;**98**:10463–8.

47. Paolocci N, Katori T, Champion HC, St John ME, Miranda KM, Fukuto JM, et al. Positive inotropic and lusitropic effects of HNO/NO- in failing hearts: independence from beta-adrenergic signaling. *Proc. Natl. Acad. Sci. USA* 2003;**100**:5537–42.
48. Sabbah HN, Tocchetti CG, Wang M, Daya S, Gupta RC, Tunin RS, et al. Nitroxyl (HNO): A novel approach for the acute treatment of heart failure. *Circ Heart Fail* 2013;**6**(6):1250–8.
49. Cheong E, Tumbev V, Abramson J, Salama G, Stoyanovsky DA. Nitroxyl triggers Ca^{2+} release from skeletal and cardiac sarcoplasmic reticulum by oxidizing ryanodine receptors. *Cell Calcium.* 2005;**37**:87–96.
50. Tocchetti CG, Wang W, Froehlich JP, Huke S, Aon MA, Wilson GM, et al. Nitroxyl improves cellular heart finction by directly enhancing cardiac sarcoplasmic reticulum Ca2 + cycling. *Circ Res.* 2007;**100**:96–104.
51. Donoso P, Sanchez G, Bull R, Hidalgo C. Modulation of cardiac ryanodine receptor activity by ROS and RNS. *Front Biosci (Landmark Ed).* 2011;**16**:553–67.
52. Kohr MJ, Kaluderic N, Tocchetti CG, Gao WD, Kass DA, Janssen PML, et al. Nitroxyl enhances myocyte Ca2 + transients by exclusively targeting SR Ca2 + -cylcing. *Front Biosci.* 2010;**E2**:614–26.
53. Lancel S, Zhang J, Evangelista A, Trucillo MP, Tong X, Siwik DA, et al. Nitroxyl activates SERCA in cardiac myocytes via glutathiolation of cysteine 674. *Circ Res.* 2009;**104**(6):720–3.
54. Gao WD, Murray CI, Tian Y, Zhong X, DuMond JF, Shen X, et al. Nitroxyl-mediated disulfide bond formation between cardiac myofilament cysteines enhances contractile function. *Circ Res.* 2012;**111**(8):1002–11.
55. Shiva S, Crawford JH, Ramachandran A, Ceaser EK, Hillson T, Brookes PS, et al. Mechanisms of the interaction of nitroxyl with mitochondria. *Biochem J.* 2004;**379**(Pt 2):359–66.
56. Queliconi BB, Wojtovich AP, Nadtochiy SM, Kowaltowski AJ, Brookes PS. Redox regulation of the mitochondrial K(ATP) channel in cardioprotection. *Biochim Biophys Acta.* 2011;**1813**(7):1309–15.
57. Arnelle DR, Stamler JS. NO +, N.O., and NO − Donation by S-Nitrosothiols: Implications for Regulation of Physiological Functions by S-Nitrosylation and Acceleration of Disulfide Formation. *Arch Biochem Biophys.* 1995;**318**:279–85.
58. Seth D, Stamler JS. The SNO-proteome: causation and classifications. *Curr Opin Chem Biol.* 2011;**15**:129–36.
59. Hess DT, Stamler JS. Regulation by S-nitrosylation of protein post-translational modification. *J Biol Chem.* 2012;**287**:4411–18.
60. Foster MW, Hess DT, Stamler JS. Protein S-nitrosylation in health and disease: a current perspective. *Trends Mol Med.* 2009;**15**:391–404.
61. Talipov MR, Timerghazin QK. Protein control of S-nitrosothiol reactivity: interplay of antagonistic resonance structures. *J Phys Chem B.* 2013;**117**:1827–37.
62. Abe K, Kimura H. The possible role of hydrogen sulfide as an endogenous neuromodulator. *J Neurosci.* 1996;**16**:1066–71.
63. Li L, Hsu A, Moore PK. Actions and interactions of nitric oxide, carbon monoxide and hydrogen sulphide in the cardiovascular system and in inflammation-a tale of three gases! *Pharmacol Ther.* 2009;**123**:386–400.
64. Filipovic MR. H2S and persulfidation (S-sulfhydration). *Handbook Exp. Pharmacol.* 2015;**230**:29–59.
65. Mathai JC, Missner A, Kugler P, et al. No facilitator required for membrane transport of hydrogen sulfide. *Proc Natl Acad Sci USA* 2009;**106**:16633–8.

66. Cuevasanta E, Denicola A, Alvarez B, Moller MN. Solubility and permeation of hydrogen sulfide in lipid membranes. *PLoS One.* 2017;**7**:e34562.
67. Filipovic MR, Miljkovic JLj, Nauser T, Royzen M, Klos K, Shubina T, et al. Chemical Characterization of the Smallest S-Nitrosothiol, HSNO; Cellular Cross-talk of H2S and S-Nitrosothiols. *J Am Chem Soc.* 2012;**134**:12016–27.
68. Rosenthal J, Lippard S. Direct Detection of Nitroxyl in Aqueous Solution using a TripodalCopper(II) BODIPY Complex. *J Am Chem Soc.* 2010;**132**:5536–7.
69. Filipovic MR, Eberhardt M, Prokopovic V, Mijuskovic A, Orescanin Dusic O, et al. Beyond H2S and N.O interplay: hydrogen sulfide and nitroprusside react directly to give nitroxyl (HNO). A new pharmacological source of HNO. *J Med Chem.* 2013;**56**:1499–508.
70. Gao Y, Toubaei A, Kong X, Wu G. Solving the 170-Year-Old Mystery About Red-Violet and Blue Transient Intermediates in the Gmelin Reaction. *Chemistry.* 2015;**21** (48):17172–7.
71. Longhi R, Ragsdale RO, Drago RS. Reactions of Nitrogen(II) Oxide with Miscellaneous Lewis Bases. *Inorg. Chem.* 1962;**1**:768–70.
72. Raihlen H, Friedolsheim A, Oswald W. Über Stickoxyd- und Kohlenoxydverbindungen des scheinbar einwertigen Eisens und Nickels. *Zugleich Erwiderung an die Herren W. Manchot und W. HieberJustus Liebigs Ann. Chem.* 1928;**465**:72–96.
73. Pryor WA, Church DF, Govindan CK, Crank G. Oxidation of thiols by nitric oxide and nitrogen dioxide: Synthetic utility and toxicological implications. *J. Org. Chem.* 1982;**47**:156–9.
74. Aravindakumar C, Ley, Ceulmans J. Kinetics of the anaerobic reaction of nitric oxide with cysteine, glutathione and cysteine-containing proteins: implication for in vivo S-nitrosation. *J. Chem. Soc. Perkin Trans.* 2002;**2**:663–9.
75. Zhao YL, Houk KN. Thionitroxides, RSNHO.: the structure of the SNO moiety in "S-nitrosohemoglobin", a possible NO reservoir and transporter. *J. Am. Chem. Soc.* 2006;**128**:1422–3.
76. Chan NL, Kavanaugh JS, Rogers PH, Arnone A. Crystallographic analysis of the interaction of nitric oxide with quaternary-T human hemoglobin. *Biochemistry.* 2004; **43**:118–32.
77. Arulsamy N, Bohle DS, Butt JA, Irvine GJ, Jordan PA, Sagan E. Interrelationships between Conformational Dynamics and the Redox Chemistry of S-Nitrosothiols. *J. Am. Chem. Soc.* 1999;**121**:7115–23.
78. Zhao YL, Bartberger MD, Goto K, Shimada K, Kawashima T, Houk KN. Theoretical evidence for enhanced NO dimerization in aromatic hosts: implication for the role of the electrophile (NO)2 in nitric oxide chemistry. *J. Am. Chem. Soc.* 2005;**127**:7964–5.
79. Yong QC, Hu LF, Wang S, Huang D, Bian JS. Hydrogen sulfide interacts with nitric oxide in the heart: possible involvement of nitroxyl. *Cardiovasc Res.* 2010;**88**:482–91.
80. Yong QC, Cheong JL, Hua F, Deng LW, Khoo YM, Lee HS, et al. Regulation of heart function by endogenous gaseous mediators-crosstalk between nitric oxide and hydrogen sulfide. *Antioxid Redox Signal.* 2011;**14**:2081–91.
81. Koppenol W. Nitrosation, thiols, and hemoglobin: energetics and kinetics. *Inorg Chem.* 2012;**51**(10):5637–41.
82. Suárez SA, Bikiel DA, Wetzler D, Martí MA, Doctorovich F. Time-resolved eelectrochemical quantification of azanone (HNO) at low nanomolar level. *Anal Chem.* 2013;**85**:10262–9.
83. Wood KC, Batchelor AM, Bartus K, Harris KL, Garthwaite G, Vernon J, et al. Picomolar nitric oxide signals from central neurons recorded using ultrasensitive detector cells. *J Biol Chem.* 2011;**286**:43172–81.

84. Coletta C, Papapetropoulos A, Erdelyi K, Olah G, Modis K, Panopoulos P, et al. Hydrogen sulfide and nitric oxide are mutually dependent in the regulation of angiogenesis and endothelium-dependent vasorelaxation. *Proc Natl Acad Sci USA* 2012;**109**: 9161–6.
85. Medani M, Collins D, Docherty NG, Baird AW, O'Connell PR, Winter DC. Emerging role of hydrogen sulfide in colonic physiology and pathophysiology. *Inflamm Bowel Dis.* 2011;**17**:1620–5.
86. Engel MA, Leffler A, Niedermirtl F, Babes A, Zimmermann K, Filipovic MR, et al. TRPA1 and substance P Mediate Colitis in Mice. *Gastroenterol* 2011;**141**:1346–58.
87. Engel MA, Khalil M, Siklosi N, Mueller-Tribbensee SM, Neuhuber WL, Neurath MF, et al. Opposite effects of substance P and calcitonin gene-related peptide in oxazolone colitis. *Dig Liv Dis* 2012;**44**:24–9.
88. Chattopadhyay M, Kodela R, Olson KR, Kashfi K. NOSH-aspirin (NBS-1120), a novel nitric oxide- and hydrogen sulfide-releasing hybrid is a potent inhibitor of colon cancer cell growth in vitro and in a xenograft mouse model. *Biochem Biophys Res Commun.* 2012;**419**(3):523–8.
89. Kodela R, Chattopadhyay M, Kashfi K. NOSH-Aspirin: A Novel Nitric Oxide-Hydrogen Sulfide-Releasing Hybrid: A New Class of Anti-inflammatory Pharmaceuticals. *ACS Med Chem Lett* 2012;**3**(3):257–62.
90. Cortese-Krott MM, Kuhnle GG, Dyson A, Fernandez BO, Grman M, DuMond JF, et al. Key bioactive reaction products of the NO/H2S interaction are S/N-hybrid species, polysulfides, and nitroxyl. *Proc Natl Acad Sci USA* 2015;**112**(34):E4651–60.
91. Seel F, Guttler H-J, Simon G, Wieckowski A. Colored sulfur species in EPD-solvents. *Pure Appl Chem.* 1977;**49**:45–54.
92. Seel F, Wagner M. The reaction of polysulfides with nitrogen monoxide in non-acqueous solvents-nitrosodisulfides. *Z Naturforsch.* 1985;**40b**:762–4.
93. Seel F, Kuhn R, Simon G, Wagner M. PNP-Perthionitrit und PNP-Monothionitrit. *Z Naturforsch.* 1985;**40b**:1607–17.
94. Seel F, Wagner M. Reaction of sulfides with nitrogen monoxide in aqueous solution. *Z Anorg Allg Chem.* 1988;**558**(3):189–92.
95. Wedmann R, et al. Does Perthionitrite (SSNO(-)) Account for Sustained Bioactivity of NO? A (Bio)chemical Characterization. *Inorg Chem.* 2015;**54**(19):9367–80.
96. Cortese-Krott MM, Fernandez BO, Santos JL, Mergia E, Grman M, Nagy P, et al. Nitrosopersulfide (SSNO(-)) accounts for sustained NO bioactivity of S-nitrosothiols following reaction with sulfide. *Redox Biol.* 2014;**2**:234–44.

Non-Heme Transition Metal Complexes of HNO

7

J.A. Olabe, S.E. Bari and L.D. Slep
Universidad de Buenos Aires, Buenos Aires, Argentina; CONICET, Buenos Aires, Argentina

Chapter Outline

Coordination to transition metals is an important chemical property of nitroxyl[1–3] given that its reactivity is significantly modified upon binding, with important implications in chemistry and biochemistry.[4,5] In pioneering studies on the redox interconversions of the nitrosyl group, Roper investigated the reactions of four coordinated (4C) $Ir(NO)(PPh_3)_3$ leading to 6C $IrCl_3(NH_2OH)(PPh_3)_2$.[6] The intermediate *cis, trans*-$IrHCl_2$ $(HNO)(PPh_3)_2$ was isolated and characterized later by Hillhouse.[7] On the other hand, a deep electrochemical mechanistic insight into the reversible transformations of coordinated NH_3 to NO_2^- was provided by Meyer through the modeling of the active site of assimilatory nitrite reductases with polypyridyl complexes $[M^{II}(bpy)(terpy)(L)]^{n+}$ (M = Ru, Os; bpy = 2,2′-bipyridine, terpy: 2,2′,2″-terpyridine; L = NH_3, NO^+, NO_2^-). Nitroxyl and other

The Chemistry and Biology of Nitroxyl (HNO). DOI: http://dx.doi.org/10.1016/B978-0-12-800934-5.00007-4

species were proposed as intermediates reacting through successive one-electron and/or proton-coupled steps in an overall six-electron process.[8a] Related results were obtained with the water-soluble complex $[Fe^{III}(H_2O)(TPPS)]^{3-}$ (H_2TPPS^{4-} = tetraanionic form of *meso*-tetrakis(*p*-sulfonatophenyl)porphyrin) through the electrocatalytic reduction of NO_2^- to NH_3 giving NH_2OH and N_2O as significant products.[8b]

The isolation of $OsCl_2(CO)(HNO)(PPh_3)_2$ [6] led to Wilson and Ibers providing for the first time X-ray structural data together with IR and crucial ^1H and ^{31}P NMR (nuclear magnetic resonance) spectroscopic information.[9] This comprehensive work allowed characterizing properly HNO as a ligand (Section 7.2). Only three X-ray structures of HNO complexes with Os, Ru, and Ir metal centers have been solved by X-ray methods. Table 7.1 shows the more significant distances and angle, together with relevant spectroscopic indicators.[7,9,10] We include the results on a more recent characterization of the first nonheme iron complex prepared in aqueous solution, $[Fe(CN)_5(HNO)]^{3-}$ [11] as well as data for the $[Ru(Me_3[9]aneN_3)(bpy)(HNO)]^{2+}$ complex affording a metal/coligand platform that allows coordination and identification of other nitrosyl species (NO^+, NO, NO^-) and NO_2^-.[12] For the sake of useful comparisons, diversity and significance, we include DFT computational results for the $[Fe(cyclam\text{-}ac)(HNO)]^{2+}$ ion,[13] for two rhenium complexes,[14] and for the heme protein derivative Mb^{II}-HNO (see chapter: Spectroscopic NMR Characterizations of HNO Adducts of Ferrous Heme Proteins).[1,2]

7.1 Synthetic methods

Different synthetic procedures in aqueous or nonaqueous solutions under anaerobic conditions proved useful for the preparation and characterization of nitroxyl complexes according to the following resumed methodologies.[1–3]

7.1.1 Chemical/electrochemical reduction of metallonitrosyls

Fig. 7.1 describes a titration experiment using two equivalents of dithionite, $S_2O_4^{2-}$, for the reduction of the nitroprusside ion $[Fe(CN)_5(NO)]^{2-}$, showing the intermediate formation of the one-electron reduced complex $[Fe(CN)_5(NO^•)]^{3-}$ and the final generation of $[Fe(CN)_5(HNO)]^{3-}$. The HNO complex shows an intense band centered at 445 nm with $\varepsilon = 4.2 \times 10^3\ M^{-1}\ cm^{-1}$ assigned to a metal-to-ligand charge transfer (MLCT) transition from Fe(II) to the π^*HNO orbital, a typical feature of the series of related $[Fe^{II}(CN)_5(L)]^{n-}$ complexes containing L chromophores absorbing in the visible region like nitrite, nitrosobenzene, pyridine, pyrazine and derivatives.[16] The robust character of the Fe–N bond (Section 7.5.1) precludes HNO release and subsequent fast N_2O formation through dehydrative dimerization,[4] a dramatic consequence of the changes in stability of HNO upon coordination.

Table 7.1 **Structural and spectroscopic (IR, NMR) data for selected 6C M-HNO complexes**

Compound	ν_{NH} $^{14}N/^{15}N$ (cm^{-1})	ν_{NO} $^{14}N/^{15}N$ (cm^{-1})	d_{M-N} (Å)	d_{N-O} (Å)	∠MNO (deg)	δ^1H/J_{15NH} (ppm/Hz)	References
cis,trans-$IrHCl_2(NHO)(PPh_3)_2$	2810/2801	1493/1465^a	1.879(7)	1.235(11)	129.8(7)	22.75/77.6^b	[7]
$OsCl_2CO(NHO)(PPh_3)_2$^{c}	–	1410/1393^d	1.915(6)	1.193(7)	136.9(6)	21.2/75^e	[9]
Ru(NHO)('py^{bu}S_4')	–	1358/1378^f	1.875(7)	1.242(9)	130.0(6)	19.56^g/ –	[10]
$[Ru(Me_3[9]aneN_3)(bpy)(NHO)]^{2+h}$		1376	1.93	1.28	127.5	–	[12]
$[Fe(CN)_5(NHO)]^{3-}$	–	1384/1352^i	1.783^h	1.249^h	137.5^h	20.0/71.14^j	[11]
$[Fe(cyclam\text{-}ac)(NHO)]^{+k}$	–	1351	1.78	1.27	126.3	–	[13]
$Mb^{II}(NHO)$	–	1385^l	1.82^m	1.24^m	131^m	14.8/72.5^n	[1,2]
$[ReCl(CO)_2(NHO)(PPh_3)_2]$	3059	1376°	–	–	–	20.66/66.2^b	[14]
$[Re(CO)_3(NHO)(PPh_3)_2]^+$	3056	1391	–	–	–	21.7/72.5^b	[15]

Abbreviations: bpy = 2,2′-bipyridine; cyclam-ac = 1,4,8,11-tetraazacyclotetradecane-1-acetate pentaanion; $Me_3[9]aneN_3$ = 1,4,7-trimethyl-1,4,7-triazacyclononane; Mb = myoglobin; PPh_3 = triphenylphosphine; 'py^{bu}S_4' = 2,6-Bis(2-mercapto-3,5-di-*tert*-butylphenylthio)dimethylpyridine dianion.
^{a}In Fluorolube/CaF_2 mull.
^{b}In CD_2Cl_2.
^{c}Value for d_{NH} = 0.94 (11) Å.
^{d}Nujol mull.
^{e}In C_6D_6.
^{f}In KBr/THF.
^{g}In $[D_8]$THF.
^{h}Data from DFT calculations, the complex was identified in water solution.
^{i}ATR/FTIR, aq pH 6.
^{j}H_2O, 25% D_2O, pH 6.
^{k}Data from DFT calculations (complex not identified in acetonitrile solution).
^{l}From resonance Raman.
^{m}From XAFS.
^{n}In H_2O.
^{o}In Fluorolube-S30 mull.
Source: Formally Re^I, Fe^{II}, Ru^{II}, Os^{II}, Ir^{III}, all singlet low-spin metal centers.

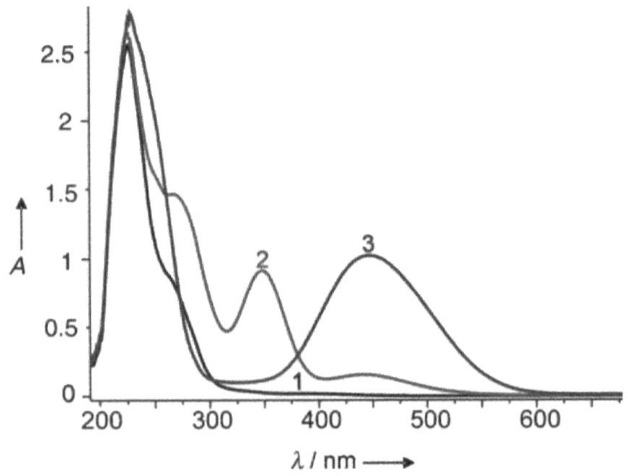

Figure 7.1 Spectra of 3×10^{-4} M $[Fe(CN)_5(NO)]^{2-}$ (1) and the one- and two-electron reduced complexes $[Fe(CN)_5(NO)]^{3-}$ (2) and $[Fe(CN)_5(HNO)]^{3-}$ (3), obtained by two sequential one-equivalent additions of $S_2O_4^{2-}$ at pH 10, $T = 25.0°C$, 6×10^{-3} M excess cyanide.[11]

In a related way the reduction of [Ru(NO)('py^{bu}S$_4$')]Br with BH_4^- in methanol yielded crystals of Ru(HNO)('py^{bu}S$_4$'). It could be stored at −20°C for 2 weeks though it decomposed in THF or CH_2Cl_2 solutions in 24 h at room temperature. Attempts to deprotonate bound nitroxyl were unsuccessful.[10] Chapter 9, Interactions of HNO With Metallated Porphyrins, Corroles, and Corrines, accounts for the preparation and properties of a Ru^{II}-HNO porphyrin complex using a hydride reductant.

The electrochemical generation of the $[Ru(Me_3[9]aneN_3)(bpy)(HNO)]^{2+}$ ion was achieved in aqueous solution at pH 2.5, starting from the nitrosonium (NO^+) derivative.[12] In a general way, cyclic voltammetry (CV) and spectroelectrochemical techniques have been used in aqueous and organic solvents for obtaining and characterizing bound HNO.[1,10] Chapter 14, Spectroscopic NMR Characterization of HNO Adducts of Ferrous Heme Proteins, deals with Mb^{II}-HNO and related globins, which also behave as robust systems in aqueous solutions in the pH range of 7–10.[1,2,17] The first ferrous heme–HNO model complex [Fe(3,5-Me-BAFP)(HNO)], (3,5-Me-BAFP^{2-} = 3,5-methyl-bis(aryloxy)-fence porphyrin dianion), was prepared through a bulk electrolytic reduction of [Fe(3,5-Me-BAFP)] in THF in the presence of NO and subsequent acidification with acetic acid. The picket-fence porphyrin character rendered a very stable HNO complex, with a half-life of ~5 h in solution, in contrast with the previous unsuccessful results with other related 5C low-spin iron complexes containing unhindered porphyrins.[18]

7.1.2 Oxidation of a metal–NH_2OH complex

The selective oxidation of *mer,trans*-$[Re(CO)_3(NH_2OH)(PPh_3)_2](SO_3CF_3)$ with $Pb(OAc)_4$ in CH_2Cl_2 at −78°C yielded *mer,trans*-$[Re(CO)_3(HNO)(PPh_3)_2](SO_3CF_3)$, which was stable under inert atmosphere in the solid state or in solution at $-78°C$ but decomposed at room temperature with interchange of HNO by triflate.[15] Crystals of Mo(CN)(HNO)(NO)(terpy) were obtained through proton abstraction from the η^2-hydroxylamido(2−) ligand in $[Mo(H_2O)(H_2NO)(NO)(terpy)]^{2+}$ in the presence of hydroxide and cyanide.[19]

7.1.3 Addition of HX (X = nucleophile) to low coordinated metallonitrosyl complexes

A number of trigonal bipyramidal (tbp) nitrosyl complexes react with anhydrous HX (X = Cl, Br, etc.) in dry organic solvents (diethylether, CH_2Cl_2, alcohols, etc.) leading to mixtures of 6C nitroxyl complexes with unavoidable side products. Reaction (Eq. 7.1) describes an example for a rhenium complex obtained in 85% yield by working with an excess of PPh_3 in Et_2O,[14] producing also $ReCl(CO)_3(PPh_3)_2$ (8%) and $ReCl_2(CO)(PPh_3)_2$ (7%).

$$Re(CO)_2(NO)(PPh_3)_2 + HCl \rightarrow ReCl(CO)_2(HNO)(PPh_3)_2 \qquad (7.1)$$

Reactions like (Eq. 7.1) consist in a formal protonation of NO and require that X behaves as a good nucleophile. In contrast, poorly coordinating anions such as triflate $SO_3CF_3^-$ (or ClO_4^-) lead to protonation at the metal center affording the hydride complex [*trans,trans*-$[Re(H)(CO)_2(NO)(PPh_3)_2](SO_3CF_3)$. It has been proposed that HCl reacts through a direct 1,2-addition across the Re−N multiple bond of the 5C-precursor, consistent with the observed stereochemistry in the product having Cl *cis* to the nitroxyl ligand. The low value of ν_{NO} for the 5C complex (1622 cm^{-1}) reflects a strong $d_\pi(Re) \rightarrow \pi^*NO$ back bonding with a relatively large amount of electron density at nitrogen suitable for proton binding. The $OsCl_2(CO)(HNO)(PPh_3)_2$ complex was early prepared by reacting gaseous HCl with tbp $OsCl(CO)(NO)(PPh_3)_2$ in dry degassed benzene,[6,9] and $IrHCl_2(HNO)(PPh_3)_2$ was obtained starting with $IrHCl(NO)(PPh_3)_2$, which was previously generated from $Ir(NO)(PPh_3)_3$ through the reaction with anhydrous HCl in CH_2Cl_2 at $-78°C$.[6,7]

Dehydrohalogenation of the nitroxyl complexes led back to the 5C precursors and could be achieved by using sufficiently strong Bronsted bases (pK_b ~7) such as $NaN(SiMe_3)_2$ in THF for the Ir-complex. Using ethanol or humid reagents avoided the isolation of the Ir−HNO complex and favored the ensuing reduction to the NH_2OH derivative.[6,7,14] Addition of bromide to tbp $[Co(das)_2(NO)]^{2+}$ (das = dimethylarsine) led to the 6C bent product $[CoBr(das)_2(NO)]^+$, which reacted with HBr (not with the weakly coordinating triflate) or through proton addition in methanolic solution giving $[CoBr(das)_2(HNO)]^{2+}$. The latter species was poorly characterized and was unstable toward the release of HNO.[20] The $RhCl_3(HNO)(PPh_3)_2$ [21] and the $[Mo(dpe)_2(F)(HNO)]PF_6$ (dppe = 1,2-bis(diphenylphosphanyl)ethane)[22] complexes have been similarly prepared.

7.1.4 Insertion of NO^+ into metal-hydride bonds

Reaction of $ReH(CO)_3(PPh_3)_2$ with $NO(SO_3CF_3)$ in CH_2Cl_2 at −78°C led to an alternative route for obtaining *mer,trans*-$[Re(CO)_3(HNO)(PPh_3)_2](SO_3CF_3)$, representative of 1,1-insertion reactions of nitrosonium salts with metal hydrides. Similarly, *cis,trans*-$OsBr(CO)_2(H)(PPh_3)_2$ led to $[OsBr(CO)_2(HNO)(PPh_3)_2]^+$.[23]

7.1.5 Coupling of hydride with a metallonitrosyl. Intramolecular H-transfer

Addition of NO to $RuCl(CO)(H)(PPh_3)_2$ in benzene yields $RuCl(CO)(H)(NO)(PPh_3)_2$, that converts subsequently into an equimolar mixture of $RuCl(CO)(NO)(PPh_3)_2$ and $RuCl(CO)(H)(HNO)(PPh_3)_2$. The second step of the reaction stands alone as a method for obtaining nitroxyl compounds from hydride-containing nitrosyl complexes. Similar reactions evolve with osmium complexes.[24]

Remarkably all the experimentally well-characterized nonheme HNO complexes are six-coordinated and contain transition metal centers from groups 6 to 9. Within a formal oxidation state approach, they can be described as diamagnetic low-spin d^6 systems containing M° (Mo), M^I(Re), M^{II}(Fe, Ru, Os), and M^{III}(Co, Rh, Ir) centers.[1–3]

7.2 Binding mode of HNO. Characterization by NMR, IR, and UV-vis spectroscopies, X-ray diffraction, and DFT calculations

The identification of HNO transition metal complexes can be best achieved by using a variety of spectroscopic techniques, DFT calculations and eventually X-ray structural data.[1–3] Currently used binding models imply considering the onset of σ-bonds of HNO to the metal within a Lewis acid–base approach, potentially through either N or O, with additional π-backbonding interactions. Only N-bound complexes have been identified. Table 7.1 shows that the MNO angles are in the range of 127–137°, pointing to an sp^2-hybridized N-atom in HNO. The properties of the N–H bond require a specific insight and NMR results have been crucial for their characterization.

Table 7.1 shows a generalized onset of values in the 20–23 ppm range for the ^1H chemical shifts (δ) that are extraordinary downfield for diamagnetic molecules. Fig. 7.2 shows the results for aqueous $[Fe(CN)_5(H^{14}NO)]^{3-}$ with the doublet obtained upon labeling with ^{15}N, which allowed obtaining the coupling constant, $^1J_{NH} = 71.14$ Hz.[11] Such values of $^1J_{NH}$ are as expected for hydrogen in a sp^2 hybridized N-bound $H^{15}NO$ ligand, given that the O-bound species should attain a substantially lower value at ~45 Hz.[15] Though without an X-ray structure, the NMR results defined unambiguously HNO as a ligand, complemented by other data in Table 7.1. NMR signatures for bound HNO were first assigned in $OsCl_2(CO)(HNO)(PPh_3)_2$ as a diagnostic tool on the basis of comparisons with the similarly low-field hydrogen resonances for metal-formyl complexes and for OsC(=S)H and OsC(=NR)H

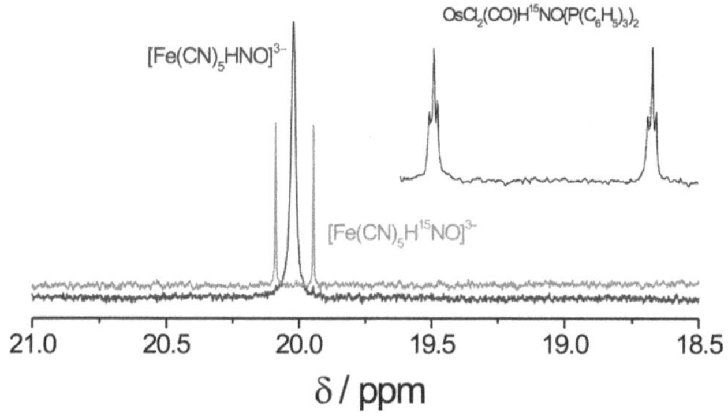

Figure 7.2 1H NMR spectrum of $[Fe(CN)_5(H^{14}NO)]^{3-}$ (in blue) (dark gray in print versions), and doublet signal centered at $\delta = 20.0$ ppm upon ^{15}N-labeling (in red) (light gray in print versions).[11] *Inset*: High-resolution view of the doublet signal centered at $\delta = 21.2$ ppm for $OsCl_2(CO)(H^{15}NO)(PPh_3)_2$, due to coupling with the phosphine ligands.[9]

systems.[9] The inset in Fig. 7.2 shows a higher resolution ^{15}N NMR spectrum for the osmium complex, which allowed detecting the coupling of hydrogen to the two equivalent phosphorus atoms of the phosphines, revealing the onset of corresponding triplets with a further 1.4-Hz splitting of the 21.2 ppm signals, also confirmed by complementary ^{31}P NMR spectra.[9] Finally, chemical shifts of ^{15}N and ^{17}O have been measured for the first time in nonheme HNO-complexes for $[Fe(CN)_5(HNO)]^{3-}$, with values at $\delta = 642.8$ ppm (relative to liquid NH_3) and 1099 ppm, respectively.[25]

X-ray results, particularly the N−O and N−H distances and MNO angles together with IR spectroscopy, are powerful tools for the characterization of nitrosyl/nitroxyl complexes. The IR measurements for the ν_{NO} stretchings are highly sensitive to bond order changes,[3,26] though care must be taken in the assignments because they are quite dependent on the geometries and binding sites of NO/HNO in the complexes, as well as on the formal charge at the metal center and the donor/acceptor abilities of the coligands. On the other hand, the values of ν_{NH}, when available, are very useful for describing the properties of the N−H bonds. In fact, the metal-HNO $\sigma-\pi$ interactions can be sensed by both ν_{NO} and ν_{NH}, and the same can be said on the values of the relevant distances, d_{N-O} and d_{N-H}, though the latter parameters are much less sensitive and/or experimentally available. Table 7.1 shows that the values of ν_{NO} are in the range of $1350-1500\ cm^{-1}$, with a decrease upon ^{15}N labeling as generally observed.[1,2] The values for the strong π-donor metals Re^{I} and M^{II} (Fe, Ru, Os) appear in a more restricted range at $\sim 1350-1400\ cm^{-1}$, and are similar to those measured in related angular Fe(II)-complexes containing the NOX ligands (X = alkyl, acyl, formamide, sulfide, thiolate), suggesting a bond order ~2 for the N−O bond in every case.[27] The upward shift of ν_{NO} for the Ir-HNO complex agrees with a poor π-donor interaction of Ir^{III} (cf. ν_{NO} at $1563\ cm^{-1}$ for free HNO). A low π-donor ability of the Co(III) center might also determine a weak Co−N bond and poor back-bonding stabilization of nitroxyl in the $[Co^{III}Br(das)_2(HNO)]^{2+}$ ion, thus accounting for the facile release of HNO in solution.[20]

Table 7.2 Calculated bond distances and N−H dissociation energies for model rhenium complexes and experimental vibrational frequencies for structurally related complexes

Compound	d_{Re-N} (Å)	d_{N-O} (Å)	d_{N-H} (Å)	ν_{NO} (cm^{-1})	ν_{NH} (cm^{-1})	D_{N-H}[a,b] kcal mol^{-1}
Free HNO	–	1.199	1.065	1563[a]	2717[a]	48.5
$[Re(CO)_5(NHO)]^{+}$ [a]	2.125	1.201	1.043	–	–	–
$[Re(CO)_3(NHO)(PH_3)_2]^{+}$ [a]	2.068	1.215	1.042	1391[c]	3056[c]	62.1
$ReCl(CO)_2(NHO)(PH_3)_2$ [a]	2.047	1.220	1.045	1376,[a,d] 1335[a,e]	3059[c]	64.8

[a]Modeling consisted in using PH_3 instead of PR_3 (R = phenyl or tricyclohexyl).[14]
[b]Calculated as: $E_{N-H} = E_{(HN=O)} - [E(H) + E(NO)]$ for free HNO, and as: $E_{N-H} = E[L_nRe(NH = O)] - \{E(H) + E[L_nRe(NO)]\}$ for the Re complexes.
[c]Experimental IR values for [*mer,trans*-$Re(CO)_3(NHO)(PPh_3)_2](SO_3CF_3)$.[15]
[d]Experimental IR values for *cis,trans*-$ReCl(CO)_2(NHO)(PPh_3)_2$.
[e]Experimental IR value for *cis,trans*-$ReCl(CO)_2(NHO)(PCy_3)_2$.

For a better understanding of the stability that metal coordination confers to nitroxyl, DFT calculations have been made with some model rhenium compounds and compared with those for free HNO.[14] The results are described in Table 7.2, and can be viewed together with those for the members of Table 7.1.

It can be seen that the Re−N bond lengths decrease (i.e., the Re−N bond order increases) when going downwards in Table 7.2, correlating with the decrease in the π-acceptor capability of the coligand set ($CO > PH_3 > Cl$). This is indicative of a competitive role of HNO with the coligands for the π-electron density and indicates an increase in π-bonding toward HNO when descending in the Table. The N−O distances increase correspondingly in comparison with the values for free HNO, in agreement with HNO behaving as a π-acceptor ligand. Further confirmation is achieved by the decreasing values of ν_{NO}, in agreement with the increasing population of the π^*_{NHO} orbital. The π-acceptor ability of HNO may be compared with that of the nitrosonium NO^+ ligand ($n = 6$), a very strong π-acceptor species[26] containing two perpendicular π^*_{NO} orbitals for accepting back-donation from the metal (*N* is sp-hybridized, compare with Fig. 7.4 below in a similar diagram for chromium, $n = 5$). Instead HNO contains only one accessible π^*_{HNO} orbital (*N* is sp^{2-} hybridized), see also Fig. 7.3A. As an illustration of the π-acceptor capability, HNO is generally oriented in the plane containing the largest number of π-acceptor ligands, as demonstrated by X-ray structural data. This allows minimizing competition for metal π-electron density and maximizing its own π-interactions.[14] A detailed DFT computational study with the heme model $[Fe^{II}(P)(HNO)(MI)]$ ($P = porphin^{2-}$, MI = 1-methylimidazole) also showed that HNO is mostly a π-acceptor ligand with weak σ-donor ability.[28]

Most significant are the trends in the values of d_{N-H}, ν_{NH} and D_{N-H} with respect to the values for free HNO. Though the changes are not much dependent on the coligand set, the decrease in d_{N-H} and the increase in ν_{NH} reveal a strengthening of the N−H bond upon coordination, also confirmed by the increase in the

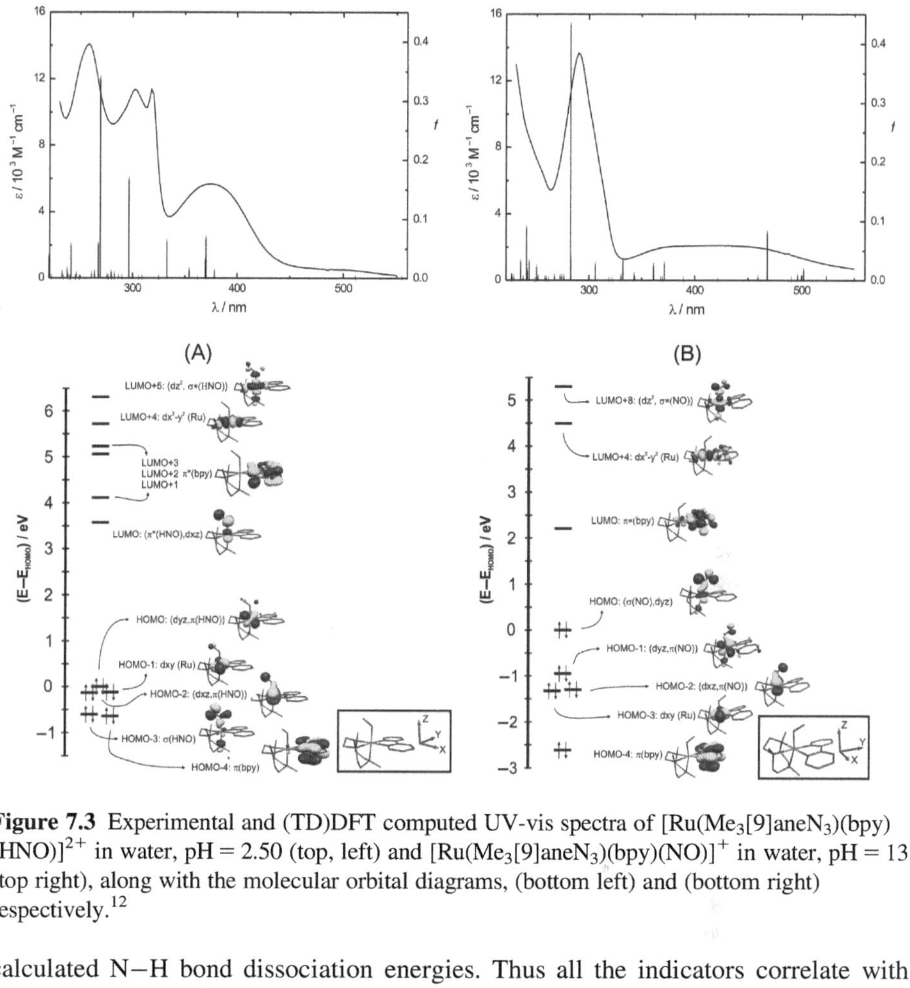

Figure 7.3 Experimental and (TD)DFT computed UV-vis spectra of $[Ru(Me_3[9]aneN_3)(bpy)(HNO)]^{2+}$ in water, pH = 2.50 (top, left) and $[Ru(Me_3[9]aneN_3)(bpy)(NO)]^{+}$ in water, pH = 13 (top right), along with the molecular orbital diagrams, (bottom left) and (bottom right) respectively.[12]

calculated N−H bond dissociation energies. Thus all the indicators correlate with the backbonding ability, as also observed when turning back to Table 7.1, in which the values of ν_{NH} (when available) show to be very sensitive to the influence of formal metal charge on the backbonding: the upward differences with respect to ν_{NH} in free HNO increase much more for the strong π-donor Re(I) complexes than for the weaker Ir(III) one.

A systematic study on the metal center effects on HNO binding has been afforded computationally while maintaining the same coligand framework, searching for the influence of electronic configuration, position in the periodic table and metal oxidation state.[29] Binding Gibbs free energies of HNO complexes with low-spin d^6 metal centers from groups 7 to 9 have been calculated. The values decrease when going to the left in the same period, and also going downwards in the same group, consistent with a stronger back donation to the π^*_{HNO} orbital. Values were calculated for the bond lengths in N−O, N−H, M−N, the computed charges in the NO-group, and the ν_{NO}'s. For the M^I centers the trends in ν_{NO} (cm^{-1}) are: 1267 (Mn) > 1248 (Tc) > 1219 (Re), while for M^{II} centers ν_{NO} decreases in the sense Fe (1399) > Ru (1376) > Os (1317). Finally the M^{III} complexes render values of 1527, 1503, and 1454 cm^{-1} for Co, Rh, and Ir, respectively. Most of these calculated numbers correspond with the experimentally available results, and agree with a generalized multiple bonding scenario within the {M(H)NO} moieties.

Fig. 7.3A shows the calculated MO diagram for $[Ru^{II}(Me_3[9]aneN_3)(bpy)(HNO)]^{2+}$ together with experimental and (TD)DFT computed UV-vis spectra in aqueous solution at pH 2.5 (top).[12] The t_{2g} set of d_{Ru} orbitals contribute to the HOMO, HOMO-1, and HOMO-2 orbitals. Though the contribution of HNO-centered orbitals to the HOMO and HOMO-1 is small (4%), the HOMO-2 comprises a large contribution from both Ru- and HNO-centered orbitals (63% and 21%) resulting from the Ru-HNO backbonding. The HOMO-3 is σ-bonding with respect to the HNO and to the axial N atom trans to the π^*_{HNO} orbital. The LUMO's feature an empty orbital mainly π^*_{HNO} in character, a set of empty bpy-centered orbitals, and the two e_g-like mostly metal-centered orbitals. The strong spectral overlap prevents a clean assignment of the whole spectrum, though the computed one is in good agreement with experiment. Accordingly the lowest energy absorption bands are mostly d_{Ru}→bpy MLCT bands. The (TD)DFT analysis also suggests the possibility of MLCT transitions involving the HNO-centered LUMO, but these bands are not clearly resolved in the experimental spectra.

From the examination of structural and spectroscopic results for non heme nitroxyl complexes and for porphyrin or heme protein analogs (see chapters: Interactions of HNO With Metallated Porphyrins, Corroles, and Corrines; Spectroscopic NMR Characterizations of HNO Adducts of Ferrous Heme Proteins) we conclude that common bonding features are established for the HNO ligand in the first coordination sphere. Second sphere effects as H-bonding to oxygen of the HNO group have been observed.[1,10] The latter interaction reinforces π-back bonding and weakens the N–O bond.[1] Nonbonding interactions of distal histidines with H and O have been studied for Mb^{II}-HNO (see chapter: Spectroscopic NMR Characterizations of HNO Adducts of Ferrous Heme Proteins).

7.3 Electronic structure of metallonitroxyl complexes containing diatomic NO^-

Deprotonation of HNO complexes lead to bound NO^-, a diatomic nitrosyl species usually called "nitroxyl anion." The electronic structure of metallonitrosyls is currently discussed in terms of the Enemark–Feltham model,[30] a molecular orbital approach focusing on the electron content of the frontier orbitals. By using the $\{MNO\}^n$ description (n is the total number of electrons occupying the metal d and/or π^*_{NO} orbitals) the gross geometrical features and main reactivity properties of a large amount of metallonitrosyls can be predicted. The influence of the coligands can be accounted separately under a fine tuning of the model.[3]

For $n \leq 6$ the pure metal d_π and π^*_{NO} orbitals are occupied and vacant, respectively. In fact their mixing allows for π-backbonding interactions, which are maximized under a linear MNO geometry. Linearity is approached by most if not all of the isolated nitrosyl compounds with $n = 6$.[3] Though the model precludes a detailed description of the electronic distribution within the M–N and N–O bonds, a goal that needs to be assessed through detailed spectroscopic measurements and/or theoretical calculations,[31] a limiting distribution M-(NO^+) is currently accepted for most of the $n = 6$ complexes, with the exception of those containing strongly σ-π donor ligands in the *trans*-position to nitrosyl like SR^- or OH^-.[1–3,31]

An illustrative example of the ambiguities arising from the application of the Enemark–Feltham model when coupled to a formal description in terms of oxidation states is given by the series of $CrL_5(NO)$ complexes (L = H_2O, CN^-, NCS^-, etc.), described as $\{CrNO\}^5$. The model predicts a linear geometry of the CrNO group as effectively found by X-ray methods for $[Cr(H_2O)_5(NO)]SO_4$ and other compounds.[32] The assignment of oxidation states in such complexes to limiting $Cr^I(NO^+)$, $Cr^{II}(NO^\bullet)$, or $Cr^{III}(NO^-)$ descriptions was highly controversial. The first one was supported by EPR spectra as well as by a value of 2.2 μ_B for the magnetic susceptibility. The value of ν_{NO} at 1747 cm^{-1}, though unable for discrimination between "NO^+" and "NO^-," was consistent with a strong backbonding picture. While DFT calculations led to oxidation states close to Cr(II) the X-ray structures and the detailed X-ray absorption measurements with several related compounds allowed choosing the $Cr^{III}NO^-$ description as the best approach, in agreement with the kinetic inertness and stability toward oxidation of the Cr(III) complexes. The parameters of the XANES spectra support the formation of a Cr–NO triple bond by overlapping two π^*_{NO} and two chromium d_π orbitals. The π-backbonding accounts for the short Cr–NO distances at ~1.68 – 1.71 Å. Fig. 7.4 shows a general MO scheme for the pseudooctahedral $CrL_5(NO)$ complexes putting in evidence the strong mixing of the d_{xz}, d_{yz}, and the π^*_{NO} orbitals and supporting an electron rich NO ligand close to NO^-.[32] Given the following discussion on the *strongly bent* MNO moieties for NO^- complexes, the message is: this is not *always* so!

In contrast with the previous analysis the one-electron and two-electron reduction of the $n = 6$ nitrosonium (NO^+) complexes afford $n = 7$ and 8 distributions. The electron content requires the occupancy of the π^*_{NO} orbitals, leading to increasingly bent geometries with MNO angles close to 140° ($n = 7$) and 120° ($n = 8$).[3,26] For $n = 7$ the electronic description in terms of limiting oxidation states is more versatile than for $n = 6$. Either $Fe^{II}(NO^\bullet)$, $Fe^{III}(NO^-)$ or $Fe^I(NO^+)$ distributions have been proposed for different nitrosyl complexes, with the first one dominantly attained under strong ligand fields. On the other hand, the weak-field "brown-ring" complex $[Fe(H_2O)_5(NO)]^{2+}$ ($n = 7$, total spin $S = 1$) is best described as an $\{Fe^{III}NO^-\}^7$ species. The reader might appreciate a detailed analysis on the spectroscopic and computational evidence for describing these electronic structures in the recent reviews.[3,26,31]

Most relevant to our aims in this Chapter, the low-spin complexes with $n = 8$ use to attain a dominant M–NO^- electronic distribution.[3] Table 7.3 contains structural and IR information for selected NO^- complexes that have been characterized as 6C

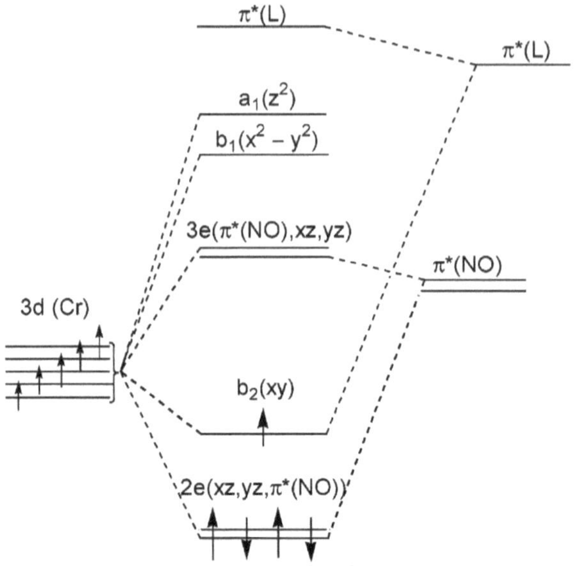

Figure 7.4 General MO orbital scheme for pseudooctahedral $Cr(NO)L_5$ complexes ($L = H_2O$, CN^-, NCS^-, etc.).[32]

or 5C species, most of them low-spin. The first structurally reported 6C compound *trans*-$[CoCl(en)_2(NO)]ClO_4$ [33] and the *trans*-$[Co(das)_2(NCS)(NO)]NCS$ analog[20] behave similarly to many other non heme Co(III) complexes as well as to the biologically relevant nitroxylcob(III)alamin derivative.[34] In agreement with the $Co^{III}NO^-$ assignment these complexes show strongly bent geometries (120−130°) and comparatively high values of ν_{NO} at ~1600 cm^{-1}, consistent with a weak back-bonding. On the other hand, the 6C $M^{II}NO^-$ complexes Fe(cyclam-ac)(NO)[13] and $[Ru(Me_3[9]aneN_3)(bpy)(NO)]^+$ [12] show lower values of ν_{NO} due to a comparatively stronger backbonding. The iron complex has been prepared in acetonitrile by electrochemical reduction of the NO^+-derivative, or using cobaltocene (Cp_2Co).[13] The DFT-optimized geometry for $[Ru^{II}(Me_3[9]aneN_3)(bpy)(NO)]^+$ shows an N−O bond length comparable to the one observed in the HNO-derivative, but the Ru−N distance is 0.02 Å shorter. Remarkably the NO^- complex has also a more pronounced RuNO bending than in the HNO-analog, close to 120°, consistent with the greater repulsions arising in the lone pair at nitrogen. The length of the Ru−N (trans to NO) bond is 0.19 Å greater that in the protonated derivative, pointing to a strong *trans*-effect exerted by the bound NO^-. The MO picture is consistent with the NO^- behaving as a donor fragment involving its HOMO electrons residing in a π orbital mostly located at the N atom, antibonding with respect to the N−O bond. For the sake of comparisons, we include data in Table 7.3 for Rh^{III} [35] and Pt^{IV} [36] compounds.

As highlighted for $[Ru^{II}(Me_3[9]aneN_3)(bpy)(NO)]^+$ the 6C Fe, Ru, and Co NO^- complexes show distinctive, significant elongations of the M−L distances (L trans to the nitroxyl group) by ~0.03−0.05 Å, as compared with the M−L distances

Table 7.3 Structural and spectroscopic IR data for selected 6C and 5C nitroxyl complexes $[M(NO)]^8$

	Compound	ν_{NO} (cm^{-1})	d_{M-N} (Å)	d_{N-O} (Å)	$\angle_{MNO}$ (deg)	References
6C	*trans*-$[CoCl(en)_2(NO)](ClO_4)$	1611	1.820(11)	1.043(17)	124.4(11)	[33]
	$[Co(das)_2(NCS)(NO)](NCS)$	1587,1561[a]	1.87(1)	1.18(2)	132.3(14)	[19]
	$NOCbl \cdot 15H_2O$	–	1.927(6)	1.14-1.20	121–117	[34]
	Fe(cyclam-ac)(NO)	1271[b]	1.752[c]	1.261[c]	122.4[c]	[13]
	$[Ru(Me_3[9]aneN_3)(bpy)NO]^+$	1315,1286[d]	1.91[c]	1.27[c]	122.9[c]	[12]
	$[Rh(NCMe)_3(PPh_3)_2(NO)]^{2+}$	–	2.026(8)	1.159(10)	118.4(6)	[35]
	$[Pt(NO_2)_4(H_2O)(NO)]^-$	1655[e]	2.10(2)	1.19	120–129	[36]
5C	$Co(LN_4)(NO)$ (sqp)	1645[f]	1.7890(11)	1.1551(15)	125.97(9)	[37]
	$[Co(12\text{-}TMC)(NO)](ClO_4)_2$ (sqp)	1712[g]	1.7844(14)	1.155(2)	128.50(13)	[38]
	$[(Cp)_2Co][Fe(TFPBr_8)(NO)]$ (sqp)	1547[h]	1.790	1.201	122.7	[39]
	$[Co(das)_2(NO)](ClO_4)_2$ (tbp)	1852[a]	1.68(3)	1.16(2)	178(2)	[19]
	$[Fe(CN)_4(NO)]^{3-}$ (tbp)[c]	1581	1.637	1.22	171.2	[40]

Abbreviations: bpy = 2,2′-bipyridine; cyclam-ac = 1,4,8,11-tetraazacyclotetradecane-1-acetate pentaanion; das = *o*-phenylenebis(dimethylarsine); en = ethylenediamine; LN_4 = diimine/dipyrrolide dianion; $Me_3[9]aneN_3$ = 1,4,7-trimethyl-1,4,7-triazacyclononane; NOCbl = nitroxylcob(III)alamine; sp = square pyramid; tbp = trigonal bipyramid; NCMe = methyl isocyanide; PPh_3 = triphenylphosphine; $TFPBr_8$ = 2,3,7,8,12,13,17,18-octabromo-5,10,15,20-[Tetrakis-(pentafluorophenyl)]porphyrin dianion; 12-TMC = 1,4,7,10-tetramethyl-1,4,7,10-tetraazacyclododecane.

[a] As Nujol mulls between KBr disks.

[b] In CD_3CN.

[c] From DFT calculations.

[d] From spectroelectrochemical reduction of the $n = 7$ complex in acetonitrile.

[e] In aqueous solution, with IR-tran cell.

[f] In KBr.

[g] In CH_3CN, −40°C.

[h] In solid film or CH_2Cl_2.

Source: M = *formally* Fe^{II}, Ru^{II}, Co^{III}, Rh^{III}, Pt^{IV}.

Table 7.4 Nitrosyl stretchings ν_{NO}, selected metric parameters and Mayer bond orders (in parenthesis) obtained from DFT computations in $[Ru(Me_3[9]aneN_3)(bpy)(L)]^n$ complexes ($L = NO^+$, NO, NO^-, NHO, and NO_2^-)

L[a]	ν_{NO} (cm^{-1})	d_{Ru-N1} (Å)	d_{Ru-N2} (Å)	d_{N-O} (Å)	$\angle_{RuNO}$ (deg)
NO^+	1934/1898[b]	1.78(1.04)	2.18(0.54)	1.18(1.82)	179.5
NO	1637/1605[b]	1.90(0.84)	2.22(0.39)	1.22(1.58)	142.4
NO^-	1315/1290[b] 1290,1282[b]	1.91(1.18)	2.41(0.20)	1.27(1.45)	122.9
NHO	1376	1.93(0.77)	2.22(0.37)	1.28(1.43)	127.5
NO_2^-	1300, 1265[c]	2.05(0.48)[d]	2.23(0.41)[e]	1.30(1.39) 1.29(1.34)[f]	120.3 120.6[g]

[a]Ref. [11].
[b]In acetonitrile, $^{14}N/^{15}N$ derivatives.
[c]In KBr.
[d]X-ray: 2.018(5) Å.
[e]2.150(6).
[f]X-ray: 1.250(6), 1.261(6).
[g]X-ray: 117.0(4).

found for complexes where nitroxyl is absent (cf. Table 7.4). The σ-*trans* effect explains in a limiting situation the abundance of 5C NO^- compounds with square pyramidal (sqp) geometries, as also shown by the Co(III) nitroxyl-porphyrinates.[41] Table 7.3 contains X-ray structural and spectroscopic data for two new nonheme 5C $Co^{III}(NO^-)$ complexes[37,38] and for the first 5C Fe complex with a substituted porphyrinate platform, $[Cp_2Co][Fe(TFPBr_8)(NO)]$, isolated from a methylene chloride solution under anaerobic conditions and well characterized through ^{15}N NMR and other signatures[39] (see chapter: HNO Generation From $NO^•$, Nitrite, Inorganic or Organic Nitrosyls, and Crosstalk With H_2S). Other thermally stable $\{FeNO\}^8$ analogs containing derivatives of a planar LN_4 ligand have been also prepared in toluene with decamethylcobaltocene as a reductant, and were isolated and spectroscopically characterized.[42] Additional ferrous heme-nitroxyl model complexes have been prepared using electron-poor porphyrin ligands as well as 3,5-Me-$BAFP^{2-}$, allowing for the establishment of correlations between the electronic structures of analogous $\{FeNO\}^7$ and $\{FeNO\}^8$ complexes.[18] All the $n = 8$ iron complexes are diamagnetic low-spin with $S = 0$ ground states and their electronic structures were described as intermediate between $Fe^{II}NO^-$ and $Fe^{I}NO^•$. As seen in Section 7.1 the tbp $[Co(das)_2(NO)]^+$ ion (allowing HBr-addition to give $[CoBr(das)_2(HNO)]^{2+}$) could be hardly described as a "nitroxyl" species,[20] given the high value of ν_{NO} (Table 7.3) and consequent electrophilic character. This could be also the case of tbp $[Fe(CN)_4(NO)]^{3-}$, not isolated nor adequately characterized yet (Table 7.3 shows DFT calculated data), though it is seemingly generated upon deprotonation and cyanide-labilization from $[Fe(CN)_5(HNO)]^{3-}$.[11]

The spin state of the nitroxyl ligand can afford a singlet or a triplet state, and this depends critically on the spin state of the metal. The electronic structures and biologically relevant activities of heme versus nonheme iron-nitroxyl $\{Fe(H)NO\}^8$ complexes have been reviewed.[43] In an effort to model the nonheme $\{Fe(H)NO\}^8$ complexes implicated in the catalytic cycles of bacterial NO reductases, the high-spin nonheme $\{FeNO\}^8$ complex $[Fe(TMG_3tren)(NO)]^+$ containing a sterically encumbering coligand has been prepared in acetonitrile solution by reducing the $n=7$ analog through an iron centered process occurring at a low redox-potential, a characteristic feature of high spin $\{FeNO\}^{7,8}$ systems. The product is a tbp 5C species with a linear NO in axial position, and has been characterized by IR, EPR, and ^{1}H NMR spectroscopies.[44] The new $Fe^{II}NO^-$ complex ($S=1$) was described as a high spin Fe(II) antiferromagnetically coupled to a triplet NO^-. Analogous electronic distributions were found for an NO adduct of the nonheme iron enzyme taurine deoxygenase[45] and for the sqp $[Fe(dithiolene)(NO)]^{3-}$.[46]

7.4 Structural and spectroscopic changes in $\{MX_5L\}$ complexes ($L = HNO$, NO_2^-, and diatomic nitrosyls: NO^+, NO, NO^-) in a common MX_5 platform. Spectroelectrochemical (UV–vis, IR) measurements

The generation and characterization of nitroxyl and related N-bound derivatives in a common metal/coligand platform is a highly desirable goal.[12] As pointed in Section 7.2, the structural and spectroscopic indicators in the MNO moieties depend on the σ-π binding properties of the coligands. Fig. 7.5 (top) shows a CV of the $[Ru^{II}(Me_3[9]aneN_3)(bpy)(NO)]^{3+}$ ion ($n=6$, formally $Ru^{II}NO^+$) in aqueous solution.

Fig. 7.5 (bottom) shows two one-electron reversible waves at 0.20 V and 0.53 V in water and acetonitrile respectively. One-electron reduction to the $n=7$ complex $[Ru^{II}(Me_3[9]aneN_3)(bpy)(NO)]^{2+}$ was achieved and monitored by UV-vis and IR spectroscopies, and the product showed stable in the absence of oxygen. A second reversible wave at -0.44 V can be appreciated in acetonitrile. By applying a potential of -0.65 V an additional one-electron reduction is onset supporting reaction (Eq. 7.2) with generation of $[Ru^{II}(Me_3[9]aneN_3)(bpy)(NO)]^+$, the $n=8$ derivative containing bound NO^-.

$$[Ru^{II}(Me_3[9]aneN_3)(bpy)(NO)]^{2+} + e^- \rightarrow [Ru^{II}(Me_3[9]aneN_3)(bpy)(NO)]^+ \quad (7.2)$$

Quite importantly the CV in water led to an electrochemically irreversible second wave (Fig. 6.5 top) though a controlled potential coulometric experiment at pH 2.5 of $[Ru^{II}(Me_3[9]aneN_3)(bpy)(NO)]^{2+}$ rendered a chemically reversible one electron conversion. In contrast with the behavior observed for the $\{RuNO\}^6/\{RuNO\}^7$

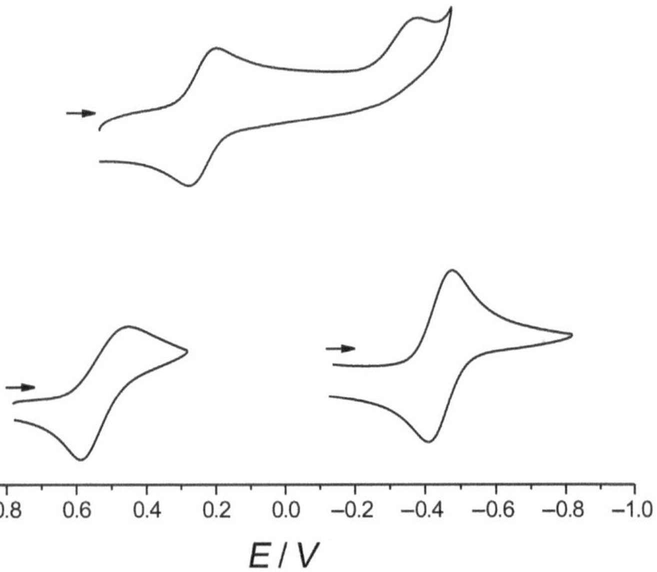

Figure 7.5 Cyclic voltammogram of $[Ru(Me_3[9]aneN_3)(bpy)(NO)]^{3+}$ in water, pH 4.5 (0.3 M acetic acid/acetate, $I = 1$ M (NaCl), (top) and in acetonitrile, $I = 1$ M ($TBAPF_6$) (bottom), Scan rate = 200 mV s^{-1}, $T = 25$°C. E in V versus AgCl (3M NaCl)/Ag°.[11]

process, the spectral changes for $\{RuNO\}^7/\{RuNO\}^8$ in aqueous acidic and in acetonitrile solutions were noticeably different. The observed pH-exploration of the redox potential showed a change of 60 mV per pH unit and allowed proposing reaction (Eq. 7.3) for the second proton-coupled one-electron reduction in aqueous acidic medium:

$$[Ru^{II}(Me_3[9]aneN_3)(bpy)(NO)]^{2+} + e^- + H^+ \rightarrow [Ru^{II}(Me_3[9]aneN_3)(bpy)(HNO)]^{2+} \tag{7.3}$$

Reactions (7.2) and (7.3) reveal the pioneering efforts for a clean generation and spectroscopic distinctive characterization in aqueous solution of the $n = 8$ nitroxyl derivatives containing bound NO^- and HNO.[12] A similar methodological approach has been used for characterizing the three redox states of $[Fe(cyclam\text{-}ac)(L)]^{2+/+/0}$ ($L = NO^+$, NO, NO^-) in acetonitrile solutions. The HNO derivative could not be generated in that solvent though DFT calculated data are shown in Table 7.1.[13] Table 7.4 shows the DFT optimized geometries and the sensitive IR stretchings ν_{NO} for the nitroxyl- and related complexes.[12]

Though the Ru–N bond lengths are slightly (and consistently) overestimated, the comparisons between experimental and theoretical bond lengths at the used level of theory have been reported for other $\{RuNO\}^6$ complexes. These computations reproduce the main features observed in the spectra of all species, and allow for a complete assignment of the vibrational modes.[12] Reaction (Eq. 7.4) for the generation of the isolated nitrite derivative has much synthetic value. The facile coordination of nitrite to the aqua-complexes allows isolating and characterizing complexes with many other MX_5 fragments as well.[3] Formation of an N-bound nitrite complex appears as the first coordination event during the activation process of the nitrite reductase enzymes,[47,48] followed by the fast proton-assisted dehydration of NO_2^-, as in reaction (Eq. 7.5). The nitrosonium (NO^+) product has a low-energy LUMO suitable for a facile subsequent reduction and release of NO.

$$[Ru^{II}(Me_3[9]aneN_3)(bpy)(H_2O)]^{2+} + NO_2^- \rightleftarrows [Ru^{II}(Me_3[9]aneN_3)(bpy)(NO_2)]^+ + H_2O \quad (7.4)$$

$$[RU^{II}(Me_3[9]aneN_3)(bpy)(NO_2)]^+ + 2H^+ \rightleftarrows [Ru^{II}(Me_3[9]aneN_3)(bpy)(NO]^{3+} + H_2O \quad (7.5)$$

Let us compare the entries in Table 7.4 for the complexes containing the triatomic angular species HNO and NO_2^-. The $Ru{-}N_1$ bond is shorter for the HNO complex, revealing a strong σ-π interaction in $[Ru^{II}(Me_3[9]aneN_3)(bpy)(HNO)]^{2+}$. The NO bond is also shorter, reflecting a greater (double bond) character for the NO group in the nitroxyl complex in comparison with the longer NO bond in the nitrite complex related to its resonance structure of single/double bonds. Remarkably the $Ru{-}N_1$ bond length in the (NO^+)-complex is the shortest of all, consistent with the strong π-backbonding.[3] The increase in $Ru{-}N_1$ and N−O bond lengths when going from the NO^+ complex to the one-electron reduced NO analog reflects the increased population of the antibonding RuNO fragment, which goes in parallel with the changes from a linear to an angular RuNO moiety, close to 140°. Similar changes can be observed when going to the two-electron reduced NO^- complex, with a RuNO angle at ~120°, smaller than for the HNO derivative. The values of ν_{NO} are quite sensitive to the electron input, decreasing in the trend $M(NO^+) > M(NO) > M(NO^-)$ upon consecutive reduction.[12]

Finally, as anticipated in Section 7.3, we compare the observed changes in the $Ru{-}N_2$ distances, trans to the nitrosyl/nitroxyl groups. The length is the shortest for the $n = 6$ compound, a result described as a "negative" *trans*-influence of the strongly electron attractive NO^+ group. These relative contractions have been observed for a large number of non heme- and heme-containing NO^+ complexes.[3] For the $n = 7$ analog the lengthening of ~0.04 Å for the $Ru{-}N_2$ distance has been traced to the population of a mixed π^*_{NO}/d_{z2} orbital by the additional electron.[3,31] The latter MO is antibonding with respect to the $Ru{-}N_2$ bond and contributes to its relative lengthening with respect to the $n = 6$ situation. Finally the $n = 8$ complex containing singlet NO^- implies an additional population of the π^*_{NO}/d_{z2} orbital, attaining a $Ru{-}N_2$ distance even much longer by ~ 0.05 Å with respect to $n = 7$ (Section 7.3, Fig. 7.3B). A pertinent comparison of the $Ru{-}N_2$ distance for the HNO complex shows it to be comparable to the one measured for the NO_2^- complex, a weak-to-moderate π acceptor, showing that the *trans*-effect has been significantly diminished for HNO by the protonation at NO^-. The *trans*-elongations account qualitatively for the occurrence of either 6C or 5C compounds, as seen In Table 7.3 for $n = 8$ but also occurring for $n = 7$.[26] Note that the great stability of the Ru complexes in Table 7.4, associated with the chelating nature of the coligands, inhibits the complete detachment of the N-arm of the tridentate ligand, a situation not generally found for the more weakly bound monodentate ligands. The quantitative evaluation of the *trans*-labilizing effects associated with the presence of bound NO, NO^- or HNO has a great bioinorganic significance,[31] related to the detailed activation mechanism of soluble guanylate cyclase following the coordination of NO to the 5C iron enzyme and further release of a *trans*-histidine ligand, see Chapter 8, The Interaction of HNO With Transition Metal Centers and Its Biological Significance. Insight Into Electronic Structure From Theoretical Calculations. The magnitude of the *trans*-effect of HNO remains a controversial issue, though the previous analysis of results in Table 7.4, together with recent theoretical calculations,[28] suggest that HNO has a minor *trans*-activating influence, in high contrast to NO^-.

7.5 Stability and reactivity in HNO complexes

We analyze the main reactivity modes of the HNO complexes in aqueous solutions dealing with: (1) the coordination ability as measured by the formation k_f and dissociation k_d specific rate constants in solution (also known as k_{on} and k_{off}, respectively), displayed in reaction (Eq. 7.6). (2) the acid–base reactivity, i.e., deprotonation forming the NO^- ligand according to reaction (Eq. 7.7). (3) the redox activity producing either more oxidized (NO, NO^+/NO_2^-) or more reduced ($NHOH/N_2$, NH_2OH, NH_3) N-bound ligands upon reaction with appropriate reagents.

$$X_5M(H_2O) + HNO \rightleftarrows X_5M(HNO) + H_2O \quad K_{st} \quad k_f/k_d \tag{7.6}$$

$$X_5M(^1HNO) \rightleftarrows X_5M(^1NO^-) + H^+ K_a \tag{7.7}$$

7.5.1 Formation and dissociation of HNO from metal centers in aqueous solution

Values of the formation rate constants k_f for reactions like (7.6) are unavailable for non heme systems because in most cases aqua-ions cannot be generated in the currently employed organic solvents. Related to the low spin complexes in Table 7.1 the reaction of the aqua-ion $[Fe(CN)_5(H_2O)]^{3-}$ could in principle lead to $[Fe(CN)_5(HNO)]^{3-}$ upon mixing with a solution of Angeli's salt containing $[HN_2O_3]^-$, which in turn releases HNO and NO_2^- to the solutions under pH-controlled conditions.[4] This is not the case and only the nitroprusside ion is formed through the reaction with nitrite. As the values of k_f for reactions like (7.6) are usually limited by the rate of water exchange for the aqua-ions, the value of k_{ex} for $[Fe(CN)_5(H_2O)]^{3-}$ is too low ($\sim 100\ s^{-1}$) for trapping HNO once it is released by the HNO donor. Therefore, reaction (7.6) is unable to compete with the much faster dehydrative dimerization of HNO into N_2O (see chapter: HNO: Redox Chemistry and Interactions With Small Inorganic Molecules). There is only one report involving Mb^{II}-HNO with an estimated value of $k_f \sim 10^4\ M^{-1}\ s^{-1}$ in aqueous solution. It was considered a lower limit to the bimolecular rate of trapping upon mixing the highly substitution-labile Mb^{II} with the HNO-donor.[1] Interestingly the vitamin B12 derivative aquacobalamin (H_2OCbl^+/HOCbl, $pK_a = 7.8$), containing Co(III), reacts with $HN_2O_3^-$ giving NOCbl, with NO^- bound to Co^{III}. The net reaction involves deprotonation of HNO, which appears as facile proton-releasing upon binding to M(III) species.[49]

The reports on the dissociation of most of the available HNO complexes (reaction 7.6, reverse) have been quite elusive to useful kinetic examinations because the compounds are frequently insoluble in aqueous media. Some of them decompose to unknown products at room temperatures.[1] In contrast the water-soluble $[Fe(CN)_5(HNO)]^{3-}$ [11,50] and $[Ru(Me_3[9]aneN_3)(bpy)(HNO)]^{2+}$ [12] remain stable toward the release of HNO for months in the absence of oxygen, as is also the case with Mb^{II}-HNO.[1] A revision of the aqueous solution behavior of $[Fe(CN)_5HNO]^{3-}$ indicates a low limit of $k_d = k_{-HNO} \sim 5 \times 10^{-7}\ s^{-1}$ at 25°C, with negligible formation of N_2O.[50] By comparing with related $[Fe^{II}(CN)_5L]^{n-}$ complexes, the order of k_{-L}'s is: $NO^+ << CO < CN^- < HNO < NOBz < NO < dmso < pz < py < NH_3$ (with the latter k_{-NH3} at $\sim 10^{-2}\ s^{-1}$). HNO fits in an intermediate position among the strong π-acceptor ligands, reflecting its ubiquity in the so-called spectrochemical series, which reflects the capability of the L ligands toward $\sigma-\pi$ bonding with the metal.[26] The three pentacyano(L)ferrate(II) complexes ($L = NO^+$, NO, HNO) afford high stability constants, generating very robust species in aqueous solutions. On the other hand, all of them are redox-active. As anticipated in Section 7.1, the preparation of the $NO^{\bullet}$ and HNO derivatives can be achieved by reducing the nitroprusside ion with dithionite.

7.5.2 Acid–base reactivity of HNO complexes

The value of pK_a for free HNO in aqueous solutions has been largely considered in the literature.[4] Insight Into Electronic Structure From Theoretical Calculations. Free NO^- displays a ground state triplet and an excited state singlet, and the pK_a for $^1HNO \rightleftarrows {}^3NO^- + H^+$ has been reevaluated as 11.5. The corresponding value for $^1HNO \rightleftarrows {}^1NO^- + H^+$ is ~ 23. Therefore, equilibrated nitroxyls might exist almost exclusively in the protonated form HNO in neutral aqueous solutions. For coordinated HNO a value of $pK_a = 7.7$ (reaction 7.7) has been assigned to $[Fe(CN)_5(HNO)]^{3-}$ through a titration NMR experiment.[11] Though the value seemed consistent, a recent NMR study of $[Fe(CN)_5(HN^{17}O)]^{3-}$ showed the persistence of the ^{17}O signal up to pH 10–11.[25] Also the spectral and ^1H NMR properties of

Mb^{II}-HNO remain unchanged in aqueous solutions in the pH range of 6–10, and a value of pK_a close to 11 has been suggested.[1,17]

A spectroelectrochemical reduction in aqueous solution allowed identifying the acid–base interconversion between bound HNO and NO^- for the $[Ru(Me_3[9]aneN_3)(bpy)(L)]^n$ complexes.[12] Fig. 7.6 shows a potential pH diagram revealing a thermodynamic picture for the predominance of five nitrogenated species with bound NO^+, NO, NO^-, HNO, and NO_2^- on the $[Ru(Me_3[9]aneN_3)(bpy)]^n$ fragment.

The diagram was built up from the combination of CV, square-wave voltammetry (SWV) and spectroelectrochemical experiments. The different stability regions for the $\{RuNO\}^n$ species are clearly defined. The line that separates the $n = 7$ from $n = 8$ compound breaks at a pH close to 10, due to reaction (Eq. 7.8).

$$[Ru(Me_3[9]aneN_3)(bpy)(HNO)]^{2+} \rightleftarrows [Ru(Me_3[9]aneN_3)(bpy)(NO)]^+ + H^+ \quad (7.8)$$

A complementary acid–base titration and global analysis of the UV-vis spectra allowed measuring a value of $pK_a = 9.78 \pm 0.15$ for reaction (7.8). The value is most reliable for M^{II}–HNO complexes, probably with some dependence on the metal center and the more so with eventual variation of the coligands. The resistance of HNO complexes to deprotonation is consistent with a strong N–H bond (Section 7.2).

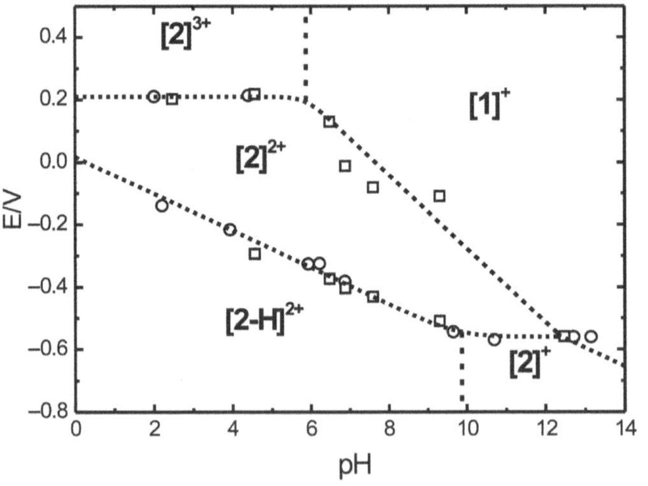

Figure 7.6 Potential ($E°$)–pH diagram showing the redox and acid–base behavior of the system based on the $[Ru(Me_3[9]aneN_3)(bpy)]^{n+}$ fragment. $[\mathbf{1}]^+ = [Ru(Me_3[9]aneN_3)(NO_2)]^+$; $[\mathbf{2}]^{n+} = [Ru(Me_3[9]aneN_3)(NO)]^{n+}$ ($n = 1,2,3$); $[\mathbf{2\text{-}H}]^{2+} = [Ru(Me_3[9]aneN_3)(HNO)]^{2+}$ Data arising from CV and SWV experiments are displayed as squares, while those originating from spectroelectrochemistry are represented as circles.[12]

It is important to remark that the speciation of M–HNO or M–NO^- complexes appears to be not only dependent on the pK_a values. As discussed in Section 7.4 the instability of $MX_5(NO^-)$ complexes toward the release of the *trans*-ligand to NO^- might define a specific scenario for a given complex, according to the following reaction sequence for $[Fe(CN)_5(HNO)]^{3-}$, Eqs. (7.9) and (7.10):

$$[Fe(CN)_5(HNO)]^{3-} \rightleftarrows [Fe(CN)_5(NO)]^{4-} + H^+ \; K_a \tag{7.9}$$

$$[Fe(CN)_5(NO)]^{4-} \rightleftarrows [Fe(CN)_4(NO)]^{3-} + CN^- \; K_{trans} \tag{7.10}$$

Theoretical evidence supports a tbp geometry for the $[Fe(CN)_4NO]^{3-}$ ion.[40] The *trans*-labilization described by reaction (7.10) has a precedent in the corresponding reaction for $[Fe(CN)_5(NO)]^{3-}$ ($n = 7$) giving the sqp $[Fe(CN)_4(NO)]^{2-}$ ion.[51,52] The putative tbp $[Fe(CN)_4(NO)]^{3-}$ product in reaction (7.10) reverts to $[Fe(CN)_5(HNO)]^{3-}$ in a high yield by reacidification to pH 6.

The controversial value of p$K_a = 7.7$ is under close scrutiny and could be revisited by considering that the measured value during the titration experiment[11] corresponds to a more complicated process including the successive reactions (7.9 and 7.10) and probably the isomerization of the initially generated sqp $[Fe(CN)_4(NO)]^{3-}$ to the more stable tbp species. On that basis, and considering related data,[1,12,25] we might assume that a value of p$K_a \sim 10$ for $[Fe(CN)_5(HNO)]^{3-}$ is the best approach. In a longer time scale a slow oxidative conversion of the deprotonated nitroxyl species to $[Fe(CN)_5(NO)]^{3-}$ has been found at pH 10, most probably associated with the decay of the strongly reducing $[Fe(CN)_4(NO)]^{3-}$, probably through water reduction.

7.5.3 Redox reactivity

There is a big difference between related HNO and NO^- nitroxyl complexes concerning the redox chemistry, given that the HNO complexes behave as mild reductants, while the NO^- species act as strong reductants. Quite revealing are the results for the reactions of the two-electron reduction products of $[Fe(CN)_5(NO)]^{2-}$ at pH 6 and 10 with poorly oxidizing reagents. With methylviologen (MV^{2+}, $E^o = -0.44$ V), no reaction is observed at pH 6 with $[Fe(CN)_5(HNO)]^{3-}$, as is also the case with free HNO.[1] On the other hand, at pH 10 the reduced viologen radical can be immediately observed after mixing.[50] Similar results follow the addition of one equivalent of the weak oxidant $[Fe(CN)_5(NO)]^{2-}$ ($E^o = 0.05$ V) to the nitroxyl complexes. While no spectral changes are observed at pH 6, the immediate formation of $[Fe(CN)_5(NO)]^{3-}$ is achieved at pH 10, revealing a quantitative onset of the comproportionation reaction (Eq. 7.11).

$$[Fe(CN)_5(NO)]^{4-} + [Fe(CN)_5(NO)]^{2-} \rightleftarrows 2[Fe(CN)_5(NO)]^{3-} \tag{7.11}$$

The results suggest a value of $E^o << -0.44$ V for $[Fe(CN)_5(NO)]^{4-}$, whilst $[Fe(CN)_5(HNO)]^{3-}$ affords an estimated value of 0.3 V, in terms of the oxidation wave measured in the CV at pH 6 and assigned to the proton-coupled redox potential, E^o(FeNO,H^+/FeHNO).[53] This is remarkable because the related potential for NO, H^+/HNO has been reported to be −0.55 V at pH 7, suggesting that aqueous free HNO is a strong reductant.

The $[Fe(CN)_5(HNO)]^{3-}$ ion reacts with the mild oxidant $[Fe(CN)_6]^{3-}$ ($E^o \sim 0.4$ V), according to reaction (Eq. 7.12):

$$[Fe(CN)_5(HNO)]^{3-} + 2[Fe(CN)_6]^{3-} \rightleftarrows [Fe(CN)_5(NO)]^{2-} + 2[Fe(CN)_6]^{4-} + H^+ \quad (7.12)$$

The rate law is first order in each reactant with $k = 70\ M^{-1}\ s^{-1}$ at 25°C.[50] This value is similar to the one reported for free HNO reacting with $[Fe(CN)_6]^{3-}$.[4] If we assume an initial slow one-electron oxidation of $[Fe(CN)_5(HNO)]^{3-}$ by $[Fe(CN)_6]^{3-}$, followed by a rapid oxidation to $[Fe(CN)_5NO]^{2-}$, the use of the Marcus "cross-reaction" model to the first step led to a value of $k_{ex} = 6 \times 10^{-3}\ M^{-1}\ s^{-1}$ for the self-exchange rate constant for the $[Fe(CN)_5(NO)]^{3-}$, H^+/$[Fe(CN)_5(HNO)]^{3-}$ redox couple. This is a very low value compared to the much higher values of k_{ex} ($\sim 10^5\ M^{-1}\ s^{-1}$) for the $[Fe^{II,III}(CN)_5(L)]^{n,(n-1)-}$ complexes that react through metal-centered mechanisms when L is nonredox active.[26] On the other hand, reaction (7.12) appears to involve a ligand-centered redox reaction involving a proton-coupled process with a high reorganization energy.[50]

By mixing the nitroprusside ion $[Fe(CN)_5(NO)]^{2-}$ with an excess of dithionite over the necessary two equivalents for attaining bound HNO, the observed decay of the intense band at 445 nm of $[Fe(CN)_5(HNO)]^{3-}$ indicates a 4-electron reduction with $[Fe(CN)_5(NH_3)]^{3-}$ as the final product.[50] The onset of consecutive ligand-centered processes with N-bound intermediates is at the heart of the behavior of assimilatory nitrite reductases.[8,47,48]

The HNO complexes react in the presence of NO leading to the same products obtained for the reaction of aqueous HNO + NO, suggesting the stoichiometry described by reaction (Eq. 7.13). This type of reactivity has been explored by Caulton's group.[24]

$$\text{M-HNO} + 2NO \rightarrow \text{M-}NO_2^- + H^+ + N_2O \quad (7.13)$$

NO and nitroxyls (both $^3NO^-$ and ^1HNO) form the hyponitrite radicals $HN_2O_2^{\bullet}/N_2O_2^{\bullet -}$ ($pK_a \sim 5.6$), which may add another NO yielding $N_3O_3^-$ and further decompose into N_2O and NO_2^- in aqueous media.[53] These radicals are potentially important species in the redox transformations of nitrogen in biological (nitrification/denitrification, assimilatory nitrite reduction) and aquatic environmental processes, suggesting the relevance of reactions like (7.13). The radicals are both strongly oxidizing and moderately reducing, and can also decay through complex redox disproportionations.[54]

As already anticipated in the Introduction of this chapter, heme complexes showed to be excellent electrocatalysts for the reduction of NO_2^- to NH_3 via HNO intermediates,[8b] as also done by multiheme cytochrome c reductases.[48] Based on the structural information and previous studies,[48a] the mechanism of the six-electron, seven-proton nitrite enzymatic reduction has been elaborated through modern DFT studies,[48b] presenting a plausible sequence of elementary steps, with a discussion on the electronic structure and reactivity of the intermediates. In the first step, NO_2^- binding is associated with N–O heterolytic bond cleavage, aided by double protonation with His_{277} as a proton donor. The second phase consists of two proton-coupled electron-transfer events, leading to an HNO intermediate. The third phase involves the formation of NH_2OH, with Arg_{114} providing the necessary proton. The second N–O bond is cleaved in the fourth phase of the mechanism, again triggered by proton transfer from His_{277}. The Tyr_{218} side chain governs the fifth and last phase, consisting of radical transfer and formation of NH_3. Thus, the mechanism implies that all conserved active-site side chains work in a concerted way in order to achieve this complex chemical transformation from NO_2^- to NH_3.

7.5.4 Reactions of M-HNO and M-NO⁻ complexes with dioxygen and reduced species

The reactivity of NO^-/HNO complexes toward O_2 is an underdeveloped field contrasting with the bioinorganic relevance, and relates to the still unclear mechanism for the autoxidation of free HNO.[4] A comprehensive set of 5C $[Co^{III}L_4(NO^-)]$ complexes (L = quadridentate or bis-bidentate platforms) react with O_2 in acetonitrile in the presence of nitrogen or phosphorus bases (B) to yield the corresponding 6C nitrocompounds *trans*-$[Co(L_4)(B)(NO_2)]$.[55] The rates increase with the basicity of B influencing the nucleophilicity of the $\{CoNO\}^8$ moieties toward O_2. The mechanism has been described as an initial equilibrium of the 5C/6C species with *only the latter* species reacting with O_2 and leading to a $\{Co(L_4)(B)N(O)O_2\}$ peroxynitrito-intermediate in the rate-determining step, followed by fast dimerization with another $\{CoL_4B(NO)\}$ unit, homolysis of the O–O bond, and final formation of the nitro-complex. Other Co, Ru, and Ir–NO^- complexes react similarly in the initial step, though with the formation of nitrato complexes as final products, presumably through the isomerization of the N-bound peroxynitrite-intermediate to the O-bound species.[3] The factors influencing one or other type of stoichiometry are not clearly understood. A kinetic and mechanistic study with aqueous NOCbl at pH 7.4 shows again that only "base-on" NOCbl (with 5,6-dimethylbenzimidazole in the *trans*-position to nitroxyl) reacts with O_2, and the reaction proceeds via an associative mechanism involving a $Co^{III}(N(O)O_2^-)$ intermediate, as described above. The latter species undergoes O–O bond homolysis and ligand isomerization to yield NO_2Cbl and H_2OCbl^+/HOCbl, respectively, via the formation of $^\bullet OH$ and $^\bullet NO_2$ intermediates arising in the homolysis.[56] The unreactivity of the 5C species toward O_2 has been recently highlighted for the $[Co^{III}(12\text{-TMC})(NO)]^{2+}$ ion, though it does react with superoxide, leading to the nitro-complex plus O_2, with the proposed intermediacy of bound peroxynitrite.[38]

The M–NO^- and M–HNO (M = Fe, Ru) complexes described in this work react under an oxygen atmosphere, either in nonaqueous or aqueous media, though no kinetic/mechanistic studies are available.[3] There is a need of disclosing the similarities and differences with the autoxidation reactions of $n = 7$ complexes like those described for $[Fe(CN)_5(NO)]^{3-}$ [57] and other related nonheme- and heme-$\{MX_5NO\}^n$ ions.[3]

References

1. Farmer PJ, Sulc FJ. Coordination chemistry of the HNO ligand with hemes and synthetic coordination complexes. *J Inorg Biochem* 2005;**339**:166–84.
2. Farmer PJ, Kumar MR, Almaraz E. The coordination chemistry of HNO: From warren roper to hemoglobin. *Comments Inorg Chem* 2010;**31**:130–43.
3. Bari SE, Olabe JA, Slep LD. Three redox states of metallonitrosyls in aqueous solution. *Adv Inorg Chem* 2015;**67**:87–144.
4. Miranda KM. The chemistry of nitroxyl (HNO) and implications in biology. *Coord Chem Rev* 2005;**249**:433–55.
5. Fukuto J, Cisneros CJ, Kinkade RL. A comparison of the chemistry associated with the biological signaling and actions of nitroxyl (HNO) and nitric oxide (NO). *J Inorg Biochem* 2013;**118**:201–8.
6. Grundy KR, Reed CR, Roper WR. Hydrogen transfer from HCl to the nitrosyl ligand. Examples of co-ordinated HNO, NHOH–, and NH_2OH. *J Chem Soc Chem Commun* 1970;**22**:1501–2.
7. Melenkivitz R, Hillhouse GL. Synthesis, structure, and reactions of a nitroxyl complex of iridium(III), cis,trans-$IrHCl_2(NH{=}O)(PPh_3)_2$. *Chem Commun* 2002;**2**:660–1.
8. (a) Murphy Jr WR, Takeuchi K, Barley MH, Meyer TJ. Mechanism of reduction of bound nitrite to ammonia. *Inorg Chem* 1986;**25**:1041–53.
   (b) Barley MH, Takeuchi KJ, Meyer TJ. Electrocatalytic reduction of nitrite to ammonia based on a water-soluble iron porphyrin. *J Am Chem Soc* 1986;**108**:5876–85.
9. Wilson RD, Ibers JA. Coordinated nitrosyl hydride: structural and spectroscopic study of $OsCl_2(CO)(HNO)(P(C_6H_5)_3)_2$. *Inorg Chem* 1979;**18**:336–43.
10. Sellmann D, Gottschalk-Gaudig T, Haussinger D, Heinemann FW, Hess BA. [Ru(HNO)('py(bu)S(4)')], the first HNO complex resulting from hydride addition to a NO complex ('py(bu)S(4)'(2-)=2,6-bis(2-mercapto-3,5-di-tert-butylphenylthio)dimethylpyridine(2-)) Part 148. Transition metal complexes with sulfur ligands. *Chem Eur J* 2001;**7**:2099–103.
11. Montenegro AC, Amorebieta VT, Slep LD, Martin DF, Roncaroli F, Murgida DH, et al. Three redox states of nitrosyl: NO^+, NO*, and NO^-/HNO interconvert reversibly on the same pentacyanoferrate(II) platform. *Angew Chem Int Ed* 2009;**48**:4213–16.
12. Osa Codesido N, Weyhermüller T, Olabe JA, Slep LD. Nitrosyl-centered redox and acid-base interconversions in $[Ru(Me_3[9]aneN_3)(bpy)(NO)]$(3,2,1+). The pKa of HNO for its nitroxyl derivative in aqueous solution. *Inorg Chem* 2014;**53**:981–97.
13. García Serres R, Grapperhaus CA, Bothe E, Bill E, Weyhermuller T, Neese F, et al. Structural, spectroscopic, and computational study of an octahedral, non-heme $[Fe\text{-}NO]^{6\text{-}8}$ Series: $[Fe(NO)(cyclam\text{-}ac)]^{2+/+/0}$. *J Am Chem Soc* 2004;**126**:5138–53.
14. Southern JS, Green MT, Hillhouse GL, Guzei IA, Rheingold AL. Chemistry of coordinated nitroxyl. Reagent-specific protonations of *trans*-Re(CO)(2)(NO)(PR_3)(2) (R = Ph, Cy) that give the neutral nitroxyl complexes cis,*trans*-ReCl(CO)(2)(NH=O)(PR_3)(2) or

the cationic hydride complex [*trans,trans*-ReH(CO)(2)(NO)(PPh_3)(2)(+)][$SO_3CF_3^-$]. *Inorg Chem* 2001;**40**:6039–46.
15. Southern JS, Hillhouse GL, Rheingold AL. Preparation of a nitroxyl (HNO) complex of rhenium by selective oxidation of coordinated hydroxylamine. *J Am Chem Soc* 1997;**119**:12406–7.
16. (a) Olabe JA. The coordination chemistry of nitrosyl in cyanoferrates. An exhibit of bioinorganic relevant reactions. *Dalton Trans* 2008;**28**:3633.
    (b) Olabe JA. Redox reactivity of coordinated ligands in pentacyano(L)ferrate complexes. *Adv Inorg Chem* 2004;**55**:61.
17. Kumar MR, Fukuto JM, Miranda KM, Farmer PJ. Reactions of HNO with heme proteins: new routes to HNO-heme complexes and insight into physiological effects. *Inorg Chem* 2010;**49**:6283–92.
18. Goodrich LE, Roy S, Ercan Alp E, Zhao J, Hu MY, Lehnert N. Electronic structure and biologically relevant reactivity of low-spin $\{FeNO\}^8$ porphyrin model complexes: new insight from a bis-picket fence porphyrin. *Inorg Chem* 2013;**52**:7766–80.
19. Wieghardt K, Holzbach J, Weiss JZ. η^2-Coordination of a Hydroxylamido(2-) Anion: The Crystal Structure of [Mo(NO)(terpy)(CN)(HNO)]$2H_2O$. *Z Naturforsch B* 1982; **37**:680–3.
20. Enemark JH, Feltham RD, Riker-Nappier J, Bizot KF. Stereochemical control of valence. IV. Comparison of the structures and chemical reactivities of five- and six-coordinate diarsine complexes of $\{CoNO\}^8$ group. *Inorg Chem* 1975;**14**:624–32.
21. Collman JP, Hoffman NW, Morris DE. Oxidative additions and catalysis of olefin hydrogenation by tris(triphenylphosphine)nitrosylrhodium. *J Am Chem Soc* 1969;**91**:5659–60.
22. Tatsumi T, Sekizawa K, Tominaga H. Preparation and properties of nitrosyl complexes of molybdenum. *Bull Chem Soc Japan* 1980;**53**:2297–303.
23. Melenkiwicz R, Southern JS, Hillhouse GL, Concolino TE, Liable-Sands LM, Rheingold AL. A new route to coordination complexes of nitroxyl (HN=O) via insertion reactions of nitrosonium triflate with transition-metal hydrides. *J Am Chem Soc* 2002;**124**:12068–9.
24. Marchenko AV, Vedernikov AN, Die DF, Pink M, Zaleski JM, Caulton KG. Reactivity of the hydrido/nitrosyl radical MHCl(NO)(CO)(P(i)Pr(3))(2), M = Ru, Os. *Inorg Chem* 2004;**43**:351–60.
25. Gao Y, Toubaei A, Kong X, Wiu G. Acidity and hydrogen exchange dynamics of iron(II)-bound nitroxyl in aqueous solution. *Angew Chem Int Ed* 2014;**53**(43):11547–51.
26. Roncaroli F, Videla M, Slep LD, Olabe JA. New features in the redox coordination chemistry of metal nitrosyls {M–NO^+; M–NO; M–NO–(HNO)}. *Coord Chem Rev* 2007;**251**:1903–30.
27. Montenegro AC, Dabrowski SG, Gutiérrez MM, Amorebieta VT, Bari SE, Olabe JA. Catalytic oxidation of hydroxyurea to bound NO^+/mediated by pentacyano (L) ferrates. Characterization of the nitroxide radical, bound C-nitrosoformamide and NO as reaction intermediates. *Inorg Chim Acta* 2011;**374**:447–52, and references therein.
28. Goodrich LE, Lehnert N. The trans effect of nitroxyl (HNO) in ferrous heme systems: implications for soluble guanylate cyclase activation by HNO. *J Inorg Biochem* 2013;**118**:179–86.
29. Yang L, Fang W, Zhang Y. Metal centre effects on HNO binding in porphyrins and the electronic origin: metal's electronic configuration, position in the periodic table, and oxidation state. *Chem Commun* 2012;**48**:3842–4.
30. Enemark JH, Feltham RD. Principles of structure, bonding and reactivity for metal nitrosyl complexes. *Coord Chem Rev* 1974;**13**:339–406.

31. Lehnert N, Scheidt WR, Wolf MW. Structure and bonding in heme-nitrosyl complexes and implications for biology. *Struct Bond* 2014;**154**:155–223.
32. Levina A, Turner P, Lay PA. X-ray absorption spectroscopic studies of chromium nitroso complexes. Crystal and molecular structure of $(Ph_4P)_3[Cr(NO)(NCS)_5].2.4(CH_3)_2CO$. *Inorg Chem* 2003;**42**:5392–8.
33. Snyder DA, Weaver DL. Crystal and molecular structure of *trans*-chloronitrosylbis(ethylenediamine)cobalt(III) perchlorate. *Inorg Chem* 1970;**9**:2760–7.
34. Hannibal L, Smith CA, Jacobsen DW, Brasch NE. Nitroxylcob(III)alamin: synthesis and X-ray structural characterization. *Angew Chem Int Ed* 2007;**46**:5140–3.
35. Kelly BA, Welch AJ, Woodward P. Molecular and crystal structure of tris(acetonitrile) nitrosylbis(triphenylphosphine)rhodium(III) dication as its hexafluorophosphate salt. *J Chem Soc Dalton Trans* 1977;2237.
36. Schaniel D, Woike T, Behrnd NR, Hauser J, Krämer KW, Todorova T, et al. Photogeneration of nitrosyl linkage isomers in octahedrally coordinated platinum complexes in the red spectral range. *Inorg Chem* 2009;**48**:11399–406.
37. Patra AK, Dube KS, Sanders BC, Papaefthymiou GC, Conradie J, Ghosh A, et al. A thermally stable $\{FeNO\}^8$ complex: properties and biological reactivity of reduced MNO systems. *Chem Sci* 2012;**3**:364–9.
38. Kumar P, Lee YM, Park YJ, Siegler MA, Karlin KD, Nam W. Reactions of Co(III)-nitrosyl complexes with superoxide and their mechanistic insights. *J Am Chem Soc* 2015;**137**:4284–7.
39. Pellegrino J, Bari SE, Bikiel DE, Doctorovich F. Successful stabilization of the elusive species $\{FeNO\}^8$ in a heme model. *J Am Chem Soc* 2009;**132**:989–95.
40. González Lebrero MC, Scherlis DA, Estiú GL, Olabe JA, Estrin DA. Theoretical investigation on the electronic structure of pentacyano(L)ferrate(II) complexes with NO(+), NO, and NO(-) ligands. Redox interconversion, protonation, and cyanide-releasing reactions. *Inorg Chem* 2001;**40**:4127–33.
41. Scheidt WR, Ellison MK. The synthetic and structural chemistry of heme derivatives with nitric oxide ligands. *Acc Chem Res* 1999;**32**:350.
42. Sanders BC, Patra AK, Harrop TC. Synthesis, properties, and reactivity of a series of non-heme $\{FeNO\}^{7/8}$ complexes: implications for Fe-nitroxyl coordination. *J Inorg Biochem* 2013;**118**:115–27.
43. Speelman AL, Lehnert N. Heme versus Non-Heme Iron-Nitroxyl $\{FeN(H)O\}^8$ Complexes: Electronic Structure and Biologically Relevant Reactivity. *Acc Chem Res* 2014;**47**:1106–16.
44. Speelman AL, Lehnert N. Characterization of a high-spin non-heme $\{FeNO\}^8$ complex: Implications for the reactivity of iron nitroxyl species in biology. *Angew Chem Int Ed* 2013;**52**:12283–7.
45. Ye S, Price JC, Barr EW, Green MT, Bollinger JM, Krebs C, et al. Cryoreduction of the NO-adduct of taurine:alpha-ketoglutarate dioxygenase (TauD) yields an elusive $\{FeNO\}^8$ species. *J Am Chem Soc* 2010;**132**:4739–51.
46. Surawatanawong P, Sproules S, Neese F, Wieghardt K. Electronic structures and spectroscopy of the electron transfer series $[Fe(NO)L_2]z$ (z = 1+, 0, 1−, 2−, 3−; L = dithiolene). *Inorg Chem* 2011;**50**:12064–74.
47. Averill BA. Dissimilatory Nitrite and Nitric Oxide Reductases. *Chem Rev* 1996;**96**:2951–64.
48. (a) Einsle O, Messerschmidt A, Huber R, Kroneck PMH, Neese F. Mechanism of the six-electron reduction of nitrite to ammonia by cytochrome c nitrite reductase. *J Am Chem Soc* 2002;**124**:11737–45.

(b) Bykov D, Neese F. Six-electron reduction of nitrite to ammonia by cytochrome c nitrite reductase: insights from density functional theory studies. *Inorg Chem* 2015;**54**:9303–16.

49. Subedi H, Hassanin HA, Brasch NE. Kinetic and mechanistic studies on the reaction of the vitamin B12 complex aquacobalamin with the HNO donor Angeli's salt: Angeli's salt and HNO react with aquacobalamin. *Inorg Chem* 2014;**53**:1570–7.
50. Montenegro AC, Bari SE, Olabe JA. Reactivity of iron(II)-bound nitrosyl hydride (HNO, nitroxyl) in aqueous solution. *J Inorg Biochem* 2013;**118**:108–14.
51. Cheney RP, Simic MG, Hoffman MZ, Taub IA, Asmus KD. One-electron reduction of pentacyanonitrosylferrate(II) ion in aqueous solution. *Inorg Chem* 1977;**16**:2187–92.
52. Schwane JD, Ashby M. FTIR investigation of the intermediates formed in the reaction of nitroprusside and thiolates. *J Am Chem Soc* 2002;**124**:6822–3.
53. (a) Poskrebyshev GA, Shafirovich V, Lymar SV. Hyponitrite radical, a stable adduct of nitric oxide and nitroxyl. *J Am Chem Soc* 2004;**126**:891–9.
    (b) Lymar SV, Shafirovich V, Poskrebyshev GA. One-electron reduction of aqueous nitric oxide: a mechanistic revision. *Inorg Chem* 2005;**44**:5212–21.
54. Poskrebyshev GA, Shafirovich V, Lymar SV. Disproportionation pathways of aqueous hyponitrite radicals ($HN_2O_2^{\bullet}/N_2O_2^{\bullet-}$). *J Phys Chem* 2008;**112**:8295–302.
55. Clarkson SG, Basolo F. Study of reaction of some cobalt nitrosyl complexes with oxygen. *Inorg Chem* 1973;**12**:1528–34.
56. Subedi H, Brasch NE. Mechanistic studies on the reaction of nitroxylcobalamin with dioxygen: evidence for formation of a peroxynitritocob(III)alamin intermediate. *Inorg Chem* 2013;**52**:11608–17.
57. Videla M, Roncaroli F, Slep LD, Olabe JA. Reactivity of reduced nitroprusside, $[Fe(CN)5NO^{\bullet}]^{3-}$, toward oxygen. *J Am Chem Soc* 2007;**129**:278–9.

The Interaction of HNO With Transition Metal Centers and Its Biological Significance. Insight Into Electronic Structure From Theoretical Calculations

8

C. Van Stappen, L.E. Goodrich and N. Lehnert
University of Michigan, Ann Arbor, MI, United States

Chapter Outline

The Chemistry and Biology of Nitroxyl (HNO). DOI: http://dx.doi.org/10.1016/B978-0-12-800934-5.00008-6

8.1 Introduction

Nitroxyl, HNO, is a very versatile molecule in biological systems as demonstrated by the many contributions to this book, which is solely devoted to this topic. Whereas NO itself was established as a signaling and immune defense agent in mammals in the 1980s,[1–6] the role of HNO in biological systems is still controversial.[7] The main reason for this is that HNO is quite unstable compared to NO, and is easily oxidized to NO and other NO_x species.[8] In addition, at higher concentrations HNO dimerizes at a diffusion limited rate to yield N_2O and water. It is therefore very difficult to directly detect HNO in biological systems,[9] and to prove beyond doubt that physiological effects observed by application of HNO donors to laboratory animals are in fact directly caused by HNO. A famous example for this is the question of whether HNO can effect vasodilation by directly binding and activating soluble guanylate cyclase (sGC) (see Section 8.2),[10] or whether this is due to a secondary effect that causes an increase in NO concentration (in the simplest case by HNO oxidation) that then induces vasodilation via the established NO-related pathway. As new detection methods and probes are developed for HNO that can be used in vivo[11] and that are selective,[12] the question of whether or not HNO is an endogenously produced signaling molecule in mammals will likely be answered in the near future. Proposals for endogenous HNO generation include reduction of NO by cyctochrome *c* oxidase and globins,[13] reaction of S-nitrosothiols with excess thiol,[14] reduction of NO by ascorbate or tyrosine,[15] or perhaps most prominently through nitric oxide synthase mediated oxidation of L-arginine in the absence of the tetrahydrobiopterin cofactor[16] (see also chapter: Hydroxylamines With Organic-Based Leaving Groups as HNO Donors, "Interaction of HNO with Small Molecules," and chapter: Global Kinetic Analysis and Singular Value Decomposition Methods Applied to Complex Multicomponent Reactions of HNO, "Is Azanone Endogenously Produced in Mammals?", in this book). Ferrous nonheme iron NO complexes show quite mild redox potentials, and could serve as platforms for HNO generation by simple one-electron reduction and protonation.[17–19] In addition, HNO has also been proposed to be an intermediate in the catalytic cycles of a number of enzymes involved in NO_x interconversion, in particular assimilatory and multiheme nitrite reductases[20–23] and P450 nitric oxide reductase.[24–26]

On the other hand, it is clear that HNO has a number of important pharmacological effects in humans that are distinct from NO, and hence, it is interesting to develop HNO donors as prodrugs that release HNO in vivo.[27] In particular, HNO has emerged as a potential drug against ischemia-reperfusion injury in heart attack and stroke.[28] One important target for HNO in biological systems are thiols. In particular, HNO is an irreversible inhibitor of mitochondrial aldehyde dehydrogenase[29] and glyceraldehyde-3-phosphate dehydrogenase[30] through cysteine modification, and an oxidant of thiols to the corresponding disulfides with formation of hydroxylamine[13] (see chapter: The Reactions of HNO With Nonheme Proteins: An Emphasis on Thiol-Containing Proteins, "Reactions of HNO with Thiol Containing Proteins," in this book).

Besides thiols, transition metal centers are thought to be the prime targets for the detection (sensing) and generation of HNO in biological systems.[9] HNO is trapped very efficiently by ferric hemes with secondary rate constants in the range of 10^5 $M^{-1}s^{-1}$, generating corresponding ferrous heme-nitrosyl complexes[31]:

$$[Fe^{III}(porphyrin)(L)] + HNO \rightarrow [Fe^{II}(porphyrin)(L)(NO)] + H^+ \tag{8.1}$$

The latter complexes are classified as $\{FeNO\}^7$ in the Enemark-Feltham notation, where the exponent "7" indicates the number of valence electrons (six d-electrons for Fe(II) plus one electron that occupies the π^* orbitals of NO).[32] Ferrous hemes coordinate HNO and form stable complexes, as demonstrated for myoglobin (Mb) and other heme proteins by Farmer and coworkers.[33] Here, the steric protection of the HNO complex in the active sites of these heme proteins is key for the stability of these species as evident from corresponding model complexes. Using simple porphyrin ligands like tetraphenyl- and octaethylporphyrin, ferrous HNO complexes are unstable and disproportionate readily to release H_2[34]:

$$2\,[Fe^{II}(porphyrin)(HNO)] \rightarrow [Fe^{II}(porphyrin)(NO)] + H_2 \tag{8.2}$$

However, the HNO complex can be stabilized by steric protection, using a bis-picket fence porphyrin,[35] which emphasizes the bimolecular nature of the decomposition reaction in Eq. (8.2). Model systems can also be used to prepare the corresponding Fe(II)-NO^- complexes, $\{FeNO\}^8$, which formally correspond to deprotonated Fe(II)-HNO complexes. These species are very basic and can be protonated with weak acids,[35] indicating that $\{FeNO\}^8$ complexes cannot exist at physiological pH.[17] Other targets for HNO are copper and nonheme iron enzymes. With respect to the latter, the first stable nonheme iron $\{FeNO\}^8$ complex was recently reported, and it was shown that this complex is readily protonated by weak acids.[19] A putative HNO complex has also been generated by cryoreduction of the $\{FeNO\}^7$ precursor in the enzyme taurine dioxygenase (TauD).[36]

This chapter focuses on the interaction of HNO with transition metal complexes, in particular hemes, using computational approaches to gain insight into the geometric and electronic structures of these complexes. As mentioned above, the question whether HNO could directly activate the mammalian NO sensor protein sGC is still a matter of heated debate, and is therefore further discussed in Sections 8.2 and 8.3 of this chapter. This includes computational work into the geometric and electronic structure of the HNO adduct of Mb. Finally, computational methods are used to estimate the pK_a of coordinated HNO in hemes and other transition metal complexes.

8.2 HNO and hemes: electronic structure and relevance for soluble guanylate cyclase

As described above, HNO has chemical properties, and thus physiological effects, that are distinct from that of NO,[9,37] yet both small molecules have been reported to activate sGC, which is a ferrous heme enzyme that serves as the primary NO sensor in mammalian physiology and that is involved in vasodilation and nerve signal transduction in the brain.[10] sGC is a ~150 kDa heterodimer that consists of three distinct domains: the N-terminal sensing domain that contains a ferrous heme, the C-terminal catalytic domain, and a dimerization domain that constitutes the interface between them. Upon NO coordination to the ferrous heme in the sensor domain, a large conformational change of the whole enzyme takes place[38] which activates the catalytic domain for the conversion of guanosine triphosphate (GTP) to the biochemical messenger cyclic guanosine monophosphate, cyclic guanosine monophosphate (cGMP).[39] Work by Marletta and coworkers has shown that sGC responds to NO in a two-step activation mechanism,[40,41] which implies that the sensing domain contains two distinct binding sites for NO. A five-coordinate, ferrous heme *b* ligated by a proximal histidine ligand, His105 in human and bovine sGC,[42] serves as the primary, high-affinity binding site for NO. Upon coordination of NO to the heme, an intermediate six-coordinate complex is formed first, which, due to the strong thermodynamic σ-*trans* effect (also called *trans* "interaction") of NO induces cleavage of the $Fe{-}N_{His105}$ bond to form the activated, five-coordinate heme-nitrosyl complex.[43–47] Release of His105 from the heme induces a conformational change in the sensing domain which activates the catalytic domain of the enzyme.[48] The second, low-affinity binding site is potentially a thiol of a cysteine residue.[49] Alternatively, the high-activity form of sGC could correspond to a five-coordinate complex where NO is bound at the proximal side of the heme.[48]

While NO is now established as the endogenous activator of sGC,[50] several reports have shown that HNO donors induce an increase in sGC-dependent vasodilation.[51–54] However, as mentioned in the Introduction, HNO can readily be oxidized to NO, e.g., by superoxide dismutases (SODs), and the NO generated in this way could then activate sGC through the previously discussed NO-dependent pathway.[55,56] Hence, the key question that needs to be addressed is whether HNO could *directly* activate sGC. Dierks and Burstyn reported initially that HNO is unable to activate sGC.[57] However, these experiments were conducted in the presence of 10 mM DTT in the buffer, and hence, it is possible that HNO was scavenged prior to reaction with sGC in these experiments. In agreement with these results, Mayer and coworkers confirmed that HNO does not activate sGC in the absence of SOD.[58] On the contrary, Miller et al. reported sGC activity in thiol and O_2-free buffer in the presence of the HNO donors 1-nitrosocyclohexyl trifluoroacetate and Angeli's salt.[59] This activity was found to be 1.9- and 3.4-fold lower than NO-induced activity at similar concentrations. Further experiments were performed in heme-free sGC, which led to decreased HNO-induced sGC activation, indicating that activation occurs predominately at the heme center. Interestingly, HNO-mediated cysteine thiol modification led to inhibition of enzyme activity.[59]

8.2.1 The thermodynamic σ-trans effect of NO[47]

With these conflicting reports in hand, the key question remains if it is chemically feasible for HNO to activate sGC. Experimentally-calibrated density functional

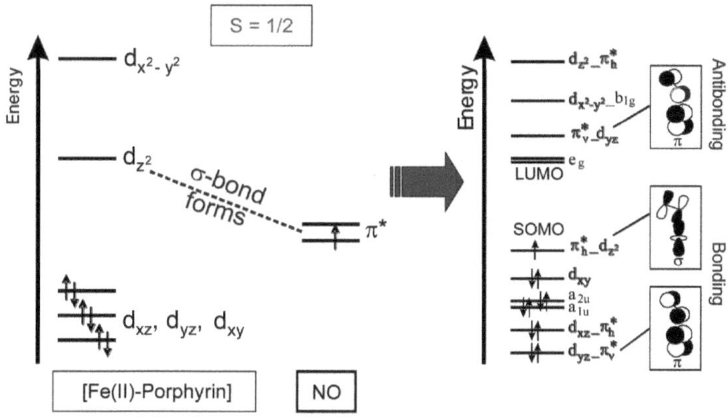

Scheme 8.1 Electronic structure of five-coordinate ferrous heme-nitrosyls.

theory (DFT) calculations show that the binding of NO to a ferrous heme leads to the formation of a strong σ bond between the singly occupied π* orbital of NO and the empty d_{z2} orbital of the low-spin iron(II) center.[60–63] If the heme is five-coordinate (5C), the resulting bonding combination, which constitutes the SOMO (singly occupied molecular orbital) of the complex, has roughly equal contributions from d_{z2} and the singly occupied π^*_h orbital of NO (h = horizontal, the π* orbital located in the Fe-N-O plane; see Scheme 8.1). This leads to a spin density distribution with 50% on Fe and 50% on NO as indicated in Fig. 8.1, left. In addition, the second π* orbital of NO, π^*_v (v = vertical, the π* orbital orthogonal to the Fe-N-O plane; see Scheme 8.1), forms a strong π-backbond with the appropriate $t_{2(g)}$ type d-orbital of iron (d_{yz} in Scheme 8.1). NO therefore acts as a "classic" σ-donor/π-acceptor ligand in ferrous heme-nitrosyl complexes. In the presence of an axial imidazole ligand, the Fe–NO bond becomes weaker due to a reduction of the covalency of the Fe–NO bond, which is directly evident from the comparison of the Fe–NO force constants in analogous five- and six-coordinate (6C) complexes (Fe–NO force constants are 2.9–3.0 mdyn $Å^{-1}$ for 5C and 2.5–2.6 mdyn $Å^{-1}$ for 6C ferrous heme-nitrosyls). This is caused by a thermodynamic σ-*trans* effect (σ-*trans* interaction) between NO and the axial imidazole ligand. This σ-*trans* interaction can be understood as a competition between the σ-donor orbitals of NO and the imidazole ligand for the d_{z2} orbital of iron. Here, NO makes the stronger (more covalent) interaction, and hence, in the SOMO of the complex, the interaction between d_{z2} of iron and π^*_h of NO is bonding, whereas the interaction between d_{z2} and the imidazole σ-donor orbital, Im(σ), is antibonding. Fig. 8.1, right, and Scheme 8.2 illustrate this point. The Fe–N_{IM} antibonding interaction in the SOMO leads to a great weakening of the Fe–imidazole bond, which is directly evident from the small binding constants of N-donor ligands to five-coordinate ferrous heme-nitrosyls in model systems.[61] Corresponding binding constants for the reaction:

$$[Fe^{II}(porphyrin)(NO)] + N\text{-donor} \rightarrow [Fe^{II}(porphyrin)(N\text{-donor})(NO)] \qquad (8.3)$$

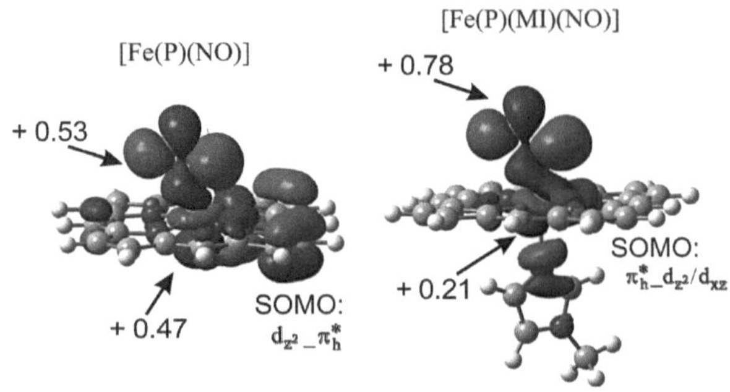

Figure 8.1 Contour plots of the SOMOs of [Fe(P)(NO)] and [Fe(P)(MI)(NO)] (P^{2-} = porphine dianion) calculated with B3LYP/LanL2DZ*. Calculated for the fully optimized structures obtained with BP86/TZVP. Spin densities on Fe and NO are indicated. *Source*: Reprinted with permission from Praneeth VKK, Näther C, Peters G, Lehnert N. Spectroscopic properties and electronic structure of five- and six-coordinate iron(II) porphyrin NO complexes: effect of the axial N-donor ligand. *Inorg Chem* 2006; 45: 2795–2811. Copyright 2006 American Chemical Society.

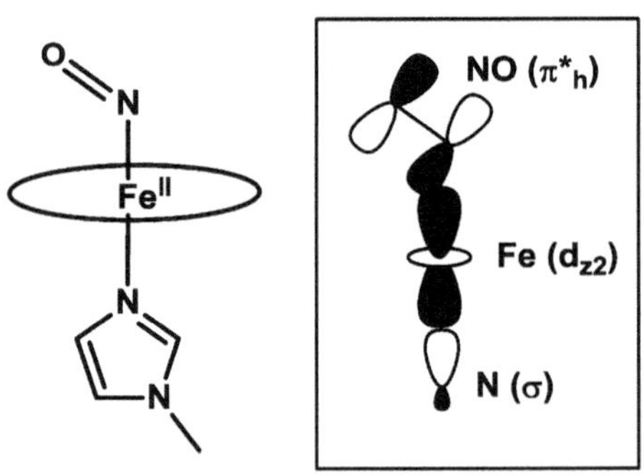

Scheme 8.2 The key Fe–NO σ-bonding orbital of six-coordinate ferrous heme nitrosyls.[46,61]

are generally found in the 1–100 M^{-1} range, compared to imidazole binding constants to ferrous hemes, which are in the 10^4–10^5 M^{-1} range.[64] It is therefore challenging to prepare six-coordinate ferrous heme-nitrosyls in the absence of a large excess of the desired N-donor ligand. This requires the application of specifically designed, "tailed" porphyrins.[65,66] Further evidence that the σ-bonding SOMO is in fact responsible for the *trans*-interaction in ferrous heme-nitrosyls was derived from spectro-electrochemical studies. One-electron reduction of ferrous heme-nitrosyls leads to a double-occupation of the SOMO, creating corresponding low-spin $\{FeNO\}^8$, or ferrous heme-nitroxyl, complexes. These compounds show dramatically increased σ-*trans* interactions and stronger Fe–NO bonds, directly supporting the electronic structure descriptions provided above, and these conclusions are further supported by DFT calculations.[35] In model complexes of six-coordinate ferrous heme-nitrosyls, the σ-*trans* interaction further manifests itself in greatly increased

Fe–N bond distances between iron and axial N-donor ligands like pyridine or imidazole, etc. For example, whereas Fe(II)-imidazole bond lengths in low-spin ferrous hemes are usually around 2.0 Å, these bond distances increase to 2.15–2.2 Å in the six-coordinate nitrosyl complexes (see Section 8.3).[67] This σ-*trans* interaction is therefore responsible for the dramatic weakening of the $Fe{-}N_{His105}$ bond upon NO binding to the sensor domain of sGC, and ultimately induces breaking of this bond and in this way, activates the catalytic domain of sGC. For detailed reviews on the geometric, spectroscopic, and electronic properties of ferrous heme-nitrosyls, see references 46,68,69.

8.2.2 CO binding to sGC

Interestingly, the weak σ-donor carbon monoxide (CO) has been shown to form a stable six-coordinate complex with the ferrous heme of sGC, where the $Fe{-}N_{His105}$ bond remains intact.[70,71] Correspondingly, CO causes only a low-level activation of sGC: 100% CO and 0.5% NO atmosphere result in 4.4- and 128-fold activation of sGC, respectively.[70] This is in agreement with the much weaker *trans*-effect of CO compared to NO. This weaker *trans*-effect of CO is also evident from spectroscopic and crystallographic data on tetraphenylporphyrin (TPP^{2-}) model complexes. First, the $Fe{-}N_{MI}$ stretching frequency, $\upsilon(Fe{-}N_{MI})$, in [Fe(TPP)(MI)(NO)] [63] (MI = 1-methylimidazole) is located at 149 cm^{-1}, whereas this mode is observed at 172/225 cm^{-1} for the analogous CO complex [Fe(TPP)(CO)(MI)] [72] and at 210–220 cm^{-1} in deoxy-Mb. Additionally, the crystal structures of these compounds show $Fe{-}N_{MI}$ bond lengths of 2.173 Å and 2.071 Å for the NO and CO complex, respectively.[67,73] By comparison, the $Fe{-}N_{MI}$ bond length of $[Fe(TPP)(MI)_2]$,[74] where no *trans*-interaction exists, is 2.014 Å. These differences originate from the fact that CO binding to ferrous hemes is dominated by strong π-backbonding, which does not give rise to a large thermodynamic *trans*-effect. Note that the low-level activation of sGC by CO has been suggested to be due to changes in heme conformation.[75–77]

8.2.3 HNO versus NO

In summary, a moderate lengthening of the $Fe{-}N_{His105}$ bond, as in the case of CO coordination, does not seem to induce the conformational change necessary for high (as in the case of NO) catalytic activity for production of cGMP. Instead, a stronger thermodynamic *trans*-effect, as in the case of NO, is required. This poses the question of whether HNO coordination to a ferrous heme could induce a strong *trans*-effect as in the case of NO. Using computational results in direct correlation with experimental findings, this point is elucidated further in the following. Based on the detailed discussion of the electronic structure of ferrous heme-nitrosyls (see Section 8.2.1), the *trans*-interaction between NO and its *trans* ligand is mediated by the SOMO of the complex, which corresponds to the σ-bonding orbital between d_{z2} of iron and π^*_h of NO (see Scheme 8.2). In the simplest approach, we can ask the question how this MO compares between analogous ferrous NO and HNO complexes. Fig. 8.2 shows a comparison of the corresponding MOs, calculated with

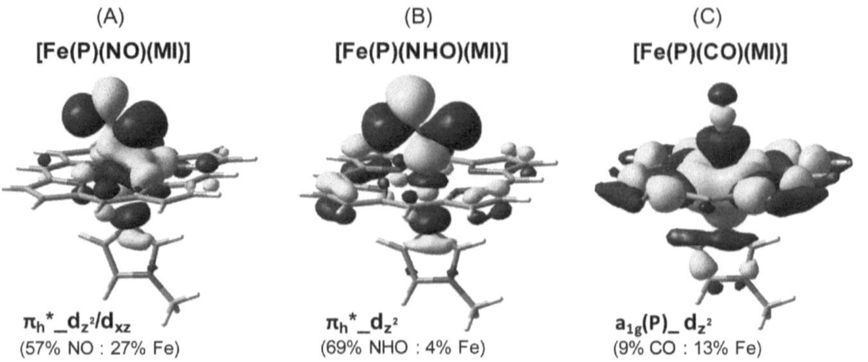

Figure 8.2 Relevant molecular orbitals of (A) [Fe^{II}(P)(NO)(MI)], (B) [Fe^{II}(P)(NHO)(MI)], and (C) [Fe^{II}(P)(CO)(MI)] which define the thermodynamic σ-*trans* effect in these ferrous porphyrin systems. Calculated using B3LYP/TZVP on the BP86/TZVP optimized structures.[78]

B3LYP/TZVP. For [Fe(P)(MI)(NO)] (P^{2-} = porphine ligand), the important Fe−NO σ bonding orbital shows 57% π^*_h and 27% Fe d_{z2}/d_{xz} contributions, corresponding to a strong σ-bond, inducing a large *trans*-interaction. In contrast, in the HNO-bound model system, the important Fe−NO σ-bonding orbital shows 69% π^*_h character *and only 4% Fe d_{z2} contribution*. This dramatic decrease in the Fe d_{z2} percentage in the HOMO of the HNO complex indicates a much smaller σ-*trans* effect for HNO, i.e., a weaker Fe−N_{MI} antibonding interaction, thus making it unlikely that this molecule could induce the cleavage of the Fe−histidine bond in sGC.[46,78] This also indicates that HNO is mostly a π-backbonding ligand (with the main π-backbonding (occupied) MO containing 48% d_{yz} and 15% π^*_v character), similar to CO.

Comparison to the ferrous CO complex, [Fe(P)(CO)(MI)], further supports this conclusion. The contour plot of the corresponding Fe−CO bonding orbital in Fig. 8.2 does not show significant CO contributions, in agreement with a weak σ-*trans* effect of this diatomic. Interestingly, this orbital contains 13% d_{z2} which is higher than that of HNO. The stronger d_{z2} contribution (compared to HNO), however, is counteracted by the fact that the overall CO contribution to this orbital is quite small, only 9%. In fact, most of the contributions to this molecular orbital originate from the porphyrin(a_{1g})_d_{z2} bonding interaction (this molecular orbital has 67% porphyrin character). Overall, we would predict HNO and CO to exert similar thermodynamic σ-*trans* effects in six-coordinate ferrous porphyrin systems.

Further support for the weakened *trans*-effect of HNO relative to NO is obtained by the BP86/TZVP optimized Fe−N_{MI} bond lengths for the six-coordinate structures [Fe(P)(MI)(X)] with X = HNO or NO,[78] where values of 2.082 Å and 2.179 Å are obtained, respectively. These values are in fact in very good agreement with the experimental data provided in Table 8.2 (see Section 8.3). This comparison indicates that HNO does not induce a *trans*-effect of the same magnitude as NO. HNO does, however, induce slightly longer Fe−N_{MI} bond lengths than both CO, 2.068 Å, and MI, 1.994 Å in the BP86/TZVP optimizations. Nevertheless, since the BP86/TZVP calculated $Fe{-}N_{MI}$ distances with either CO or HNO differ by only ~0.01 Å, one would expect that the *trans*-interaction induced by these ligands is quite similar, indicating that HNO will behave much more similarly to CO as compared to NO when bound to a ferrous heme.[78] This clearly suggests that HNO cannot chemically induce the cleavage of the $Fe{-}N_{His105}$ bond in sGC.

8.2.4 Mb-HNO structure

Most of the structural information currently available about ferrous heme–HNO complexes is based on the HNO adduct of ferrous Mb, which was first reported by Farmer and coworkers.[79,80] Unfortunately, crystallographic information is not available for this species, but based on EXAFS and other spectroscopic data the basic structural features of the six-coordinate HNO adduct of Mb have been determined.[81] The overall, proposed structure of the Fe–NHO unit is illustrated in Fig. 8.3. Here, the nitroxyl ligand is bound through its N-atom to ferrous heme and is also N-protonated, with Fe–N and N–O distances of 1.82 and 1.24 Å and an Fe-N-O angle of 131°. The $Fe{-}N_{His}$ distance is 2.09 Å, which compares well with the value of 2.082 Å calculated for the simple model system [Fe(P)(MI)(NHO)] as discussed in Section 8.2.3. Further computational work has shown that HNO ligands are generally N-protonated when bound to a ferrous heme, with the O-protonated form being higher in energy by >20 kcal mol^{-1}.[25] The measured N–O stretching frequency of 1385 cm^{-1} for the Mb(II)-HNO adduct[81] is also a direct reflection of N-protonation, whereas the O-protonated form has been predicted to

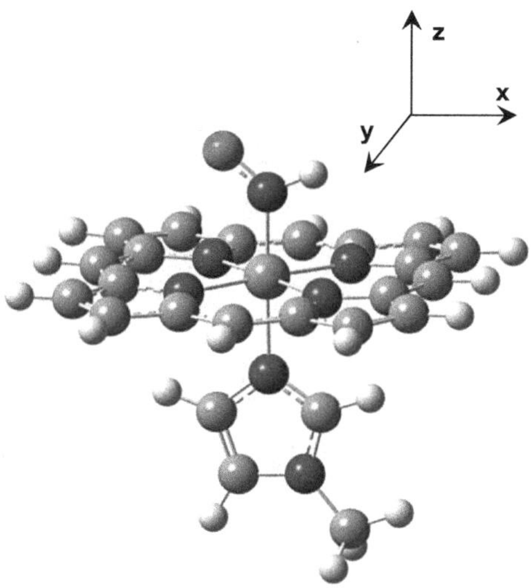

Figure 8.3 The model system [Fe(P)(MI)(X)], where P = porphine^{2-}, MI = 1-methylimidazole, and X = NHO, and applied coordinate system. The structure shown is calculated with BP86/TZVP.[78]

have a much lower N–O stretch.[25] Finally, nuclear magnetic resonance (NMR) spectroscopy also directly shows N-protonation of the HNO ligand.[33]

One interesting question is how the structure of the Mb(II)-NHO adduct is further influenced by second coordination sphere effects, in particular hydrogen bonding to the distal His of Mb. This aspect was studied in detail using computational methods by Zhang and coworkers.[82,83] It has been reported previously, on the basis of NMR observations, that several distinct isomers of the Mb(II)-NHO complex may exist, depending on the protonation state of the distal histidine residue, leading to differences in second coordination sphere interactions.[33] The formation of these isomers and their stabilities have been closely studied from a computational perspective[81] in order to improve previous models[33,84] and better fit the experimentally observed spectroscopic parameters of the system, such as the ^{15}N-NMR and ^{1}H-NMR chemical shifts of bound HNO, as well as the observed N–O vibrational frequency, ν(N–O). These spectroscopic parameters are particularly useful due to their high sensitivity to the electronic structure of the HNO complexes, with characteristic peaks for Mb(II)-NHO found at 14.9 and 550 ppm in the ^{1}H- and ^{15}N-NMR spectra, and an N–O stretching frequency of 1385 cm^{-1}.[33,81,85]

In order to determine the exact structures of the potential isomers, a series of active site models, based on Mb(II)-HNO, were set forth which vary in the number of hydrogen bond participants, the hydrogen bonding mode of the distal histidine, and the extent of histidine backbone inclusion into the model. Two isomers were generated, one in which the N_ε-position of histidine is protonated (**1**), resulting in a (HisH•••ONH) interaction, and one where the N_δ-position of histidine is protonated (**2**), resulting in a (His•••HNO) hydrogen bonding interaction; a comparison of these computed structures is provided in Fig. 8.4. Initial structures were formulated based on the Mb(II)-CO X-ray crystal structure (PDB code: 1DWR).[86] Peripheral substituents of the heme molecule were annotated to the porphyrin ring, and histidines were annotated to their α, β, or γ-carbon pending on the model employed. On the basis of their previous investigation into the

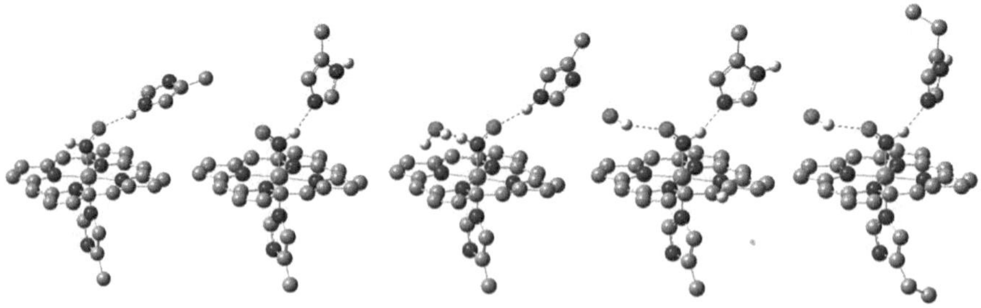

1 2 3 4 4′

Figure 8.4 The Mb(II)-NHO model systems employed by Zhang and coworkers[81] in which the distal histidine may be either N_ε or N_δ protonated. Hydrogens of the porphyrin ring and of histidine residues have been excluded for clarity, with the exception of the N_ε-position of the distal histidine as this is necessary for illustrating the hydrogen bonding in the system.

electronic structure of Mb(II)-NHO,[83] Zhang and coworkers performed geometry optimizations and frequency calculations utilizing the mPWVWN method with the Wachters basis set for Fe, the 6-311++G** basis for HNO and the first coordination shell atoms of iron, and the 6-31G(d) basis for the remaining atoms. Additional computational details may be found in the supplement of their paper.[82]

For diagnostic purposes, the calculated binding energies of HNO to Mb(II) were compared to those found previously for singlet oxygen, $^1O_2$, since it has been proposed by Farmer and Sulc that these two isoelectronic molecules share analogous hydrogen bonding interactions, yet HNO has a much higher binding affinity than $^1O_2$.[87] Previous studies have shown $^1O_2$ binding to be stabilized by 7.1 kcal mol^{-1} through hydrogen bonding with the distal histidine.[88] However, the models **1** and **2** were found to be 3.8 and 4.3 kcal mol^{-1} less stabilized than the $^1O_2$ complex. In order to obtain more reasonable estimates of the hydrogen bond stabilization in the HNO complex, it is important to recognize that HNO, unlike $^1O_2$, can be simultaneously involved in two hydrogen bonding interactions. This kind of interaction is well recognized[83,89] and with the addition of a water molecule to models **1** and **2**, models **3** and **4** were generated, leading to 4–5 kcal mol^{-1} greater stabilization than that found for $^1O_2$. In model **3**, water acts as a hydrogen bond acceptor, while in model **4** water acts as a hydrogen bond donor. Model **4′** in Fig. 8.4 constitutes a further refinement of model **4**, where a portion of the amino acid backbone is included to give a better representation of the protein environment.

With the double hydrogen bonding models in hand, further comparisons to the experimentally determined spectroscopic data can then be made. The presence of two isomers was initially conjectured on the basis of ^{1}H-NMR data where Mb(II)-NHO prepared via a nitrite/borohydride mixture produced two proton resonances at 14.93 and 14.87 ppm, referred to as isomers A and B. Under alternative reaction conditions, only a single peak for the A isomer was observed at 14.93 ppm. Additionally, in the analogous hemoglobin complex, the ^{15}N-NMR shift was observed at 560 ppm for the A isomer, which is shifted slightly downfield to 556 ppm for the B isomer.[33] Thus, a trend of a downfield shift for both the proton and nitrogen NMR HNO signals in the B isomer is evident from experiment. Comparing calculated values, model **3** in which the distal histidine is N_ε-protonated is consistently shifted downfield in both ^{1}H- and ^{15}N-NMR resonances relative to the N_δ-protonated model **4** (regardless of slight variations in the model), as documented in Table 8.1. While these types of calculations may not provide extremely accurate values relative to experiment, they are usually internally consistent, which strongly indicates that the identity of the A isomer is in fact the N_δ-protonated (His•••HNO•••H_2O) structure represented by model **4**, while the B isomer matches well with the N_ε-protonated (H_2O•••HNO•••HHis) model **3**. In general, such computational studies have proven particularly useful in examining specific protein active site interactions and structural isomers that may be very difficult to directly observe in experiment, particularly when X-ray crystal structural data are unavailable.[90–92] For example, Capece and coworkers have used extensive QM/MM calculations to explore distal pocket effects in globins (with a focus on Mb), in particular with respect to the binding of diatomic ligands such as O_2, NO, and

Table 8.1 Experimental and calculated N–O stretching frequencies and NMR chemical shifts of Mb(II)-NHO and of complementing computational models (P^{2-} = porphine dianion)[82]

	System	υ (cm^{-1})	δ_H ppm	δ_N ppm
	Mb(II)-NHO A	1385	14.93	661
	Mb(II)-NHO B	–	14.87	–
1	Fe(P)(His)(HNO•••His)	1374	14.88	605
2	Fe(P)(His)(His•••HNO)	1400	15.55	665
3	Fe(P)(His)(H_2O•••HNO•••His)	1384	15.03	649
4	Fe(P)(His)(His•••HNO•••H_2O)	1380	15.10	664
4′	Fe(P)(His)(His•••HNO•••H_2O)	1384	15.13	658

CO.[93] Such investigations demonstrate the ability of computational methods to model the behavior of even relatively large protein systems, and infer the nature and properties of ligand-binding sites within them.

8.3 Calculation of accurate binding constants using DFT methods: application to soluble guanylate cyclase activation by NO, HNO, and CO

To better understand the difference in sGC activation by NO, HNO, and CO, we employed DFT calculations to investigate the thermodynamic σ-*trans* effect induced by these ligands on 1-methylimidazole (MI) in the ferrous heme model complexes [Fe(P)(MI)(X)] (X = NO, NHO, CO, MI).[78] For this purpose, we used DFT total energy calculations to evaluate the binding of MI to five-coordinate ferrous porphyrins in *trans* position to NO, HNO, CO, and MI as a way to (a) systematically assess the relative strength of the *trans*-interaction induced by each of these small molecules and (b) to calibrate DFT methods for the accurate calculation of weak binding constants in ferrous heme complexes. To decrease computational cost, the porphine approximation was applied and MI was used as a model for histidine ligation to the ferrous heme center. The optimized structure of the corresponding HNO complex, [Fe(P)(MI)(NHO)], obtained with BP86/TZVP, is shown in Fig. 8.3. Note that BP86/TZVP is a reliable method to predict the geometric structures of ferrous heme complexes with diatomics like NO, which is further illustrated in Table 8.2.[46,78]

8.3.1 Spin-state energetics

Prior to the calculation of binding energies, the ground states of all reactants need to be considered. Five- and six-coordinate ferrous heme-nitrosyl complexes have S = ½ ground states.[68] For X = NHO, CO, and MI the ferrous six-coordinate

Table 8.2 **Experimental and calculated geometric parameters of [Fe(P)(X)] and [Fe(P)(MI)(X)], where X = NO, HNO, CO, and MI**

Complex		Method	Geometric parameters [Å] [°]					References
			ΔFe–X	ΔX–O	ΔFe–N_{MI}	ΔFe–N_{porph}	<Fe-X-O	
[Fe(P)(NO)]	$S = ½$	*exp. [Fe(TPP)(NO)]*	1.739	1.163	–	2.000	144	[94]
		BP86/TZVP	1.704	1.179	–	2.019	146	
[Fe(P)(MI)(NO)]	$S = ½$	*exp. [Fe(TPP)(MI)(NO)]*	1.750	1.182	2.173	2.008	138	[95]
		BP86/TZVP	1.734	1.186	2.179	2.021	140	
		B3LYP/6-31G*	1.759	1.177	2.105	2.019	140	
		B3LYP/LanL2DZ*	1.787	1.172	2.105	2.030	140	
		mPWVWN/ 6-311++G**	1.745	1.188	2.518	2.045	141	
		OLYP/TZVP	*Fe–N_{MI} bond broken during optimization*					
[Fe(P)(NHO)]	$S = 0$	*exp.*	–	–	–	–	–	
		BP86/TZVP	1.741	1.235	–	2.004	131	
		B3LYP/6-31G*	1.732	1.226	–	2.002	131	
	$S = 2$	*exp.*	–	–	–	–	–	
		BP86/TZVP	1.953	1.246	–	2.097	131	
		B3LYP/6-31G*	1.953	1.246	–	2.097	131	
[Fe(P)(MI) (NHO)]	$S = 0$	*exp. Mb-NHO* (EXAFS)	1.82	1.24	2.09	2	131	[81]
		BP86/TZVP	1.789	1.236	2.082	2.016	132	
		B3LYP/6-31G*	1.789	1.226	2.060	2.017	132	
[Fe(P)(CO)]	$S = 0$	*exp.*	–	–	–	–	–	
		BP86/TZVP	1.705	1.166	–	2.001	180	
		B3LYP/6-31G*	1.716	1.154	–	2.001	180	
	$S = 2$	*exp.*	–	–	–	–	–	
		BP86/TZVP	1.999	1.149	–	2.093	180	
		B3LYP/6-31G*	2.371	1.133	–	2.070	180	

(*Continued*)

Table 8.2 (Continued)

Complex		Method	Geometric parameters [Å] [°]					References
			ΔFe–X	ΔX–O	ΔFe–N_{MI}	ΔFe–N_{porph}	<Fe-X-O	
[Fe(P)(MI)(CO)]	*S* = *0*	*exp. [Fe(TPP)(MI)(CO)]*	1.793	1.061	2.071	2.003	179	[73]
		BP86/TZVP	1.756	1.162	2.068	2.018	180	
		B3LYP/6-31G*	1.774	1.151	2.063	2.019	180	
[Fe(P)(MI)]	*S* = *0*	*exp.*	–	–	–	–	–	
		BP86/TZVP	1.900	–	–	1.995	–	
		B3LYP/6-31G*	1.924	–	–	2.002	–	
	S = 2	*exp. [Fe(TPP) (2-MeHIm)]*	2.127	–	–	2.073	–	[96]
		BP86/TZVP	2.146	–	–	2.086	–	
		B3LYP/6-31G*	2.158	–	–	2.086	–	
[Fe(P)(MI)(MI)]	*S* = *0*	*exp.* $[Fe(TPP)(MI)_2]$	2.014	–	2.014	1.997	–	[74]
		BP86/TZVP	1.994	–	1.994	2.071	–	
		B3LYP/LanL2DZ*	2.036	–	2.036	2.028	–	
		B3LYP/6-31G*	2.011	–	2.011	2.013	–	
		mPWVWN/ 6-311++G**	2.067	–	2.067	2.033	–	
		OLYP/TZVP	2.047	–	2.047	2.010	–	

complexes were computed as low-spin (S = 0), but for the corresponding five-coordinate complexes both the S = 0 and 2 spin states were included to determine the lowest energy spin state in each case. Whereas DFT predicts quite clearly that [Fe(P)(X)] with X = CO, NHO to be low-spin (experimentally not known for X = NHO), the spin state of [Fe(P)(MI)] is very ambiguous.[97] Experimentally, the latter complex is high-spin. The reliable prediction of the spin states of transition metal complexes is very challenging for DFT,[98,99] and as will be evident from the calculated binding constants, much of the problems that DFT is facing in the accurate prediction of binding constants goes back to incorrectly calculated spin state energetics. *Ligand binding that is accompanied by a spin state change is therefore particularly difficult to describe quantitatively by DFT.*

8.3.2 Method calibration: calculation of binding constants for 1-methylimidazole ligation to [Fe(P)(NO)] and [Fe(P)(MI)]

As a measure of the σ-*trans* interaction exerted by each of the four small molecules HNO, CO, NO, and MI in ferrous heme complexes, we have evaluated the binding constants (K_{eq}) of MI to the five-coordinate species [Fe(P)(X)] (for X = NHO, NO, CO, MI), which are obtained from the calculated ΔG values at 298.15K for the reaction:

$$[\text{Fe(P)(X)}] + \text{MI} \rightleftarrows [\text{Fe(P)(MI)(X)}] \tag{8.4}$$

Due to the lack of experimental binding constants for both X = NHO and CO, the NO and MI complexes were utilized first to assess the accuracy of DFT to calculate these binding energies (Table 8.3) and K_{eq} values (Table 8.4). Since gradient-corrected functionals generally overestimate metal-ligand covalencies, and also show strong preference for low-spin states, accurate metal-ligand binding energies are often times only available from hybrid functionals.[46] Given that BP86/TZVP generates good structures at low computational cost, we used these structures and then calculated binding energies, ΔE, with a large number of methods as listed in Table 8.3. Calculated basis set superposition errors, usually around 2 kcal mol^{-1}, are included in the ΔE values. Thermal and entropic corrections (taken from the BP86/TZVP calculations) are further applied to the calculated binding energies, resulting in the free energies, ΔG at 298.15K, listed in Table 8.3. Importantly, this thermal correction is essentially method independent (with typical errors of less than 5%) and, as a result, we can apply the BP86/TZVP thermal corrections to energies calculated with alternate methods.[78]

The experimental ΔG value for binding MI to [Fe(TPP)(NO)][61] *is* -1.9 *kcal mol*$^{-1}$ *and, using a calculated thermal correction of about 14.0 kcal mol*$^{-1}$*, the binding energy can be estimated around* -16 *kcal mol*$^{-1}$. The B3LYP/TZVP calculated binding energy for MI ligation to [Fe(P)(NO)] (Table 8.3) is, however, only -3.5 kcal mol^{-1}; underestimated by >10 kcal mol^{-1}. Similarly, ΔG for binding MI to [Fe(TPP)(MI)] is -6.7 kcal mol^{-1} experimentally. After applying a 18.9 kcal mol^{-1} thermal correction,

Table 8.3 Reaction energies (kcal mol^{-1}) for $[Fe(P)(X)] + MI \rightleftarrows [Fe(P)(X)(MI)]$ at 298.15K

	SP method	X = NO	X = NHO[b]		X = CO[b]		X = MI[b]	
		$S = ½$[c]	$S = 0$[c]	$S = 2$[c]	$S = 0$[c]	$S = 2$[c]	$S = 0$[c]	$S = 2$[c]
ΔE ΔG	*Experimental*[a]	**(−16)** **−1.9**						**(−26)** **−6.7**
BP86/TZVP optimized geometries								
ΔE	BP86/TZVP	−2.4	**−11.7**	−41.0	**−12.5**	−49.3	**−17.7**	−31.0
	B3LYP/TZVP	−3.5	**−11.9**	−13.7	**−12.2**	−34.8	−13.7	**−5.1**
	B3LYP*/TZVP	−3.8	**−12.3**	−20.5	**−12.6**	−29.0	−15.2	**−14.8**
	B3LYP-D/TZVP	−16.8	**−26.2**	−29.2	**−26.2**	−37.1	−28.6	**−26.7**
	B3LYP*-D/TZVP	−17.3	**−26.7**	−36.9	**−26.8**	−44.9	**−30.2**	−34.5
	OLYP/TZVP	−17.3	**0.3**	−11.6	**−0.7**	−23.1	−5.2	**0.6**
	O3LYP/TZVP	−2.2	**−12.7**	−20.9	**−13.3**	−31.8	−17.0	**−11.3**
	O3LYP*/TZVP	−0.9	**−13.2**	−15.6	**−13.8**	−27.1	−16.8	**−6.5**
	O3LYP-D/TZVP	−17.6	**−29.1**	−39.6	**−29.5**	−50.0	−34.2	**−33.9**
	O3LYP*-D/TZVP	−16.3	**−29.6**	−34.4	**−30.1**	−45.2	−34.0	**−29.0**
Thermal correction		*14.0*	*13.8*	*19.6*	*13.7*	*19.9*	*14.8*	*18.9*
ΔG	BP86/TZVP	11.6	2.1	−21.4	1.2	−29.4	−2.9	−12.1
	B3LYP/TZVP	10.5	1.9	5.9	1.5	−14.9	1.1	13.8
	B3LYP*/TZVP	10.2	1.5	−0.9	1.1	−9.1	−0.4	4.1
	B3LYP-D/TZVP	−2.8	−12.4	−9.6	−12.5	−17.2	−13.8	−7.8
	B3LYP*-D/TZVP	−3.3	−12.9	−17.3	−13.1	−25.0	−15.4	−15.6
	OLYP/TZVP	−3.3	14.1	8.0	13.0	−3.2	9.6	19.5
	O3LYP/TZVP	11.8	1.1	−1.3	0.4	−11.9	−2.2	7.6
	O3LYP*/TZVP	13.1	0.6	4.0	−0.1	−7.2	−2.0	12.4
	O3LYP-D/TZVP	−3.6	−15.3	−20.0	−15.8	−30.1	−19.4	−15.0
	O3LYP*-D/TZVP	−2.3	−15.8	−14.8	−16.4	−25.3	−19.2	−10.1

[a]Determined for the corresponding TPP complexes [Fe(TPP)(MI)(NO)] and $[Fe(TPP)(MI)_2]$. ΔE is predicted using experimental ΔG values and BP86/TZVP predicted thermal and entropic corrections (see text).
[b]For [Fe(P)(CO)] and [Fe(P)(NHO)] the spin state of the complex is not known and could either be high-spin ($S = 2$) or low-spin ($S = 0$). Hence, both spin states were considered here. For completion, we also included the low-spin state of [Fe(P)(MI)]. In all of these cases, ΔE values printed in bold represent the ones for which the spin state of the five-coordinate precursor is predicted by DFT to have the lowest energy.
[c]Spin state of the five-coordinate precursor, the six-coordinate product (after MI binding) is always low-spin.

Table 8.4 Binding constants (M^{-1}) for [Fe(P)(X)] + MI ⇄ [Fe(P)(X)(MI)] at 298.15K. K_{eq} are calculated using the listed method with BP86/TZVP geometries

SP method	X = NO	X = NHO		X = CO		X = MI	
	$S = ½$[a]	$S = 0$[a]	$S = 2$[a]	$S = 0$[a]	$S = 2$[a]	$S = 0$[a]	$S = 2$[a]
Experimental	**26**						$\mathbf{7.8 \times 10^{4}}$
BP86	3.1×10^{-9}	2.9×10^{-2}	5.0×10^{15}	1.3×10^{-1}	3.5×10^{21}	1.4×10^{2}	7.2×10^{8}
B3LYP	1.9×10^{-8}	4.0×10^{-2}	5.0×10^{-5}	7.6×10^{-2}	7.9×10^{10}	1.7×10^{-1}	7.4×10^{-11}
B3LYP*	3.2×10^{-8}	7.8×10^{-2}	4.2	1.7×10^{-1}	4.5×10^{6}	1.8	9.2×10^{-4}
B3LYP-D	$\mathbf{1.1 \times 10^{2}}$	$\mathbf{1.1 \times 10^{9}}$	1.2×10^{7}	$\mathbf{1.5 \times 10^{9}}$	4.1×10^{12}	1.3×10^{10}	$\mathbf{5.6 \times 10^{5}}$
B3LYP*-D	2.7×10^{2}	2.7×10^{9}	4.5×10^{12}	4.3×10^{9}	2.0×10^{18}	1.9×10^{11}	2.7×10^{11}
OLYP	2.7×10^{2}	4.7×10^{-11}	1.3×10^{-6}	2.9×10^{-10}	2.1×10^{2}	8.5×10^{-8}	5.2×10^{-15}
O3LYP	2.2×10^{-9}	1.5×10^{-1}	8.7	5.2×10^{-1}	5.3×10^{8}	39	2.7×10^{-6}
O3LYP*	2.4×10^{-10}	3.7×10^{-1}	1.2×10^{-3}	1.2	1.9×10^{8}	30	7.5×10^{-10}
O3LYP-D	4.7×10^{2}	1.7×10^{11}	4.9×10^{14}	4.1×10^{11}	1.1×10^{22}	1.5×10^{14}	9.4×10^{10}
O3LYP*-D	**53**	$\mathbf{4.1 \times 10^{11}}$	6.9×10^{10}	$\mathbf{9.7 \times 10^{11}}$	3.5×10^{18}	1.2×10^{14}	$\mathbf{2.6 \times 10^{7}}$

[a]Spin state of the five-coordinate precursor, the six-coordinate product (after MI binding) is always low-spin.

ΔE can be predicted around $-26\ \text{kcal mol}^{-1}$. B3LYP/TZVP significantly underestimates this value by $>20\ \text{kcal mol}^{-1}$, calculating a ΔE value of $-5.1\ \text{kcal mol}^{-1}$. Accordingly, the B3LYP/TZVP K_{eq} values (calculated from ΔG) for MI ligation to [Fe(P)(NO)] and [Fe(P)(MI)] show significant errors compared to reported experimental values. For [Fe(TPP)(NO)] and [Fe(TPP)(MI)],[64] MI binding constants of 26 and $7.8 \times 10^4\ \text{M}^{-1}$ are reported, whereas the calculations yield 1.1×10^{-6} and $4.2 \times 10^{-9}\ \text{M}^{-1}$, respectively (Table 8.4).

Since the calculated binding constants are seven (or more) orders of magnitude underestimated using B3LYP/TZVP, ΔE was recalculated for X = NO and MI using a modified B3LYP functional where HF exact exchange is reduced to 15% (B3LYP*), in combination with the TZVP basis set. While using the B3LYP* functional favorably increased the binding energy of MI to [Fe(P)(MI)] to $-14.8\ \text{kcal mol}^{-1}$, ΔE for [Fe(P)(NO)] is essentially unaffected at $-3.8\ \text{kcal mol}^{-1}$. This translates to a free energy for MI binding to [Fe(P)(MI)] of $4.1\ \text{kcal mol}^{-1}$, $10\ \text{kcal mol}^{-1}$ higher than the experimental ΔG of $-6.7\ \text{kcal mol}^{-1}$ for [Fe(TPP)(MI)]. Correspondingly, the predicted K_{eq} value for [Fe(P)(MI)] of $9.2 \times 10^{-4}\ \text{M}^{-1}$ is eight orders of magnitude too low. The calculated free energy, ΔG, for MI binding to [Fe(P)(NO)] is still predicted $\sim 10\ \text{kcal mol}^{-1}$ too high compared to the experimental value of $-1.9\ \text{kcal mol}^{-1}$ for [Fe(TPP)(NO)]; translating to a calculated K_{eq} value of $3.2 \times 10^{-8}\ \text{M}^{-1}$ (nine orders of magnitude too low, see Table 8.4).

Finally, recent computational work by Siegbahn and coworkers has shown that inclusion of van der Waals interactions is important for the accurate determination of metal-ligand binding constants.[100] If van der Waals interactions are included in the B3LYP functional (B3LYP-D/TZVP), ΔE is now predicted for [Fe(P)(NO)] to be $-16.8\ \text{kcal mol}^{-1}$, only $0.8\ \text{kcal mol}^{-1}$ from the "experimental" MI binding energy of $-16\ \text{kcal mol}^{-1}$ as shown in Table 8.3. K_{eq} of the NO complex increases to $1.1 \times 10^2\ \text{M}^{-1}$, now only overestimating MI affinity for [Fe(P)(NO)] by one order of magnitude, or $0.9\ \text{kcal mol}^{-1}$ in terms of ΔG. Additionally, with B3LYP-D we predict ΔE of $-26.7\ \text{kcal mol}^{-1}$ for [Fe(P)(MI)] which is also within $1\ \text{kcal mol}^{-1}$ of the "experimental" value for the tetraphenylporphyrin complex. The B3LYP-D/TZVP calculated MI binding constant for [Fe(P)(MI)] is $5.6 \times 10^5\ \text{M}^{-1}$, again only one order of magnitude from the experimental value of $7.8 \times 10^4\ \text{M}^{-1}$ ($1.1\ \text{kcal mol}^{-1}$ in terms of ΔG). Excitingly, the ΔG values obtained with B3LYP-D/TZVP are well within the generally accepted error of DFT ($\sim 2\ \text{kcal mol}^{-1}$).

In addition, the newer functional OLYP was recommended by Radon and Pierloot[101] and others in recent studies for the calculation of binding constants in heme systems, so we also investigated how this method performs for the scientific problem investigated here. OLYP/TZVP energies (calculated from BP86/TZVP structures) predict ΔE for ligation of MI to [Fe(P)(NO)] quite well at $-17.3\ \text{kcal mol}^{-1}$, which is within $2\ \text{kcal mol}^{-1}$ of the estimated experimental MI binding energy for [Fe(TPP)(NO)] of $\Delta E \sim -16\ \text{kcal mol}^{-1}$. However, OLYP actually predicts stronger MI binding to the five-coordinate NO complex than to the corresponding five-coordinate MI complex, as shown in Table 8.4, with a predicted binding energy of only $0.6\ \text{kcal mol}^{-1}$ for [Fe(P)(MI)]. This is in stark contrast to experimental findings and likely due to the problems in predicting good spin-state energies in the case of OLYP. Combined with the problems of

OLYP/TZVP to determine an accurate structure for [Fe(P)(MI)(NO)] (the six-coordinate structure is unstable upon geometry optimization with OLYP; see Table 8.2), this clearly renders this method unusable for the system under study here.

On the other hand, the O3LYP functional with both the reduced HF exchange (15%) and the van der Waals correction included (O3LYP*-D) actually predicts the best binding energies, ΔE, for X = NO and MI, -16.3 and -29.0 kcal mol^{-1}, respectively. As a result, the calculated ΔG value for X = NO is -2.3 kcal mol^{-1} with O3LYP*-D, only overestimating the experimental value by 0.4 kcal mol^{-1}. The free energy for X = MI is -10.1 kcal mol^{-1}, overestimated compared to experiment by 3.4 kcal mol^{-1}. Therefore, the O3LYP*-D binding constants are the most accurate for any modified O3LYP functional investigated here (see Fig. 8.5) with K_{eq} values for MI binding of 53 and 2.6×10^7 M^{-1} for X = NO and MI, respectively. This now provides a quite accurate prediction of MI binding constants to ferrous heme-nitrosyl systems, although binding of MI to [Fe(P)(MI)] is overestimated by three orders of magnitude.

In summary, in comparison to experimental values for the corresponding ferrous tetraphenylporphyrin complexes, B3LYP-D/TZVP predicts the most accurate MI binding constants to [Fe(P)(X)] (where X = NO and MI) of all the methods tested here. Excitingly, the K_{eq} values predicted by this method are well within the accepted error of density functional theory calculations. B3LYP-D/TZVP is followed closely in accuracy by binding constants for O3LYP*-D/TZVP as shown in Tables 8.3 and 8.4.

To examine the effect of alternate geometries on the predicted ΔE values, B3LYP-D/TZVP and O3LYP*-D/TZVP energies (most accurate for the BP86/TZVP structures) were recalculated using B3LYP/6-31G* fully optimized structures. Importantly, calculated binding energies (ΔE) vary by less than 0.5 kcal mol^{-1} when using the BP86/TZVP and B3LYP/6-31G* structures.[78] This is found for both the B3LYP-D and the O3LYP*-D functional used in combination

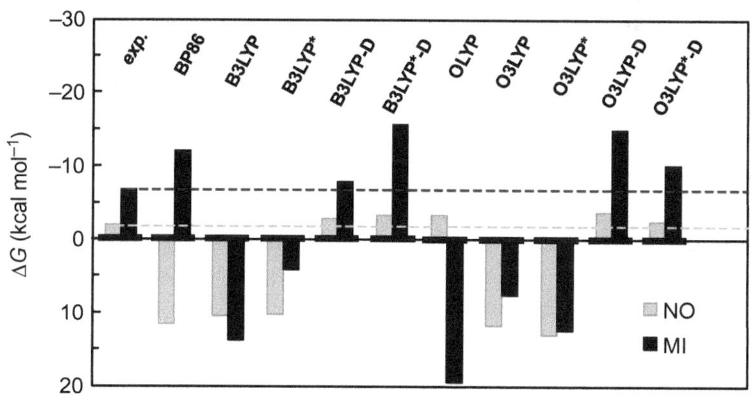

Figure 8.5 Experimental and DFT-calculated free energies (kcal mol^{-1}) for the reaction: [Fe(P)(X)]+MI ⇄ [Fe(P)(MI)(X)] where X = NO and MI at 298.15K. All calculations were performed on BP86/TZVP-optimized structures.[78]

with the TZVP basis set. This indicates that the most important parameter in determining accurate *absolute* binding energies and K_{eq} values is the method by which the single-point energies are calculated, as long as the structures are reasonable.

In summary, prediction of accurate MI binding constants, K_{eq}, in ferrous heme systems poses a serious challenge due to the fact that computational errors in ligand binding energies are of the same magnitude as the actual binding energies that we are trying to calculate. Several previous computational studies[100,101] have discussed the difficulties in obtaining accurate binding energies for heme systems and our work suggests similar conclusions. Therefore, *it is crucial that all theoretical binding constants reported use experimental values as a calibration for method accuracy.*

8.3.3 Examination of the thermodynamic σ-trans effect of HNO in sGC model systems using calculated binding constants

Based on the results from the method calibration studies described above, inclusion of van der Waals interactions is crucial for the calculation of MI binding constants to [Fe(P)(X)] where X = NO and MI. Specifically, the B3LYP-D and O3LYP*-D functionals give particularly accurate K_{eq} values (Table 8.4) for our system. While the absolute values of the calculated binding constants for many of the tested methods (Table 8.4) are significantly in error compared to experimental data, the binding constants of MI to [Fe(P)(NHO)] and [Fe(P)(CO)] are generally predicted to be within the same order of magnitude, indicating that HNO and CO exhibit a comparable *trans*-effect. This does not appear to depend much on the applied computational method. In addition, the best method from our method calibration study, B3LYP-D/TZVP, predicts MI binding constants of 1.1×10^9 and $1.5 \times 10^9\ M^{-1}$ for HNO (five-coordinate, S = 0) and CO (five-coordinate, S = 0) complexes, respectively. K_{eq} values for the five-coordinate HNO and CO complexes are approximately seven orders of magnitude larger than that of the corresponding NO complex. Oddly, however, B3LYP-D/TZVP predicts these binding constants to be even larger than that for MI binding to [Fe(P)(MI)] (S = 2), which is likely caused by the difficulty DFT methods are facing with respect to the required spin-state change in the latter case. While CO should have a relatively small *trans*-effect, it is still thought to have a larger *trans*-effect than MI, based on crystallographic Fe—N_{MI} bond lengths and Fe—N_{MI} stretching frequencies, see Section 8.2. A similar situation is observed for O3LYP*-D/TZVP binding constants where X = NHO and CO are predicted to have larger MI binding constants than X = MI. Conversely, utilization of B3LYP*-D/TZVP, where the HF exact exchange is lowered to 15%, yields MI binding constants for the five-coordinate HNO and CO complexes (S = 0) that are smaller than that for X = MI (S = 2); i.e., B3LYP*-D predicts that both CO and HNO exert a stronger *trans*-interaction in ferrous heme systems than MI. In accordance with the strong σ-*trans* effect of NO, the binding constant of MI to [Fe(P)(NO)] obtained with B3LYP*-D/TZVP is much lower than that for X = CO, HNO, and MI and is predicted to be $2.7 \times 10^2\ M^{-1}$ (see Table 8.4). Therefore, although the absolute MI binding constant to [Fe(P)(MI)] is significantly in error with B3LYP*-D, lowering the HF exact exchange in B3LYP to 15% appears to aid in the prediction of good *relative* MI binding constants for all complexes considered here. While DFT's ability to accurately predict binding constants is in question, it appears to consistently predict similar MI binding constants for both HNO and CO complexes, which are at least six orders of magnitude larger than that of NO. Hence, NHO and CO should have similar properties as ligands to ferrous hemes. This indicates that the thermodynamic σ-*trans* effect of NO is much stronger than that of CO and HNO, suggesting again that HNO cannot *directly* activate sGC through cleavage of the $Fe{-}N_{His105}$ bond, as observed experimentally for CO.

8.3.4 Alternate methods of sGC activation by HNO

Although HNO does not induce a significant thermodynamic σ-*trans* effect, we have shown that NO^-, the deprotonated form of HNO, actually has an even stronger σ-*trans* effect than NO.[35] DFT optimized structures predict an $Fe{-}N_{MI}$ bond length of 2.44 Å for $[Fe(P)(MI)(NO)]^-$, which corresponds to a very weak interaction (close to non-bonding).[25] This is significantly longer than the calculated $Fe{-}N_{MI}$ bond length of 2.18 Å in [Fe(P)(MI)(NO)]. Therefore, it may be feasible for HNO to activate sGC if the distal pocket contains a strong enough hydrogen bond accepting amino acid, e.g., histidine as shown in Scheme 8.3. As the pK_a of heme-bound HNO is unknown, it is not known if this partial "deprotonation" to give this ligand more NO^- character is biologically relevant. DFT calculations suggest that heme-bound NO^- is very basic (in ferrous heme-thiolate complexes), suggesting that biological hydrogen bonds are not strong enough to cause a significant deprotonation, and hence, an increase in *trans*-effect of HNO (see also Section 8.4.4).

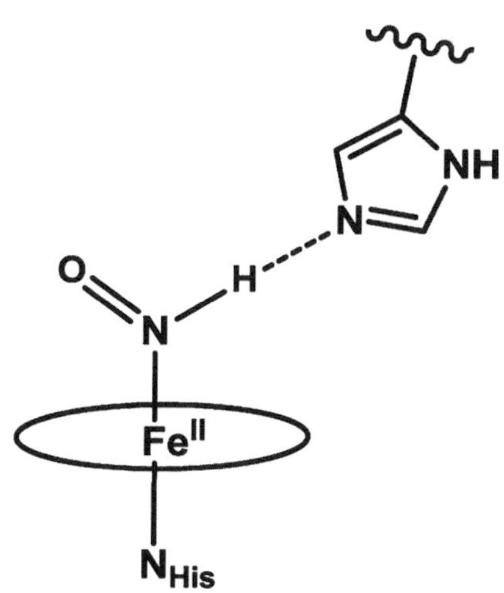

Scheme 8.3 Possible route for sGC activation by HNO through strong hydrogen bonding from HNO to an amino acid side chain.

Interestingly, however, a recent study by Montenegro et al. reports the pK_a for the nonheme $[Fe(CN)_5(HNO)]^{3-}$ complex to be 7.7,[102] significantly lower than that of free HNO. At such a low pK_a, HNO would potentially be prone to significant hydrogen bonding to distal pocket residues. This possibility is further addressed in Section 8.4.

One completely different possibility how HNO could activate sGC by conversion into NO was recently proposed by Farmer and coworkers.[103] Here, HNO would bind to the ferrous heme of sGC first, forming a six-coordinate ferrous–HNO complex. This species would not be able to activate sGC, but could react with a second molecule of HNO, following the equation:

$$\text{Fe(II)-NHO} + \text{HNO} \rightarrow \text{Fe(II)-NO} + \text{``}H_2NO\text{''} \qquad (8.5)$$

The resulting sGC–NO complex then activates sGC in the usual way, and the H_2NO radical would be released into solution. This is certainly a very interesting possibility that could explain the conflicting results when sGC is reacted with different HNO donors under different experimental conditions, and deserves further study.

8.4 The pK_a's of transition metal HNO complexes

Despite the great breadth of research that has been performed to investigate the biological and physiological roles of HNO, little is still known regarding the pK_a of HNO when bound to biologically relevant transition metal centers such as found in the active sites of Mb, sGC, and Cytochrome P450. In Mb, attempts have been made to determine the pK_a of HNO when bound to the iron center; however, studies observing the ^{1}H-NMR resonance of the HNO proton in the pH range of 6.5–10 have not observed significant changes, indicating the pK_a of Mb(II)-NHO to be greater than 10.[33] Unfortunately, the protein itself is unstable at higher pH levels, making these experiments difficult and unsuitable for NMR.[104] Therefore, it is highly useful to utilize computational methods to overcome experimental limitations as will be demonstrated in this section.

One of the great difficulties in comparing the pK_a of free HNO with those found in metal-HNO complexes is the required spin flip for the release of a proton, H^+, from free HNO, generating a kinetic barrier and making the proton transfer process slow.[8,13,87] This also provides us with two possible pK_a values, one for the formation of singlet $^1NO^-$, and the second one for triplet $^3NO^-$. The pK_a of HNO in solution was originally reported as 4.7 in the 1970s,[105] though this has been revised in more recent years and is now estimated as ~11.6 for triplet $^3NO^-$ and ~23 for the singlet $^1NO^-$ species.[106,107]

$$^1HNO \rightarrow {}^{3/1}NO^- + H^+ \qquad (8.6)$$

However, metal-bound NO^- remains as a singlet species in most cases, eliminating the need for a change in spin multiplicity and making this process similar to any regular acid-conjugate base pair:

$$[M^{X+}-{}^{1}NHO] \rightarrow [M^{X+}-{}^{1}NO^{-} \Leftrightarrow M^{(X-1)+}-{}^{2}NO]^{-} + H^{+} \tag{8.7}$$

It has been shown that metal assisted stabilization of the singlet $^1NO^-$ species can act to significantly lower the pK_a of HNO, since the Lewis-acidic metal is able to accept electrons from NO^-. This also implies that the pK_a of metal-bound NHO should be compared to the value of 23 for the free HNO/$^1NO^-$ couple, rather than that of 11.6 for the HNO/$^3NO^-$ couple as is usually done. A classic example of this behavior is seen in the nitroprusside $[Fe^{II}(CN)_5(NHO)]^{3-}$ complex where Olabe and coworkers utilized ^{1}H-NMR to propose a pK_a of 7.7 for the equilibrium provided in the equation:[102]

$$[Fe^{II}(CN)_5(HNO)]^{3-} \Leftrightarrow [Fe^{II}(CN)_5(NO)]^{4-} + H^{+} \tag{8.8}$$

Reduced nitroprusside forms an extremely simplified analog to biologically relevant systems such as Mb, as it contains a six-coordinate (6C) low-spin Fe(II) center with bound HNO/NO^-. Additional work by Slep, Olabe and coworkers includes the synthesis of a very well characterized 6C Ru(II) system capable of binding HNO and undergoing deprotonation:[108]

$$[Ru^{II}(Me_3[9]aneN_3)(bpy)(NHO)]^{2+} \Leftrightarrow [Ru^{II}(Me_3[9]aneN_3)(bpy)(NO)]^{1+} + H^{+} \tag{8.9}$$

Based upon three independent extrapolations, the pK_a of this ruthenium-based complex was accurately determined as 9.78 ± 0.15. This is significantly lower than the pK_a of free HNO (to yield singlet $^1NO^-$), understandably so since the Lewis-acidic Ru(II) center allows for the stabilization of the conjugate-base $^1NO^-$. Interestingly, this pK_a is higher than that of the significantly more negatively charged prusside species $[Fe^{II}(CN)_5(NHO)]^{3-}$. This is counter-intuitive, considering the conjugate base, $[Fe^{II}(CN)_5(NO)]^{4-}$, may be considered less capable of stabilization due to the greater negative charge of the complex; hence, the bound HNO would be expected to be less acidic in this case, leading to a higher pK_a compared to the Ru(II) complex. On the other hand, it has been speculated that *trans*-labilization and loss of CN^- in $[Fe^{II}(CN)_5(NO)]^{4-}$ could lead to a lower pK_a, a complication which is avoided in the chelating $Me_3[9]aneN_3$ ligand.[108] This aspect is further discussed in Section 8.4.3.

8.4.1 Computational approach

In order to obtain a better grasp of the pK_a of the Mb(II)-NHO system, or more generally, of HNO bound to heme active sites containing a proximal histidine such as sGC, we have conducted a computational investigation using techniques calibrated to the above-mentioned ruthenium complex for which the pK_a is well known. Following Eq. (8.7), the pK_a of HNO may be estimated by calculation of the change in Gibbs free energy (ΔG), since

$$\Delta G = -\text{RT} \ln (K_{\text{eq}}) \qquad (8.10)$$

In order to model the gain and loss of a proton, we can utilize a conjugate base pair; in this case we have utilized the ammonia/ammonium couple. Another option may be to model a proton in an aqueous environment; however, these calculations can become very tedious when trying to properly describe the hydrogen bonding network which forms around a proton in a water cluster. With our approach, the chemical reaction that we model involves the protonation of ammonia, and the formation of NO^-:

$$\text{M-NHO} + \text{NH}_3 \Leftrightarrow \text{M-NO}^- + \text{NH}_4^+ \qquad (8.11)$$

Here, ΔG can be estimated as the difference of the individual Gibb's free enthalpies of the products and the reactants of the system:

$$\Delta G = (G_{\text{M-NO}-} + G_{\text{NH4+}}) - (G_{\text{M-NHO}} + G_{\text{NH3}}) \qquad (8.12)$$

We may also write the chemical Eq. (8.11) in terms of the equilibrium constant, K_{eq}:

$$K_{\text{eq}} = \frac{[\text{M-NO}^-][\text{NH}_4^+]}{[\text{M-NHO}][\text{NH}_3]} = \frac{K_a(\text{M-NHO})}{K_a(\text{NH}_4^+)} \qquad (8.13)$$

From Eq. (8.13), it can be seen that, using the logarithmic definition of the pK_a value:

$$-\log(K_{\text{eq}}) = pK_a(\text{M-NHO}) - pK_a(\text{NH}_4^+) \rightarrow pK_a(\text{M-NHO}) = pK_a(\text{NH}_4^+) - \log(K_{\text{eq}}) \qquad (8.14)$$

In this way, the Gibb's free enthalpy and K_{eq} for Eq. (8.11) can be calculated, and using the known experimental pK_a of ammonia (9.25 at 25°C) in conjunction with Eq. (8.14), the pK_a for a given M-NHO system can be obtained computationally.

We have previously calibrated a computational method for the accurate description of a 6C ferrous heme—NHO complex with proximally bound methylimidazole, [Fe(P)(MI)(NHO)]; showing that the BP86/TZVP functional/basis set combination in conjunction with the COSMO solvation model as employed by ORCA[110] provides reliable geometries (see Section 8.2.3 and Table 8.2). Additional investigations showed only a subtle dependence of ΔG on molecular geometry with respect to the functional utilized in the geometry optimization as long as reasonable structures are obtained (see Section 8.3.2). Much more importantly, however, the predicted ΔG values in this study are *highly* dependent on the functional utilized in the single-point energy calculations (see Section 8.3.2 and Table 8.4).[78] It is therefore a reasonable approximation in such systems to utilize the less expensive BP86/TZVP approach, along with the appropriate solvation model such as COSMO or PCM, to optimize molecular geometries, and then employ appropriate functionals to more accurately predict ΔG and, in turn, pK_a values as presented in the following.[110]

8.4.2 Method calibration: calculated pK_a's for $[Ru^{II}(Me_3[9]aneN_3)(bpy)(NHO)]^{2+}$

Since the pK_a of $[Ru^{II}(Me_3[9]aneN_3)(bpy)(NHO)]^{2+}$ has been well investigated, this complex provides a suitable model to gauge the accuracy of various DFT functionals to generally predict the pK_a of HNO in 6C metal complexes. Optimized structures of $[Ru^{II}(Me_3[9]aneN_3)(bpy)(NHO)]^{2+}$ and $[Ru^{II}(Me_3[9]aneN_3)(bpy)(NO)]^{+}$ are provided in Fig. 8.6; as noted previously, the potentially complex hydrogen bonding network which may form in aqueous solution has been omitted in favor of the simpler combination of COSMO treatment in conjunction with the ammonium proton donor/acceptor pair. The strong σ-*trans* effect of NO^- (see Section 8.3.4) is evident in the optimized geometry of $[Ru^{II}(Me_3[9]aneN_3)(bpy)(NO)]^{+}$ (see Table 8.5), where the trans-bond to NO^- increases by a distance of 0.2 Å relative to $[Ru^{II}(Me_3[9]aneN_3)(bpy)(NHO)]^{2+}$. The p$K_a$ of this complex is well represented with several of the employed functionals, including B3LYP,

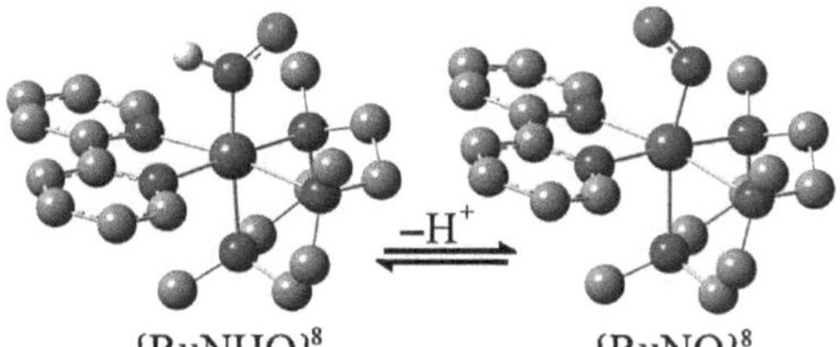

Figure 8.6 BP86/TZVP-optimized structures of $[Ru^{II}(Me_3[9]aneN_3)(bpy)(NHO/NO^-)]^{2+/1+}$ complexes reported by Olabe, Slep, and coworkers, referred to as Ru-NHO (left) and Ru-NO^- (right).[111]

Table 8.5 Calculated bond distances and angles of the indicated Ru^{II}-NHO and Ru^{II}-NO^- complexes (all $S = 0$; optimized using BP86/TZVP)

Complex	Parameter				
	Fe−NO (Å)	**Fe−N(trans) (Å)**	**N−O (Å)**	**Fe-N-O (deg)**	**(trans)N-Fe-N(O) (deg)**
$[Ru(Me_3[9]aneN_3)(bpy)(NHO)]^{2+}$	1.93	2.23	1.25	132	177
$[Ru(Me_3[9]aneN_3)(bpy)(NO)]^{1+}$	1.88	2.43	1.23	127	166

Table 8.6 Calculated ΔG (in kcal mol^{-1}) (using Eq. 8.12) and pK_a's of $[Ru(Me_3[9]aneN_3)(bpy)(NHO)]^{2+}$

Method	ΔG	pK_a
exp.	**0.72**	**9.78**
BP86	−4.76	5.76
B3LYP	**0.35**	**9.51**
B3LYP-D	5.03	12.94
B3LYP*	**−0.64**	**8.78**
B3LYP*-D	4.41	12.48
OLYP	− 6.72	4.32
O3LYP	**−2.46**	**7.45**
O3LYP-D	2.59	11.15
O3LYP*	**−1.45**	**8.19**
O3LYP*-D	6.19	13.79
TPSSH	**−0.48**	**8.90**
ωB97X-D	−4.56	5.91

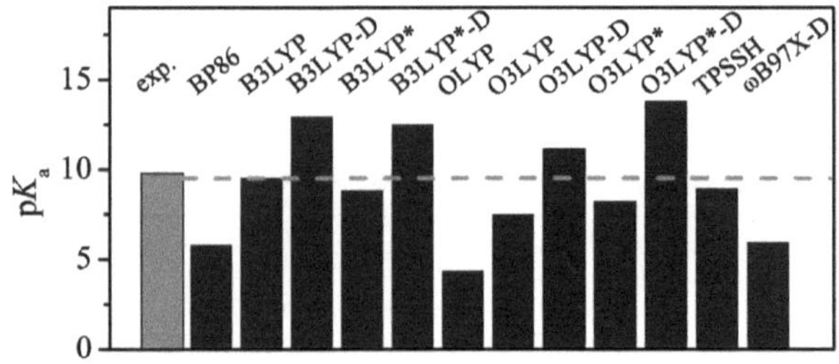

Figure 8.7 Calculated pK_a values for the $[Ru^{II}(Me_3[9]aneN_3)(bpy)(NHO)]^{2+}$ complex.[111] The dashed line indicates the experimentally determined pK_a of 9.78 of this species.

B3LYP*, O3LYP*, and TPSSH. The calculated values are compared with the experimentally determined pK_a of 9.78 in Table 8.6 and Fig. 8.7. Interestingly, the addition of the three-body dispersion term, D3, tends to cause an overestimation of the pK_a, with the exception of ωB97X-D where the value is severely underestimated. These issues are seen in all applications of the dispersion term for the calculation of pK_a values described in this chapter.

8.4.3 Prediction of the pK_a for the reduced nitroprusside complex, $[Fe(CN)_5(NHO)]^{3-}$

As mentioned above, previous studies have proposed an unusually low pK_a for the $[Fe(CN)_5(NHO)]^{3-}$ complex, particularly considering this value (7.7) is proposed to be significantly lower than that of $[Ru^{II}(Me_3[9]aneN_3)(bpy)(NHO)]^{2+}$ (9.78). To investigate this, we have applied our calibrated DFT methods to a series of

Table 8.7 Calculated pK_a's and ΔG (in kcal mol^{-1}) of nitroprusside complexes under investigation (all $S = 0$). For abbreviation, complexes are labeled numerically as follows: (a) $[Fe(CN)_5(NO)]^{4-}$, (b) $[Fe(CN)_4(NO)]^{3-}$, (c) $[Fe(CN)_4(NHO)]^{2-}$, and (d) $[Fe(CN)_5(NHO)]^{3-}$

Method	D→A pK_a	A→B ΔG	C→B pK_a	C→D ΔG
BP86	31.68	−26.4	2.85	−12.9
B3LYP	**33.04**	**−21.8**	**11.04**	**−8.3**
B3LYP-D	49.14	−90.1	1.76	39.1
B3LYP*	**33.16**	**−23.2**	**8.93**	**−9.9**
B3LYP*-D	49.21	−91.5	−0.36	37.5
OLYP	34.35	−34.4	5.92	−4.4
O3LYP	**34.63**	**−24.1**	**6.08**	**−14.8**
O3LYP-D	51.85	−93.6	−3.32	31.9
O3LYP*	**35.57**	**−22.5**	**7.82**	**−15.3**
O3LYP*-D	52.67	−92.2	−1.60	31.7
TPSSH	**32.75**	**−20.3**	**6.73**	**−15.2**

Table 8.8 Calculated bond distances and angles for the nitroprusside complexes under investigation (all $S = 0$)

	Complex	Parameter					
		Fe–NO (Å)	Fe–CN (trans) (Å)	N–O (Å)	Fe–CN (Å)	Fe-N-O (deg)	(trans) CN-Fe-N (O) (deg)
A	$[Fe(CN)_5(NO)]^{4-}$	1.87	2.03	1.26	1.93	123	175
B	$[Fe(CN)_4(NO)]^{3-}$	1.66	–	1.21	1.92	178	–
C	$[Fe(CN)_4(NHO)]^{2-}$	1.74	–	1.24	1.93	123	–
D	$[Fe(CN)_5(NHO)]^{3-}$	1.83	1.95	1.25	1.93	134	177
E	$[Fe(CN)_4(CNH)(NO)]^{3-}$	1.79	1.83	1.23	1.93	123	179

nitroprusside related complexes. Initial calculations for the pK_a according to Eq. (8.8) are presented in Table 8.7. To our surprise, these pK_a's are astonishingly high, around 31–35. Even consideration of a generous margin of error would imply that Eq. (8.8) is highly unfavorable. It is also known that NO^- exhibits a significant thermodynamic σ-*trans* effect, which is not particularly well reproduced in our calculations, as noted in Table 8.8. Due to this *trans*-effect, we suspected that $[Fe(CN)_5(NO)]^{4-}$ may be an unstable complex, and the possibility of labilization of the *trans*-CN^- arises. For this to occur, CN^- may either dissociate independently, or dissociate following protonation; the latter behavior has been observed

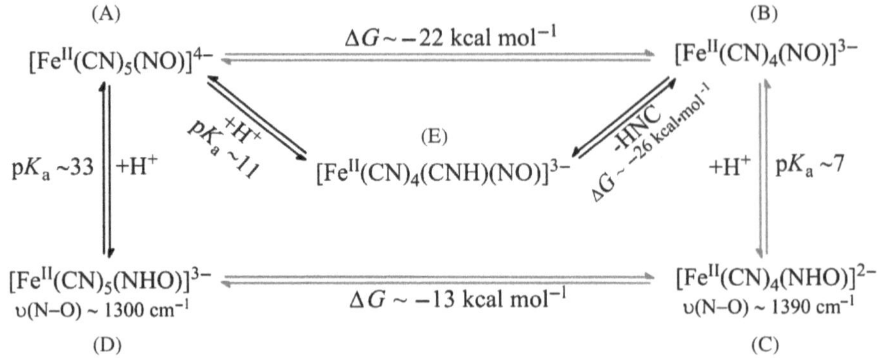

Scheme 8.4 Possible routes for the protonation of the reduced nitroprusside complex $[Fe^{II}(CN)_5(NO)]^{4-}$. Reported values for pK_a's, ΔG, and υ(N−O) are based upon an average of several DFT methods.[111]

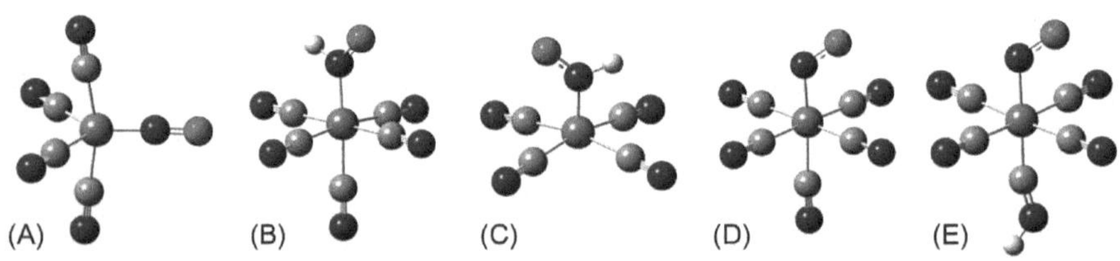

Figure 8.8 BP86/TZVP-optimized nitroprusside complexes and complementing species. (A) $[Fe(CN)_5(HNO)]^{3-}$. (B) $[Fe(CN)_5(NO)]^{4-}$. (C) $[Fe(CN)_4(CNH)(NO)]^{3-}$. (D) $[Fe(CN)_4(NO)]^{3-}$. (E) $[Fe(CN)_4(NHO)]^{2-}$ [111] (all $S = 0$).

previously in hydrogenase model complexes.[109] This will in turn result in the $[Fe^{II}(CN)_4(NO)]^{3-}$ complex. From here, $[Fe^{II}(CN)_4(NO)]^{3-}$ can be protonated to form $[Fe^{II}(CN)_4(NHO)]^{2-}$. A free energy cycle describing this process is provided in Scheme 8.4, with the structures involved summarized in Fig. 8.8. It can been seen in Fig. 8.9 that the calculated pK_a's for the 5C complex match quite favorably with the experimentally observed pK_a of 7.7, especially compared to calculated pK_a's for the 6C complex, which are in the 31−35 range. Interestingly, once protonated, the $[Fe^{II}(CN)_4(NHO)]^{2-}$ complex may favorably bind CN^- once again, but only if the CN^- ligand is not protonated under the experimental conditions. This would generate the complex $[Fe^{II}(CN)_5(NHO)]^{3-}$, which is the species that was proposed to be formed in the initial report of the experimental data.[102]

In summary, one-electron reduction of the $[Fe^{II}(CN)_5(NO)]^{3-}$ precursor complex results in the labilization of the *trans*-CN^- ligand, and the resulting complex $[Fe^{II}(CN)_4(NO)]^{3-}$ can then undergo protonation of NO^- to form the five-coordinate NHO complex. From here, the complex could in theory reform a 6C species, resulting in the proposed $[Fe^{II}(CN)_5(NHO)]^{3-}$ complex. It is important to note that the experimentally observed stretching frequency υ(N−O) for the HNO complex was found to

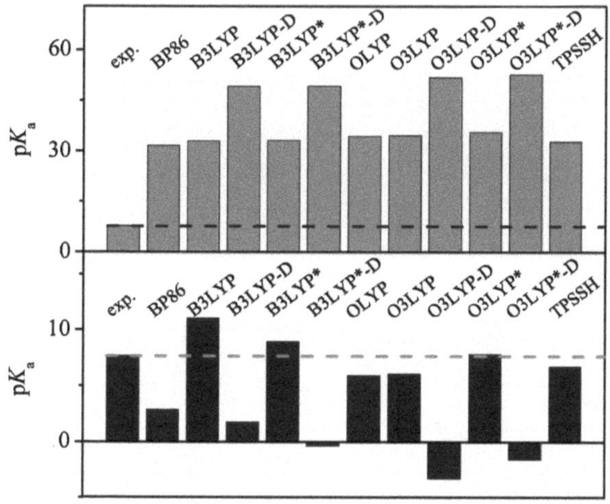

Figure 8.9 Comparison of calculated pK_a values for $[Fe(CN)_5(NHO)]^{3-}$ (top) and $[Fe(CN)_4(NHO)]^{2-}$ (bottom). The dashed horizontal line in each panel is indicative of the experimentally observed pK_a of 7.7.[111]

be 1380 cm^{-1}, which falls close to that calculated for the 5C complex $[Fe^{II}(CN)_4(NHO)]^{2-}$ (1390 cm^{-1}). In contrast, for the 6C complex $[Fe^{II}(CN)_5(NHO)]^{3-}$, υ(N–O) is predicted around 1300 cm^{-1}. While one must be cautious in drawing direct conclusions from DFT calculated frequency values alone, these results may be viewed in conjunction with the thermodynamic results which ultimately indicate that the experimentally observed HNO complex in ref.102 is actually $[Fe^{II}(CN)_4(NHO)]^{2-}$. It could be imagined that under very acidic conditions, the system could bifurcate between the direct protonation and the labilization pathway; however, due to the experimental conditions, the labilized CN^- pathway is likely what was observed in ref.102 Here we see once again the significance of the σ-*trans* effect of NO^-, and the utility of DFT in probing possible interpretations of experimental observations.

8.4.4 The pK_a of the Mb(II)-NHO complex

A substantial amount of work has been put forth to determine the pK_a's of small molecules bound to Mb and other heme proteins such as hemoglobin, sGC, and cytochrome P450. A notable limitation in this work has been the stability of these proteins at the pH levels which are required to observe the acid-dissociation of H^+. Previous studies by Farmer and coworkers have shown no observable changes in the ^{1}H-NMR resonance of the HNO proton over the pH range of 6.5–10 in Mb(II)-NHO, and proposed that the pK_a lies close to 11.[33,112] As demonstrated by Zhang and coworkers in their computational study, His64 plays an integral role in the stabilization of HNO in the Mb(II)-NHO complex primarily through the hydrogen bonding interaction [His•••HNO•••H_2O], referred to previously as isomer A. An interesting consequence of the His-N_ε interaction is that upon deprotonation to form

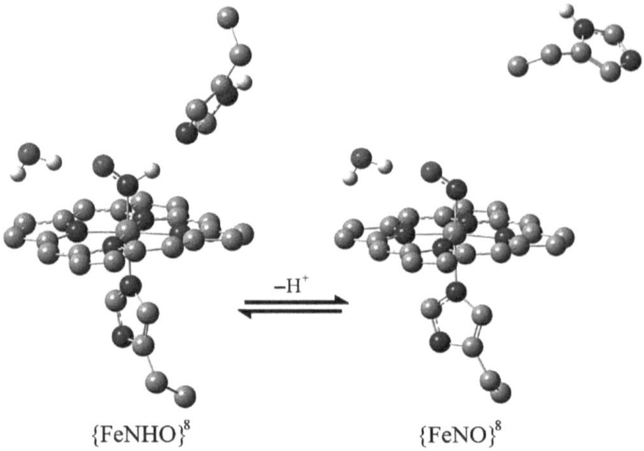

Figure 8.10 BP86/TZVP-optimized models of Mb(II)-NHO (left) and Mb(II)-NO^- (right) based on previous work by Zhang and coworkers for the Mb(II)-NHO A isomer (see text). Hydrogens of the porphyrin ring and of histidines have been omitted for clarity, with the exception of the N_δ-hydrogen of the distal histidine.[111]

Table 8.9 Calculated bond distances and angles of the Fe^{II}-NO^-/NHO porphyrin model complexes utilizing the BP86/TZVP method (all $S = 0$; cf. Fig. 8.10)

	Complex	Parameter				
		Fe–NO (Å)	Fe–N(trans) (Å)	N–O (Å)	Fe-N-O (deg)	(trans)N-Fe-N(O) (deg)
1	Mb(II)-(HNO)	1.80	2.10	1.26	131	177
2	Mb(II)-(NO^-)	1.74	2.39	1.24	125	176

NO^-, the lone pairs present on His64-N_ε and NO^- now repel each other, shifting the imidazole side-chain away to minimize this interaction. This is pronounced in the optimized model of Mb(II)-NO^-, shown in Fig. 8.10, where the distal imidazole rotates away to maximize distance. This is not unusual in such calculations since the side chain is fixed by the α-carbon to its crystallographically relevant position, yet additional residues which may prevent such an extreme motion are omitted due to computational cost. Potential hydrogen bonding through the His-N_δH was also considered; however, due to the geometric constraint imposed to keep the distal imidazole close to the observed crystallographic position, a new hydrogen bond from His-N_δH to NO^- was not observed to occur. Key geometric parameters are provided in Table 8.9. The calculated pK_a's for the Mb(II)-NO^-/NHO models are provided in Fig. 8.11 and Table 8.10. It can be seen that the calculated values lie on average around 14, somewhat higher than the previously proposed pK_a of 11,

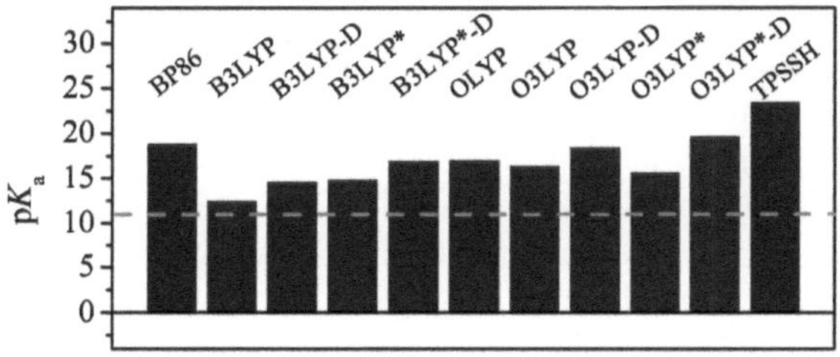

Figure 8.11 Calculated pK_a's of Mb(II)-NHO; the dashed line represents the *lower limit* for the pK_a value (about 11) as determined experimentally (see text).[111]

Table 8.10 Calculated ΔG (in kcal mol^{-1}) (using Eq. 8.12) and pK_a's of Mb(II)-NHO

Method	ΔG	pK_a
BP86	13.01	18.79
B3LYP	**4.32**	**12.42**
B3LYP-D	7.17	14.51
B3LYP*	**7.53**	**14.77**
B3LYP*-D	10.38	16.86
OLYP	10.48	16.93
O3LYP	**9.61**	**16.29**
O3LYP-D	12.45	18.38
O3LYP*	**8.68**	**15.61**
O3LYP*-D	14.12	19.60
TPSSH	**19.33**	**23.42**

but still within a reasonable range considering that the experimental pK_a is not exactly known. As described above, current experimental efforts spanning out to very alkaline pH ranges have not been able to determine the exact pK_a of Mb(II)-NHO. This again demonstrates the value of supplementing such problems through carefully performed computational work.

8.5 Conclusions

In conclusion, modern DFT methods are very useful to analyze the spectroscopic properties and electronic structures of transition metal complexes, as demonstrated here for NO and HNO complexes of hemes. However, if not carefully performed and calibrated against experimental results, DFT methods can also be misleading, and this is particularly obvious for the calculated metal-ligand bind constants (free energies) presented here. The results are highly functional dependent as shown in Table 8.4 and Fig. 8.5, and the accuracy of the results decreases further if a change in spin state is involved in ligand binding. Depending on the choice of functional, binding constants for MI to the 5C ferrous heme NO and MI complexes vary by 10 orders of magnitude! Without proper calibration to known experimental binding constants, there is no way to know which functional to trust. The calculation of pK_a values for transition metal-NHO complexes shows more consistency overall, but again, a method calibration to choose a proper functional is mandatory if trustable results are to be obtained.

In this chapter, DFT methods were applied to shed light on the properties of ferrous heme-NHO complexes. First, we discuss whether HNO could impose a suitable thermodynamic *trans*-effect on the proximal His ligand in sGC, and hence, whether HNO could induce breaking of the Fe$-$N$_{His105}$ bond upon binding to the heme and in this way, directly activate sGC, the mammalian NO sensor protein. Calculations on the electronic structure of the ferrous heme-NHO complex and careful analysis of the electronic structure show that this is not the case. By comparison of the calculated imidazole binding constants, it becomes further evident that HNO should behave similarly to CO, which is known to bind to sGC without being able to induce cleavage of the proximal Fe$-$N$_{His105}$ bond. This further supports the electronic-structural argument that HNO does not have a strong enough *trans*-effect to activate sGC. The best K_{eq} values for MI binding to [Fe(P)(MI)(X)] were obtained with B3LYP-D/TZVP energies on BP86/TZVP geometries. The calculated MI binding constants for X = NO and MI are 110 and 5.6×10^5 M^{-1}, predicted only one order of magnitude higher than experimentally determined values of 26 and 7.8×10^4 M^{-1} for ferrous tetraphenylporphyrin complexes.

Next, we investigated the properties of the best characterized heme$-$NHO complex, which is the HNO adduct of Mb, Mb(II)-NHO. DFT studies by Zhang and coworkers have elucidated how HNO is bound in the Mb active site, forming a dual hydrogen bond to the distal His and a water molecule. In addition, our results allow for an estimate of the pK_a of the HNO ligand in Mb(II)-NHO, which is calculated to be around 14. This result is based on a carefully performed method calibration study against the known pK_a value of HNO in the Ru(II) complex $[Ru^{II}(Me_3[9]aneN_3)(bpy)(NHO)]^{2+}$. Here, B3LYP, B3LYP*, O3LYP, O3LYP*, and TPSSH single point calculations on BP86/TZVP-optimized structures deliver the best results, with a calculated pK_a value of about nine for the Ru(II) complex (experiment: 9.78). All these results demonstrate that DFT calculations, if carefully performed and calibrated against experimental data, are able to supplement and extend experimental results, and elucidate experimental findings that have remained mysterious. In this regard, our calculations show that for the nitroprusside system, the experimentally obtained pK_a value (7.7) is actually for the five-coordinate complex $[Fe^{II}(CN)_4(NHO)]^{2-}$ (calculated pK_a: $\sim$7) and not $[Fe^{II}(CN)_5(NHO)]^{3-}$ (calculated pK_a: $\sim$33) as initially proposed.

Acknowledgment

This work was supported by the National Science Foundation (CHE 1305777 and CHE 1464696 to NL).

References

1. Culottta E, Koshland Jr. DE. NO news is good news. *Science* 1992;**258**:1862–5.
2. Bredt DS, Snyder SH. Nitric oxide: a physiologic messenger molecule. *Annu Rev Biochem* 1994;**63**:175–95.
3. Moncada S, Palmer RM, Higgs EA. Nitric oxide: physiology, pathophysiology, and pharmacology. *Pharmacol Rev* 1991;**43**:109–42.
4. Butler AR, Williams DLH., The physiological role of nitric oxide, *Chem Soc Rev* 1993; 233–41.
5. Ignarro L. *Nitric oxide: biology and pathobiology*. San Diego: Academic Press; 2000.
6. Palmer RM, Ferrige AG, Moncada S. Nitric oxide release accounts for the biological activity of endothelium-derived relaxing factor. *Nature* 1987;**327**:524–6.
7. Doctorovich F, Farmer PJ. Preface: the intricate puzzle of HNO chemistry. *J Inorg Biochem* 2013;**118**:107.
8. Fukuto JM, Cisneros CJ, Kinkade RL. A comparison of the chemistry associated with the biological signaling and actions of nitroxyl (HNO) and nitric oxide (NO). *J Inorg Biochem* 2013;**118**:201–8.
9. Doctorovich F, Bikiel D, Pellegrino J, Suárez SA, Larsen A, Martí MA. Nitroxyl (azanone) trapping by metalloporphyrins. *Coord Chem Rev* 2011;**255**:2764–84.
10. Miller TW, Cherney MM, Lee AJ, Francoleon NE, Farmer PJ, King SB, et al. The effects of nitroxyl (HNO) on soluble guanylate cyclase activity. Interactions at ferrous heme and cysteine thiols. *J Biol Chem* 2009;**284**:21788–96.
11. Rosenthal J, Lippard SJ. Direct detection of nitroxyl in aqueous solution using a tripodal copper(II) BODIPY complex. *J Am Chem Soc* 2010;**132**:5536–7.
12. Suárez SA, Fonticelli MH, Rubert AA, de La Llave E, Scherlis D, Salvarezza RC, et al. A surface effect allows HNO/NO discrimination by a cobalt porphyrin bound to gold. *Inorg Chem* 2010;**49**:6955–66.
13. Paolocci N, Jackson MI, Lopez BE, Miranda K, Tochetti CG, Wink DA, et al. The pharmacology of nitroxyl (HNO) and its therapeutic potential: not just the Janus face of NO. *Pharmacol Ther* 2007;**113**:442–58.
14. Arnelle DR, Stamler JS. NO^+, NO, and NO^- donation by S-nitrosothiols: implications for regulation of physiological functions by S-nitrosylation and acceleration of disulfide formation. *Arch Biochem Biophys* 1995;**318**:279–85.
15. Suarez SA, Neuman NI, Muñoz M, Álvarez L, Bikiel DE, Brondino CD, et al. Nitric oxide is reduced to HNO by proton-coupled nucleophilic attack by ascorbate, tyrosine, and other alcohols. A new route to HNO in biological media? *J Am Chem Soc* 2015;**137**:4720–7.
16. Schmidt HH, Hofmann H, Schindler U, Shutenko ZS, Cunningham DD, Feelish M. No•NO from NO synthase. *Proc Natl Acad Sci USA* 1996;**93**:14492–7.
17. Speelman A, Lehnert N. Heme versus non-heme iron-nitroxyl $\{FeN(H)O\}^8$ complexes: electronic structure and biologically relevant reactivity. *Acc Chem Res* 2014;**47**:1106–16.
18. Berto TC, Hoffman MB, Murata Y, Landenberger KB, Alp EE, Zhao J, et al. Structural and electronic characterization of non-heme Fe(II)-nitrosyls as biomimetic models of the Fe_B center of bacterial nitric oxide reductase. *J Am Chem Soc* 2011;**133**:16714–17.
19. Speelman AL, Lehnert N. Characterization of a high-spin non-heme $\{FeNO\}^8$ complex: implications for the reactivity of iron nitroxyl species in biology. *Angew Chem Int Ed* 2013;**52**:12283–7.

20. Lui SM, Soriano A, Cowan JA. Enzymatic reduction of inorganic anions. Pre-steady-state kinetics analysis of the dissimilatory sulfite reductase (Desulfoviridin) from *Desulfovibrio vulgaris* (Hildenborough). Mechanistic implications. *J Am Chem Soc* 1993;**115**:10483–6.
21. Crane BR, Seigel LM, Getzoff ED. Probing the catalytic mechanism of sulfite reductase by X-ray crystallography: structures of the *Escherichia coli* hemoprotein in complex with substrates, inhibitors, intermediates, and products. *Biochemistry* 1997;**36**:12120–37.
22. Einsle O, Messerschmidt A, Huber R, Kroneck PMH, Neese F. Mechanism of the six-electron reduction of nitrite to ammonia by cytochrome c nitrite reductase. *J Am Chem Soc* 2002;**124**:11737–45.
23. Bykov D, Neese F. Substrate binding and activation in the active site of cytochrome c nitrite reductase: a density functional study. *Inorg Chem* 2015;**54**:9303–16.
24. Shiro Y, Fujii M, Iizuka T, Adachi S, Tsukamoto K, Nakahara K, et al. Spectroscopic and kinetic studies on reaction of cytochrome P450nor with nitric oxide. Implication for its nitric oxide reduction mechanism. *J Biol Chem* 1995;**270**:1617–23.
25. Lehnert N, Praneeth VKK, Paulat F. Electronic structure of Fe(II)-porphyrin nitroxyl complexes: molecular mechanism of fungal nitric oxide reductase (P450nor). *J Comp Chem* 2006;**27**:1338–51.
26. Riplinger C, Bill E, Daiber A, Ullrich V, Shoun H, Neese F. New insights into the nature of observable reaction intermediates in cytochrome P450 NO reductase by using a combination of spectroscopy and quantum mechanics/molecular mechanics calculations. *Chem Eur J* 2014;**20**:1602–14.
27. Nakagawa H. Controlled release of HNO from chemical donors for biological applications. *J Inorg Biochem* 2013;**118**:187–90.
28. Ma XL, Gao F, Liu G-L, Lopez BL, Christopher TA, Fukuto JM, et al. Opposite effects of nitric oxide and nitroxyl on postischemic myocardial injury. *Proc Natl Acad Sci USA* 1999;**96**:14617–22.
29. Nagasawa HT, DeMaster EG, Redfern B, Shirota FN, Goon DJW. Evidence for nitroxyl in the catalase-mediated bioactivation of the alcohol deterrent agent cyanamide. *J Med Chem* 1990;**33**:3120–2.
30. Lopez BE, Wink DA, Fukuto JM. The inhibition of glyceraldehyde-3-phosphate dehydrogenase by nitroxyl (HNO). *Arch Biochem Biophys* 2007;**465**:430–6.
31. Miranda KM, Paolocci N, Katori T, Thomas DD, Ford E, Bartberger MD, et al. A biochemical rationale for the discrete behavior of nitroxyl and nitric oxide in the cardiovascular system. *Proc Natl Acad Sci USA* 2003;**100**:9196–201.
32. Enemark JH, Feltham RD. Principles of structure, bonding, and reactivity for metal nitrosyl complexes. *Coord Chem Rev* 1974;**13**:339–406.
33. Kumar MR, Fukuto JM, Miranda KM, Farmer PJ. Reactions of HNO with heme proteins: new routes to HNO-heme complexes and insight into physiological effects. *Inorg Chem* 2010;**49**:6283–92.
34. Choi I-K, Liu Y, Feng D, Paeng K-J, Ryan MD. Electrochemical and spectroscopic studies of iron porphyrin nitrosyls and their reduction products. *Inorg Chem* 1991;**30**:1832–9.
35. Goodrich LE, Saikat R, Alp EE, Zhao J, Hu MY, Lehnert N. Electronic structure and biologically relevant reactivity of low-spin $\{FeNO\}^8$ porphyrin model complexes: new insight from a bis-picket Fence porphyrin. *Inorg Chem* 2013;**52**:7766–80.
36. Ye S, Price JC, Barr EW, Green MT, Bollinger JM, Krebs C, et al. Cryoreduction of the NO-adduct of taurine: α-Ketoglutarate dioxygenase (TauD) yields an elusive $\{FeNO\}^8$ species. *J Am Chem Soc* 2010;**132**:4739–51.
37. Miranda KM. The chemistry of nitroxyl (HNO) and implications in biology. *Coord Chem Rev* 2005;**249**:433–55.

38. Campbell MG, Underbakke ES, Potter CS, Carragher B, Marletta MA. Single-particle EM reveals the higher-order domain architecture of soluble guanylate cyclase. *Proc Natl Acad Sci USA* 2014;**111**:2960–5.
39. Ignarro LJ. Haem-dependent activation of cytosolic guanylate cyclase by nitric oxide: a widespread signal transduction mechanism. *J Biochem Soc Trans* 1992;**20**:465–9.
40. Cary SPL, Winger JA, Marletta MA. Tonic and acute nitric oxide signaling through soluble guanylate cyclase is mediated by nonheme nitric oxide, ATP and GTP. *Proc Natl Acad Sci USA* 2005;**102**:13064–9.
41. Cary SPL, Winger JA, Derbyshire ER, Marletta MA. Nitric oxide signaling: no longer simply On or Off. *Trends Biochem Sci* 2006;**31**:231–9.
42. Zhao Y, Marletta MA. Localization of the heme binding region in soluble guanylate cyclase. *Biochemistry* 1997;**36**:15959–64.
43. Traylor TG, Sharma VS. Why NO? *Biochemistry* 1992;**31**:2847–9.
44. Yu AE, Hu S, Spiro TG, Burstyn JN. Resonance raman spectroscopy of soluble guanylyl cyclase reveals displacement of distal and proximal heme ligands by NO. *J Am Chem Soc* 1994;**116**:4117–18.
45. Zhao Y, Brandish PE, Ballou DP, Marletta MA. A molecular basis for nitric oxide sensing by soluble guanylate cyclase. *Proc Natl Acad Sci USA* 1999;**96**:14753–8.
46. Goodrich LE, Paulat F, Praneeth VKK, Lehnert N. Electronic structure of heme-nitrosyls and its significance for nitric oxide reactivity, sensing, transport, and toxicity in biological systems. *Inorg Chem* 2010;**49**:6293–316.
47. Hunt AP, Lehnert N. Heme-nitrosyls: electronic structure implications for function in biology. *Acc Chem Res* 2015;**48**:2117–25.
48. Herzik Jr. MA, Jonnalagadda R, Kuriyana J, Marletta MA. Structural insights into the role of iron–histidine bond cleavage in nitric oxide-induced activation of H-NOX gas sensor proteins. *Proc Natl Acad Sci USA* 2014;**111**:E4156–64.
49. Fernhoff NB, Derbyshire ER, Marletta MA. A nitric oxide/cysteine interaction mediates the activation of soluble guanylate cyclase. *Proc Natl Acad Sci USA* 2009;**106**:21602–7.
50. Kimura H, Mittal CK, Murad F. Activation of guanylate cyclase from rat liver and other tissues by sodium azide. *J Biol Chem* 1975;**250**:8016–22.
51. Paolocci N, Saavedra WF, Miranda KM, Martignani C, Isoda T, Hare JM, et al. Nitroxyl anion exerts redox-sensitive positive cardiac inotropy in vivo by calcitonin gene-related peptide signaling. *Proc Natl Acad Sci USA* 2001;**98**:10463–8.
52. Ellis A, Li CG, Rand MJ. Differential actions of L-cysteine on responses to nitric oxide, nitroxyl anions and EDRF in the rat aorta. *Br J Pharmacol* 2000;**129**:315–22.
53. Favaloro JL, Kemp-Harper BK. The nitroxyl anion (HNO) is a potent dilator of rat coronary vasculature. *Cardiovas Res* 2007;**73**:587–96.
54. Fukuto JM, Chiang K, Hszieh R, Wong P, Chaudhuri G. The pharmacological activity of nitroxyl: a potent vasodilator with activity similar to nitric oxide and/or endothelium-derived relaxing factor. *J Pharmacol Exp Ther* 1992;**263**:546–51.
55. Fukuto JM, Hobbs AJ, Ignarro LJ. Conversion of nitroxyl (HNO) to nitric oxide (NO) in biological systems: the role of physiological oxidants and relevance to the biological activity of HNO. *Biochem Biophys Res Commun* 1993;**196**:707–13.
56. Murphy ME, Sies H. Reversible conversion of nitroxyl anion to nitric oxide by superoxide dismutase. *Proc Natl Acad Sci USA* 1991;**88**:10860–4.
57. Dierks EA, Burstyn JN. Nitric oxide (NO), the only nitrogen monoxide redox form capable of activating soluble guanylyl cyclase. *Biochem Pharmacol* 1996;**51**:1593–600.

58. Zeller A, Wenzl MV, Beretta M, Stessel H, Russwurm M, Koesling D, et al. Mechanisms underlying activation of soluble guanylate cyclase by the nitroxyl donor Angeli's salt. *Mol Pharmacol* 2009;**76**:1115–22.
59. Miller TW, Cherney MM, Lee AJ, Francoleon NE, Farmer PJ, King SB, et al. The effects of nitroxyl (HNO) on soluble guanylate cyclase activity. Interactions at ferrous heme and cysteine thiols. *J Biol Chem* 2009;**284**:21788–96.
60. Praneeth VKK, Neese F, Lehnert N. Spin density distribution in five- and six-coordinate iron(II)–porphyrin NO complexes evidenced by magnetic circular dichroism spectroscopy. *Inorg Chem* 2005;**44**:2570–2.
61. Praneeth VKK, Näther C, Peters G, Lehnert N. Spectroscopic properties and electronic structure of five- and six-coordinate iron(II) porphyrin NO complexes: effect of the axial N-donor ligand. *Inorg Chem* 2006;**45**:2795–811.
62. Lehnert N, Galinato MGI, Paulat F, Richter-Addo GB, Sturhahn W, Xu N, et al. Nuclear resonance vibrational spectroscopy applied to [Fe(OEP)(NO)]: the vibrational assignments of five-coordinate ferrous heme-nitrosyls and implications for electronic structure. *Inorg Chem* 2010;**49**:4133–48.
63. Lehnert N, Sage JT, Silvernail NJ, Scheidt WR, Alp EE, Sturhahn W, et al. Oriented single-crystal nuclear resonance vibrational spectroscopy of [Fe(TPP)(MI)(NO)]: quantitative assessment of the trans effect of NO. *Inorg Chem* 2010;**49**:7197–215.
64. Portela CF, Magde D, Traylor TG. The ortho effect in ligation of iron tetraphenylporphyrins. *Inorg Chem* 1993;**32**:1313–20.
65. Collman JP, Yang Y, Dey A, Decreau RA, Ghosh S, Ohta T, et al. A functional nitric oxide reductase model. *Proc Natl Acad Sci USA* 2008;**105**:15660–5.
66. Berto TC, Praneeth VKK, Goodrich LE, Lehnert N. Iron-porphyrin NO complexes with covalently attached N-donor ligands: the formation of a stable six-coordinate species in solution. *J Am Chem Soc* 2009;**131**:17116–26.
67. Wyllie GRA, Schulz CE, Scheidt WR. Five- to six-coordination in (nitrosyl)iron(II) porphyrinates: effects of binding the sixth ligand. *Inorg Chem* 2003;**42**:5722–34.
68. Lehnert N, Berto TC, Galinato MGI, Goodrich LE. In: Kadish KM, Smith KM, Guilard R, editors. *The handbook of porphyrin science*, vol. 14. New Jersey: World Scientific; 2011, p. 1–247, (Chapter 63).
69. Lehnert N, Scheidt WR, Wolf MW. Structure and bonding in heme-nitrosyl complexes and implications for biology. *Struct Bond* 2014;**154**:155–224.
70. Stone JR, Marletta MA. Soluble guanylate cyclase from bovine lung: activation with nitric oxide and carbon monoxide and spectral characterization of the ferrous and ferric states. *Biochemistry* 1994;**33**:5636–40.
71. Burstyn JN, Yu AE, Dierks EA, Hawkins BK, Dawson JH., Studies of the Heme Coordination and Ligand Binding Properties of Soluble Guanylyl Cyclase (sGC): Characterization of Fe(II)sGC and Fe(II)sGC(CO) by Electronic Absorption and Magnetic Circular Dichroism Spectroscopies and Failure of CO To Activate the Enzyme, Biochemistry 1995; 34: 5896–5903.
72. Leu BM, Silvernail NJ, Zgierski MZ, Wyllie GRA, Ellison MK, Scheidt WR, et al. Quantitative vibrational dynamics of iron in carbonyl porphyrins. *Biophys J* 2007;**92**:3764–83.
73. Salzmann R, Ziegler CJ, Godbout N, McMahon MT, Suslick KS, Oldfield E. Carbonyl complexes of iron (II), ruthenium (II), and osmium (II) 5, 10, 15, 20-tetraphenylporphyrinates: A comparative investigation by X-ray crystallography, solid-state NMR spectroscopy, and density functional theory. *J Am Chem Soc* 1998;**120**:11323–34.
74. Li J, Nair SM, Noll BC, Schulz CE, Scheidt WR. Relative axial ligand orientation in bis (imidazole)iron(II) porphyrinates: are "picket fence" derivatives different? *Inorg Chem* 2008;**47**:3841–50.

75. Erbil WK, Price MS, Wemmer DE, Marletta MA. A structural basis for H-NOX signaling in *Shewanella oneidensis* by trapping a histidine kinase inhibitory conformation. *Proc Natl Acad Sci USA* 2009;**106**:19753–60.
76. Ma X, Sayed N, Bbeuve A, van den Akker F. NO and CO differentially activate soluble guanylyl cyclase via a heme pivot-bend mechanism. *EMBO J* 2007;**26**:578–88.
77. Muralidharan S, Boon EM. Heme flattening is sufficient for signal transduction in the H-NOX family. *J Am Chem Soc* 2012;**134**:2044–6.
78. Goodrich LE, Lehnert N. The trans effect of nitroxyl (HNO) in ferrous heme systems: Implications for soluble guanylate cyclase activation by HNO. *J Inorg Biochem* 2013;**118**:179–86.
79. Bayachou M, Lin R, Cho W, Farmer PJ. Electrochemical reduction of NO by myoglobin in surfactant film: characterization and reactivity of the nitroxyl (NO−) adduct. *J Am Chem Soc* 1998;**120**:9888–93.
80. Sulc F, Immoos CE, Pervitsky D, Farmer PJ. Efficient trapping of HNO by deoxymyoglobin. *J Am Chem Soc* 2004;**126**:1096.
81. Immoos CE, Sulc F, Farmer PJ, Czarnecki K, Bocian DF, Levina A, et al. Bonding in HNO-myoglobin as characterized by X-ray absorption and resonance raman spectroscopies. *J Am Chem Soc* 2005;**127**:814–15.
82. Yang L, Ling Y, Zhang Y. HNO binding in a heme protein: structures, spectroscopic properties, and stabilities. *J Am Chem Soc* 2011;**133**:13814–17.
83. Ling Y, Mills C, Weber R, Yang L, Zhang Y. NMR, IR/Raman, and structural properties in HNO and RNO (R = alkyl and aryl) metalloporphyrins with implication for the HNO-myoglobin complex. *J Am Chem Soc* 2010;**132**:1583–91.
84. Sulc F, Fleischer E, Farmer PJ, Ma DJ, La Mar GN. 1H NMR structure of the heme pocket of HNO-myoglobin. *J Biol Inorg Chem* 2003;**8**:348–52.
85. Kumar MR, Pervitsky D, Chen L, Poulos T, Kundu S, Hargrove MS, et al. Nitrosyl hydride (HNO) as an O2 analogue: long-lived HNO adducts of ferrous globins. *Biochemistry* 2009;**48**:5018–25.
86. Chu K, Vojtechovsky J, McMahon BH, Sweet RM, Berendzen J, Schlichting I. Structure of a ligand-binding intermediate in wild-type carbonmonoxy myoglobin. *Nature* 2000;**403**:921–3.
87. Farmer PJ, Sulc F. Coordination chemistry of the HNO ligand with hemes and synthetic coordination complexes. *J Inorg Biochem* 2005;**99**:166–84.
88. Ling Y, Zhang Y. In: Wheeler RA, editor. *Annual reports in computational chemistry*, vol. 6. New York: Elsevier; 2010, p. 65–77.
89. Liu Y, Liu W, Li H, Liu J, Yang Y. Theoretical study of hydrogen bonding interaction in nitroxyl (HNO) dimer: interrelationship of the two N − H· · ·O blue-shifting hydrogen bonds. *J Phys Chem A* 2006;**110**:11760–4.
90. Radoul M, Bykov D, Rinaldo S, Cutruzzola F, Neese F, Goldfarb D. Dynamic hydrogen-bonding network in the distal pocket of the nitrosyl complex of *Pseudomonas aeruginosa* cd_1 nitrite reductase. *J Am Chem Soc* 2011;**133**:3043–55.
91. Bykov D, Neese F. Substrate binding and activation in the active site of cytochrome *c* nitrite reductase. A density functional study. *J Biol Inorg Chem* 2011;**16**:417–30.
92. Greco C, Bruschi M, De Gioia L, Ryde U. A QM/MM investigation of the activation and catalytic mechanism of Fe-only hydrogenases. *Inorg Chem* 2007;**46**:5911–21.
93. Capece L, Boechi L, Perissinotti LL, Arroyo-Mañez P, Bikiel DE, Smulevich G, et al. Small ligand-globin interactions: reviewing lessons derived from computer simulation. *Biochim Biophys Acta* 2013;**1834**:1722–38.
94. Silvernail NJ, Olmstead MM, Noll BC, Scheidt WR. Tetragonal to triclinic—a phase change for [Fe(TPP)(NO)]. *Inorg Chem* 2009;**48**:971–7.

95. Wyllie GRA, Schulz CE, Scheidt WR. Five- to six-coordination in (Nitrosyl)iron(II) porphyrinates: effects of binding the sixth ligand. *Inorg Chem* 2003;**42**:5722–34.
96. Ellison MK, Schulz CE, Scheidt WR. Structure of the deoxymyoglobin model [Fe(TPP)(2-MeHIm)] reveals unusual porphyrin core distortions. *Inorg Chem* 2002;**41**:2173–81.
97. Jensen KP, Ryde U. Importance of proximal hydrogen bonds in haem proteins. *Mol Phys* 2003;**101**:2003–18.
98. Houghton BJ, Deeth RJ. Spin-state energetics of Fe^{II} complexes—the continuing voyage through the density functional minefield. *Eur J Inorg Chem* 2014;4573–80.
99. Reiher M, Salomon O, Hess BA. Reparameterization of hybrid functionals based on energy differences of states of different multiplicity. *Theor Chem Acc* 2001;**107**:48–55.
100. Siegbahn PEM, Blomberg MRA, Chen S-L. Significant van der waals effects in transition metal complexes. *J Chem Theor Comput* 2010;**6**:2040–4.
101. Radon M, Pierloot K. Binding of CO, NO, and O_2 to heme by density functional and multireference ab initio calculations. *J Phys Chem A* 2008;**112**:11824–32.
102. Montenegro AC, Amorebieta VT, Slep LD, Martin DF, Roncaroli F, Murgida DH, et al. Three redox states of nitrosyl: NO^+, NO*, and NO^-/HNO interconvert reversibly on the same pentacyanoferrate(II) platform. *Angew Chem Int Ed* 2009;**48**:4213–16.
103. Zapata AL, Kumar MR, Pervitsky D, Farmer PJ. A singular value decomposition approach for kinetic analysis of reactions of HNO with myoglobin. *J Inorg Biochem* 2013;**118**:171–8.
104. Nagao S, Hirai Y, Suzuki A, Yamamoto Y. ^{19}F NMR characterization of the thermodynamics and dynamics of the acid – alkaline transition in a reconstituted sperm whale metmyoglobin. *J Am Chem Soc* 2005;**127**:4146–7.
105. Grätzel M, Taniguchi S, Henglein A., Pulsradiolytische Untersuchung kurzlebiger Zwischenprodukte der NO-Reduktion in wässriger Lösung, *Ber Bunsen-Ges Phys Chem* 1970; 1003–10.
106. Bartberger MD, Liu W, Ford E, Miranda KM, Switzer C, Fukuto JM, et al. The reduction potential of nitric oxide (NO) and its importance to NO biochemistry. *Proc Natl Acad Sci USA* 2002;**99**:10958–63.
107. Shafirovich V, Lymar SV. Nitroxyl and its anion in aqueous solutions: Spin states, protic equilibria, and reactivities toward oxygen and nitric oxide. *Proc Natl Acad Sci USA* 2002;**99**:7340–5.
108. Codesido NO, Weyhermüller T, Olabe JA, Slep LD. Nitrosyl-centered redox and acid–base interconversions in $[Ru(Me3[9]aneN3)(bpy)(NO)]^{3,2,1+}$. The pKa of HNO for its nitroxyl derivative in aqueous solution. *Inorg Chem* 2014;**53**:981–97.
109. Gloaguen F, Lawrence JD, Rauchfuss TB, Bénard M, Rohmer M-M. Bimetallic carbonyl thiolates as functional models for Fe-only hydrogenases. *Inorg Chem* 2002; **41**:6573–82.
110. Models of nitroprusside-NHO, $[Ru^{II}(Me^3[9]aneN^3)(bpy)(HNO)]^{2+}$, and Mb(II)-NHO were generated in the program GaussView5; the initial geometries of nitroprusside-NHO, ammonia, and ammonium were constructed manually, while the $[Ru^{II}(Me_3[9]aneN_3)(bpy)(HNO/NO^-)]^{2+/1+}$ and Mb(II)-(NHO/NO^-) models used were based on previous DFT studies by Slep and Zhang, respectively [82,108]. These structures were reoptimized using the BP86/TZVP method; further single-point and frequency calculations were performed using the ORCA 3.0.2 quantum-chemical computing package [F. Neese, ORCA 3.0.2; Max-Planck Institut fuer chemische Energiekonversion, Muelheim/Ruhr, Germany, 2011].
111. C. Van Stappen, N. Lehnert, unpublished results.
112. Sulc F. *Dissertation*. Irvine: University of California; 2006.

Interactions of HNO With Metallated Porphyrins, Corroles, and Corrines

9

M.A. Morales Vásquez[1], *J. Pellegrino*[1], *L. Álvarez*[1], *N.I. Neuman*[1,2], *F. Doctorovich*[1] *and M.A. Martí*[1]
[1]Universidad de Buenos Aires and CONICET, Buenos Aires, Argentina,
[2]Universidad Nacional del Litoral, Santa Fe, Argentina

Chapter Outline

9.1 General reactivity of Fe, Mn, and Co porphyrins with HNO

9.1.1 Background on porphyrin nitrosyl complexes

One of the main targets of NO and HNO in biological systems is heme-proteins, thus there are a wide range of structure/reactivity studies using models, i.e., metalloporphyrins (MP), mainly of iron, but also manganese, cobalt, and ruthenium. The preferred (almost exclusively) binding mode of NO to the metal atom is through the nitrogen atom. Once bound, its character ranges (formally) from that of NO^+ to that of NO^-.[1] In recognition of the covalent nature of the M—NO bond, Enemark and Feltham long ago proposed using the following notation for the corresponding metal nitrosyls: $\{MNO\}^n$, where M designates the metal, and n stands for the total number of electrons in the d orbital of the metal ion plus those in NO π^* orbitals. Its relevance relies mostly on the fact that it allows rationalization of the structure (specifically the M—N—O angle, as shown in Fig. 9.1) of a wide range of different complexes.[2] We will begin describing NO (as well as NO^-) studies of iron

The Chemistry and Biology of Nitroxyl (HNO). DOI: http://dx.doi.org/10.1016/B978-0-12-800934-5.00009-8

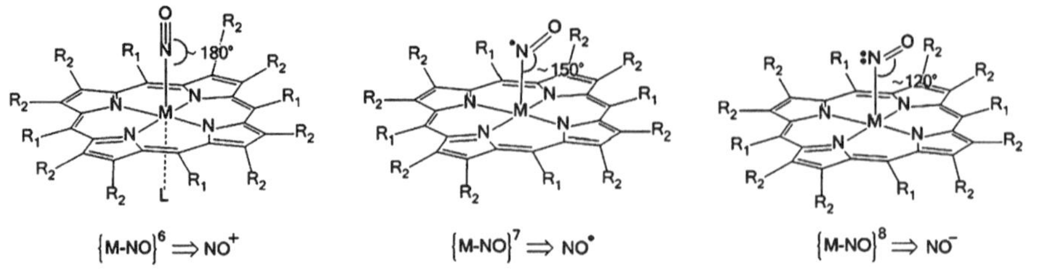

Figure 9.1 $\{MNO\}^n$ complexes with different values of n.

porphyrins, since they are possibly the most studied systems because of their biological relevance as "heme models."

9.1.2 $\{FeNO\}^n$ complexes

Firstly, it is interesting to point out that for iron nitrosyl complexes, there is good correspondence between the Enemark and Feltham description, i.e., $\{FeNO\}^{6/7/8}$ and their formal description as iron Fe^{II} coordinated to NO^+, $NO^\bullet$ or NO^-/HNO, respectively. For example, $\{FeNO\}^6$ complexes have a linear Fe—N—O geometry and NO stretching mode $\nu_{NO} \approx 1900 cm^{-1}$, consistent with its NO^+ character, while $\{FeNO\}^7$ complexes are considered ferrous nitrosyls with an NO• character, a bent Fe—N—O angle and $\nu_{NO} \approx 1670 cm^{-1}$, as observed in several works that include crystallographic characterization.[3] Most importantly, iron porphyrins also yield $\{FeNO\}^8$ complexes, which are supposed to have NO^- (nitroxyl) character, and stretching frequencies $\nu_{NO} \approx 1500 cm^{-1}$.[1,4]

Concerning their stability, the $\{FeNO\}^7$ complexes are by far the most stable ones; and they are readily obtained (see section below) from the reaction of HNO with the corresponding ferric porphyrin, or by bubbling $NO_{(g)}$ through the iron (II or III) porphyrin solution. Given their stability, is not surprising that the first iron nitrosyl porphyrin complex reported was the $\{FeNO\}^7$ complex Fe(TPP)NO, in 1975.[5] Since then, many reports on $\{FeNO\}^7$ complexes were published, dealing with the electronic structure, reactivity and biological implications of this relevant species. One salient feature of the $\{FeNO\}^7$ complexes is the *trans*-effect exerted by the NO ligand, which is directly related to, and possible responsible for, the soluble guanylate cyclase activation mechanism. It has been proposed that due to the NO *trans*-effect, the iron center loses the distal histidine coordination which in turn triggers enzyme activation.[6] This effect is supposed to be even stronger in $\{FeNO\}^8$ but not $\{Fe(H)NO\}^8$ adducts, a fact that could have important physiological implications.[7] Another important issue revealed by several studies concerns the dependence of the electronic structure of the $\{FeNO\}^7$ complexes on the presence of a sixth ligand. While the five-coordinate complexes are best described as intermediate between $Fe^{I}NO^+$ and $Fe^{II}NO^\bullet$, six-coordinate complexes are mostly represented as $Fe^{II}NO^\bullet$, as concluded by EPR, IR, and DFT calculations.[8]

Regarding its redox behavior, $\{FeNO\}^7$ porphyrin complexes exhibit one-electron conversions to the oxidized and reduced $\{FeNO\}^{6/8}$ forms at accessible potentials within certain range, depending on the presence of withdrawing or donating groups in the macrocycle. The pioneering spectroelectrochemical studies by Kadish and coworkers revealed the spectroscopic features of both $\{FeNO\}^{6/8}$ species,[9] and many years later, these redox forms could be properly prepared and isolated.

Consistently with its higher reactivity, the first isolated $\{FeNO\}^6$ complexes were prepared and crystallographically characterized almost 10 years after the first $\{FeNO\}^7$ complex.[10] These complexes are more prone to NO loss or reaction with nucleophiles, converting to the $\{FeNO\}^7$ form in the presence of excess NO (through a reductive nitrosylation) reaction.[11] Surprisingly, the analogous $\{RuNO\}^6$ proved to be a much more stable species; these complexes do not easily lose NO or convert to the $\{RuNO\}^7$ form. The weakness of the Fe—N bond in the $\{FeNO\}^6$ adducts is still a matter of debate, and it has been suggested to be an intrinsic feature that relates to the properties of a high-spin $Fe^{III}NO^{\bullet}$ ($S = 2$) state that appears at low energy and is dissociative regarding the Fe—NO bond.[12] This state would be much higher in energy in the corresponding $\{RuNO\}^6$ complexes and this fact would account for the stability of these systems. For a complete review of the structure and reactivity of $\{FeNO\}^{6/7}$ heme nitrosyls and its implications in biology, the reader is referred to an article by Lehnert and coworkers.[13]

Regarding the $\{FeNO\}^8$, or HNO/nitroxyl adducts, the first complexes obtained by chemical reduction were reported in 1991 by Ryan's group with the widely used TPP (tetraphenyl) and OEP (octaethyl) porphyrins. The adducts could be characterized by UV—Vis and Raman spectroscopies but none of them could be isolated.[14] Recently, an increasing number of reports on $\{FeNO\}^8$ complexes have contributed to the spectroscopic characterization of this initially elusive species, but still much more research is clearly needed to completely understand their nature and implications in biological systems.[15]

Since previously reported $\{FeNO\}^8$ adducts proved impossible to isolate, mainly due to their susceptibility to oxidation to the more stable $\{FeNO\}^7$ species,[14] it could be expected that electron-poor porphyrins would stabilize it. By using this strategy the stable $\{FeNO\}^8$ complex $[Co(C_5H_5)_2]^+[Fe(TFPPBr_8)NO]^-$ was obtained through the reduction of the corresponding $\{FeNO\}^7$ complex with cobaltocene. This product could be isolated and characterized by UV—Vis, IR, and ^{15}NMR.[16] Intriguingly, while the perhalogenatedporphyrin resulted an appropriate platform to stabilize the $FeNO^-$ moiety, attempts to protonate it to yield the expected $\{Fe(H)NO\}^8$ complex were unsuccessful, yielding instead the stable $\{FeNO\}^7$ complex.

It is interesting to note that the $\{Fe(H)NO\}^8$ adduct was indeed stabilized in the distal pocket of several globins as described by Farmer and coworkers in Chapter 14, Spectroscopic NMR Characterizations of HNO Adducts of Ferrous Heme Proteins[17] and also recently using a bis-picket fence porphyrin.[18] Finally, and also noteworthy are the observations that the nonporphyrin $[Fe(CN)_5(HNO)]^{3-}$ species has been reported as a stable $\{Fe^{II}HNO\}^8$ complex in aqueous solutions (see chapter: Non-Heme Transition Metal Complexes of HNO),[19] as well as a fairly stable $\{RuHNO\}^8$ complex, [Ru(TTP)(HNO)(1-MeIm)].[20] In these systems we can neither invoke the supplied stabilization by amino acids as in the protein complexes nor the steric protection provided by the bulky substituents in the picket fence adduct. Clearly there must be some other factors influencing the stability of the $\{Fe(H)NO\}^8$ moiety, perhaps hydrogen bonding with the solvent.

9.1.3 $\{CoNO\}^8$ and $\{MnNO\}^6$ complexes

Interestingly, the first nitrosyl porphyrin complex reported was not an iron but a cobalt complex, namely Co(TPP)NO, in 1973.[21] Since then, several other cobalt NO complexes were prepared (all of them $\{CoNO\}^8$) and thoroughly studied from the spectroscopic, kinetic, and particularly electrochemical perspectives.[22] $\{CoNO\}^8$ adducts are best described as $Co^{III}NO^-$, and all of them are five-coordinated due to the strong *trans*-effect exerted by the NO^- ligand, as previously mentioned for iron nitrosyls. A notable difference from their iron analogs, however, is that $\{CoNO\}^8$ adducts cannot be protonated to yield the HNO complexes, i.e., CoHNO, since the protonation occurs in the macrocyclic ligand.[23] In contrast with the iron systems, all reported stable cobalt adducts are $\{CoNO\}^8$ complexes; the oxidation of the complexes does not give the $\{CoNO\}^{6/7}$ forms, since the porphyrin ligand is the electron transfer site.[22a,b] The complexes can also be reduced, and again the electron is mainly added to the porphyrin ring, the same as found for the $\{FeNO\}^8$ complexes,[14,24] and the reduced cobalt complexes are prone to NO loss.

For manganese porphyrins, all nitrosyl complexes are $\{MnNO\}^6$ systems and can be described as Mn^INO^+. Differently to what is observed for the $\{FeNO\}^6$ adducts, Mn nitrosyls are quite sensitive to oxidation by O_2, which suggests an important $Mn^{II}NO^{\bullet}$ character of the adducts.[25] The first reported crystal structure of a $\{MnNO\}^6$ adduct was for Mn(TPP)(NO)(4-MePip) in 1974.[26] The list of complexes increased later, in both five- and six-coordinate systems, with extensive redox characterization.

Again, as in the cobalt systems, and differently from that observed for iron, the Enemark and Feltham parameter "n" could not be altered, since the reductions of the $\{MnNO\}^6$ systems occur in the porphyrin ring, and there is not access to the $\{MNO\}^{7/8}$ forms. Moreover, the reduced species are not stable, also due to labilization of the NO ligand.[27]

In summary while iron porphyrins allow several type of nitrosyl adducts, yielding $\{FeNO\}^{6,7,8}$ type of complexes, cobalt porphyrins yield exclusively a $\{CoNO\}^8$ complex, and manganese an $\{MnNO\}^6$ one. All of them can be readily obtained from the corresponding reaction of the free porphyrin in the appropriate redox state and HNO, as will be described below.

9.1.4 Kinetics and mechanistic aspects of HNO and donor reactions with metalloporphyrins

The reaction mechanism and kinetics of $Fe^{III}P$ and $Mn^{III}P$ (where P is the porphyrin) with two common azanone donors, sodium trioxidinitrate (Angeli's Salt, AS) and toluensulfohydroxamic acid (TSHA), were extensively studied from a kinetic perspective.[28–30] Results show that efficient conversion to the corresponding nitrosyl adduct $\{FeNO\}^7$ or $\{MnNO\}^6$ is observed, as evidenced by the UV–Vis spectral change. The shift in the Soret band is significantly larger for $Mn^{III}P$, (see Table 9.1), compared to $Fe^{III}P$ (shifts around 5 nm), and that is one of the reasons why $Mn^{III}P$ can be used for HNO quantification.

Table 9.1 Spectral changes for the Mn porphyrins and their NO complexes

Porphyrin	λ Soret absorption band for the porphyrin, in nm (ε, $cm^{-1}M^{-1}$)		λ Soret absorption band for the NO-complexes, in nm (ε, $cm^{-1}M^{-1}$)
$[Mn^{III}HematoP]^-$	368 (69,300)	461 (30,400)	414 (108,000)
$[Mn^{III}ProtoP]^-$	371 (82,500)	466 (34,500)	414 (100,000)
$[Mn^{III}TSPP]^{3-}$	401 (40,000)	468 (70,000)	426 (140,000)
$[Mn^{III}Br_8TCPP]^{3-}$	394 (43,000)	493 (78,000)	380[a] (38,000) 460[a] (83,000)
$[Mn^{III}$ T-2-PyP$]^+$	374 (25,000)	462 (70,000)	421 (95,000)
$[Mn^{III}T(TriMA)P]^{5+}$	379 (27,000)	464 (68,700)	423 (89,000)

[a]Correspond to the two maxima in the UV–vis spectrum.

Most strikingly, analysis of the reaction kinetics for both tested donors (AS and Piloty's Acid, PA) revealed two markedly different behaviors, depending on the porphyrin reduction potential (see Fig. 9.1). For "oxidizing" porphyrins, i.e., those with a positive reduction potential (>100 mV), reaction with the donor is extremely fast (on the order of seconds); full conversion to the nitrosyl porphyrin is obtained using stoichiometric donor:porphyrins ratios, and reaction kinetics is first order in donor concentration, obeying the following overall rate law:

$$\frac{d[\mathrm{MPNO}]}{dt} = k_{\mathrm{donor}}[\mathrm{donor}][\mathrm{Porphyrin}]$$

where $d[\mathrm{MPNO}]/dt$ is the corresponding nitrosyl formation rate as determined from the UV–Vis spectra. On the other hand, for "reducing" porphyrins, with a negative reduction potential (< −160 mV), a large excess of donor is required to drive the reaction, which is slow (half-life of over 100 min for stoichiometric donor:porphyrin ratios). Moreover, the rate does not increase linearly with donor concentration, and instead, the HNO and donor concentrations follow a typical square-root trend. On the contrary, the reaction is first order in [HNO], strongly suggesting HNO trapping by the porphyrin, and obeying the following rate law:

$$\frac{d[\mathrm{MPNO}]}{dt} = k_{\mathrm{on}}[\mathrm{HNO}][\mathrm{Porphyrin}]$$

where [HNO] is estimated using the following equation:

$$[\mathrm{HNO}]_{\mathrm{ss}} = \sqrt{\frac{k_d[\mathrm{donor}]_0}{2k_h}}$$

In other words, while oxidizing porphyrins interact directly with the donor, promoting NO release which is "concertedly" trapped by the reduced porphyrin to

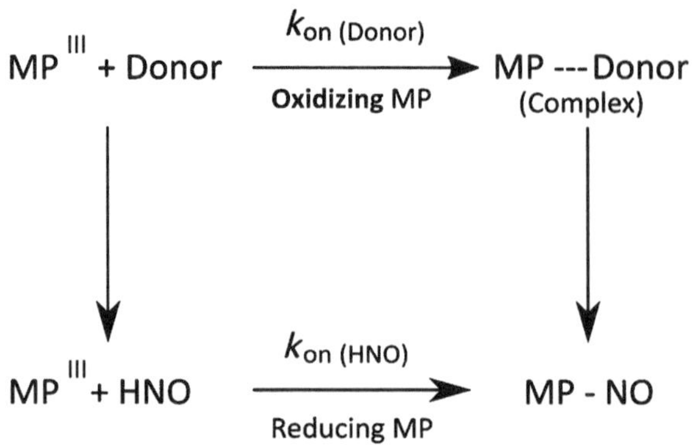

Scheme 9.1 Reaction mechanisms between metalloporphyrins and HNO donors.

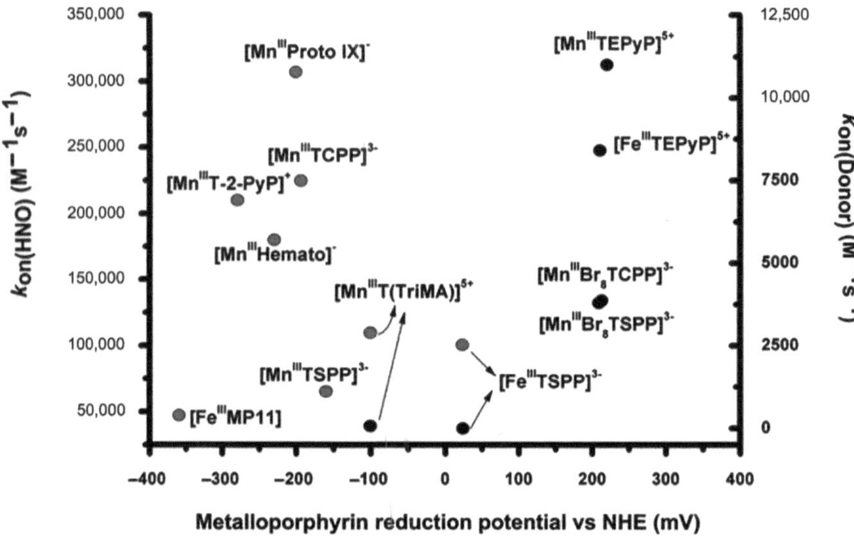

Figure 9.2 Values of k_{on} (HNO) ($M^{-1}s^{-1}$) (red, left axis) and k_{on} (Donor) ($M^{-1}s^{-1}$) (black, right axis) for Fe and Mn-porphyrins plotted against the reduction potential for the Fe^{III}/Fe^{II} or Mn^{III}/Mn^{II} couples.

yield the resulting $\{MeNO\}^7$ complex, reducing Fe and Mn porphyrins trap the free HNO produced by the spontaneous decomposition of the donor (see Scheme 9.1). Results for oxidizing porphyrins also highlight a key factor of donor decomposition mechanism to yield azanone (at least for NONOates and sulfohydroxamic acids represented by AS and PA, respectively) since they show that donor oxidation possibly yield NO (and not HNO) (Fig 9.2).

Finally, it is interesting to compare the kinetic results obtained for the Mn and Fe porphyrins with those obtained for HNO and NO binding to ferrous and ferric

Table 9.2 Fe^{3+}/Fe^{2+} reduction potentials and NO/HNO rate constants for MP and heme-proteins

Target	$E_{1/2}$ vs NHE (mV)	Reactant	k_{on} ($M^{-1}s^{-1}$)	References
[a]$[Fe^{III}MP11]^{3-}$	−360	HNO	3.1×10^4	31
		NO	1.1×10^6	7
[b]$[(OH)Fe^{III}TSPP]^{4-}$	+23	NO	4.5×10^5	32
$[Fe^{II}TSPP]^-$	−300	NO	1.5×10^9	32
[c]Cyt-c(Fe^{III})	+250	HNO	4×10^4	33
Catalase(Fe^{III})	−226	HNO	3×10^5	33
[d]metMb(Fe^{III})	+120	HNO	8×10^5	33
[e]Mb(Mn^{III})	−320	HNO	3.4×10^5	25
Mb(Fe^{II})	−430	HNO	1.4×10^4	33
		NO	2.0×10^7	34
O_2	-------	HNO	3×10^3	34
NO	-------	HNO	5.8×10^6	35

[a]MP11 = Microperoxidase 11.
[b]TSPP = meso-Tetrakis(*p*-sulfonatophenyl) porphyrin.
[c]Cyt-c = Cytochrome c.
[d]metMb = met-Myoglobin.
[e]Mb = Myoglobin.

heme-proteins, which values are shown in Table 9.2. For all studied porphyrins, as well as most proteins, HNO association rate constants to ferric hemes fall in the 1×10^4 to $1 \times 10^5\ M^{-1}s^{-1}$ range, which suggests a minor role for the protein matrix. These values are significantly smaller compared to those of NO binding to ferrous porphyrins or heme-proteins, which are greater than $1 \times 10^7\ M^{-1}s^{-1}$, and only slightly smaller than those reported for NO binding to Mn^{III} and Fe^{III} porphyrins. A possible explanation for this fact is, as proposed by Lorkovic and Ford,[29] that for NO (and possibly also HNO) binding to ferric hemes, the release of a weakly coordinated water/hydroxide ligand is rate limiting.

The obtained rates can also be considered in the context of their possible HNO targets (like O_2, NO or thiols). HNO trapping by porphyrins is 10−100 times faster than reaction with oxygen, thus even for physiological "microaerobic" conditions, $Mn^{III}P$ and $Fe^{II}P$ can be used as efficient HNO traps, and the reaction of HNO with ferric heme-proteins is expected to occur. On the other hand, reaction of MP with NO is fast, and thus NO can be considered an HNO sink.

In summary, the reactions of HNO donors with Mn and Fe porphyrins proceed by two alternative mechanisms. The first one, in the case of oxidizing porphyrins ($E_{1/2}$ >100 mV), which catalyze the oxidative decomposition of the donor, and the second one with reducing porphyrins ($E_{1/2} < -160$ mV), requires spontaneous decomposition of the donor to yield free HNO, which is subsequently trapped via a bimolecular reaction with rate constants in the 4×10^4 to $4 \times 10^5\ M^{-1}s^{-1}$ range. These studies also provide a rationale for selecting adequate Mn-porphyrins to be used in HNO trapping, detection, and quantification experiments, which should be reducing, and most stable to oxygen-mediated decay.[25,30]

9.2 Metallocorroles and metallocorrines

9.2.1 *Nitrosyl complexes of metallocorroles*

Corroles are contracted porphyrinoids constituted by four pyrrole groups linked by three *meso* carbons and one direct β-β'pyrrole–pyrrole bond (Fig. 9.3A). The first corrole syntheses were performed in the 1960s motivated by natural occurrence of cobalamin (vitamin B_{12}, see Fig. 9.3B), which is a corrin (a reduced type of corrole). But it was not until 1999, when two seminal papers were published by Paolesse et al.[36] and Gross et al.[37], reporting facile syntheses for 5,10,15-*tris*(phenyl)corrole and 5,10,15-*tris*(pentafluorophenyl)corrol (tpfc), respectively, that the field began to grow exponentially. Later Paolesse[38] and Gryko[39] perfected various synthetic approaches allowing the collection of a large set of corroles giving important yields. Compared to porphyrins that act as metal dianionic ligands, corroles act as metal trianionic ligands and display a smaller central cavity; therefore, they are able to stabilize coordinated metals in higher oxidation states.[40] These properties are reflected in the observed differences between the electronic and chemical properties of the corresponding nitrosyl metal correlates compared to nitrosyl metal porphyrins. However, it is important to keep in mind that only a few works have been published on nitrosyl-corrolates.

Among key nitrosylcorrole studies, Sinha et al.,[41] e.g., synthesized several iron corrolates with NO apical ligands and performed spectroelectrochemistry experiments to obtain both oxidized and reduced species. The oxidized species showed a single EPR signal compatible with a corrole centered radical (and thus with NO^+ character). Furthermore, there is a small shift of the νNO stretching frequency, supporting the fact that the redox process is not {FeNO} centered. Instead, the reduced species presents an EPR spectra with ^{14}N hyperfine splitting ((^{14}N) = 15.4 G),

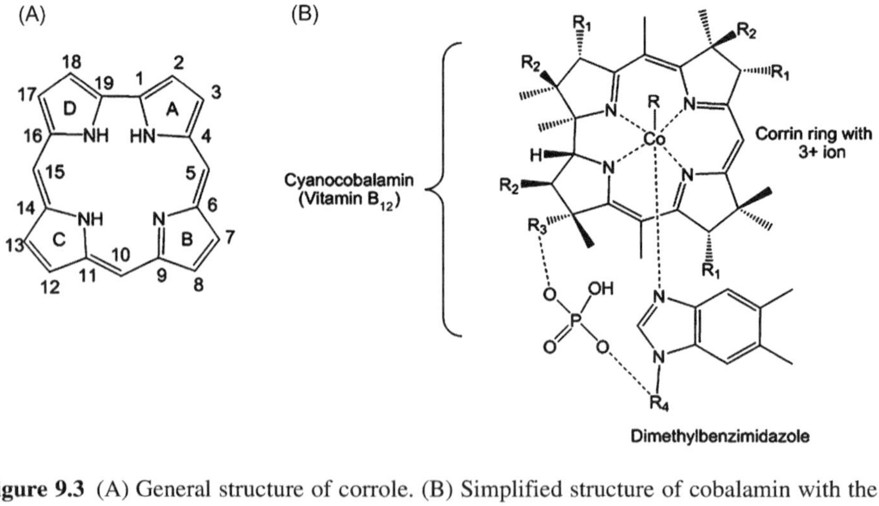

Figure 9.3 (A) General structure of corrole. (B) Simplified structure of cobalamin with the upper axial ligand (R = CN^-, OH^-, -CH_3, Ado^- [adenosyl]).

showing that the unpaired electron is not corrole but {FeNO} centered. Moreover νNO stretching frequency shifts from $1785\ cm^{-1}$ in the neutral species to $1604\ cm^{-1}$ in the reduced {FeNO}(Corr). Sinha et al.[41] also noted that the obtained results of electrochemistry, spectroelectrochemistry measurements, and DFT calculations clearly showed the tendency of the $\{FeNO\}^6$ adduct to be reduced and the opposite tendency of the corrolate adduct to get oxidized. They realized that the species obtained had a $[(Corr)^{2\cdot -}\{Fe(NO)\}^6]^{\cdot +}$ structure.

From the above observations, they established that the EPR signals were mainly originated at the metal center from a low-spin ferric iron. To our knowledge there are no published studies of nitrosyl complexes of cobalt and manganese corrolates. We now turn our attention to cobalamin nitroxyl complexes, which on the other hand have been extensively studied.

9.2.2 Nitroxyl complexes of cobalamin

Cobalamins (Cbl)—vitamin B_{12}—are derivatized corrines (macrocycles with a corrole framework and reduced 1,2,3,7,8,12,13,17,18,19 carbon atoms, as shown in Fig. 9.3B), which act as monobasic ligands for $Co^{II/III}$. Hassanin et al.[42] have studied the nitroxyl complex of cobalamin both in solution using NMR and UV−vis techniques[43] and in their crystallized forms.[44] Their studies show that the Co−N = O angle is about 120°, compatible with a Co^{III}−NO^- description of the corresponding complex. The bent orientation is consistent with the presence of a lone pair in the N atom. The N−O distances in NOCbl are 1.12−1.14 Å, and assigned to an N = O double bond. In comparison, for the nitrite adducts, NO_2−Cbl, the N−O distances are in the 1.22−1.32 Å range. UV−visible spectra show that the visible bands in NO−Cbl are noticeably blue-shifted compared to those of CN−Cbl and NO_2−Cbl and more similar to base-off protonated Me−Cbl ($Me = CH_3^-$). Computational studies by Surducan et al.[45] show that HNO can bind to either formally Co^{II}Cbl and Co^{III}Cbl. They also suggest that NO−Co^{II}Cbl is possibly better described as NO^-−Co^{III}Cbl. Concerning HNO association kinetics, Subedi et al.[46] studied the reaction between $H_2OCo^{III}Cbl^+$/$HOCo^{III}$Cbl and Angeli's salt at several pH's. They found that the main reaction product is the corresponding, and previously described, NO^-−Co^{III} Cbl nitroxyl complex (NOCbl) and that the reaction is fast, although no kinetic parameters have been determined. Clearly studies of NO and HNO with iron (or other metal) corroles as well as cobalamin are just beginning and a lot of future work is needed to understand its reactivity patterns and physiological relevance.[47,48]

9.3 Conclusions

Azanone reactions with iron, manganese, and cobalt porphyrins have been extensively studied in the last decades. The most straightforward and well-characterized reaction is that of HNO with a metal porphyrin in an appropriate redox state, which allows to obtain the corresponding most stable nitrosyl complexes $\{FeNO\}^7$,

$\{MnNO\}^6$, and $\{CoNO\}^8$. Interestingly, while Mn and Co do not yield stable reduced or oxidized states, iron is able to form as well stable $\{FeNO\}^8$ and $\{FeHNO\}^8$ (actually HNO/NO^- coordinated) complexes. Mechanistically speaking, HNO reactions with porphyrins follow the expected bimolecular kinetics with rate constants similar to those observed for other small ligands (like NO) with the same porphyrins. Care should be taken, however, with "oxidizing" porphyrins that are able to directly interact with the donor, since they are prone to oxidize it and promote its decomposition. Physiological as well as technological aspects of these reactions will be further analyzed in Chapter 7, Non-Heme Transition Metal Complexes of HNO and Chapter 13, Detection of HNO by Membrane Inlet Mass Spectrometry, respectively.

References

1. McCleverty JA. Chemistry of nitric oxide relevant to biology. *Chem Rev* 2004;**104**: 403–18.
2. Enemark JH, Feltham RD. Principles of structure, bonding, and reactivity for metal nitrosyl complexes. *Coord Chem Rev* 1974;**13**:339–406.
3. Wyllie GRA, Scheidt WR. Solid-state structures of metalloporphyrin NO_x compounds. *Chem Rev* 2002;**102**:1067–89.
4. Harrop TC. Chapter five—new insights on $\{FeNO\}^n$ ($n = 7$, 8) systems as enzyme models and HNO donors. *Adv Inorg Chem* 2015;**67**:243–63.
5. Scheidt WR, Frisse ME. Nitrosylmetalloporphyrins. II. Synthesis and molecular stereochemistry of nitrosyl-.alpha, beta, gamma, delta-tetraphenylporphinatoiron(II). *J Am Chem Soc* 1975;**97**:17–21.
6. Stone JR, Marletta MA. Spectral and kinetic studies on the activation of soluble guanylate cyclase by nitric oxide. *Biochemistry* 1996;**35**:1093–9.
7. Goodrich LE, Lehnert N. Principles of structure, bonding, and reactivity for metal nitrosyl complexes. *J Inorg Biochem* 2013;**118**:179–86.
8. (a) Pranneth VKK, Näther C, Peters G, Lehnert N. Spectroscopic properties and electronic structure of five- and six-coordinate iron(II) porphyrin NO complexes: effect of the axial N-donor ligand. *Inorg Chem* 2006;**45**:2795–811.
   (b) Berto TC, Praneeth VKK, Goodrich LE, Lehnert N. Iron-porphyrin NO complexes with covalently attached N-donor ligands: formation of a stable six-coordinate species in solution. *J Am Chem Soc* 2009;**131**:17116–26.
   (c) Cheng L, Powell DR, Khanand MA, Richter-Addo GB. The first unambiguous determination of a nitrosyl-to-nitrite conversion in an iron nitrosyl porphyrin. *Chem Commun* 2000;**23**:2301–2.
9. Lancon D, Kadish KM. Electrochemical and spectral characterization of iron mono- and dinitrosyl porphyrins. *J Am Chem Soc* 1983;**105**:5610–17.
10. Scheidt WR, Lee YJ, Hatano K. Preparation and structural characterization of nitrosyl complexes of ferric porphyrinates. Molecular structure of aquonitrosyl (meso-tetraphenylporphinato) iron(III) perchlorate and nitrosyl (octaethyl-porphinato) iron(III) perchlorate. *J Am Chem Soc* 1984;**106**:3191–8.
11. Ford PC. Reactions of NO and nitrite with heme models and proteins. *Inorg Chem* 2010;**49**:6226–39.

12. Praneeth VKK, Paulat F, Berto TC, DeBeer George S, Näther C, Sulok CD, et al. Electronic structure of six-coordinate iron(III) – porphyrin NO adducts: the elusive iron (III) – NO(radical) state and its influence on the properties of these complexes. *J Am Chem Soc* 2008;**130**:15288–303.
13. Goodrich LE, Paulat F, Praneeth VKK, Lehnert N. Electronic structure of heme-nitrosyls and its significance for nitric oxide reactivity, sensing, transport, and toxicity in biological systems. *Inorg Chem* 2010;**49**:6293–316.
14. Choi IK, Liu Y, Feng D, Paeng KJ, Ryan MD. Electrochemical and spectroscopic studies of iron porphyrin nitrosyls and their reduction products. *Inorg Chem* 1991;**30**: 1832–9.
15. (a) Speelman AL, Lehnert N. Heme versus non-heme iron-nitroxyl $\{FeN(H)O\}^8$ complexes: electronic structure and biologically relevant reactivity. *Acc Chem Res* 2014; **47**(4):1106–16.
    (b) Doctorovich F, Bikiel DE, Pellegrino J, Suárez SA, Martí MA. Reactions of HNO with metal porphyrins: underscoring the biological relevance of HNO. *Acc Chem Res* 2014;**47**(10):2907–16.
16. Pellegrino J, Bari SE, Bikiel DE, Doctorovich F. successful stabilization of the elusive species $\{FeNO\}^8$ in a heme model. *J Am Chem Soc* 2010;**132**:989–95.
17. (a) Kumar MR, Pervitsky D, Chen L, Poulos T, Kundu S, Hargrove MS, et al. Nitrosyl hydride (HNO) as an O_2 analogue: long-lived HNO adducts of ferrous globins. *Biochemistry* 2009;**48**:5018–25.
    (b) Lin R, Farmer PJ. The HNO adduct of myoglobin: synthesis and characterization. *J Am Chem Soc* 2000;**122**:2393–4.
18. Goodrich LE, Roy S, Alp EE, Zhao J, Hu MY, Lehnert N. Electronic structure and biologically relevant reactivity of low-spin $\{FeNO\}^8$ porphyrin model complexes: new insight from a bis-picket fence porphyrin. *Inorg Chem* 2013;**52**:7766–80.
19. Montenegro AC, Amorebieta VT, Slep LD, Martín DF, Roncaroli F, Murgida DH, et al. Three redox states of nitrosyl: NO^+, NO·, and NO^-/HNO interconvert reversibly on the same pentacyanoferrate(II) platform. *Angew Chem Int Ed* 2009;**48**:4213–16.
20. Lee J, Richter-Addo GB. A nitrosyl hydride complex of a heme model [Ru(ttp)(HNO)(1-MeIm)] (ttp = tetratolylporphyrinato dianion). *J Inorg Biochem* 2004;**98**:1247–50.
21. Sheidt WR, Hoard JL. Stereochemistry of low-spin cobalt porphyrins. I. Structure and bonding in a nitrosylcobalt porphyrin and their bearing on one rational model for the oxygenated protoheme. *J Am Chem Soc* 1973;**95**:8281–8.
22. (a) Kelly S, Lancon D, Kadish KM. Electron-transfer and ligand-addition reactions of (TTP)Mn(NO) and (TTP)Co(NO) in a nonaqueous media. *Inorg Chem* 1984; **23**:1451–8.
    (b) Kadish KM, Mu XH, Lin XQ. Microvoltammetry and in situ FTIR, ESR, and UV-visible spectroelectrochemical studies of nitrosyl(tetrophenylporphyrinato)cobalt oxidation/reduction in dichloromethane. *Inorg Chem* 1988;**27**:1489–92.
    (c) Richter-Addo GB, Hodge SJ, Yi G-B, Khan MA, Ma T, et al. Synthesis, characterization, and spectroelectrochemistry of cobalt porphyrins containing axially bound nitric oxide. *Inorg Chem* 1996;**35**:6530–8.
    (d) Kini AD, Washington J, Kubiak CP, Morimoto BH. Spectroelectrochemical characterization of substituted cobalt nitrosyl porphyrins. *Inorg Chem* 1996;**35**:6904–6.
    (e) Kadish KM, Ou Z, Tan X, Boschi T, Monti D, Fares V, et al. Synthesis and electrochemistry of cobalt β-halogenated meso-tetraphenylporphyrins containing a nitrosyl axial ligand. Crystal structure of $(TPPBr_4NO_2)Co(NO)$. *J Chem Soc, Dalton Trans* 1999;1595–602. Available from: http://dx.doi.org/10.1039/A809719D.

23. Rhine MA, Rodrigues AV, Bieber Urbauer RJ, Urbauer JL, Stemmler TL, Harrop TC. Proton-Induced Reactivity of NO^- from a $\{CoNO\}^8$ Complex. *J Am Chem Soc* 2014; **136**:12560–3.
24. Pellegrino J, Hübner R, Doctorovich F, Kaim W. Spectroelectrochemical evidence for the nitrosyl redox siblings NO^+, $NO^{\cdot}$, and NO^- coordinated to a strongly electron-accepting Fe^{II} Porphyrin: DFT calculations suggest the presence of high-spin states after reduction of the $Fe^{II}{-}NO^-$ complex. *Chem Eur J* 2011;**17**:7868–74.
25. Borón I, Suárez SA, Doctorovich F, Martí MA, Bari SE. A protective protein matrix improves the discrimination of nitroxyl from nitric oxide by Mn^{III} protoporphyrinate IX in aerobic media. *J Inorg Biochem* 2011;**105**:1044–9.
26. Piciulo PG, Rupprecht G, Scheidt WR. Stereochemistry of nitrosylmetalloporphyrins. Nitrosyl- alpha, beta, gamma, delta-tetraphenylporphinato(1-methylimidazole) iron and nitrosyl- alpha, beta, gamma, delta-tetraphenylporphinato(4-methylpiperidine)manganese. *J Am Chem Soc* 1974;**96**:5293–5.
27. Zahran ZN, Shaw MJ, Khan MA, Richter-Addo GB. Fiber-optic infrared spectroelectrochemical studies of six-coordinate manganese nitrosyl porphyrins in nonaqueous media. *Inorg Chem* 2006;**45**(6):2661–8.
28. Álvarez L, Suarez SA, Bikiel DE, Reboucas JS, Batinić-Haberle I, Martí MA, et al. Redox potential determines the reaction mechanism of HNO donors with Mn and Fe porphyrins: defining the better traps. *Inorg Chem* 2014;**53**(14):7351–60.
29. Bari SE, Martí MA, Amorebieta VT, Estrin DA, Doctorovich F. Fast nitroxyl trapping by ferric porphyrins. *J Am Chem Soc* 2003;**125**(50):15272–3.
30. Martí MA, Bari S, Estrin D, Doctorovich F. Discrimination of nitroxyl and nitric oxide by water-soluble Mn(III) porphyrins. *J Am Chem Soc* 2005;**127**:4680–4.
31. Suárez SA, Martí MA, De Biase PM, Estrin DA, Bari SE, Doctorovich F. HNO trapping and assisted decomposition of nitroxyl donors by ferric hemes. *Polyhedron* 2007;**26**: 4673–9.
32. Laverman LE, Ford PC. Mechanistic studies of nitric oxide reactions with water soluble iron(II), cobalt(II), and iron(III) porphyrin complexes in aqueous solutions: implications for biological activity. *J Am Chem Soc* 2001;**123**(47):11614–22.
33. Miranda KM, Nims RW, Thomas DD, Espey MG, Citrin D, Bartberger MD, et al. Comparison of the reactivity of nitric oxide and nitroxyl with heme proteins: a chemical discussion of the differential biological effects of these redox related products of NOS. *J Inorg Biochem* 2003;**93**:52–60.
34. Scott EE, Gibson QH, Olsen JS. Mapping the pathways for O_2 entry into and exit from myoglobin. *J Biol Chem* 2001;**276**:5177–88.
35. Lymar SV, Shafirovich V, Poskrebyshev GA. One-electron reduction of aqueous nitric oxide: a mechanistic revision. *Inorg Chem* 2005;**44**:5212–21.
36. Paolesse R, Di Natale C, Macagnano A, Sagone F, Scarselli MA, Chiaradia P, et al. Langmuir – blodgett films of a manganese corrole derivative. *Langmuir* 1999;**15**(4): 1268–74.
37. Gross Z, Galili N, Saltsman I. The first direct synthesis of corroles from pyrrole. *Angew Chem Int* 1999;**38**(10):1427–9 1999.
38. Paolesse R, Nardis S, Sagone F, Khoury RG. Synthesis and functionalization of *meso*-aryl-substituted corroles. *J Org Chem* 2001;**66**(2):550–6.
39. (a) Gryko DT, Koszarna B. Refined methods for the synthesis of meso-substituted A_3- and trans-A_2B-corroles. *Org Biom Chem* 2003;**1**(2):350–7.
    (b) Koszarna B, Gryko DT. Efficient synthesis of meso-substituted corroles in a H_2O – MeOH mixture. *J Org Chem* 2006;**71**(10):3707–17.

40. Aviv-Harel I, Gross Z. Aura of corroles. *Chemistry* 2009;**15**(34):8382–94.
41. Sinha W, Deibel N, Agarwala H, Garai A, Schweinfurth D, Purohit CS, et al. Synthesis, spectral characterization, structures, and oxidation state distributions in [(corrolato) $Fe^{III}(NO)]^n$ ($n = 0, +1, -1$) complexes. *Inorg Chem* 2014;**53**(3):1417–29 2014.
42. Hassanin H, Hannibal L, Jacobsen DW, Brown KL, Marques HM, Brasch NE. NMR spectroscopy and molecular modelling studies of nitrosylcobalamin: further evidence that the deprotonated, base-off form is important for nitrosylcobalamin in solution. *Dalton Trans* 2009;**3**:424–33.
43. Hassanin H, Hannibal L, Jacobsen DW, El-Shahat MF, Hamza MS, Brasch NE. Mechanistic studies on the reaction between R_2N-NONOates and aquacobalamin: evidence for direct transfer of a nitroxyl group from R_2N-NONOates to cobalt(III) centers. *Angew Chem* 2009;**48**(47):8909–13.
44. Hassanin H, El-Shahat MF, DeBeer S, Smith C, Brasch NE. Redetermination of the X-ray structure of nitroxylcobalamin: base-onnitroxyl cobalamin exhibits a remarkably long Co–N (dimethylbenzimidazole) bond distance. *Dalton Trans* 2010;**39**(44): 10626–30.
45. Surducan M, Makarov SV, Silaghi-dumitrescu R. Redox and linkage isomerism with ligands relevant to oxidative and nitrosative stress in cobalamin. *Polyhedron* 2014;**78**: 72–84.
46. Subedi H, Hassanin HA, Brasch NE. Kinetic and mechanistic studies on the reaction of the vitamin B_{12} complex aquacobalamin with the HNO donor angeli's salt: angeli's salt and HNO react with aquacobalamin. *Inorg Chem* 2014;**53**:1570–7.
47. Lim MD, Lorkovic IM, Ford PC. Kinetics of the oxidation of triphenylphosphine by nitric oxide. *Inorg Chem* 2002;**41**(4):1026–8.
48. Sulc F, Immoos CE, Pervitsky D, Farmer PJ. Efficient trapping of HNO by deoxymyoglobin. *J Am Chem Soc* 2004;**126**:1096–101.

Fluorescent Probes for HNO Detection

10

M. Ren, B. Dong and W. Lin
University of Jinan, Jinan, Shandong, People's Republic of China

Chapter Outline

10.1 Introduction

HNO has attracted intense attention since the late 1900s.[1,2] A significant body of studies have demonstrated that HNO may confer critical functions in various physiological processes. For example, HNO can enhance the contractility of heart cells,[3] elicit vasorelaxation in muscle cells,[4] and deregulate platelet aggregation.[5] Furthermore, it has also been reported that HNO may facilitate the relaxation of resistant-like arteries by triggering the voltage-dependent K^+ channel.[6,7] However, oxidation of HNO may afford strong oxidizing species, which can induce damage in DNA and result in the depletion of cellular thiols.[4,8,9] On the other hand, HNO may act as a valid therapeutic agent. For instance, HNO may be employed for the treatment of heart failure. In light of its strong ability to inhibit platelet aggregation, HNO is superior to the traditional nitrovasodilators for the treatment of atherothrombotic syndromes.[10] Thus, it is of high significance to develop powerful methods for the tracking and detection of HNO in living systems.

A number of analytical techniques, such as colorimetric methods,[11–13] EPR,[14–16] HPLC,[17,18] mass spectrometry,[19] and electrochemical analysis,[20] have been formulated for HNO detection. However, the shortcomings of these methods include that they are time-consuming and require destruction of cells and tissues. Thus, these procedures are not suitable for tracking and detecting HNO in the context of living systems. By sharp contrast, recently, fluorescence sensing and

The Chemistry and Biology of Nitroxyl (HNO). DOI: http://dx.doi.org/10.1016/B978-0-12-800934-5.00010-4

imaging has emerged as a powerful means for studying bio-species in the living samples in light of its unparalleled advantages such as high sensitivity and spatio-temporal resolution, real-time detection, in particular, compatible with the living systems. Thus, the fluorescent probes for HNO are highly desirable.

Toward this end, in the past few years, a wide variety of fluorescent HNO probes have been engineered and applied for imaging of HNO in the living systems. Thus, the aim of this chapter is to highlight the recent advances in this area. We have classified the fluorescent HNO probes into three types based on their distinct interaction mechanisms. The first type refers to the fluorescent HNO probes operating by the HNO-induced reduction of Cu(II) to Cu(I) (*Copper-based fluorescent HNO probes*).[21–28] For the second type, we will describe the fluorescent HNO probes designed based on the interactions of HNO with nitroxide radicals such as 2,2,6,6-tetramethyl-1-piperidinyloxy (TEMPO) (*TEMPO-based fluorescent HNO probes*).[29] Finally, we will summarize the fluorescent HNO probes working by the reaction of HNO with triarylphosphines (*Phosphine-based fluorescent HNO probes*).[30–39] The reaction mechanism of the third type of fluorescent probes will be covered on Chapter 11, Phosphine-Based HNO Detection. Overall, this chapter focuses on the design strategies, fluorescence response, reaction mechanisms, and the biological applications of some representative fluorescent HNO probes.

10.2 Copper-based fluorescent HNO probes

Reductive nitrosylation of oxidized metals such as Mn(III), Fe(III), and Cu(II) is one of the important reactions mediated by HNO. In 2005, Doctorovich and coworkers developed the colorimetric probes for HNO based on Mn (III) porphyrins. As HNO and NO have the different reductibility for Mn(III) porphyrins, the probes can be used to distinguish HNO from NO. The presence of HNO can induce the conversion of Mn(III) porphyrins to Mn(II)(Por)NO leading to a large shift in the UV–Vis Soret band.[12] Building on this work, Lippard's group constructed a few probes for fluorescent detection of nitroxyl based on the reductive nitrosylation of Cu(II). In 2007, they introduced the first fluorescent HNO probe based on a polymer scaffold, bithiophene-substituted poly(*p*-phenyleneethynylene) derivative (CP1) complexed with Cu(II).[21] This complex displayed an enhancement in the emission upon exposure to excess NO in unbuffered solutions; however, it showed a small decrease in the emission under simulated biological conditions (pH 7.4 HEPES/KCl buffer), likely due to precipitation of the probe.[21] By contrast, treatment of the probe with AS (a stable source of HNO) could elicit a twofold fluorescence increase. Thus, this probe can selectively sense HNO over NO in a buffered aqueous solution.

A survey of the recent literature reveals that at least five small-molecule Cu-based fluorescent HNO probes have been reported from 2010 to 2014 (Fig. 10.1). Interestingly, the probes reported in 2010–12 exhibit emission in the visible-light range. By contrast, the probe illustrated in 2013 has emission in the far-red light

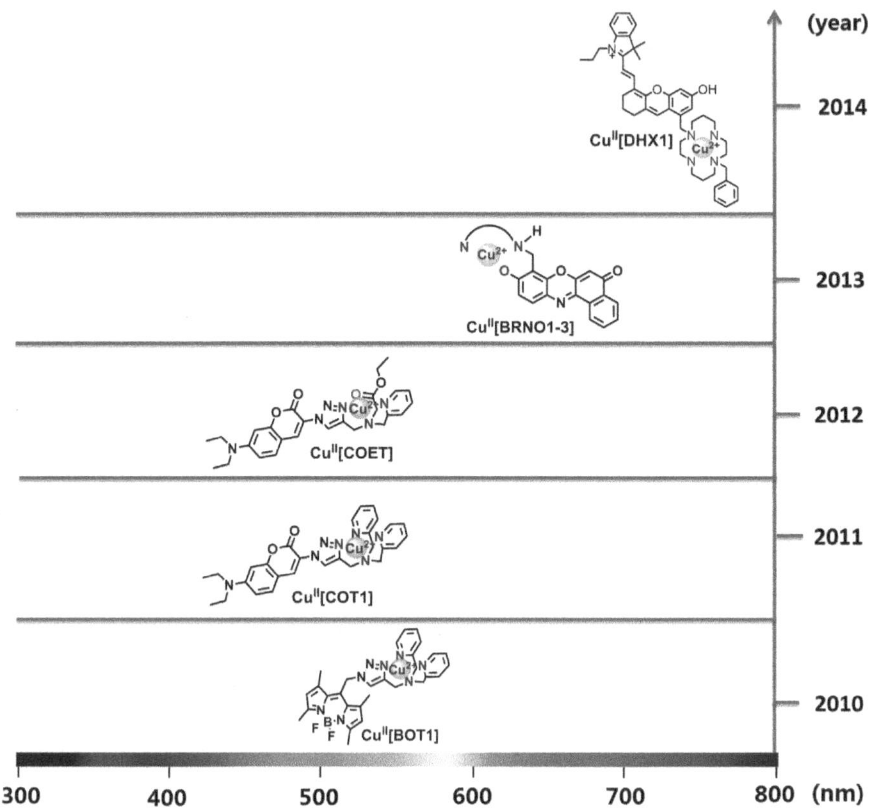

Figure 10.1 Structures and emission wavelengths of the Cu-based fluorescent HNO probes reported in the last 5 years.

range, and the one described in 2014 displays near infrared (NIR) fluorescence. Thus, we can observe a trend: the emission wavelengths of the probes are enhanced over the years. For convenience, in this section, the Cu-based fluorescent HNO probes are divided into two classes: visible light and NIR fluorescent probes.

10.2.1 Visible-light fluorescent HNO probes

In 2010, Lippard and coworkers [22] developed the first small-molecule fluorescent probe for HNO, Cu^{II}[BOT1], which is composed of a BODIPY dye and a Cu^{II}-binding site (Scheme 10.1). The BODIPY fluorophore has an absorption band in the visible region centered at 518 nm, and thus, it is well suited for cellular imaging applications. The Cu^{II}-binding site displays a tripodal coordination environment provided by an *N*-(triazolylmethyl)-*N,N*-dipicolyl framework. The dye and the binding site are connected by a rigid spacer, a triazole bridge, which minimizes the distance between them, thereby assuring strong fluorescence quenching to the dye

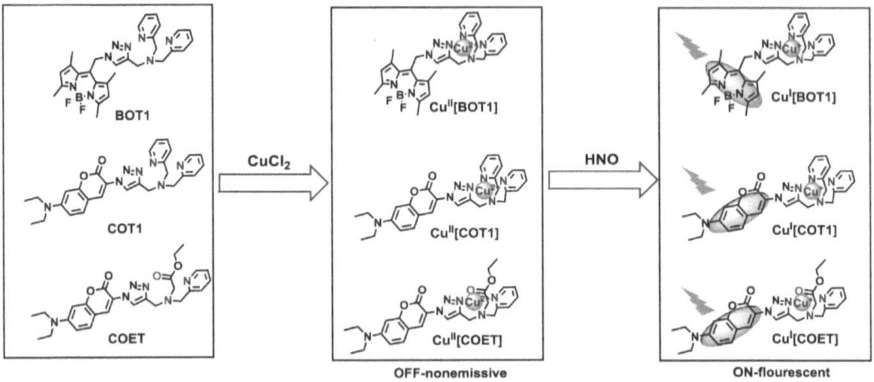

Scheme 10.1 Structures and proposed mechanisms of the fluorescent HNO probes with the emission in the visible range.

upon complexation with Cu^{II}. In other words, from the fluorescence point of view, the rigid spacer is critical for the free probe to be in the off-state. The metal-free compound, BOT1, showed emission at 526 nm with a $\Phi_{fl} = 0.12$ upon excitation at 518 nm. However, by coordination to Cu(II), the resulting complex Cu^{II}[BOT1] exhibited weak emission with a $\Phi_{fl} = 0.01$. This decrease in the emission can be recovered in the presence of HNO. When the probe, Cu^{II}[BOT1], was treated with excess AS in a buffered solution, an approximately fourfold increase in the emission was observed, suggesting HNO detection with a marked turn-on signal under physiologically relevant conditions. By contrast, other RNS and ROS such as NO, NO_3^-, $ONOO^-$, H_2O_2, and OCl^- induced almost no enhancement in the emission. The complex was applied for imaging applications in HeLa cells. Treatment of both the probe and AS to the cells led to marked red fluorescence inside the cells over the course of 10 min, in accordance with the HNO-induced fluorescence turn-on response. In 2013, Lippard's group further investigated the properties and X-ray crystal structure of Cu^{II}[BOT1].[23] By electrochemical analysis, they found that the reduction of Cu(II) to Cu(I) in the complex Cu^{II}[BOT1] by HNO is thermodynamically favored. In addition, characterization of the sensitivity of the probe to HNO indicates that Cu^{II}[BOT1] is more desirable for sensing HNO in the 0.5–5 mM concentration range.

Yao's group developed a Cu(II)-coumarin conjugate Cu^{II}[COT1] (see Scheme 10.1) acting as a dual probe for HNO by both fluorescence and EPR signals in 2011.[24] In comparison to Cu^{II}[BOT1],[22] Yao et al. selected coumarin instead of BODIPY as the fluorophore in the design of Cu^{II}[COT1]. Prior to complexation to Cu(II), COT1 has strong emission at 499 nm with a $\Phi_{fl} = 0.63$. However, the Φ_{fl} dropped to 0.03 upon coordination to Cu(II). Treatment of AS with the probe Cu^{II}[COT1] elicited a 17-fold fluorescence enhancement, which is much larger than that observed for Cu^{II}[BOT1].[22] By contrast, other RNS and ROS such as NO_3^-, NO_2^-, $ONOO^-$, ROO•, H_2O_2, and OCl^- failed to induce any marked emission enhancement of the probe. However, the addition of NO to the probe could provide only a 3.2-fold enhancement signal, much less than that in the presence of HNO. This distinction in the fluorescence enhancement could be exploited as the basis for the discrimination between NO and HNO.

By modifications on Cu^{II}[COT1], in the next year, the same group engineered the probe Cu^{II}[COET] (see Scheme 10.1), in which one pyridine moiety was replaced by an ester group to facilitate both the binding and cell membrane permeability.[25] The reaction of Cu^{II}[COET] with HNO could lead to the formation of Cu^{I}[COET], resulting in a 6.9-fold enhancement in the fluorescence intensity. To exclude the potential interference from Cys, a reducing agent, they added *N*-ethylmaleimide (NEM) as a masking reagent in the assay solution. The probe and AS were incubated with NIH3T3 cells (mouse embryonic fibroblast cells), and a significant fluorescence was noted indicating that the probe could detect the exogenous HNO in the cells. They further applied the probe for in vivo imaging of HNO in zebra fish. Only in the presence of Cu^{II}[COET], the zebra fish showed no obvious fluorescence. By contrast, when the zebra fish was further treated with AS, a significant fluorescence signal was observed in the two zygomorphic areas around the yolk and yolk extension (see Fig. 10.2).

The possible shortcomings of the above BODIPY- and coumarin-based HNO fluorescent probes include high-energy absorption and emission, and relatively short Stokes shifts of about 30 nm. These may cause light scattering, which may further lead to a low signal-to-noise ratio. To address these potential problems, in 2013, Lippard and coworkers designed and synthesized three conjugates, BRNO1 − 3 (see Fig. 10.3),[26] in which benzoresorufin was chosen as the fluorophore, as it can provide an emission at 625 nm, which is favorable for biological imaging studies. When a series of metal ions such as Na^{+}, Mg^{2+}, Ca^{2+}, Ba^{2+}, and Mn^{2+} were added to BRNO1 − 3, only a small decrease in the fluorescence intensity was noted. However, treatment of Cu^{2+} could elicit significant fluorescence quenching. Using BRNO-1 as an example, in the Cu(II)-bound form, the complex could respond to HNO and NO with a 4.8- or 1.5-fold fluorescence enhancement, respectively. In other words, the probe is more effective for sensing HNO than NO. The mechanistic studies revealed that the reduction of Cu(II) to Cu(I) is required for HNO or NO turn-on response. In addition, they found that the redox potentials of the complexes are largely influenced by the fluorescent dye instead of the different amine chelating moieties. They further demonstrated that the probes could be employed to image HNO in the different cells lines including HeLa and Raw 264.7 (see Fig. 10.3). The co-localization imaging experiments suggest that Cu^{II}[BRNO-1] and Cu^{II}[BRNO-3] were localized in the endoplasmic reticulum (ER) with a Pearson correlation coefficient of 0.61 for Cu^{II}[BRNO-1] and 0.74 for Cu^{II}[BRNO-3], respectively, when compared with an ER dye indicator.

10.2.2 NIR fluorescent HNO probes

Although the above probes Cu^{II}[BRNO1−3] have a long fluorescence emission wavelength (625 nm), their selectivity is relatively poor. To circumvent this problem, in 2014, Lippard's group developed an NIR fluorescent HNO probe,

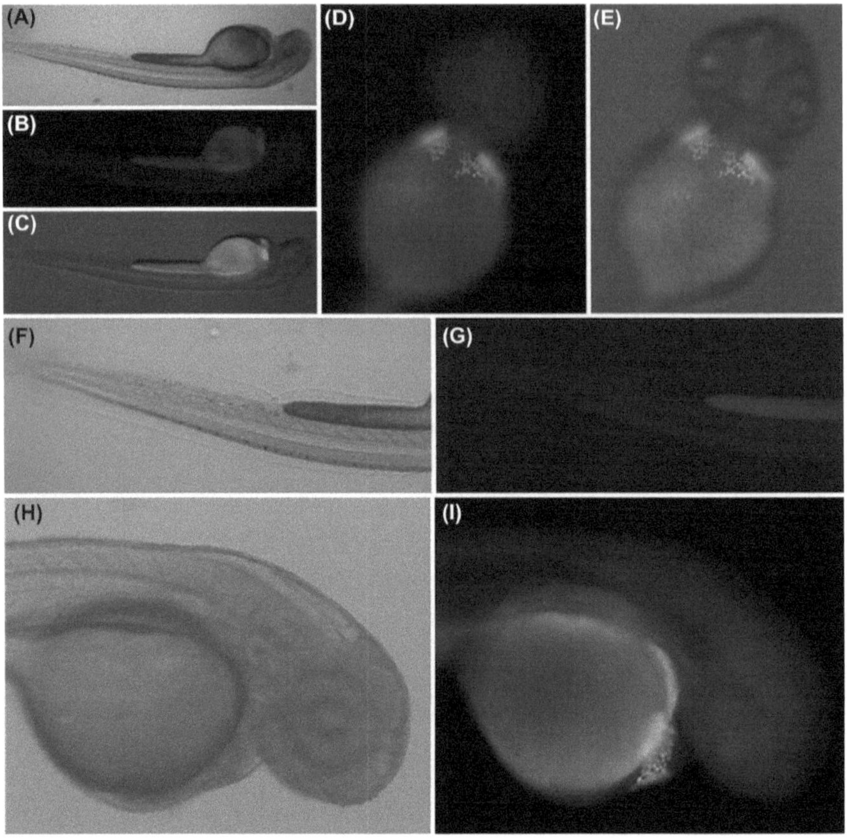

Figure 10.2 Fluorescence images of AB/Tubingen larvae zebrafish incubated in E3 embryo medium at 28.5°C. 1-Phenyl-2-thiourea (PTU, 0.003%) was added to depress the development of pigment after 8 h of incubation. (A) Brightfield image of a zebrafish supplemented with 5 μM Cu^{II}[COET] for 1 h and then further incubated with 200 μM Angeli's salt for 0.5 h; (B) fluorescence image of (A); (C) overlay between (A) and (B); (D)–(I) high magnifications revealing that the fluorescent signals mainly present in the yolk and yolk extension.
Source: Reprinted from reference [25] Zhou Y, Yao YW, Li JY, Yao C, Lin BP. Nitroxyl induced fluorescence enhancement via reduction of a copper(II) coumarin-ester complex: its application for bioimaging in vivo. *Sensor Actuat B: Chem* 2012;**174**, 414–420. Copyright (2012), with permission from Elsevier.

Cu^{II}[DHX1] (see Fig. 10.4).[27] This probe can not only detect HNO with a high selectivity but also emit in the NIR region, highly desired for in vivo imaging. The HD NIR platform reported by our group[46] was selected as the fluorophore for Cu^{II}[DHX1], because it is a bright NIR emitter with excellent biocompatibility. The absorption and emission wavelengths of compounds **3** and DHX1 are in the NIR region. The brightness of DHX1 is comparable with Indocyanine Green, a benchmark NIR fluorophore, for in vivo imaging, suggesting that DHX1 is highly suitable for in vivo applications. The quantum yield of DHX1 was decreased from

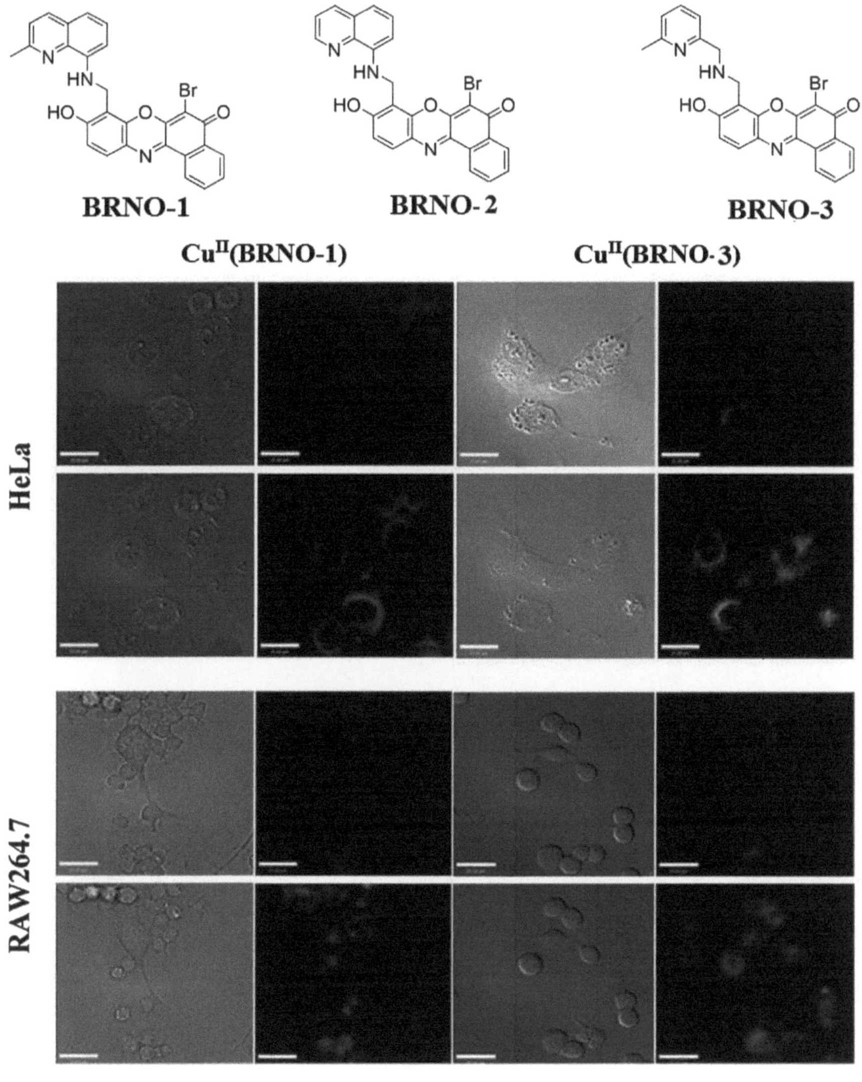

Figure 10.3 Structures of BRNO1 − 3 and fluorescence imaging of HNO by the probes Cu^{II}[BRNO-1] and Cu^{II}[BRNO-3] in HeLa and Raw 264.7 cells. For each set, the top image corresponds to the treatment of the cells with the fluorescent probe. The bottom image corresponds to the cells treated with both the probe and Angeli's salt. (Left) DIC images. (Right) Fluorescence images.

Source: Adapted from reference [26] Zhou Y, Liu K, Li JY, Fang Y, Zhao TC, Yao C. Visualization of nitroxyl in living cells by a chelated copper(II) coumarin complex. *Org Lett* 2011;**13**, 1290−1293 with permission from Copyright (2013) American Chemical Society.

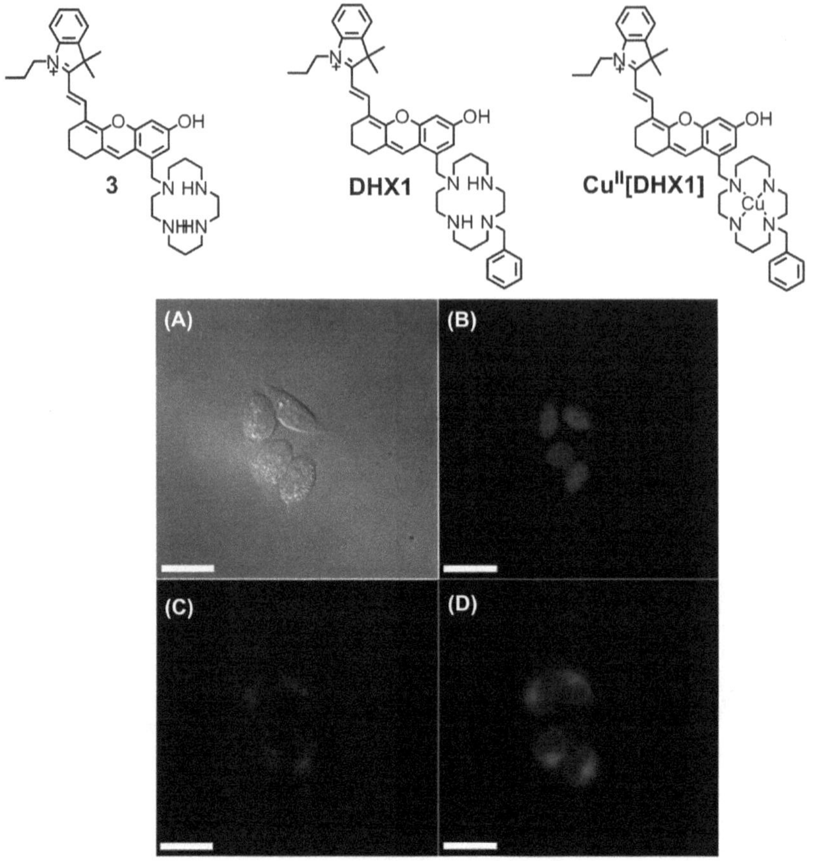

Figure 10.4 Structures of compounds 3, DHX1, and Cu^{II}[DHX1] and the images of the cells incubated with Cu^{II}[DHX1] before and after addition of Angeli's salt: (A) Differential interference contrast (DIC) image; (B) blue channel showing the nuclei; (C) NIR channel before addition of Angeli's salt; and (D) NIR channel after treatment with 1.5 mM Angeli's salt for 5 min.
Source: Adapted from reference [27] Rosenthal J, Lippard SJ. Direct detection of nitroxyl in aqueous solution using a tripodal copper(II) BODIPY complex. *J Am Chem Soc* 2010; **132**, 5536–5537 with permission from Copyright (2014) American Chemical Society.

0.048 to 0.0027, by more than 1 order of magnitude, upon coordination with Cu(II), a consequence of the paramagnetic quenching induced by the transition metal.

Interestingly, unlike Cu^{II}[BRNO1–3], the probe Cu^{II}[DHX1] has good selectivity for HNO over NO and reduced sulfur species. Upon addition of 100 equiv. of Angeli's salt (AS) to a solution of Cu^{II}[DHX1], a fivefold fluorescence enhancement was detected in about 2 min. However, addition of 5000 equiv. of NO induced almost no detectable changes in the fluorescence intensity, suggesting that Cu^{II}[DHX1] could selectively detect HNO over NO. In addition, treatment of

Cu^{II}[DHX1] with 100 equiv. of $CaCl_2$, $MgCl_2$, NaCl, $ZnCl_2$, KNO_3, H_2O_2, NaClO, sodium ascorbate, NaONOO, L-(+)-cysteine, GSH, or methionine elicited no observable fluorescence enhancement. It is worth noting that the major advantage of the probe Cu^{II}[DHX1] is that it can exclude the interference of the reduced sulfur species. By sharp contrast, the previously developed Cu-based probes all suffered from the interference of the biological thiols.[22–25]

The probe Cu^{II}[DHX1] is capable of detecting HNO in live cells with high selectivity. The cells incubated with Cu^{II}[DHX1] and AS exhibited a bright signal, about threefold increase in the fluorescence intensity, in the NIR channel (see Fig. 10.4). However, in the presence of NO and the probe, no marked fluorescence enhancement in the cells was noted. Furthermore, treatment of the cells with the probe and Na_2S only induced a very small fluorescence enhancement, implying that Cu^{II}[DHX1] has a high selectivity for HNO over H_2S in the context of the living cells.

10.2.3 Reaction mechanism

The above illustrated Cu-based fluorescent HNO probes have essentially the same reaction mechanism for HNO. For a representative example, in the probe Cu^{II}[BOT1], the role of paramagnetic Cu(II) is to quench the fluorescence by a PET pathway. However, the reduction of Cu(II) into diamagnetic Cu(I) by HNO results in the recovery of the fluorescence (see Scheme 10.2).

In 2014, Zhang's group used theoretical calculations to support the above mechanism,[41,42] and they further proposed two pathways for the HNO-induced reduction of Cu^{II}[BOT1] into Cu^{I}[BOT1] (see Scheme 10.3). Based on the theoretical model, they illustrated a number of ways to improve the reactivity of the metal-based HNO probes: (1) A strong oxidizing power of the metal center is favorable for the electron transfer; (2) A proton-accepting ligand may facilitate the proton transfer; (3) Electron-deficient substituents may be helpful for electron transfer to the metal center. Other approaches such as using the ligands with stronger metal binding capability and fluorescence response may also be beneficial.[41]

In general, the favorable properties of the Cu-based fluorescent HNO probes include good selectivity over other RNS and downstream NO oxidation products, high compatibility with the living biological samples, suitable water solubility, and

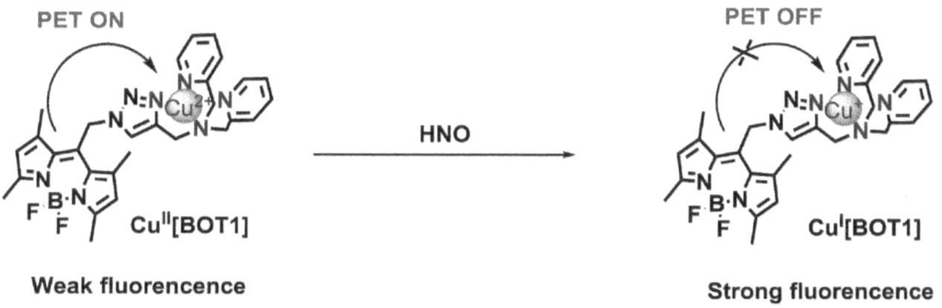

Scheme 10.2 Typical sensing mechanism of Cu-based fluorescent HNO probes.

efficient membrane permeability. In addition, the fluorescent probe Cu^{II}[DHX1] designed by Lippard et al. has superior photophysical properties with the absorption and emissions in the NIR region to avoid the unintended cellular damage by high-energy radiation and to minimize the intrinsic biological auto-fluorescence.[27] However, the Cu-based fluorescent HNO probes are associated with several problems. The sensitivity of this type of probes is not very high due to the limited fluorescence enhancement. Furthermore, most of this type of probes tend to localize in the endoplasmic reticulum (ER), and it is quite difficult to develop the HNO probes for localizing in other cellular organelles. These shortcomings may be addressed by the development of other types of fluorescent HNO probes, in particular, the nonmetal-based probes operating by a distinct mechanism.

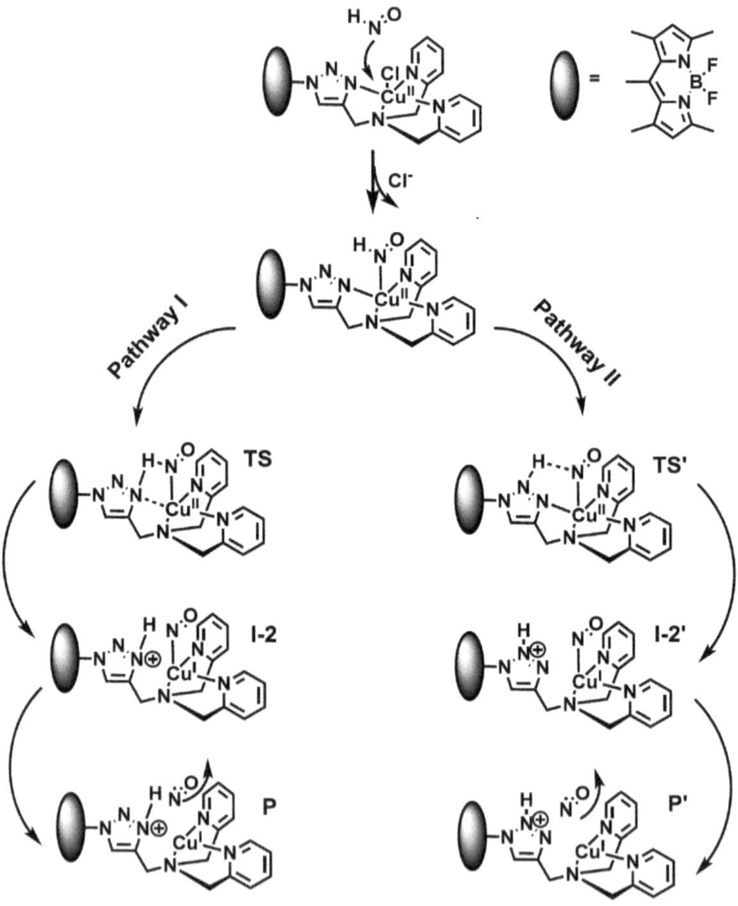

Scheme 10.3 Pathways proposed for the reduction of Cu^{II}[BOT1] by HNO.

10.3 TEMPO-based fluorescent HNO probes

The H–N bond energy of HNO is only 49.9 kcal mol^{-}1,[43] which is much smaller than 90 kcalmol^{-}1, the average H–N bond energy. Therefore, the H–N bond in HNO is prone to fracture and easy to be attacked by the protophilic reagents. In 2003, Miranda and coworkers studied the reaction of HNO with TEMPO, and found that the estimated rate constant of this reaction was about 8×10^4 M^{-1} s^{-1}. The reaction provides the products TEMPOL-H and NO via a net hydrogen abstraction from HNO to the nitroxide radical.[44]

In 2011, Toscano's group took advantage of the above reaction to design a fluorescent HNO probe, named TEMPO-9-AC (Scheme 10.4).[29] The probe TEMPO-9-AC exhibited weak fluorescence. However, the reaction product of TEMPO-9-AC with HNO, TEMPO-9-AC-H, displayed strong fluorescence. Since NO has no available proton, the probe is not able to react with it. Based on this strategy, the fluorescent probe TEMPO-9-AC can distinguish between HNO and NO with a turn-on fluorescence signal. Although this class of probes are capable of selectively sensing HNO, they are not desirable for bio-applications due to the complexity of the radical chemistry in the living systems.

10.4 Phosphine-based fluorescent HNO probes

In 2009, King's group reported for the first time the reaction of HNO with organic phosphines to produce the corresponding phosphine oxide and azaylide.[45] Please refer to Chapter 11, Phosphine-Based HNO Detection, for the detailed reaction mechanism. In 2011, King and coworkers further investigated the compatibility of the phosphine-based (P-based) method for HNO detection in vitro and in vivo.[18]

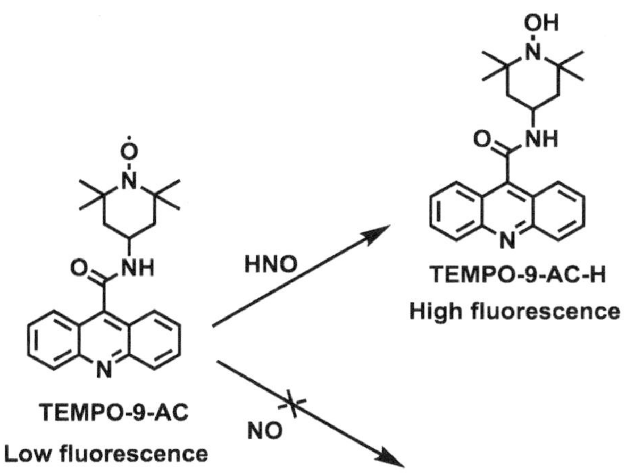

Scheme 10.4 Fluorescence sensing of HNO over NO by the probe TEMPO-9-AC.

Since then, this strategy, HNO-induced deprotection of 2-(diphenylphosphino)benzoate-caged moiety, has been well adapted for the development of the fluorescent HNO probes for biological imaging applications (see Fig. 10.5). These probes function by different sensing mechanisms including ring-opening of xanthene, FRET, ICT, and ESIPT. Since most of the P-based fluorescent HNO probes will be covered on Chapter 11, Phosphine-Based HNO Detection, herein we will only briefly describe some complementary examples of the probes including CF, GCTPOC-1, and HNO-HBT.

When compared to the turn-on probes, the Förster resonance energy transfer (FRET)-based ratiometric ones allow the measurement of fluorescence intensities at two wavelengths, which may afford an internal correction for environmental effects. In 2015, Sun's group designed and synthesized a ratiometric fluorescent HNO probe named CF based on a FRET approach (see Fig. 10.6).[30] The probe consisted of a FRET dye pair, coumarin-fluorescein. The studies on the spectral properties of the probe indicate that the probe exhibited an emission peak at 470 nm, which is the characteristic emission of the coumarin donor. However, when the probe was excited at 415 nm, the fluorescence intensity at 517 nm (the feature emission of the

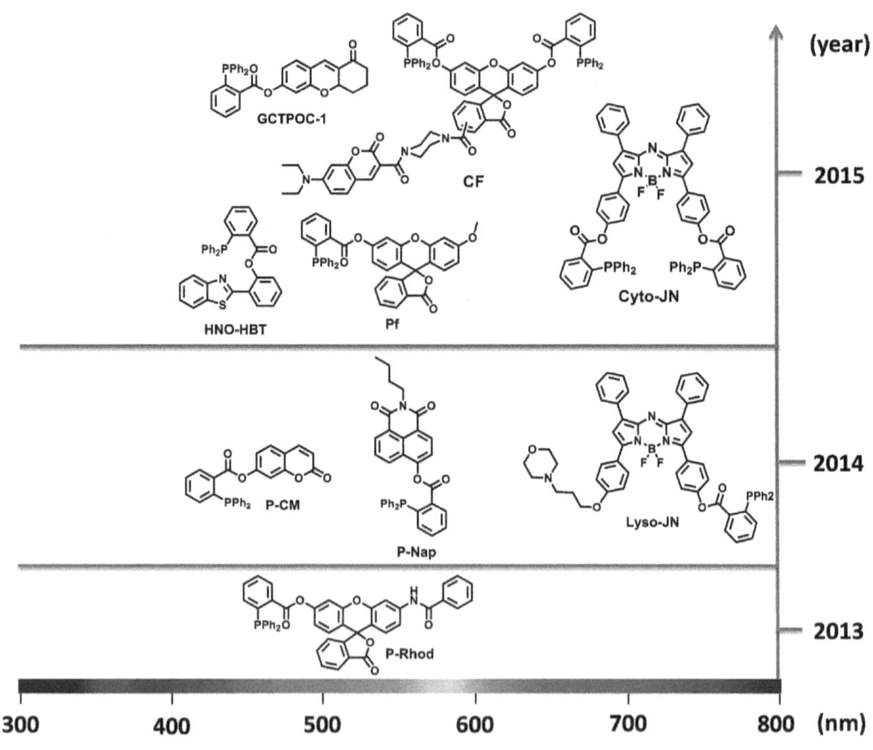

Figure 10.5 Structures and emission wavelengths of the phosphine-based fluorescent HNO probes reported in the past 3 years.

fluorescein acceptor) gradually increased accompanied by the decrease of the emission at 470 nm with the addition of AS. CF displayed a quick response toward AS with a detection limit of 1.4 μM. In addition, they demonstrated that the probe showed excellent selectivity for HNO over reactive oxygen species (ROS), reactive nitrogen species (RNS), reactive sulfur species (RSS), and biologically related metal ions tested. The HeLa cells incubated with the probe CF displayed strong blue fluorescence and weak green fluorescence suggesting that the probe is cell permeable. By contrast, the cells treated with both the probe and AS exhibited a significant fluorescence enhancement in the green channel and almost no fluorescence in the blue channel, implying that the probe could be employed to detect HNO in living cells.

Based on the ESIPT mechanism, Zhang's and Yin's group independently developed a fluorescent HNO probe, named HNO-HBT, using HBT as the fluorophore and a 2-(diphenylphosphino) benzoate moiety as the recognition receptor.[31,32] With the addition of AS to the solution of the probe HNO-HBT, a 40-fold enhancement in the fluorescence intensity at 459 nm was noted. The response rate of the probe is relatively fast, and the fluorescence signal could reach the plateau within 5 min in the presence of 10 equiv. of HNO. The favorable properties of the probe include

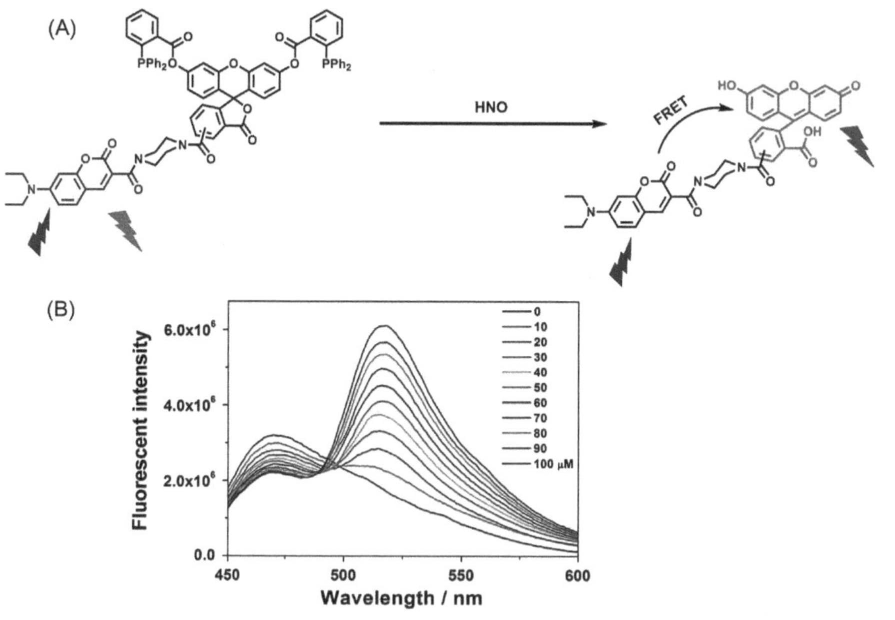

Figure 10.6 (A) Fluorescence sensing mechanism of the probe CF. (B) Fluorescence spectra of CF (10 μM) with the addition of various concentrations of AS.
Source: Adapted from reference [30] Zhang H, Liu R, Tan Y, Xie WH, Lei H, Cheung HY, et al. A FRET-based ratiometric fluorescent probe for nitroxyl detection in living cells. *Acs Appl Mater Interfaces* 2015; **7**, 5438–5443 with permission from Copyright (2015) American Chemical Society.

high sensitivity, good selectivity, low cytotoxicity, and capability for imaging HNO in living cells.

Two-photon microscopy (TPM) is advantageous over the conventional one-photon fluorescence microscopy in several aspects such as three-dimensional imaging of living tissues, reduced photodamage to biosamples, increased tissue penetration, and lower background fluorescence. Our group reported the first two-photon fluorescent HNO probe, GCTPOC-1 (see Fig. 10.7),[33] which was constructed based on the GCTPOC two-photon fluorescent platform also developed by us.[40] The probe showed weak fluorescence in the caged form. However, in the presence of HNO, the probe exhibited a 49-fold fluorescence enhancement at 512 nm. We further demonstrated that the probe was stable in a wide range of pH values (5.0–8.5) and suitable for detecting HNO at physiological pH. The notable feature of the probe GCTPOC-1 is that it is capable of three-dimensional fluorescence imaging of HNO in thick living samples. When the tissue slices were incubated with GCTPOC-1, and then treated with AS, relatively significant fluorescence was observed up to 150 μm in the two-photon mode (see Fig. 10.7).

Compared with the Cu-based HNO fluorescent probes, the P-based probes display several advantages such as higher sensitivity and better selectivity. In addition, by molecular design, a fluorescent HNO probe with localization in the lysosomal was obtained. Thus, the P-based probes are essentially complementary to the Cu-based HNO fluorescent probes.

10.5 Conclusions

In summary, the representative examples of small-molecule fluorescent HNO probes were discussed. Based on the distinction in the recognition sites, these probes were classified as Cu-based, TEMPO-based, and P-based probes. Each type of fluorescent HNO probes have their own advantages and limitations. Improvement in the properties of the probes by the judicial molecular design is still needed. In general, the selectivity of the probes for HNO over other bio-species is relatively insufficient. In addition, the detection of endogenous HNO in living samples remains an unmet challenge. We believe that with the joint efforts of the chemists and biologists in different areas these issues will be addressed in the near future. We expect that these fluorescent HNO probes and their improved next-generation ones will act as powerful molecular tools for further studies of both the chemistry and biology of HNO in the context of living systems.

Acknowledgments

This work was financially supported by NSFC (21172063, 21472067, 21502067), the Natural Science Foundation of Shandong Province, China (ZR2014BP001), Taishan Scholar Foundation (TS 201511041), and the startup fund of University of Jinan.

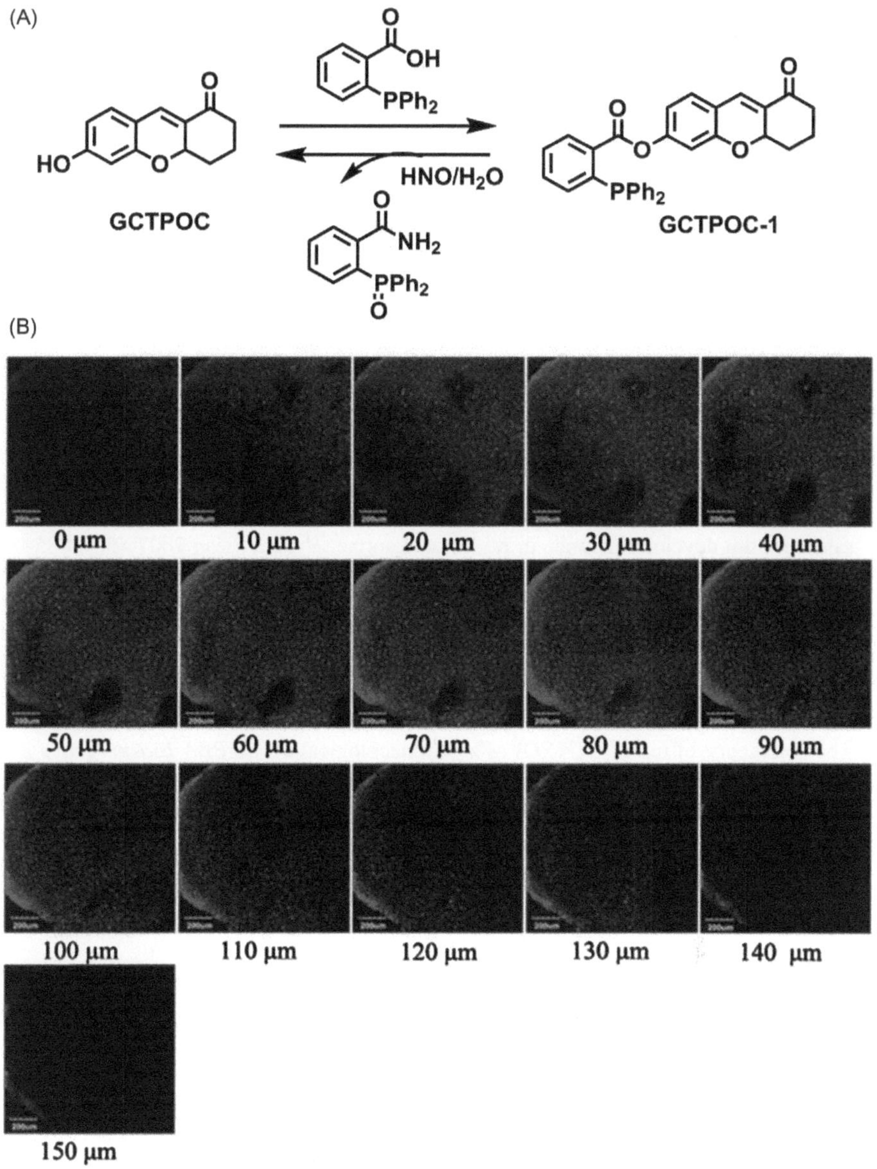

Figure 10.7 (A) Design and synthesis of GCTPOC-1 as a two-photon fluorescent HNO probe. (B) Two-photon fluorescence images of a fresh rat liver slice pretreated with GCTPOC-1 and then with AS incubated at the depths of approximately 0–150 μm, excitation at 780 nm.

Source: Reproduced from reference [33] Zheng K, Lin W, Cheng D, Chen H, Liu Y, Liu K. A two-photon fluorescent turn-on probe for nitroxyl (HNO) and its bioimaging application in living tissues. *Chem Commun* 2015; **51**, 5754–5757 with permission from the Royal Society of Chemistry.

References

1. Dalby FW. The spectrum and structure of the HNO molecule. *Can. J. Phys* 1958;**36**: 1336–71.
2. Salotto AW, Burnelle L. Potential energy curves for the HNO molecule. *Chem. Phys. Lett* 1969;**3**:80–3.
3. Bullen ML, Miller AA, Andrews KL, Irvine JC, Ritchie RH, Sobey CG, et al. Nitroxyl (HNO) as a vasoprotective signaling molecule. *Antioxid. Redox. Sign* 2010;**14**:1675–86.
4. Lopez BE, Rodriguez CE, Pribadi M, Cook NM, Shinyashiki M, Fukuto JM. Inhibition of yeast glycolysis by nitroxyl (HNO): A mechanism of HNO toxicity and implications to HNO biology. *Arch. Biochem Biophys* 2005;**442**:140–8.
5. Dautov RF, Ngo DTM, Licari G, Liu S, Sverdlov AL, Ritchie RH, et al. The nitric oxide redox sibling nitroxyl partially circumvents impairment of platelet nitric oxide responsiveness. *Nitric Oxide* 2013;**35**:72–8.
6. Yuill KH, Yarova P, Kemp-Harper BK, Garland CJ, Dora KA, Novel A. Role for HNO in local and spreading vasodilatation in rat mesenteric resistance arteries. *Antioxid. Redox. Sign* 2010;**14**:1625–35.
7. Irvine JC, Favaloro JL, Widdop RE, Kemp-Harper BK. Nitroxyl anion donor, angeli's salt, does not develop tolerance in rat isolated aortae. *Hypertension* 2007;**49**(4):885–92.
8. Lopez BE, Shinyashiki M, Han TH, Fukuto JM. Antioxidant actions of nitroxyl (HNO). *Free. Radical. Biol Med* 2007;**42**:482–91.
9. Zeng B-B, Huang J, Wright MW, King SB. Nitroxyl (HNO) release from new functionalized N-hydroxyurea-derived acyl nitroso-9,10-dimethylanthracene cycloadducts. *Bioorg. Med. Chem Lett* 2004;**14**:5565–8.
10. Switzer CH, Flores-Santana W, Mancardi D, Donzelli S, Basudhar D, Ridnour LA, et al. The emergence of nitroxyl (HNO) as a pharmacological agent. *BBA-Bioenergetics* 2009; **1787**:835–40.
11. Bari SE, Martí MA, Amorebieta VT, Estrin DA, Doctorovich F. Fast nitroxyl trapping by ferric porphyrins. *J Am Chem Soc* 2003;**125**:15272–3.
12. Martí MA, Bari SE, Estrin DA, Doctorovich F. Discrimination of nitroxyl and nitric oxide by water-soluble Mn(III) porphyrins. *J Am Chem Soc* 2005;**127**:4680–4.
13. Dobmeier KP, Riccio DA, Schoenfisch MH. Xerogel optical sensor films for quantitative detection of nitroxyl. *Anal Chem* 2008;**80**:1247–54.
14. Mitchell DG, Quine RW, Tseitlin M, Eaton SS, Eaton GR. X-band rapid-scan EPR of nitroxyl radicals. *J Magn Reson* 2012;**214**:221–6.
15. Benial AMF, Utsumi H, Ichikawa K, Murugesan R, Yamada K-i, Kinoshita Y, et al. Dynamic nuclear polarization studies of redox-sensitive nitroxyl spin probes in liposomal solution. *J Magn Reson* 2010;**204**:131–8.
16. Samuni U, Samuni Y, Goldstein S. On the distinction between nitroxyl and nitric oxide using nitronyl nitroxides. *J Am Chem Soc* 2010;**132**:8428–32.
17. Donzelli S, Espey MG, Thomas DD, Mancardi D, Tocchetti CG, Ridnour LA, et al. Discriminating formation of HNO from other reactive nitrogen oxide species. *Free. Radical Biol Med* 2006;**40**:1056–66.
18. Reisz JA, Zink CN, King SB. Rapid and selective nitroxyl (HNO) trapping by phosphines: kinetics and new aqueous ligations for HNO Detection and quantitation. *J Am Chem Soc* 2011;**133**:11675–85.
19. Cline MR, Tu C, Silverman DN, Toscano JP. Detection of nitroxyl (HNO) by membrane inlet mass spectrometry. *Free. Radical Biol Med* 2011;**50**:1274–9.

20. Suárez SA, Bikiel DE, Wetzler DE, Martí MA, Doctorovich F. Time-resolved electrochemical quantification of azanone (HNO) at low nanomolar level. *Anal Chem* 2013; **85**:10262–9.
21. Loas A, Radford RJ, Deliz Liang A, Lippard SJ. Solid-phase synthesis provides a modular, lysine-based platform for fluorescent discrimination of nitroxyl and biological thiols. *Chem Sci* 2015;**6**:4131–40.
22. Wrobel AT, Johnstone TC, Deliz Liang A, Lippard SJ, Rivera-Fuentes P. A fast and selective near-infrared fluorescent sensor for multicolor imaging of biological nitroxyl (HNO). *J Am Chem Soc* 2014;**136**:4697–705.
23. Royzen M, Wilson JJ, Lippard SJ. Physical and structural properties of [Cu(BOT1)Cl] Cl, a fluorescent imaging probe for HNO. *J Inorg Biochem* 2013;**118**:162–70.
24. Apfel UP, Buccella D, Wilson JJ, Lippard SJ. Detection of nitric oxide and nitroxyl with benzoresorufin-based fluorescent sensors. *Inorg Chem* 2013;**52**:3285–94.
25. Zhou Y, Yao YW, Li JY, Yao C, Lin BP. Nitroxyl induced fluorescence enhancement via reduction of a copper(II) coumarin-ester complex: its application for bioimaging in vivo. *Sensor. Actuat. B: Chem* 2012;**174**:414–20.
26. Zhou Y, Liu K, Li JY, Fang Y, Zhao TC, Yao C. Visualization of nitroxyl in living cells by a chelated copper(II) coumarin complex. *Org. Lett* 2011;**13**:1290–3.
27. Rosenthal J, Lippard SJ. Direct detection of nitroxyl in aqueous solution using a tripodal copper(II) BODIPY complex. *J Am Chem Soc* 2010;**132**:5536–7.
28. Tennyson AG, Do L, Smith RC, Lippard SJ. Selective fluorescence detection of nitroxyl over nitric oxide in buffered aqueous solution using a conjugated metallopolymer. *Polyhedron* 2007;**26**:4625–30.
29. Cline MR, Toscano JP. Detection of nitroxyl (HNO) by a prefluorescent probe. *J Phys Org Chem* 2011;**24**:993–8.
30. Zhang H, Liu R, Tan Y, Xie WH, Lei H, Cheung HY, et al. A FRET-based ratiometric fluorescent probe for nitroxyl detection in living cells. *Acs Appl. Mater. Interfaces* 2015; **7**:5438–43.
31. Liu C, Cao Z, Wang Z, Jia P, Liu J, Wang Z, et al. A highly sensitive and reductant-resistant fluorescent chemodosimeter for the rapid detection of nitroxyl. *Sensor. Actuat. B: Chem* 2015;**220**:727–33.
32. Lv HM, Yi C, Lei J, Au CT, Yin SF. An ESIPT-based ratiometric fluorescent probe for the imaging of nitroxyl in living cells. *Anal. Methods.-UK* 2015;**7**:3883–7.
33. Zheng K, Lin W, Cheng D, Chen H, Liu Y, Liu K. A two-photon fluorescent turn-on probe for nitroxyl (HNO) and its bioimaging application in living tissues. *Chem Commun* 2015;**51**:5754–7.
34. Kawai K, Ieda N, Aizawa K, Suzuki T, Miyata N, Nakagawa H. A reductant-resistant and metal-free fluorescent probe for nitroxyl applicable to living cells. *J Am Chem Soc* 2013;**135**:12690–6.
35. Jing X, Yu F, Chen L. Visualization of nitroxyl (HNO) in vivo via a lysosome-targetable near-infrared fluorescent probe. *Chem Commun* 2014;**50**:14253–6.
36. Liu C, Wu H, Wang Z, Shao C, Zhu B, Zhang X. A fast-response, highly sensitive and selective fluorescent probe for the ratiometric imaging of nitroxyl in living cells. *Chem Commun* 2014;**50**:6013–16.
37. Mao G-J, Zhang X-B, Shi X-L, Liu H-W, Wu Y-X, Zhou L-Y, et al. A highly sensitive and reductant-resistant fluorescent probe for nitroxyl in aqueous solution and serum. *Chem Commun* 2014;**50**:5790–2.
38. Liu P, Jing X, Yu F, Lv C, Chen L. A near-infrared fluorescent probe for the selective detection of HNO in living cells and in vivo. *Analyst* 2015;**140**:4576–83.

39. Miao Z, Reisz JA, Mitroka SM, Pan J, Xian M, King SB. A selective phosphine-based fluorescent probe for nitroxyl in living cells. *Bioorg. Med Chem Lett* 2015;**25**:16–19.
40. Yuan L, Lin W, Chen H, Zhu S, He L. A Unique family of rigid analogues of the GFP Chromophore with tunable two-photon action cross-sections for biological imaging. *Angew. Chem Int Ed* 2013;**52**:10018–22.
41. Michael MA, Pizzella G, Yang L, Shi Y, Evangelou T, Burke DT, et al. HNO/NO conversion mechanisms of Cu-based HNO probes with implications for Cu, Zn-SOD. *J Phys Chem Lett* 2014;**5**:1022–6.
42. Yang W, Chen X, Su H, Fang W, Zhang Y. The fluorescence regulation mechanism of the paramagnetic metal in a biological HNO sensor. *Chem Commun* 2015;**51**:9616–19.
43. Dixon RN. The heats of formation of HNO and of DNO. *J. Chem. Phys* 1996;**104**: 6905–6.
44. Miranda KM, Paolocci N, Katori T, Thomas DD, Ford E, Bartberger MD, et al. A biochemical rationale for the discrete behavior of nitroxyl and nitric oxide in the cardiovascular system. *Proc. Natal. Acad Sci USA* 2003;**100**:9196–201.
45. Reisz JA, Klorig EB, Wright MW, King SB. Reductive phosphine-mediated ligation of nitroxyl (HNO). *Org Lett* 2009;**11**:2719–21.
46. Yuan L, Lin W, Zhao S, Gao W, Chen B, He L, et al. A unique approach to development of near-infrared fluorescent sensors for in vivo imaging. *J Am Chem Soc* 2012; **134**:13510–23.

Phosphine-Based HNO Detection

11

Z. Miao and S. Bruce King
Wake Forest University, Winston-Salem, NC, United States

Chapter Outline

11.1 Introduction

In the past, azanone (nitroxyl, HNO) has occupied the position of an unstable and highly reactive nitrogen oxide that participates in various nitrogen-based chemistries. The discovery of the biochemical synthesis of nitric oxide (NO) and the further description of its numerous signaling properties led to the examination of the biological activities of compounds related to NO through redox chemistry (nitrite and nitrate, NO_2^-, NO_3^-, and HNO).[1–6] Over the last 20 years, extensive investigations of HNO reveal its distinct chemical and biological profiles compared to NO.[7–13] Indeed, various lines of research identify HNO as a potential therapeutic for congestive heart failure, cancer, or directly implicate HNO in biological processes.[14–19] Given the fact that HNO rapidly dimerizes and dehydrates to yield nitrous oxide (N_2O),[20] a full chemical and biological understanding of HNO depends on fast, reliable, and selective detection. Earlier methods (the identification of N_2O, trapping with thiols or heme proteins) either lack the sensitivity or the specificity for HNO providing somewhat ambiguous results regarding HNO's involvement in both chemical and biological processes.[9] These limitations necessitate the development of new and mechanistically different HNO detection strategies that fully exploit HNO's reactivity. Over the past few years, robust, selective, and sensitive copper-based fluorescent probes and electrochemical methods have been developed and used to clarify the presence and role of HNO.[21–26] Based on HNO's well-documented properties as both an electrophile and oxidant (HNO reacts with soft nucleophiles such as thiols),[27] our group and others have explored the reactions of HNO with

The Chemistry and Biology of Nitroxyl (HNO). DOI: http://dx.doi.org/10.1016/B978-0-12-800934-5.00011-6

phosphines as an alternative method of HNO detection.[28–34] This chapter focuses on the chemistry of HNO phosphine trapping and its application to HNO chemistry and biology.

11.2 Reactions of phosphines with nitroso compounds

C-Nitroso compounds demonstrate electrophilic properties and react with various nucleophiles at the polar N=O bond. Nitrosobenzene (PhNO) reacts with triphenylphosphine (TPP) to give various products in a process that also yields chemiluminescence.[35–37] Triphenylphosphine also adds to α-nitroso-β-napthol to give triphenylphosphine oxide and an azaylide species ($R_3P{=}NR$, Scheme 11.1).[38] In 1972, Haake reported the reaction of triphenylphosphine with trityl *S*-nitroso thiol in benzene produces a yellow precipitate, which was identified as the *S*-azaylide ($Ph_3P{=}N{-}SR$, Scheme 11.1) and triphenylphosphine oxide in equal amounts.[39] Recent X-ray crystallographic and ^{31}P NMR studies confirm the structure of the *S*-azaylide and the reactivity of phosphines with *S*-nitrosothiols.[40] Such results suggest that phosphines can exploit the native electrophilicity of the nitroso group as a potential trapping approach for numerous nitroso-containing substrates. By viewing H–N=O as the simplest nitroso compound and its known reactivity with soft nucleophiles,[27] we predicted that phosphine nucleophiles will react with HNO to yield chemically distinct products.

11.3 Reactions of phosphines with HNO

Initial experiments reveal that organophosphines react with HNO similarly to *C*- and *S*-nitroso compounds. Triphenylphosphine reacts with HNO generated from Angeli's salt (2:1 ratio) to give equal amounts of triphenylphosphine oxide (**1a**) and azaylide (**2a**), identified by comparison to an authentic standard, in an organic/aqueous mixed solvent (Scheme 11.2).[28] While these results confirm the reaction of HNO with organophosphines, the azaylide (**2a**), a Staudinger reduction intermediate, quickly hydrolyzes to **1a**, which may also arise from phosphine oxidation. To test the reactivity of HNO with a phosphine under completely aqueous conditions, a

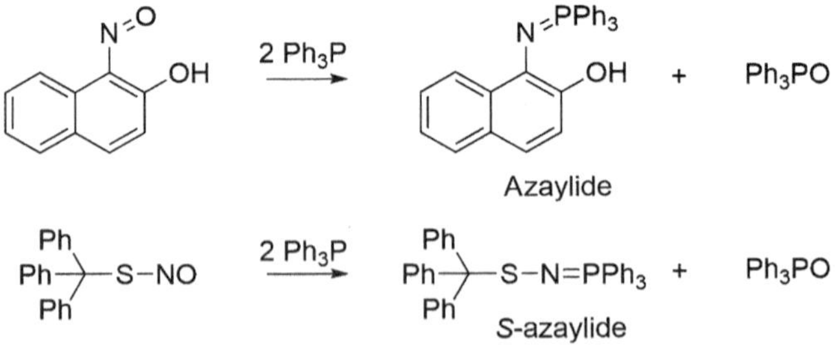

Scheme 11.1 Reaction of nitroso compounds with phosphines.

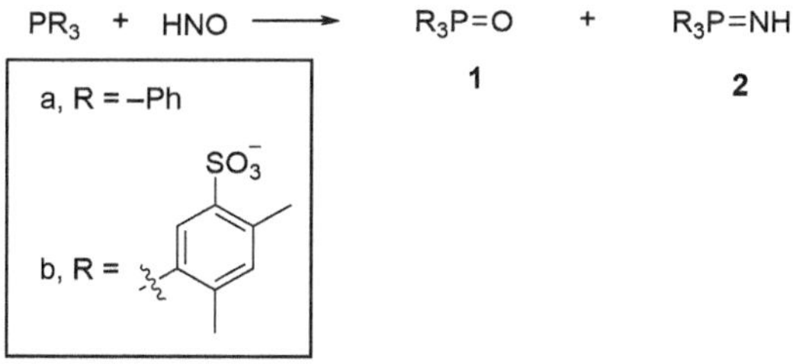

Scheme 11.2 Reaction of HNO with phosphines.

similar reaction of Angeli's salt-derived HNO and tris(4,6-dimethyl-3-sulfonatophenyl)phosphine trisodium salt hydrate (TXPTS), a water-soluble phosphine, in Tris buffer (100 mM, pH 7.6) also gives an azaylide as determined by mass spectrometry (MS) and ^{31}P nuclear magnetic resonance (NMR) spectroscopy (**2b**, Scheme 11.2) and the phosphine oxide (**1b**).[28] Unlike **2a**, the TXPTS-derived azaylide (**2b**) demonstrates stability for up to 6 days as judged by ^{31}P NMR spectroscopy making **2b** a potential marker for HNO.[28] Structural differences in the substituents on the aromatic groups attached to the phosphine appear as key factors influencing the stability of **2a** vs **2b**.[28] The addition of TXPTS to a solution of Angeli's salt also quenches over 90% of N_2O formation providing another line of evidence for the reaction of phosphines and HNO.[28]

TXPTS also reacts with *S*-nitrosothiols (RSNOs), including *S*-nitroso glutathione (GSNO), *S*-nitroso cysteine (CySNO), and *S*-nitrosated proteins to yield phosphine oxide (**1b**), azaylide (**2b**), and *S*-alkyl phosphonium adducts that contain an $S{-}P^+$ connectivity.[40] A combination of NMR and MS methods verify the structure of these products, which indicates a potential competition of phosphine probes between HNO and RSNOs.[40] A mechanism for the formation of these products follows one earlier presented by Haake and illustrated for HNO in Scheme 11.3.[28,39,40]

Phosphines trap HNO regardless of the source. Both TPP and TXPTS quickly trap HNO derived from *p*-bromo Piloty's acid to yield azaylides (**2a**–**b**) and experiments with ^{15}N-*p*-bromo Piloty's acid show the nitrogen atom of the azaylide derives from the Piloty's acid derivative.[28] TXPTS also traps enzymatically generated HNO.[29] Incubation of horseradish peroxidase (HRP) with hydrogen peroxide in the presence of hydroxylamine generates HNO and addition of TXPTS to this reaction forms the azaylide (**2b**).[29,41] TXPTS also reduces the amount of N_2O produced by this system as expected.[29] Similarly, introduction of TXPTS to the aerobic incubation of bovine hepatic catalase with cyanamide in the presence of a glucose/glucose oxidase peroxide generating system, another known HNO-producing system,[42] yields the azaylide (**2b**) and reduces N_2O formation.[29]

Scheme 11.3 shows a proposed mechanism of phosphine oxide (**1**) and azaylide (**2**) formation from the reaction of HNO with a phosphine.[28] Similar to postulated mechanisms with other nitroso compounds,[39] the reaction of HNO with phosphines may occur via a hetero three-membered ring intermediate (**5**, Scheme 11.3). Addition of phosphine to HNO at either the nitrogen or the oxygen atom would give the zwitterionic species **3** or **4**, respectively, which exist in resonance with

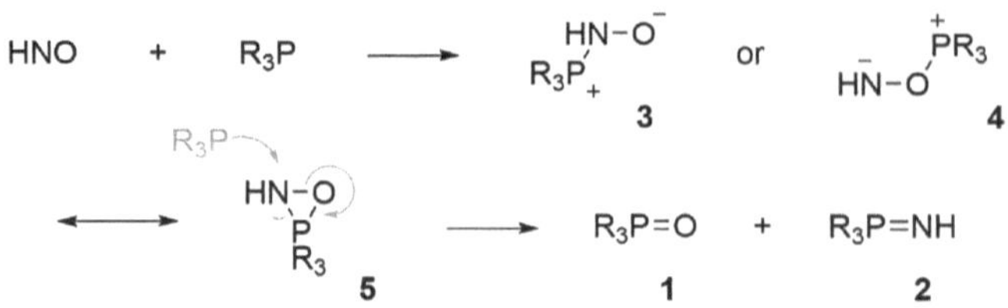

Scheme 11.3 Proposed mechanism for the reaction of HNO with phosphines.

5 (Scheme 11.3). Addition of a second phosphine to **5** (or **3** or **4**) would produce the corresponding phosphine oxide and azaylide in a 1:1 ratio (Scheme 11.3). While the oxidation of phosphines with H_2O_2 only generates phosphine oxide, the reliable formation of an azaylide (**2**) from the reaction with HNO distinguishes HNO's reactivity from other species and can be used as a selective marker for HNO and a means for HNO detection.

The inherent and rapid dimerization of HNO limits the lifespan of HNO and feasible detection and quantification chemistry must compete with dimerization and cross-reactivity with other species. Initial experimental data confirms that phosphines react with HNO fast enough to serve as qualified traps as the reaction of phosphines with HNO generally appears complete in less than 30 minutes.[28,29] The reaction of TXPTS (1 eq.) with HNO in the presence of the physiologically relevant HNO scavenger glutathione (GSH, 1 eq.) produces azaylide (**2b**), suggesting the reaction of HNO with phosphine is faster than with GSH.[28] Direct kinetic measurements for HNO reactions remain challenging due to the requirement of HNO donors (that have their own release kinetics) and its quick dimerization. Computational analysis of the reaction of tris(2-carboxylethyl)phosphine (TCEP) with HNO by Fukuto gives a second-order rate constant of $8.4 \times 10^6\ M^{-1}\ s^{-1}$.[43] ^{31}P NMR competition experiments with GSH show the second-order rate constant of the reaction of TXPTS with HNO to be $9 \times 10^5\ M^{-1}\ s^{-1}$.[29] This order of magnitude difference in rate constants may be due to the difference in kinetic methods used to measure these rates or the steric and electronic differences between these phosphines.

In addition to favorable kinetics, robust HNO detection requires selectivity over other redox species including physiologically relevant nitrogen oxides. As reported, the reaction of a phosphine with HNO produces the corresponding phosphine oxide, which can form from direct oxidation, and azaylide, which specifically indicates HNO formation. At this time, various triaryl phosphines have been screened for their specific reaction with HNO.[28–34] In general, exposure of TXPTS and other phosphine HNO probes to nitrate or nitrite yields no reaction.[28–34] Addition of diethylammonium (*Z*)-1-(*N*, *N*-diethylamino) diazen-1-ium-1, 2-diolate (DEA/NO, a common NO donor) to TXPTS generates **1b**, as previously described.[29] Incubation of TXPTS with peroxynitrite or hydrogen peroxide gives phosphine oxide.[29] Addition of AS to TXPTS under basic conditions, which do not yield HNO, does not produce any new phosphorus containing products.[29] Organophosphines resist reaction with biological reductants such as glutathione (GSH), ascorbic acid, hydroxylamine, or hydrogen sulfide (H_2S).[30–34] In this regard, phosphine-based trapping of HNO detection provides a chemical alternative to the Cu^{II}-based probes, which may yield positive results in the presence of various excess reductants.[21–24] As mentioned, organophosphines, including TPP and TXTPS, react with *S*-nitrosothiols to yield phosphine oxides and *S*-azaylide species and this reactivity must be seriously considered when using phosphines to detect HNO.[39,40,44]

Recent work shows the use of phosphines for confirming the intermediacy of HNO in chemical processes and the identification and quantification of HNO release from new HNO donors. For example, addition of TPP and subsequent identification of the phosphine oxide and azaylide (**1–2a**) supports HNO formation from $\{CoNO\}^8$ complexes via proton-induced reactions.[45] Trapping with water-soluble phosphines provides support for HNO formation in the iron-heme model reduction of nitrite that was confirmed with an HNO-selective electrode as control experiments show these phosphines react over time with nitrite at low pH.[46] TXPTS trapping verifies the presence and detection of HNO (but not NO) by membrane inlet mass spectrometry as a new method of HNO detection.[47] TXPTS also provides a useful molecular tool to quantify and kinetically evaluate (by ^{31}P NMR spectroscopy) the release of HNO from new structurally diverse donors.[48,49] Together, these studies show the ability of phosphine trapping to play an important role in the identification and characterization of HNO from various sources.

11.4 Reductive ligation

Azaylides produced from reaction of phosphines with HNO are nucleophilic species and intermediates in the Staudinger reduction of azides to amines. Considering the mechanism of the Staudinger ligation reveals the possible capture of these reactive intermediates forming the basis of a new HNO detection strategy.[50,51] Addition of an electrophilic ester adjacent to the phosphine allows the nucleophilic azaylide intermediate to react intramolecularly and produce a stable and unique HNO-derived products that can be further identified for the purpose of HNO detection and quantification.

Scheme 11.4 shows HNO trapping through reductive Staudinger ligation by a triaryl phosphine with an appropriately positioned electrophilic group.[28] The phosphine first reacts with HNO to yield an equal amount of phosphine oxide and azaylide (Scheme 11.3). The reactive azaylide nucleophilically attacks the adjacent ester group forming a tetrahedral intermediate that decomposes with alcohol release and a phosphonium-like product (**6**, Scheme 11.4). Hydrolysis of **6** provides the benzamide phosphine oxide (**7**) as the final product (Scheme 11.4). The liberation of an alcohol in this process makes feasible the further design and development of colorimetric and fluorescent HNO probes. The benzamide (**7**) directly arises from the HNO-derived azaylide with HNO serving as the nitrogen source for **7**, which acts as a stable and chemically distinct marker of HNO.

S-Nitrosothiols (RSNOs) represent an important class of posttranslational modifications that amplify and modulate the actions of NO and influence enzyme

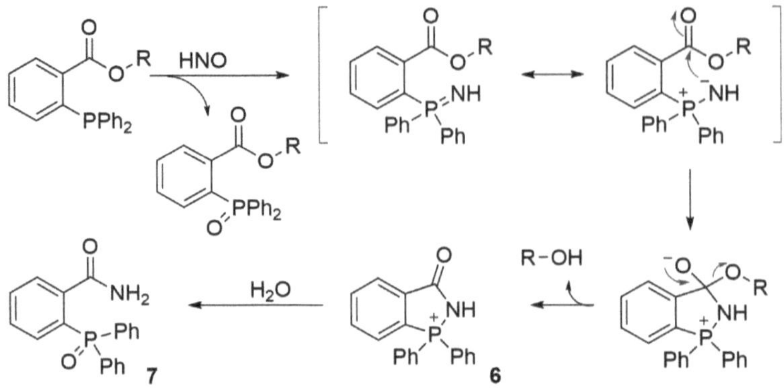

Scheme 11.4 Staudinger ligation of HNO with phosphines.

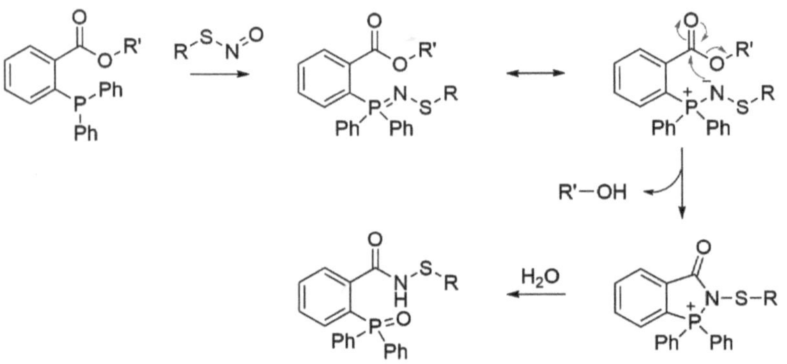

Scheme 11.5 Reductive ligation of RSNOs with phosphines.

activity.[52–54] As mentioned, phosphines react with *S*-nitrosothiols to yield 1:1 mixtures of the corresponding phosphine oxide and the *S*-azaylide and much recent attention has been focused on developing this chemistry for the reliable detection of RSNOs in biological systems.[39,40] Xian has thoroughly examined the reaction of phosphines with RSNO including the ability of these compounds to undergo reductive ligation reactions, similar to HNO (Scheme 11.5).[44,55,56] A variety of organic and fully protected *S*-nitroso thiol derivatives of L-cysteine undergo successful ligation reactions in organic/aqueous solvents suggesting reductive phosphine ligations may be useful for RSNO detection (Scheme 11.5).[44,55,56] A phosphine-derived RSNO probe, which also reacts with HNO, based on reductive ligation chemistry has been reported.[57] Despite the earlier reports of RSNO's undergoing reductive ligation,[44,55,56] more recent high-pressure liquid chromatography/mass spectrometry (HPLC–MS) and ^{31}P NMR studies show that depending on the structure of the trapping phosphine, GSNO does not act as a reliable ligation partner in completely aqueous conditions.[58] Phosphines containing an ester as the

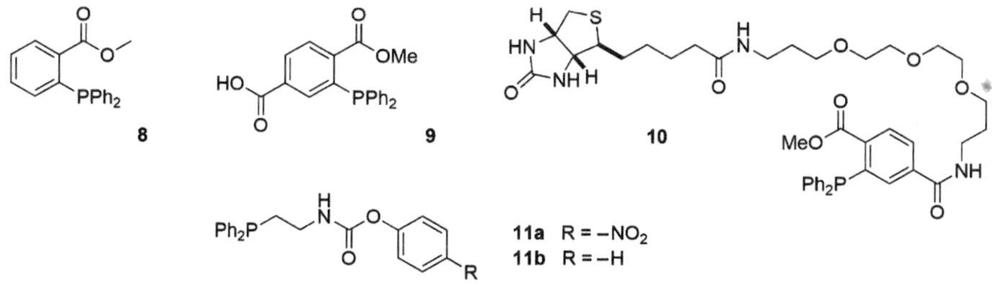

Scheme 11.6 Various phosphine-based HNO traps.

electrophilic group for *S*-azaylide addition did not undergo ligation indicating that while this approach may not permit RSNO labeling it reveals a difference in reactivity between RSNO and HNO that permits selective HNO detection.[58]

Methyl 2-(diphenylphosphino)benzoate (**8**, Scheme 11.6) represents the first substrate to undergo HNO-mediated reductive Staudinger ligation.[28] As expected, treatment of **8** with Angeli's salt in a mixture of acetonitrile/Tris buffer gives a 1:1 mixture of the benzamide (**7**) and the corresponding phosphine oxide in less than 20 minutes.[28] Control experiments indicate the selectivity of **8** for HNO and studies with ^{15}N-AS show the nitrogen atom of **7** derives from HNO.[28] These results confirm the initial formation of an azaylide as intermediate and its successful participation in a Staudinger ligation and form the basis to design new HNO detection methods to in biological systems.[28] Other modified phosphine esters (Scheme 11.6) and their ability to trap HNO have been reported.[29] As **8** suffers from poor water solubility, the carboxylated analog (**9**) was designed to improve the aqueous solubility for potential HPLC detection and quantification.[29] Comparison to authentic standards shows **9** traps over 90% of generated HNO yielding the corresponding benzamide product that can be separated and quantified by HPLC.[29] Recent work applies **9** for determining the efficiency of a new HNO donor by HPLC.[59] Commercially available phosphine (**10**), which contains a biotin moiety, is fully water soluble and traps AS or enzymatically derived HNO to yield the corresponding phosphine oxide and amide.[29]

Toward the development of a real-time HNO detection method for both in vitro and in vivo situations, phosphines (**11a–b**) containing carbamates as the electrophile for ligation were investigated.[29] Unlike esters that may hydrolyze, especially with cellular esterase catalysis, carbamates should provide a stable linkage relatively inert to hydrolysis. Phosphines **11a–b** contain the phenolate and *p*-nitrophenolate leaving groups that provide a convenient way for real-time HNO detection and quantification by UV–vis spectroscopy. Incubation of **11a–b** with AS in 7:3 acetonitrile/ammonium formate buffer produces an HNO-derived urea and corresponding phosphine oxide.[29] The release of *p*-nitrophenol from **11a** upon reaction with HNO enables a rapid colorimetric detection approach. Addition of AS to **11a** results in a bright yellow solution with a sharp increase in absorption at 400 nm and **11a** can detect HNO at the micromolar level without interference from other nitrogen oxides.[29] While this reaction may serve as a potential qualitative and quantitative

HNO detection method, **11a** suffers from self-catalyzed hydrolysis thus limiting its use in HNO detection.[29]

11.5 Fluorescent HNO probes

Recent design and synthetic efforts have yielded fluorescent HNO probes (depicted in Scheme 11.7) capable of HNO detection in living cells.[30–34] The successful development of these fluorescent probes provides a reliable and robust approach for better establishing HNO's biological profile and for identifying endogenous HNO sources. While these probes contain various fluorophores, they share structural similarities and the same basic chemical concept as **8**. Structurally, these probes embed a fluorophore in the ester group adjacent to the phosphine and introduction of HNO generates the azaylide that reacts with the ester releasing the fluorophore via Staudinger ligation. Regardless of the identity of the fluorophore, each of these probes produces the same phosphine oxide benzamide (**7**, Scheme 11.7) as a distinct molecular marker for HNO formation.

Nakagawa et al. reported the first fluorescent probe (P-Rhod, Scheme 11.7) for azanone in 2013.[30] This probe was based on rhodol (Ex/Em = 491/526 nm), a hybrid of fluorescein and rhodamine with excellent photophysical properties that can be prepared in four steps from commercially available fluorescein.[30] Direct esterification of rhodol with 2-(diphenylphosphino)benzoic acid yields P-Rhod and quenches the fluorescence.[30] Incubation of P-Rhod with excess AS in buffer leads to the liberation of the fluorophore and a significant fluorescence increase in less

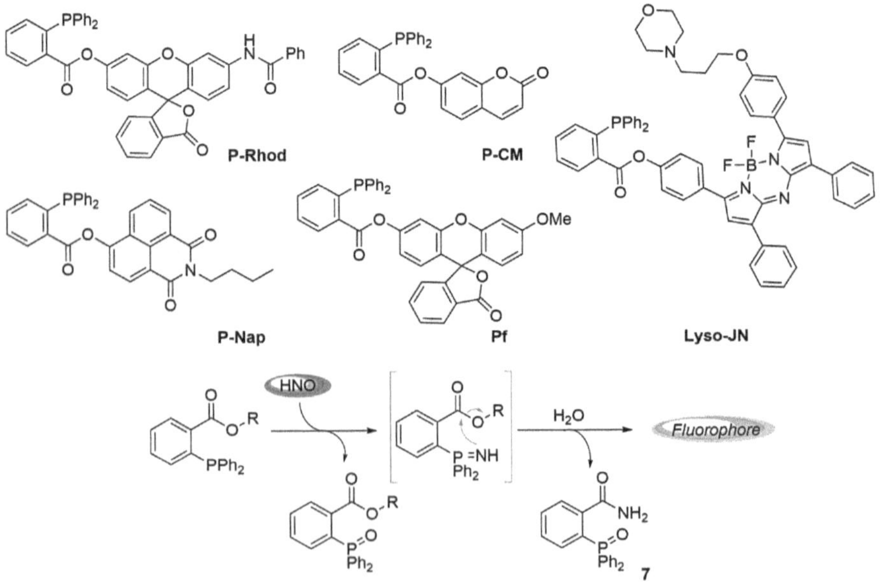

Scheme 11.7 Fluorescent HNO probes and mechanism of fluorescence generation.

than 20 minutes.[30] Control experiments demonstrate the probe's selectivity for HNO over other biological species including reactive nitrogen/oxygen species (RNS/ROS) and physiological oxidants like $ONOO^-$, NO_2^-, NO_3^-, and H_2O_2.[30] Biological reductants also do not produce any fluorescence change.[30] While earlier studies indicate *S*-nitrosothiols react with phosphines through ligation pathways, GSNO only elicits a limited fluorescence response (many fold weaker than that of HNO) indicating GSNO may not undergo efficient ligation required for fluorophore formation.[30]

P-Rhod also demonstrates the ability to detect HNO generated in a more biological system. Addition of P-Rhod to the NH_2OH/H_2O_2/HRP-based enzymatic HNO generation system induces a dramatic increase in fluorescence intensity and decreases the amount of N_2O generated.[30] Experiments with human alveolar basal epithelial A549 adenocarcinoma cells show the ability of P-Rhod to detect HNO in living cells (Fig. 11.1).[30] Incubation of P-Rhod with these cells for 3 hours produces little fluorescence increase, indicating the compound does not undergo any type of ester hydrolysis, which would yield a fluorescence response, further confirming the feasibility of these compounds for cellular applications.[30] Addition of AS to the P-Rhod loaded cells immediately yields fluorescence detectable by fluorescence confocal microscopy.[30] Addition of other redox species including an NO donor, GSNO, NaOCl or excess GSH does not yield a fluorescence response in the cells.[30] These results show that P-Rhod and other phosphine-based probes serve as promising metal-free fluorescent probes capable of HNO detection in living cells.

Following the report of P-Rhod, four other phosphorus-based fluorescent HNO probes based on similar Staudinger ligation chemistry have been described and thoroughly evaluated.[31–34] Similar to P-Rhod, these probes demonstrate selectivity for HNO over other relevant physiological species including biological oxidants and reductants.[31–34] Again, GSNO only induces a weak fluorescence response compared to HNO further distinguishing the reactivity between these nitroso compounds and demonstrating the inability of RSNO to reliably participate

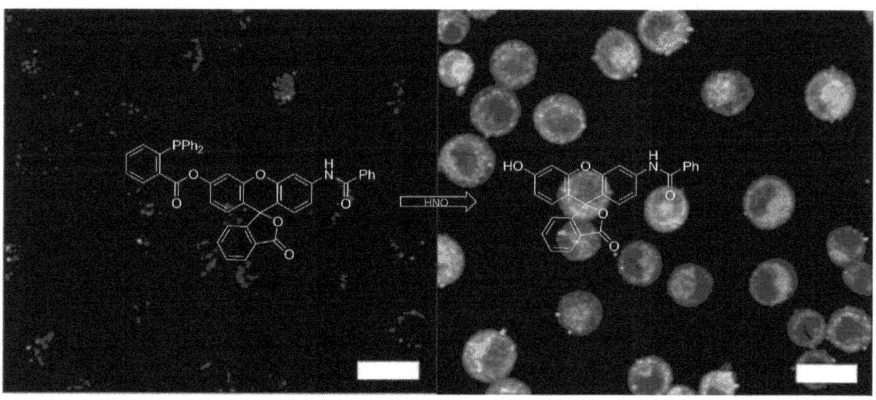

Figure 11.1 HNO-induced fluorescence images of A549 cells stained with P-Rhod before (left) and 20 min (right) after treatment with AS.

in the fluorescence generating ligation reaction.[30,31] The coumarin-based fluorescent probe (P-CM, Scheme 11.7) can be prepared in a one step from commercially available umbelliferone.[33] P-CM has an improved sensitivity toward AS (0.05 μM) and the fluorescence intensity increases linearly at low AS concentrations (0.05–10 μM). P-CM successfully detects HNO derived from AS in buffered (pH 7.4) aqueous 20% bovine serum solution demonstrating its ability to detect HNO in a complex biological matrix.[33] The 1,8-naphthalimide-derived probe (P-Nap, Scheme 11.7) changes fluorescence from blue (418 nm) to green (546 nm) upon introduction of AS, which allows the ratiometric detection of HNO.[32] The excellent linearity between the fluorescence intensity ratio and AS concentration makes P-Nap a good candidate for HNO detection. The addition of AS to live RAW 264.7 macrophage cells previously incubated with P-Nap results in a distinct fluorescence change and confocal microscopy provides high-resolution images indicating HNO detection.[32] A lysosomal-directed near-infrared HNO probe (Lyso-JN, Scheme 11.7) based on the same chemical principles has been prepared.[31] This probe shows organelle-specific HNO detection in response to addition of HNO in RAW 264.7 macrophages demonstrating the introduction of an organelle-directing chemical groups permits organelle-specific HNO detection.[32]

Our group reported a fluorescent probe based on modified fluorescein (Pf, Scheme 11.7, Ex/Em = 465/520 nm) that can be easily obtained by the straightforward coupling of fluorescein monomethyl ether with 2-(diphenylphosphino)benzoic acid in high yield.[34] Similar to the other described probes, the addition of AS-derived HNO to Pf results in a rapid increase in fluorescence intensity (Fig. 11.2) and Pf selectively fluoresces in the presence of HNO rather than other biologically relevant oxidants/reductants.[30–34] Treatment of Pf with excess amounts of GSNO and

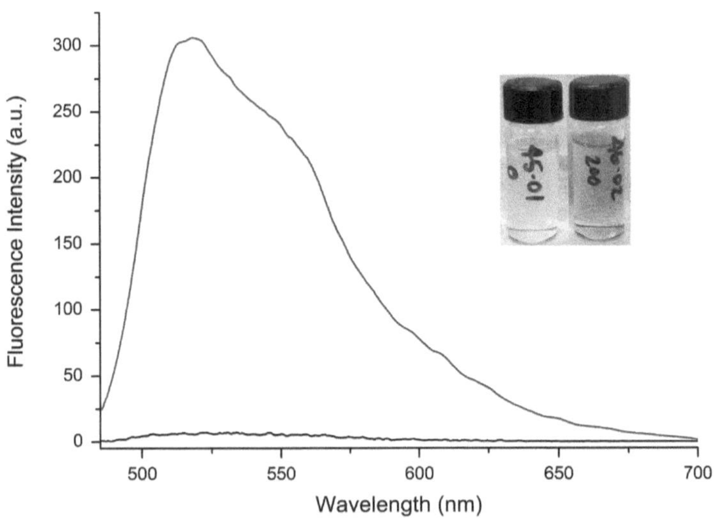

Figure 11.2 Fluorescence spectra of Pf before (black) and 60 min (red) (gray in print version) after treatment with AS. *Insert*: Color change induced by AS addition.

S-nitrosocysteine (CysNO, 200 eq.) produces a much weaker fluorescence response than a similar amount of AS.[34] Also, the fluorescence response of Pf to equal amounts of GSNO/CysNO (2 eq.) appears ~20–30 fold weaker than the response to HNO (2 eq.) further confirming the observation that phosphine probes react selectively with HNO compared to RSNO.[34] Further work remains to explain the observed differences in reactivity between HNO and RSNO with these probes. Pf also successfully detects HNO generated from AS in living HeLa cells by confocal fluorescence microscopy.[34] HeLa cells treated with Pf and no HNO do demonstrate a weak time-dependent fluorescence likely indicating slow hydrolysis of Pf within the cell.[34] LC–MS studies following extraction of the HeLa cell lysates treated with HNO show the formation of the phosphine oxide benzamide (**7**), which further confirms both HNO formation and phosphine-mediated reductive ligation.[34]

11.6 Conclusion

The phosphorus-mediated Staudinger ligation provides a convenient approach to HNO detection and quantification. Since the time of the first description of using phosphines as HNO traps in 2009,[28] this area has rapidly grown with the development of various probes that now allow HNO characterization in chemical and biological processes as well as imaging in live cells. Together with the copper-based fluorescent probes and the new electrochemical methods, HNO researchers now possess three mechanistically different means for HNO detection/measurement that were not available just a few years ago. Already these methods are better defining the role of HNO in various chemical and biological processes and support the development of HNO-based therapies. Much work remains to better understand the basic chemistry of these phosphine-mediated ligations that should lead to more diversified tools for HNO (and potentially RSNO) detection. Further development of these probes should better answer questions regarding the biology of HNO and finally permit a more complete description of this nitrogen oxide.

Acknowledgments

Work in the author's laboratory was supported by the National Institutes of Health (HL62198) and Wake Forest University.

References

1. Palmer RMJ, Ferrige AG, Moncada S. Nitric oxide release accounts for the biological activity of endothelium-derived relaxing factor. *Nature* 1987;**327**:524–6.
2. Moncada S, Palmer RMJ, Higgs EA. Nitric oxide: physiology, pathophysiology, and pharmacology. *Pharmacol Rev* 1991;**43**:109–42.

3. Stamler JS, Singel DJ, Loscalzo J. Biochemistry of nitric oxide and its redox-activated forms. *Science* 1992;**258**:1898–902.
4. Arnelle DR, Stamler JS. NO^+, NO, and NO^- donation by *S*-nitrosothiols: implications for regulation of physiological functions by *S*-nitrosylation and acceleration of disulfide formation. *Arch Biochem Biophys* 1995;**318**:279–85.
5. Bogdan C. Nitric oxide and the immune response. *Nat Immunol* 2001;**2**:907–16.
6. Lundberg JO, Weitzberg E, Gladwin MT. The nitrate–nitrite–nitric oxide pathway in physiology and therapeutics. *Nat Rev Drug Discov* 2008;**7**:156–67.
7. Gao WD, Murray CI, Tian Y, Zhong X, DuMond JF, Shen X, et al. Nitroxyl-mediated disulfide bond formation between cardiac myofilament cysteines enhances contractile function. *Circ Res* 2012;**111**:1002–11.
8. Fukuto JM, Bartberger MD, Dutton AS, Paolocci N, Wink DA, Houk KN. The physiological chemistry and biological activity of nitroxyl (HNO): the neglected, misunderstood, and enigmatic nitrogen oxide. *Chem Res Toxicol* 2005;**18**:790–801.
9. Miranda KM. The chemistry of nitroxyl (HNO) and implications in biology. *Coord Chem Rev* 2005;**249**:433–55.
10. Paolocci N, Jackson MI, Lopez BE, Miranda K, Tocchetti CG, Wink DA, et al. The pharmacology of nitroxyl (HNO) and its therapeutic potential: not just the janus face of NO. *Pharmacol Ther* 2007;**113**:442–58.
11. Irvine JC, Ritchie RH, Favaloro JL, Andrews KL, Widdop RE, Kemp-Harper BK. Nitroxyl (HNO): the Cinderella of the nitric oxide story. *Trends Pharmacol Sci* 2008;**29**: 601–8.
12. Fukuto JM, Bianco CL, Chavez TA. Nitroxyl (HNO) signaling. *Free Rad Biol Med* 2009;**47**:1318–24.
13. Norris AJ, Sartippour MR, Lu M, Park T, Rao JY, Jackson MI, et al. Nitroxyl inhibits breast tumor growth and angiogenesis. *Int J Cancer* 2008;**122**:1905–10.
14. Tocchetti CG, Stanley BA, Murray CI, Sivakumaran V, Donzelli S, Mancardi D, et al. Playing with cardiac "redox switches": the "HNO way" to modulate cardiac function. *Antioxid Redox Sign* 2011;**14**:1687–98.
15. Lopez BE, Rodriguez CE, Pribadi M, Cook NM, Shinyashiki M, Fukuto JM. Inhibition of yeast glycolysis by nitroxyl (HNO): mechanism of HNO toxicity and implications to HNO biology. *Arch Biochem Biophys* 2005;**442**:140–8.
16. Lopez BE, Wink DA, Fukuto JM. The inhibition of glyceraldehyde-3-phosphate dehydrogenase by nitroxyl (HNO). *Arch Biochem Biophys* 2007;**465**:430–6.
17. Eberhardt M, Dux M, Namer B, Miljkovic J, Cordasic N, Will C, et al. H_2S and NO cooperatively regulate vascular tone by activating a neuroendocrine HNO–TRPA1–CGRP signalling pathway. *Nat Commun* 2014;5.
18. Flores-Santana W, Salmon DJ, Donzelli S, Switzer CH, Basudhar D, Ridnour L, et al. The specificity of nitroxyl chemistry is unique among nitrogen oxides in biological systems. *Antioxid Redox Signal* 2011;**14**:1659–74.
19. Filipovic MR, Miljkovic JL, Nauser T, Royzen M, Klos K, Shubina T, et al. Chemical characterization of the smallest *S*-nitrosothiol, HSNO; cellular cross-talk of H_2S and *S*-nitrosothiols. *J Am Chem Soc* 2012;**134**:12016–27.
20. Bonner FT, Ravid B. Thermal decomposition of oxyhyponitrite (sodium trioxodinitrate (II)) in aqueous solution. *Inorg Chem* 1975;**14**:558–63.
21. Rosenthal J, Lippard SJ. Direct detection of nitroxyl in aqueous solution using a tripodal copper(II) BODIPY complex. *J Am Chem Soc* 2010;**132**:5536–7.
22. Zhou Y, Liu K, Li JY, Fang YA, Zhao TC, Yao C. Visualization of nitroxyl in living cells by a chelated copper(II) coumarin complex. *Org Lett* 2011;**13**:1290–3.

23. Zhou Y, Yao YW, Li JY, Yao C, Lin BP. Nitroxyl induced fluorescence enhancement via reduction of a copper(II) coumarin–ester complex: its application for bioimaging in vivo. *Sensor Actuat B Chem* 2012;**174**:414–20.
24. Wrobel AT, Johnstone TC, Liang AD, Lippard SJ, Rivera-Fuentes P. A fast and selective near-infrared fluorescent sensor for multicolor imaging of biological nitroxyl (HNO). *J Am Chem Soc* 2014;**136**:4697–705.
25. Suárez SA, Bikiel DE, Wetzler DE, Martí MA, Doctorovich F. Time-resolved electrochemical quantification of azanone (HNO) at low nanomolar level. *Anal Chem* 2013;**85**:10262–9.
26. Suarez SA, Fonticelli MH, Rubert AA, de la Llave E, Scherlis D, Salvarezza RC, et al. A surface effect allows HNO/NO discrimination by a cobalt porphyrin bound to gold. *Inorg Chem* 2010;**49**:6955–66.
27. Bartberger MD, Fukuto JM, Houk KN. On the acidity and reactivity of HNO in aqueous solution and biological systems. *Proc Natl Acad Sci USA* 2001;**98**:2194–8.
28. Reisz JA, Klorig EB, Wright MW, King SB. Reductive phosphine-mediated ligation of nitroxyl (HNO). *Org Lett.* 2009;**11**:2719–21.
29. Reisz JA, Zink CN, King SB. Rapid and selective nitroxyl (HNO) trapping by phosphines: kinetics and new aqueous ligations for HNO detection and quantitation. *J Am Chem Soc* 2011;**133**:11675–85.
30. Kawai K, Ieda N, Aizawa K, Suzuki T, Miyata N, Nakagawa H. A reductant-resistant and metal-free fluorescent probe for nitroxyl applicable to living cells. *J Am Chem Soc* 2013;**135**:12690–6.
31. Jing XT, Yu FB, Chen LX. Visualization of nitroxyl (HNO) in vivo via a lysosome-targetable near-infrared fluorescent probe. *Chem Commun* 2014;**50**:14253–6.
32. Liu CY, Wu HF, Wang ZK, Shao CX, Zhu BC, Zhang XL. A fast-response, highly sensitive and selective fluorescent probe for the ratiometric imaging of nitroxyl in living cells. *Chem Commun* 2014;**50**:6013–16.
33. Mao GJ, Zhang XB, Shi XL, Liu HW, Wu YX, Zhou LY, et al. A highly sensitive and reductant-resistant fluorescent probe for nitroxyl in aqueous solution and serum. *Chem Commun* 2014;**50**:5790–2.
34. Miao Z, Reisz JA, Mitroka SM, Pan J, Xian M, King SB. A selective phosphine-based fluorescent probe for nitroxyl in living cells. *Bioorg Med Chem Lett* 2015;**25**:16–19.
35. Cadogan JIG. Reduction of nitro- and nitroso-compounds by tervalent phosphorus reagents. *Q Rev Chem Soc* 1968;**22**:222–51.
36. Chainikova EM, Teregulova AN, Shamukaev VA, Safiullin RL. Chemiluminescence during deoxygenation of nitrosobenzene with triphenylphosphine. *High Energ Chem* 2009;**43**:147–8.
37. Khursan VS, Shamukaev VA, Chainikova EM, Khursan SL, Safiullin RL. Kinetics and mechanism of the nitrosobenzene deoxygenation by trivalent phosphorous compounds. *Russ Chem Bull* 2013;**62**:2477–86.
38. Sidky MM, Soliman FM, Shabana R. Organophosphorus compounds XXIV. The reaction of triphenylphosphine and trialkyl phosphites with α-nitroso-β-napthol. *Egypt J Chem* 1978;**21**:29–36.
39. Haake M. Zur desoxygenierung von tritylthionitrit. *Tetrahedron Lett* 1972;**13**:3405–8.
40. Bechtold E, Reisz JA, Klomsiri C, Tsang AW, Wright MW, Poole LB, et al. Water-soluble triarylphosphines as biomarkers for protein s-nitrosation. *ACS Chem Biol* 2010;**5**:405–14.
41. Donzelli S, Espey MG, Flores-Santana W, Switzer CH, Yeh GC, Huang JM, et al. Generation of nitroxyl by heme protein-mediated peroxidation of hydroxylamine but not *N*-hydroxy-L-arginine. *Free Rad Biol Med* 2008;**45**:578–84.

42. Nagasawa HT, Demaster EG, Redfern B, Shirota FN, Goon JW. Evidence for nitroxyl in the catalase-mediated bioactivation of the alcohol deterrent agent cyanamide. *J Med Chem* 1990;**33**:3120–2.
43. Jackson MI, Han TH, Serbulea L, Dutton A, Ford E, Miranda KM, et al. Kinetic feasibility of nitroxyl reduction by physiological reductants and biological implications. *Free Rad Biol Med* 2009;**47**:1130–9.
44. Wang H, Xian M. Fast reductive ligation of *S*-nitrosothiols. *Angew Chem Int Ed* 2008;**47**:6598–601.
45. Rhine MA, Rodrigues AV, Urbauer RJB, Urbauer JL, Stemmler TL, Harrop TC. Proton-induced reactivity of NO^- from a {CoNO}(8) complex. *J Am Chem Soc* 2014;**136**:12560–3.
46. Heinecke JL, Khin C, Pereira JCM, Suárez SA, Iretskii AV, Doctorovich F, et al. Nitrite reduction mediated by heme models. Routes to NO and HNO? *J Am Chem Soc* 2013;**135**:4007–17.
47. Cline MR, Tu CK, Silverman DN, Toscano JP. Detection of nitroxyl (HNO) by membrane inlet mass spectrometry. *Free Radical Biol Med* 2011;**50**:1274–9.
48. Guthrie DA, Ho A, Takahashi CG, Collins A, Morris M, Toscano JP. "Catch-and-Release" of HNO with pyrazolones. *J Org Chem* 2015;**80**:1338–48.
49. Guthrie DA, Nourian S, Takahashi CG, Toscano JP. Curtailing the hydroxylaminobarbituric acid-hydantoin rearrangement to favor HNO generation. *J Org Chem* 2015;**80**:1349–56.
50. Köhn M, Breinbauer R. The staudinger ligation—a gift to chemical biology. *Angew Chem Int Ed* 2004;**43**:3106–16.
51. Lin FL, Hoyt HM, van Halbeek H, Bergman RG, Bertozzi CR. Mechanistic investigation of the staudinger ligation. *J Am Chem Soc* 2005;**127**:2686–95.
52. Deem S, Kim SS, Min JH, Eveland R, Moulding J, Martyr S, et al. Pulmonary vascular effects of red blood cells containing *S*-nitrosated hemoglobin. *Am J Physiol-Heart C* 2004;**287**:H2561–8.
53. Hess DT, Matsumoto A, Kim SO, Marshall HE, Stamler JS. Protein *S*-nitrosylation: rapurview and pameters. *Nat Rev Mol Cell Bio* 2005;**6**:150–66.
54. Janssen-Heininger YMW, Mossman BT, Heintz NH, Forman HJ, Kalyanaraman B, Finkel T, et al. Redox-based regulation of signal transduction: principles, pitfalls, and promises. *Free Radical Bio Med* 2008;**45**:1–17.
55. Zhang J, Wang H, Xian M. An unexpected bis-ligation of *S*-nitrosothiols. *J Am Chem Soc* 2009;**131**:3854–5.
56. Zhang JM, Wang H, Xian M. Exploration of the "traceless" reductive ligation of *S*-nitrosothiols. *Org Lett* 2009;**11**:477–80.
57. Zhang D, Chen W, Miao Z, Ye Y, Zhao Y, King SB, et al. A reductive ligation based fluorescent probe for *S*-nitrosothiols. *Chem Commun* 2014;**50**:4806–9.
58. Seneviratne U, Godoy LC, Wishnok JS, Wogan GN, Tannenbaum SR. Mechanism-based triarylphosphine-ester probes for capture of endogenous RSNOs. *J Am Chem Soc* 2013;**135**:7693–704.
59. Bhardwaj A, Huang ZJ, Kaur J, Yang FH, Seubert JM, Knaus EE. A diazen-1-ium-1,2-diolate analog of 7-azabenzobicyclo[2.2.1]heptane: synthesis, nitric oxide and nitroxyl release, in vitro hemodynamic, and anti-hypertensive studies. *Bioorg Med Chem Lett* 2013;**23**:2769–74.

Electrochemical Detection of Azanone

12

S.A. Suarez, M. Muñoz, D.E. Bikiel, M.A. Marti and F. Doctorovich
Universidad de Buenos Aires and CONICET, Buenos Aires, Argentina

Chapter Outline

12.1 Background

One of the key factors in HNO research is to be able to detect and quantify this molecule unambiguously, a difficult task given its short lifetime and overlapping reactivity with NO and other reactive nitrogen and oxygen species (RNOS). As described in previous chapters, the last decade has seen a range of developments in this field,[1–3] of which the present description of an electrochemical HNO sensor is a paradigmatic example.

As described in Chapter 5, HNO Generation From $NO^{\bullet}$, Nitrite, Inorganic or Organic Nitrosyls, and Crosstalk With H_2S, a number of studies of nitroxyl reactivity have been performed with metalloporphyrin complexes. Although cobalt porphyrins (CoPs) were not studied as much as their iron or manganese analogs, they also gained prominent attention for the binding and activation of nitric oxide. The NO adducts of cobalt substituted myoglobin and hemoglobin have been characterized by EPR and UV-vis,[4,5] and several kinetic studies have been done on the NO reaction with Co^{II} and Co^{III} porphyrins. The specific association rate constants of

The Chemistry and Biology of Nitroxyl (HNO). DOI: http://dx.doi.org/10.1016/B978-0-12-800934-5.00012-8

NO with Co^{II} porphyrins were estimated to be $k_{on} \sim 10^9 \, M^{-1} \, s^{-1}$, while the low value of $k_{off} \sim 10^{-4} \, s^{-1}$ shows the stability of the $Co^{III}NO^{-}$ complex.[6] Interestingly the association rate of NO with Co^{III} porphyrins is notably smaller.[6] The reactions of HNO with metalloporphyrins of iron, manganese, and cobalt were also known, showing association rates of $k_{on} \sim 10^5 \, M^{-1} \, s^{-1}$.[7–9] Generally speaking whereas NO reacts with Co^{II} and it does not react with Co^{III}, HNO reacts with Co^{III} and not with Co^{II}. This is an advantage of Co^{III} over Fe^{III} porphyrins, which react both with NO and HNO.[10] In addition, the resulting $Co^{III}NO^{-}$ is stable; however, it loses NO upon oxidation. This fact is the foundation of the method for electrochemical detection of HNO, as will be described in the following sections.

Owing to their widespread use in technical applications when coupled to a surface, monolayers of porphyrins and metalloporphyrins adsorbed on an inert metal surface have been used both as model systems for studying their reactivity and for the development chemical sensors.[11–13] In situ electrochemical detection in real time is a commonly used analytical technique that allows identification and quantification of many elusive chemical species,[14–16] including nitric oxide. The most important aspects to take into account for azanone detection are selectivity (against other RNOS) and sensitivity, because HNO concentrations are usually quite low due to its high reactivity. Metalloporphyrins are good candidates for achieving this task, since they can be chosen to be selective toward azanone and are able to trap it at very small concentrations. Development of the electrochemical sensor thus started by selecting an adequate metal-porphyrin, which would trap azanone but not NO. *Based on the above-described properties, a CoP was selected as a possible candidate for an HNO-selective sensor.*

12.2 Design of an amperometric HNO sensor

12.2.1 Electrochemical properties of cobalt porphyrins and surface adsorption

During the last 40 years, the electrochemistry of cobalt and nitrosyl CoP has been studied.[6,17,18] Typically, between −1 and 1 Volt, the CoPs have five redox couples involving one electron (three oxidations and two reductions process), while the nitrosyl porphyrin undergoes only two of each one.[6,8,17,18] Stable complexes of $[(TPP)Co(NO)]^{-}$ and $[(TPP)Co(NO)]^{+}$ [TPP = tetraphenylporphyrin] could both be obtained at the electrode surfaces[19] and redox potentials for Co^{II}(*p*-OCH_3)TPP have been reported in several solvents.[20]

The most successful methods for surface-molecule linking are based on the establishment of gold-sulfur bonds, and a large number of porphyrin monomers bearing thiols have been prepared and attached to gold electrodes by this method. In order to avoid the problems associated with disulfide formation in solution, the thiol is best handled in a protected form (e.g., with an *S*-acetyl protecting group), that undergoes in situ *S-acetyl bond* cleavage on the surface. The cobalt—5,10,15,20-tetrakis [3-(*p*-acetylthiopropoxy)phenyl] porphyrin [Co(P)], shown in

Scheme 12.1, was selected due to the presence of four sulfur anchor points that allow efficient immobilization of the porphyrin on a gold surface. In order to avoid the signal due to the direct reaction of NO with the gold electrode, the holes remaining on the surface after attaching the porphyrin was occupied by aliphatic thiols.

Direct confirmation of Co(P) adsorption was performed by electrochemical measurements of the "Co(P) electrodes" built. A reversible oxidation process is observed in the cyclic voltammograms of the Co(P)-modified gold electrode in aqueous solution, within an $E_{1/2}$ value of about 0.4 V, versus Saturated calomel electrode (SCE). The reaction corresponds to the Co^{III}/Co^{II} couple, and surprisingly the value is shifted ca. 400 mV to lower potentials compared to values in solution (see Table 12.1). These results show that Co(P) adsorption on the gold electrode facilitates Co^{II} oxidation. The same results are observed in organic solvent, showing that this is not a solvent but a surface related effect.[8] Cyclic voltammograms of "$Co^{III}(P)NO^{-}$ electrodes" in aqueous solution show a clear peak corresponding to one-electron reduction at 0.8 V versus SCE, a value close to the one observed in solution, thus lacking surface effect. Overall, these results (summarized in Table 12.1) show evidence of electron transfer from the metal surface to the Co metal center of the

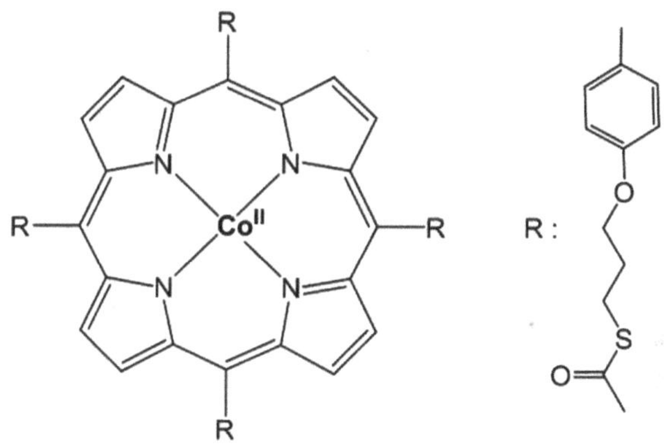

Scheme 12.1 Structure of Co(II)-5,10,15,20-tetrakis[3-(*p*-acetylthiopropoxy)phenyl] porphyrin.

Table 12.1 Measured Co^{III}/Co^{II} redox potentials for Co(P) and $Co^{III}(P)NO^{-}$

Porphyrin	Solvent	Condition	$E_{1/2}$ (volt vs SCE)
Co(P)	CH_2Cl_2	Solution	**0.79**
		Adsorbed	**0.44**
	Water	Adsorbed	**0.40**
$CoIII(P)NO^{-}$	CH_2Cl_2	Solution	**0.89**
		Adsorbed	**0.86**
	Water	Adsorbed	**0.83**

adsorbed porphyrin, which is suppressed by axial coordination of NO to the Co porphyrin.[8,21,22] The shift observed on the reduction potential is, as shown below, key for the HNO-sensing scheme.

12.2.2 CoP modified electrodes reactivity toward NO and HNO

The adsorbed CoP maintains its selectivity. As expected, when the porphyrin is immobilized on the surface Co^{III}(P), it reacts efficiently with HNO, while it does not react with NO donors or NO gas. This is evidenced by the lack of electrochemical signal when the porphyrin is maintained in Co^{III} state, at a potential compatible with the $Co^{III}(P)NO/Co^{III}(P)NO^-$ couple (about 0.8 V vs SCE) and the current intensity is monitored after addition of NO donors. On the other hand, NO reacts rapidly with Co^{II}(P) modified electrodes, while HNO does not, i.e., the $Co^{III}(P)NO/Co^{III}(P)NO^-$ couple is observed after adding NO donor to Co(P) bound electrode maintained in the Co^{II}(P) state. However, no change is observed when HNO donors are used (Eqs. 12.1–12.4). This behavior is summarized below:

$$Co^{II}(P) + NO \leftrightarrow Co^{III}(P)NO^- \qquad (12.1)$$

$$Co^{II}(P) + HNO \leftrightarrow \text{No reaction} \qquad (12.2)$$

$$[Co^{III}(P)]^+ + HNO \leftrightarrow Co^{III}(P)NO^- + H^+ \qquad (12.3)$$

$$[Co^{III}(P)]^+ + NO \leftrightarrow \text{No reaction} \qquad (12.4)$$

12.2.3 HNO amperometric sensing scheme

Based on the above-described behavior,[7,8] Scheme 12.2 was proposed as a method to detect HNO amperometrically. According to this scheme, the resting state electrode potential is set to 0.8 V,[8] a value where the porphyrin is stable as Co^{III}P and no current is observed. Reaction with HNO yields the $Co^{III}PNO^-$ complex that under the previously described conditions is oxidized to Co^{III}PNO. Due to the lability of the resulting Co^{III}PNO complex, it releases the NO ligand quickly yielding Co^{III}P, allowing the catalytic cycle to restart. Since each electrode is covered with a surface concentration of only 4×10^{-12} mol cm^{-2}, the total amount of trapped HNO, and thus released NO is small, so the reaction between released nitric oxide and HNO is not relevant.

The Co(P)-modified electrode response was tested by measuring the current versus time plot at 0.8 V, showing that few seconds after the addition of an HNO donor (Angeli's Salt, AS) the current intensity increases (Fig. 12.1), and it is maintained for several minutes due to the continuous HNO production from the donor. When the donor concentration starts to decrease, the corresponding signal consistently decays.

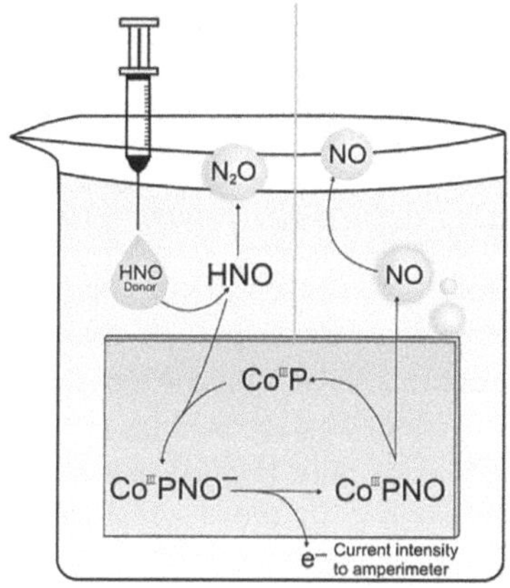

Scheme 12.2 Mechanistic scheme considered for HNO production from the decomposition of an HNO donor (Angeli's Salt, AS), plus the reactions occurring at the electrode surface. HNO reacts with the electrode surface producing a current intensity, which is proportional to the bulk concentration of HNO in solution at a given time.

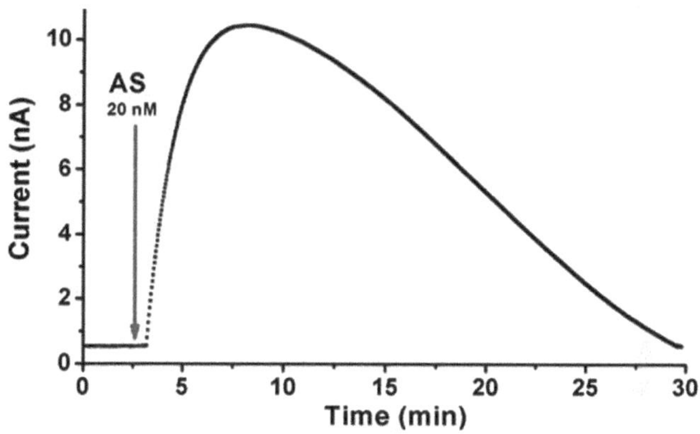

Figure 12.1 Intensity versus time plot for Co(P) modified electrodes. The arrow indicates addition of AS.

12.3 Analytical properties of the azanone detecting electrode

The HNO-sensing electrode was characterized in several aspects related to its potential use in chemical and biological studies involving azanone:[7]

- Dynamic range and linear response to HNO
- Kinetic analysis of the time-resolved electrode response in the presence of well-characterized HNO donors.

- Selectivity and media compatibility: possible occurrence of spurious signals (or interferences) due to the presence of oxygen and other RNOS; Performance of the electrode in the presence of living cells.

12.3.1 Dynamic range and linear response to HNO

In order to detect and quantify the maximum concentration of azanone present in a given system at any time, the electrode response (peak current) has to be calibrated with known [HNO] in solution, as determined from the corresponding initial donor concentration ($[HNO] = (k_d.[donor]/k_{dim})^{1/2}$). Fig. 12.2 shows the corresponding peak current (ΔC) as a function of bulk [HNO]. As it can be observed in Fig. 12.2, the electrode has a linear response in the 1–150 nM range and thus allows the detection and quantification of azanone at low nanomolar concentrations.[7] The above-described inverse square relation between donor and peak [HNO] and the apparent upper limit to azanone bulk concentration are clearly evidenced in inset of Fig. 12.2, where the measured ΔC is plotted against initial AS concentration.

12.3.2 Kinetic analysis of the time-resolved electrode response toward HNO

To analyze the electrode signal from a kinetic point of view, the equations that relate bulk [HNO] with the detected current intensity (ΔC) were derived, using a typical two-layer model. Briefly, near the electrode surface, a thin layer is established where the flow velocity is essentially zero. Mass transport takes place by diffusion alone, and thus a continuous flow is required to maintain the surface concentrations demanded by the Nernst's equation. Convection maintains a constant supply of HNO at the outer edge of this diffusion layer, which is proportional to the HNO concentration in bulk solution. The observed current will thus depend on the HNO diffusion rate, the thickness of the diffusion layer, and the HNO concentrations at the

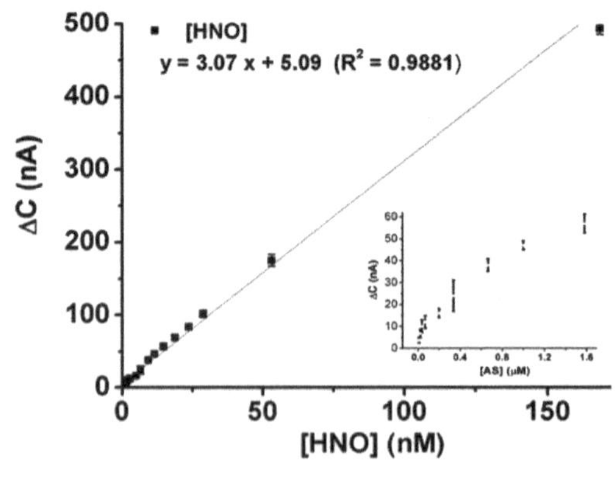

Figure 12.2 Calibration curve for HNO. *Inset*: ΔC intensity signal (nA) detected by a Co(P)-modified gold electrode as a function of AS concentration.

electrode surface and in bulk solution. Assuming that Co(P) is wired to the electrode, that HNO binding to Co(P) is fast (k_{on} is expected to be on the order of $10^4\ M^{-1}\ s^{-1}$) for a thin layer, and given that the concentration of HNO at the electrode approaches zero (since it binds and is consumed by the Co(P)), the resulting current intensity (i) is directly proportional to [HNO] in the bulk solution (Eq. 12.5).

$$i_I = \frac{n.F.A.D_{HNO}}{\delta} C_{HNO} = k_{eHNO} C_{HNO} \tag{12.5}$$

where n is the number of electrons per mol of HNO, F is the Faraday's constant, A is the electrode surface area (cm^2), D_{HNO} is the diffusion coefficient for azanone, and δ the thickness of the diffusion layer. Since $n.F.A.D_{HNO}$ and δ are constants for a given system, they can be grouped in the above-mentioned system-dependent effective rate k_{eHNO}. Therefore, for any given system, the corresponding chemical kinetic scheme can be solved using the electrode current signal to determine time-dependent bulk HNO concentrations.

The kinetic model was in the first place tested for known HNO-producing reactions (such as AS decomposition), allowing the determination of the electrode-dependent rates (mainly Co oxidation and NO release from the CoP nitrosyl complex). These rates can subsequently be used to determine unknown HNO-related rate constants (see below). Table 12.2 lists all the kinetic rate constants required by the model.

In the second place, in order to analyze the performance of the electrode (validation), the associated differential set of equations and obtained rate constants were tested. The electrode response to a solution of known methyl-derived Piloty's acid (TSHA) concentration was analyzed. The resulting plot (shown in Fig. 12.3) was analyzed using the above-described kinetic model and unknown rate constant (Table 12.2) leaving it free to adjust (and thus determine) the TSHA decomposition rate constant. The obtained value[7] of $5.5 \times 10^{-4}\ s^{-1}$ at 25°C is in excellent agreement with the reported value of $4.4 \times 10^{-4}\ s^{-1}$,[28] showing that the Co(P) electrode and associated kinetic model is appropriate to analyze reactions involving HNO donors (first-order rate constants) and determine the associated reaction rate constants.[7,29] Kinetic analysis of more complex reactions will be described in Section 12.4.2.

Table 12.2 Reactions and rate constants that were used in the simulations for HNO production as a function of time

Reaction	k (s^{-1})	References
Angeli's Salt→HNO	8.9×10^{-4}	[23–25]
Co(III)P + HNO→Co(III)PNO$^-$ + H+	$3.1 \times 10^4\ M^{-1}$	[24]
Co(III)PNO$^-$→Co(III)PNO + e$^-$	8.9×10^{-2}	[7]
Co(III)PNO→Co(III)P + NO	5.2×10^{-3}	[18]
HNO + NO→$N_2O_2^-$ + H^+	$5.8 \times 10^6\ M^{-1}$	[26]
2HNO→N_2O + H_2O	$8 \times 10^6\ M^{-1}$	[26,27]

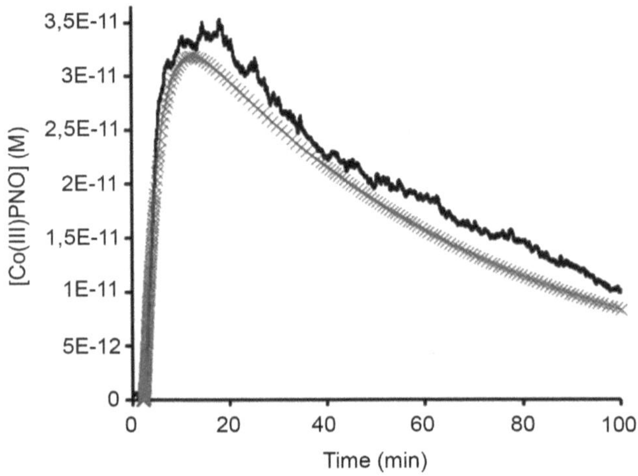

Figure 12.3 Electrode response to a solution of 3.4 μM TSHA (black: experimental; red (gray in print version): simulated).

Moreover, these results show that the electrode signal reflects the actual azanone concentration in a time-resolved fashion, thus allowing the performance of kinetic measurements of HNO production/consumption.[7] As a final comment, it is important to note that since the HNO reaction with the modified electrode is very efficient, and a small amount of current can be detected, the amount of azanone that reacts with the electrode is usually negligible compared to the total amount of HNO in the system (less than 1%), and it does not perturb significantly the HNO producing and consuming reactions, thus becoming a valuable tool in mechanistic and kinetic studies where azanone plays a key role.[7]

12.3.3 Selectivity and media compatibility

To analyze the electrode selectivity and biological media compatibility our analysis of the electrode response in the presence of RNOS and other species in complex biological media showed that although the electrode response is affected, it can still detect HNO at nM levels. The HNO concentration is completely independent of the presence of either of the tested species (KNO_3, KNO_2, NH_2OH, and H_2O_2), in both anaerobic and aerobic media. As shown in Fig. 12.4, the electrode response in cell culture with intact cells and with lysed cells, is only minimally altered.[7] The most dramatic effect on the electrode signal is due to the presence of oxygen, which, as expected, scavenges bulk HNO. Thus, azanone can still be detected and quantified in the presence of high concentrations of oxygen, such as those found in ambient air.[7,30]

12.4 Applications of the HNO sensor

As shown in the previous sections, the electrochemical sensor is able to quantitatively detect HNO at very low concentrations in real time. Moreover, it can be used

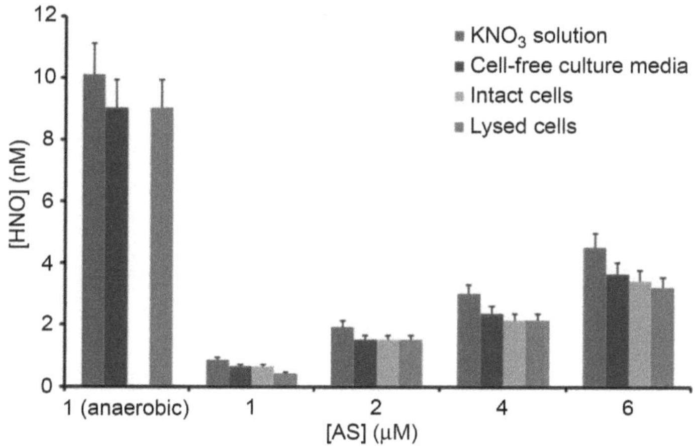

Figure 12.4 [HNO] produced by AS decomposition in the presence of *Xenopus* melanophores cell culture. The experiments were performed under air atmosphere unless otherwise noted.

in biological media without loss of sensitivity. Therefore, it can provide valuable information in a number of applications related to chemistry, biology, and medicine. Some of these applications will be described in the following sections.

12.4.1 Characterization of HNO donors

The conditions needed for HNO liberation by donors can be easily determined by the use of the electrochemical azanone sensor. For example, in the case of pH-sensitive donors, the pH at which the donation of HNO starts was determined by using the HNO-selective electrode.[29] Starting with a solution of the donor at a pH where the compound is stable, the pH was raised slowly by addition of diluted NaOH(aq.), while pH and current intensity values were measured simultaneously (Fig. 12.5A). A large increase in the current indicates that HNO is being released; the current reaches a plateau when the maximum rate of HNO production is obtained. For example, Fig. 12.5B shows the titration of methyl-Piloty's Acid where a clear increase in the signal, and thus HNO release, is evidenced above pH 9.

12.4.2 Detection and quantification of HNO in chemical and biochemical systems

12.4.2.1 HNO derived from the reaction of porphyrins with nitrite

The water-soluble ferriheme model Fe^{III}(TPPS) mediates oxygen atom transfer from inorganic nitrite to a water-soluble phosphine (Tris(3-sulfophenyl)phosphine).

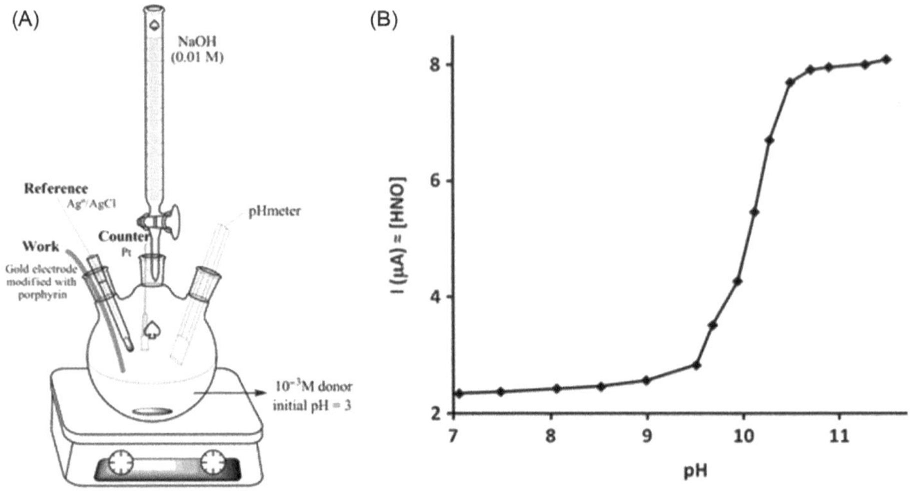

Figure 12.5 (A) Three-electrode system used for pH-dependent measurements. (B) Current intensity versus pH for methyl-Piloty's acid donor.

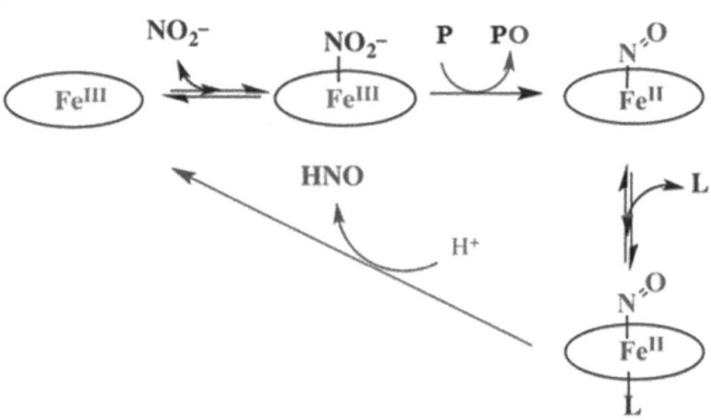

Scheme 12.3 Reactions described for aqueous pH 5.81solutions of Fe(TPPS), nitrite, and phosphine as substrate.

The nitrosyl complex Fe^{II}(TPPS)(NO) is the dominant iron species while excess substrate is present. However, in slightly acidic media (pH ~6), the system does not stop at this very stable ferrous nitrosyl. Instead, it displays a matrix of redox transformations linking spontaneous regeneration of Fe^{III}(TPPS) and the formation of HNO (Scheme 12.3).[31]

For the above-mentioned process, the maximum amount of HNO produced could be quantified. To perform this, the signal obtained in the study reaction was compared to the Angeli's Salt calibration curve made under similar conditions. In this case, a 67% of HNO was generated, which is in accordance with the amount of N_2O measured by GC.

12.4.2.2 Obtaining kinetics parameters of HNO-producing reactions

In the last few years, a new route of HNO generation was investigated. It was suggested that NO can be reduced to HNO by biologically relevant alcohols with moderate reducing capacity, such as ascorbate or tyrosine,[32,33] according to the following global reaction (which actually does not occur through direct H· atom abstraction):

$$NO + ROH \rightarrow RO^{\bullet} + HNO \quad (12.6)$$

Taking advantage of the amperometric HNO sensor, it was possible to detect azanone and also to obtain the corresponding kinetics parameters, even though in this case the reaction mechanism was not known and/or the HNO source does not follow first-order kinetics (unlike HNO donors). From the measurements of HNO concentration versus time, the corresponding initial HNO production rates (v_i) could be obtained under different reaction conditions. Fig. 12.6A shows that v_i versus [ROH] follows a linear function. From the slope of these plots, an effective bimolecular reaction rate constant (k_{eff}), corresponding to reaction five can be obtained.

$$v = k_{\text{eff}}[\text{ROH}][\text{NO}]$$

Also important, Fig. 12.6B shows that Log(v_i) versus Log[ROH] is linear, with a slope close to one, confirming that the reaction is first order in this reactant.

These results show that the HNO sensor can be used to kinetically characterize reactions in which HNO is involved.

12.4.3 Measurements in live organisms

Measurements of [HNO] as a function of time in cell culture opens up a great number of possibilities concerning the production of HNO in living systems. HNO

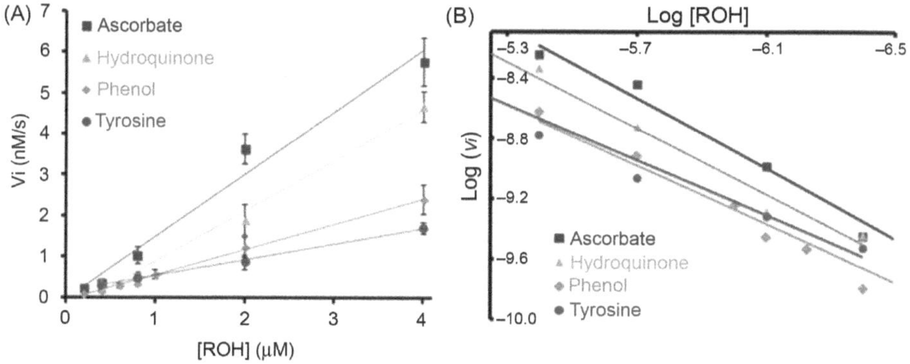

Figure 12.6 (A) v_i versus [ROH]. (B) Log(v_i) versus Log[ROH]. [NO] = 0.2 mM ROH = 0.1−4 μM. Square: ascorbate; triangle: hydroquinone; diamond: phenol; circle: tyrosine.

donors, as well as HNO formation from other sources (such as NO or produced by enzymes) can be easily tested by immersing the three-electrode system in the cell culture. In this case, the observed signal corresponds to HNO released to the cell culture medium.

12.4.3.1 *In vitro cell measurements*

The HNO electrode was used to measure azanone production in extracellular medium. The ability of ascorbate to reduce endogenously generated NO from macrophages was tested. Macrophages were stimulated with lipopolysaccharide/interferon gamma to stimulate inducible nitric oxide synthase to produce NO. After the addition of ascorbate, an immediate rise in the signal was observed, showing azanone formation (Fig. 12.7).[32]

It is known that the range of physiological concentrations of $NO^{\bullet}$ goes from 100 pM to ca. 150 nM [34–38]. Some of the reducing compounds such as ascorbate (vit C) display in vivo circulating concentrations at the low mM range. This would result in HNO produced at pM to low nM levels, which are large enough to produce a biological response, and can be detected by the HNO sensor. Therefore, HNO produced from these sources could be detected in living systems.

12.4.3.2 *In vivo measurements*

The amperometric sensor is able to detect [HNO] for example in rat[39] or Guinea pig live animals. Fig. 12.8 shows two setups that are used to perform different in vivo measurements, such as soft tissues (heart/eye/peritoneum) or vascular systems (aorta/jugular vein). In all cases, the three electrodes (work, reference, and

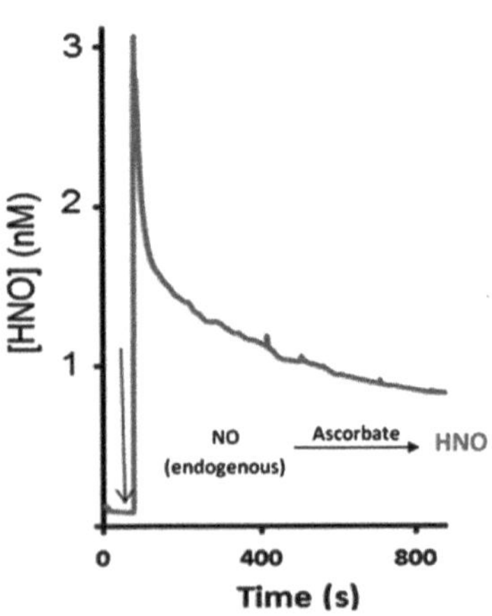

Figure 12.7 HNO formation after the addition of ascorbate to immunostimulated macrophages. The HNO sensor was immersed into 10^6 cell mL^{-1} suspension of immunostimulated macrophages in Dulbecco's Modified Eagle's Medium (DMEM). 1 mM ascorbate was added and the current was monitored.

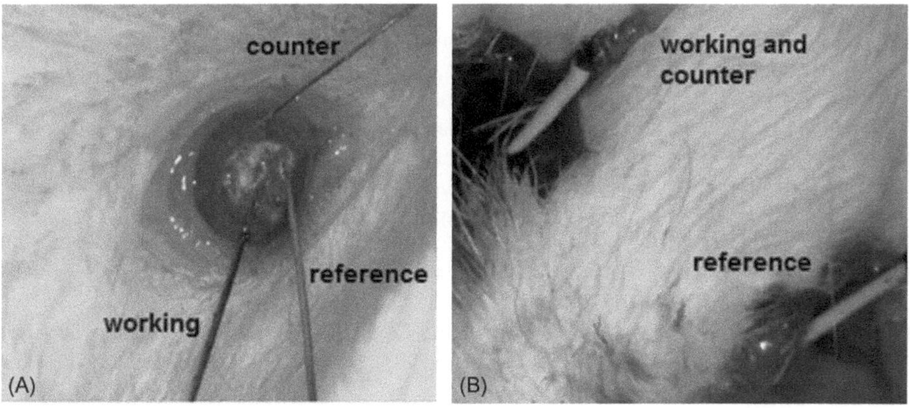

Figure 12.8 (A) Soft-tissue (example: eye). (B) Vascular system (example: jugular vein).

counter) are placed inside the respective tissue. Since the geometry of the metal electrodes can be flat or cylindrical, and their sizes can be changed as needed, down to sub-millimetric scale, the three-electrode measuring system can be introduced to the living animal directly, or through cannulas.

In summary, the Co(P) electrode can be used in studies involving HNO-related chemical and biological reactions, as well as in vitro and in vivo assays.[29,31,40,41] Preliminary results indicate that the sensor will be in the near future an important tool for the endogenous detection of HNO.

References

1. Dobmeier KP, Riccio DA, Schoenfisch MH. Xerogel optical sensor films for quantitative detection of nitroxyl. *Anal Chem* 2008;**80**(4):1247–54.
2. Tennyson AG, Do L, Smith RC, Lippard SJ. Selective fluorescence detection of nitroxyl over nitric oxide in buffered aqueous solution using a conjugated metallopolymer. *Polyhedron* 2007;**26**(16):4625–30.
3. Donzelli S, Graham M, Flores-Santana W, Switzer CH, Yeh GC, Huang J, et al. Generation of nitroxyl by heme protein-mediated peroxidation of hydroxylamine but not *N*-hydroxy-L-arginine. *Free Radic Biol Med* 2008;**45**(5):578–84.
4. Scheidt WR, Hoard JL. Stereochemistry of low-spin cobalt porphyrins. I. Structure and bonding in a nitrosylcobalt porphyrin and their bearing on one rational model for the oxygenated protoheme. *J Am Chem Soc* 1973;**95**(25):8281–8.
5. Hori H, Ikeda-Saito M, Leigh JS, Yonetani T. Single-crystal electron paramagnetic resonance studies of photolyzed oxy- and nitric oxide-cobalt myoglobins. *Biochemistry* 1982;**21**:1431–7.
6. Roncaroli F, Van Eldik R. Mechanistic analysis of reductive nitrosylation on water-soluble cobalt(III)-porphyrins. *J Am Chem Soc* 2006;**128**:8042–53.
7. Suárez S, Bikiel D, Wetzler D, Martí MA, Doctorovich F. Time-resolved E electrochemical quantification of azanone (HNO) at low nanomolar level. *Anal Chem* 2013;**85**: 10262–9.

8. Suárez SA, Fonticelli MH, Rubert AA, de la Llave E, Scherlis D, Salvarezza RC, et al. A surface effect allows HNO/NO discrimination by a cobalt porphyrin bound to gold. *Inorg Chem* 2010;**49**(15):6955–66.
9. Doctorovich F, Bikiel DE, Suárez SA, Marcelo A. 2014, **58**: 145–183.
10. Ford PC, Fernandez BO, Lim MD. Mechanisms of reductive nitrosylation in iron and copper models relevant to biological systems. *Chem Rev* 2005;**105**(6):2439–55.
11. Williams FJ, Vaughan OPH, Knox KJ, Bampos N, Lambert RM. First observation of capping/uncapping by a ligand of a Zn porphyrin adsorbed on Ag (100). *Chem Commun (Camb)* 2004;**44**(15):1688–9.
12. Guillaud G. Metallophthalocyanines: gas sensors, resistors and field effect transistors. *Coord Chem Rev* 1998;**178–180**:1433–84.
13. Mochida I, Tsuji K, Suetsugu K, Fujitsui H, Takeshita K. Modified electronic structure and enhanced catalytic activity of cobalt tetraphenylporphyrin supported by titanium dioxide. *J Phys Chem* 1980;3159–62.
14. Kudo S, Bourassa JL, Boggs SE, Sato Y, Ford PC. In situ nitric oxide (NO) measurement by modified electrodes: NO labilized by photolysis of metal nitrosyl complexes. *Anal Biochem* 1997;**247**(2):193–202.
15. Chi Y, Chen J, Aoki K. Electrochemical generation of free nitric oxide from nitrite catalyzed by iron meso-tetrakis(4-N-methylpyridiniumyl)porphyrin. *Inorg Chem* 2004;**43** (26):8437–46.
16. Wei J, Qiu J, Li L, Ren L, Zhang X, Chaudhuri J, et al. A reduced graphene oxide based electrochemical biosensor for tyrosine detection. *Nanotechnology* 2012;**23**(33):335707.
17. Owens JW, Perry M, Seybert DW. *Science* 1998; **80**: 1693.
18. Laverman LE, Ford PC. J Am Chem Soc 2001; **123**(47): 11614–22.
19. Kelly S, Lancon D, Kadish KM. Electron-transfer and ligand-addition reactions of (TTP)Mn(NO) and (TTP)Co(NO) in a nonaqueous media. *Inorg Chem* 1984;**23**:1451–8.
20. Walker FA. *Transition* 1976, No. 4.
21. Flechtner K, Kretschmann A, Steinru H, Gottfried JM. Ii C. Langmuir 2007: 12110–11.
22. Hieringer W, Flechtner K, Kretschmann A, Seufert K, Auwärter W, Barth JV, et al. The surface trans effect: influence of axial ligands on the surface chemical bonds of adsorbed metalloporphyrins. *J Am Chem Soc* 2011;**133**(16):6206–22.
23. Miranda KM, Dutton AS, Ridnour LA, Foreman CA, Ford E, Paolocci N, et al. Mechanism of aerobic decomposition of Angeli's salt (sodium trioxodinitrate) at physiological pH. *J Am Chem Soc* 2005;**127**(2):722–31.
24. Martí MA, Bari SE, Estrin DA, Doctorovich F. Discrimination of nitroxyl and nitric oxide by water-soluble Mn(III) porphyrins. *J Am Chem Soc* 2005;**127**(13):4680–4.
25. Angeli A. Chem Zentralbl 1902; **73**: 691.
26. Shafirovich V, Lymar SV. Nitroxyl and its anion in aqueous solutions: spin states, protic equilibria, and reactivities toward oxygen and nitric oxide. *Proc Natl Acad Sci USA* 2002;**99**(11):7340–5.
27. Bonner FT, Hughes MN, Poole RK, Scott RI. Kinetics of the reactions of trioxodinitrate and nitrite ions with cytochrome d in Escherichia coli. *Biochim Biophys Acta, Bioenerg* 1991;**1056**(2):133–8.
28. Bonner FT, Ko YH. Kinetic, isotopic, and N-15 NMR-study of *N*-hydroxybenzenesulfonamide decomposition—an HNO source reaction. *Inorg Chem* 1992;**31**:2514–19.
29. Sirsalmath K, Suárez SA, Bikiel DE, Doctorovich F. The pH of HNO donation is modulated by ring substituents in Piloty's acid derivatives: azanone donors at biological pH. *J Inorg Biochem* 2013;**118**:134–9.

30. Drew B, Leeuwenburgh C. Aging and the role of reactive nitrogen species. *Ann NY Acad Sci* 2002;**959**:66–81.
31. Heinecke JL, Khin C, Pereira JCM, Suárez SA, Iretskii AV, Doctorovich F, et al. Nitrite reduction mediated by heme models. Routes to NO and HNO? *J Am Chem Soc* 2013;**135**(10):4007–17.
32. Suarez SA, Neuman N, Muñoz M, Alvarez L, Brondino C, Bikiel DE, et al. J Am Chem Soc 2015; **137**: 4720–27.
33. Loas A, Radford RJ, Deliz Liang A, Lippard SJ. Solid-phase synthesis provides a modular, lysine-based platform for fluorescent discrimination of nitroxyl and biological thiols. *Chem Sci* 2015;**6**:4131–40.
34. Al-Nimer MSM, Al-Obaidi SAH, Al-Dulaimi KS. Serum nitric oxide and peroxynitrite levels in adult sero-positive rheumatoid arthritis treated with disease modifying antirheumatic drugs: a preliminary report. *Turk J Med Sci* 2010;**40**(2):191–7.
35. Coleman JW. Nitric oxide in immunity and inflammation. *Int Immunopharmacol* 2001;**1**(8):1397–406.
36. Wink DA, Kasprzak KS, Maragos CM, Elespuru RK, Misra M, Dunams TM, et al. DNA deaminating ability and genotoxicity of nitric oxide and its progenitors. *Science* 1991;**254**(5034):1001–3.
37. Ridnour LA, Thomas DD, Mancardi D, Espey MG, Miranda KM, Paolocci N, et al. The chemistry of nitrosative stress induced by nitric oxide and reactive nitrogen oxide species. Putting perspective on stressful biological situations. *Biol Chem* 2004; **385**:1–10.
38. Hall CN, Garthwaite J. What is the real physiological NO concentration in vivo? *Nitric Oxide* 2009;**21**(2):92–103.
39. Doctorovich F, Rosenstein R. Personal Communication.
40. Álvarez L, Suarez SA, Bikiel DE, Reboucas JS, Batinić-Haberle I, Martí MA, et al. Redox potential determines the reaction mechanism of HNO donors with Mn and Fe porphyrins: defining the better traps. *Inorg Chem* 2014;**53**:7351–60.
41. Eberhardt M, Dux M, Namer B, Miljkovic J, Cordasic N, Will C, et al. H2S and NO cooperatively regulate vascular tone by activating a neuroendocrine HNO-TRPA1-CGRP signalling pathway. *Nat Comm* 2014;**5**:4381.

Detection of HNO by Membrane Inlet Mass Spectrometry

13

T.A. Chavez and J.P. Toscano
Johns Hopkins University, Baltimore, MD, United States

Chapter Outline

13.1 Introduction

Membrane inlet (or introduction) mass spectrometry (MIMS) is a technique that has been in use since the early 1960s. MIMS was first reported by Hoch and Kok as a method for the detection of hydrophobic gases dissolved in aqueous solution by mass spectrometry.[1] Analytes partition from the liquid phase through a membrane and into the vapor phase via pervaporation.[2] Pervaporation is a process that occurs in three stages: First, the species of interest is adsorbed at the surface of the membrane, it then permeates through the membrane, and finally desorbs into vacuum where it is detected by mass spectrometry. Factors that influence MIMS sensitivity include the diffusability and solubility of the gas in the membrane material, membrane surface area, and membrane thickness.

The MIMS technique has been used in a variety of applications including the monitoring of biological reactions and industrial processes.[2,3] The application of MIMS for the detection of the small hydrophobic gas, nitric oxide (NO),[3,4] suggested the possibility to use MIMS to detect the closely related gas, azanone (HNO, nitroxyl). HNO, the protonated one-electron reduced form of NO, has been recently recognized for its unique biological activity, especially as a potential therapeutic for heart failure.[5–9] Detection of HNO is difficult due to its facile reactivity with itself to produce hyponitrous acid (HON = NOH, $k = 8 \times 10^6\ M^{-1}s^{-1}$), which

The Chemistry and Biology of Nitroxyl (HNO). DOI: http://dx.doi.org/10.1016/B978-0-12-800934-5.00013-X

subsequently dehydrates to form nitrous oxide (N_2O) (Scheme 13.1).[10] As a result of this reactivity, HNO must be generated in situ by the use of donor molecules. It is hoped that the development of the MIMS technique for HNO detection and investigation of the aqueous chemistry of this potentially new gaseous signaling agent will ultimately shed light on its potential mechanisms of action in vivo.

13.2 Membrane inlet design and methods

The MIMS setup originally used for HNO detection,[11] based on a design reported by Silverman and coworkers,[3] consists of a membrane probe that includes a piece of silastic tubing attached to glass tubing at one end and sealed at the other by a glass bead (Fig. 13.1A). The glass tubing is attached via an ultra-torr Swagelok connection to an external vacuum chamber containing a dosing line that leads to a quadrupole mass spectrometer where ions are detected following electron ionization (EI). The membrane probe is immersed in an aqueous solution in a sealable 4-mL glass sample cell fit with a sample injection port.

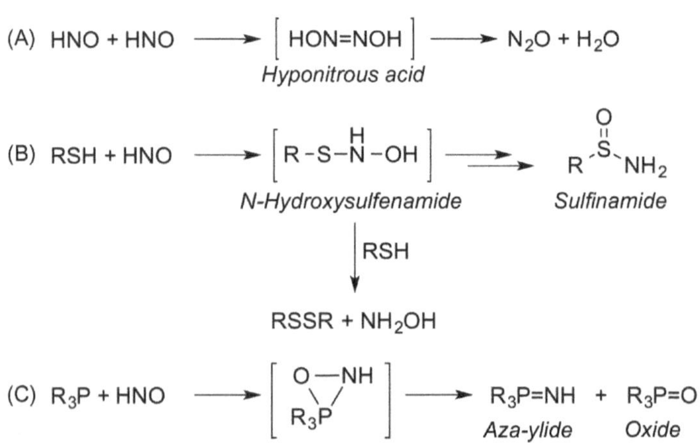

Scheme 13.1 Reactivity of HNO.

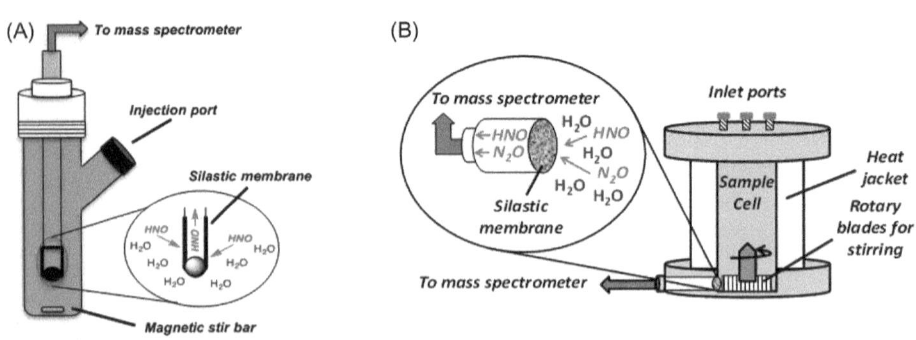

Figure 13.1 Schematic representation of MIMS sample cells and membrane probes based on (A) Silverman and coworkers' design and (B) the Hiden HPR-40 system.

Recently, our laboratory has begun to use a Hiden HPR-40 MIMS system with a sample cell and membrane probe that have been optimized to detect gases dissolved in aqueous solution. The modified cell contains rotary blades with an imbedded magnet for efficient stirring past the silastic membrane. The sample cell is fit with an interior heating/cooling jacket for temperature regulation, as well as multiple ports for sample injection (Fig. 13.1B). As described below, with this optimized setup we can reliably detect the HNO *m/z* 30 NO^+ fragment ion signal at precursor concentrations as low as 25 nM. Given the first-order rate constant for HNO donor decomposition ($t_{1/2}$ at 25°C = ca. 15 min, $k_d = 7.7 \times 10^{-4}\ s^{-1}$) and the bimolecular rate constant for HNO dimerization, we estimate a detection limit of approximately 1 nM HNO.[12]

13.3 Detection of HNO by MIMS

MIMS has been used to examine HNO production from a variety of donors, including disodium diazen-1-ium-1,2,2-triolate (Angeli's salt, AS), *N*-hydroxybenzenesulfonamide (Piloty's acid, PA), and 2-bromo-*N*-hydroxybenzenesulfonamide (2-BrPA) (Fig. 13.2). Typical MIMS traces observed at *m/z* 30 (NO^+), 31 (HNO^+/$^{15}NO^+$), and 44 (N_2O^+) after addition of AS to argon-purged phosphate-buffered saline (PBS) at pH 7.4 containing the metal chelator, diethylenetriaminepentaacetic acid (DTPA) are shown in Fig. 13.3. Aliquots of HNO donor solutions are injected after a flat baseline is established and ion current intensities are measured as a function of time.

Although the *m/z* 44 signal is easily assigned to N_2O, the ultimate product of HNO dimerization, assignments of the *m/z* 30 and 31 signals can be more complicated. Depending on the donor, its concentration, and conditions of the MIMS experiment, the *m/z* 30 signal can contain contributions from NO, HNO, and/or N_2O. The EI mass spectra of NO, HNO, and N_2O all contain strong NO^+ signals at *m/z* 30 due to either the parent ion (NO) or major fragment ions (HNO and N_2O). The HNO parent ion *m/z* 31 signal is very weak with an intensity less than 2.8% that of the *m/z* 30 fragment ion in the reported EI mass spectrum.[13] Due to this weak intensity, the detection of HNO^+ can also be complicated by natural abundance $^{15}NO^+$. For example, small *m/z* 31 MIMS signals from the NO donor, DEA/NO, and from standard N_2O gas can be observed (Fig. 13.4). The intensity ratio of the *m/z* 31 to *m/z* 30 signals in each of these cases matches well with that expected from natural abundance ^{15}N (0.4%). In light of the potential complications

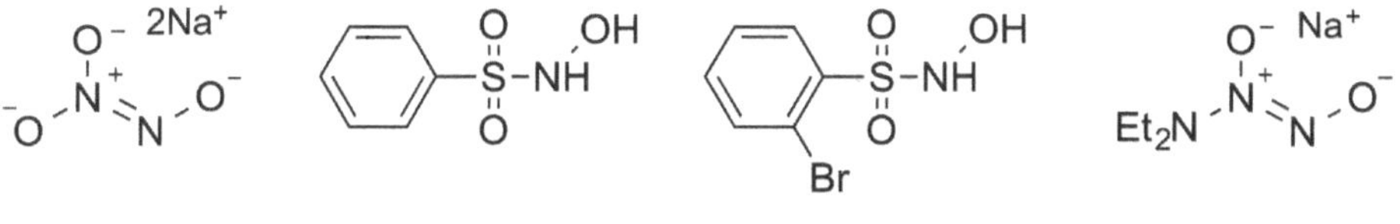

Figure 13.2 The HNO donors, AS, PA, and 2-BrPA, and the NO donor, DEA/NO.

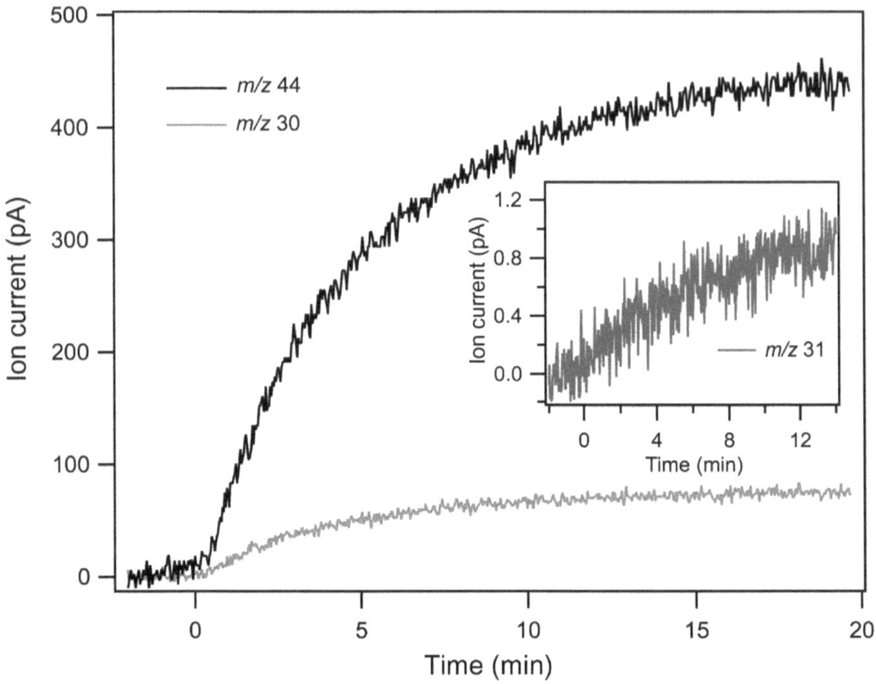

Figure 13.3 MIMS signals observed at *m/z* 30, 31, and 44 following the addition of 50 μM AS to an argon-purged 0.1M PBS solution containing 100 μM DTPA at pH 7.4 and 37°C.

surrounding identification of the *m/z* 30 and 31 MIMS signals, additional experiments must be performed to determine the relative contributions of HNO, NO, and N_2O to these signals.

13.4 Differentiating HNO and NO MIMS signals

HNO and NO can be differentiated by their reactivity with chemical traps. For example, HNO reacts readily with thiols (e.g., glutathione (GSH)) and phosphines (e.g., tris-(4,6-dimethylphenyl)phosphine-3,3′,3″-trisulfonic acid trisodium salt (TXPTS)) under both anaerobic and aerobic conditions (Scheme 13.1).[14,15] NO, on the other hand, is quite unreactive with thiols under anaerobic conditions, but in the presence of oxygen undergoes rapid autoxidation to produce nitrogen dioxide (NO_2) and/or dinitrogen trioxide (N_2O_3), both of which react rapidly with thiols.[14,16–20] In addition, NO is relatively unreactive with phosphines, both in the presence or absence of oxygen. However, at high TXPTS concentrations NO can react to produce ultimately N_2O.[21–24]

Consistent with the above reactivity, the *m/z* 30 signal generated from the NO donor, sodium 1-(*N*,*N*-diethylamino)diazen-1-ium-1,2-diolate (DEA/NO, Fig. 13.2) is unaffected by the addition of 250 μM GSH under anaerobic conditions; however,

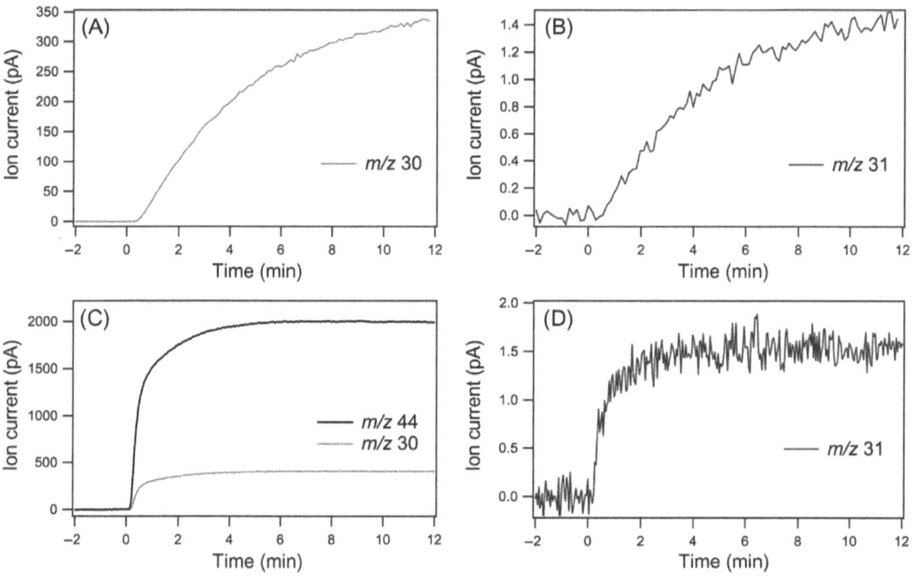

Figure 13.4 MIMS signals observed at (A) *m/z* 30 and (B) *m/z* 31 following the addition of 100 μM DEA/NO to an argon-purged 0.1 M PBS solution containing 100 μM DTPA at pH 7.4 and 37°C, and at (C) *m/z* 44, 30, and (D) 31 following the addition of 100 μL of N_2O (g) to an argon-purged 0.1 M PBS solution containing 100 μM DTPA at pH 7.4 and 37°C.

when the solution is purged with oxygen, the signal is diminished.[11] In addition, the DEA/NO *m/z* 30 MIMS signal is unaffected by up to 250 μM TXPTS. In comparison, the *m/z* 30 signal produced from the HNO donor, 2-BrPA, is completely quenched by 250 μM GSH under anaerobic conditions or by 250 μM TXPTS under aerobic or anaerobic conditions.

Another method that has been used to distinguish MIMS signals originating from NO from those due to HNO and/or N_2O is the use of liquid nitrogen cold trap.[25] Although the boiling point of HNO is unknown, it is likely higher than that of NO (bp = −152°C at 1 atm). If this is the case, it should be possible to trap HNO and N_2O (bp = −88°C at 1 atm) selectively using an appropriate cold trap. Since the MIMS system is under vacuum, a liquid nitrogen (bp = −196°C at 1 atm) cold trap was found to trap HNO (derived from 2-BrPA) and N_2O gas effectively, but not NO (produced from DEA/NO).[25] In addition, the *m/z* 31 signal ($^{15}NO^+$) from DEA/NO (Fig. 13.4B) persists in the presence of a liquid nitrogen cold trap, indicating that this signal is not due to HNO.

The contribution of N_2O to the *m/z* 30 MIMS signal observed following decomposition of HNO donors can be determined by analysis of the *m/z* 44 to *m/z* 30 signal intensity ratio. In MIMS experiments using relatively high donor concentrations (typically greater than 100 μM), dimerization is favored and primarily N_2O is observed. Under these conditions, the *m/z* 30 signal originates solely from N_2O, and the 44:30 ratio is identical to that of a standard N_2O sample. However, at lower donor concentrations (typically less than 5 μM), the *m/z* 44 signal is not observed, and the *m/z* 30 signal is due to the NO^+ fragment ion of HNO alone. Between these two extremes, the *m/z* 30 signal arises from contributions from both HNO and N_2O. The relative contribution of these two species can be calculated using the 44:30 ratio observed for a standard N_2O sample.[11]

13.5 HNO donor comparison

MIMS has been used to compare the HNO donors, AS, PA, and 2-BrPA.[11] The *m/z* 30 MIMS signals produced from either 50 μM AS or 2-BrPA were examined as function of GSH and TXPTS concentration. For 2-BrPA, the *m/z* 30 signal was completely quenched with 50 μM and above GSH or TXPTS. In the case of AS, although the intensity of the *m/z* 30 signal was reduced in presence of either GSH or TXPTS, it was not completely quenched even at high concentrations of trap (100−250 μM). Since N_2O is not observed at *m/z* 44 and all the HNO produced should be quenched at high trap concentrations, the residual *m/z* 30 signal was attributed to NO. The intensity of this residual signal arising from 50 μM AS was identical to that observed from 2 μM DEA/NO. Because the NO-forming pathways of both DEA/NO and AS produce 2 moles of NO from each mole of precursor, this result indicates that approximately 2% of AS decomposes to NO at pH 7.4. This minor production of NO would normally not be observed since HNO is an effective trap of NO,[10] and is only revealed when the HNO is removed by a selective trap. The generation of a small amount of NO by AS at neutral pH is a possibility that has recently been considered.[26]

The decomposition of PA at physiological pH is slow, leaving it susceptible to oxidation, which ultimately leads to NO production.[27,28] This oxidation requires both trace metals and oxygen.[29] The TXPTS quenching of the *m/z* 30 MIMS signal derived from PA was examined at pH 7.4 under aerobic conditions in the presence or absence of the metal chelator, DTPA.[11] In the presence of DTPA, slow production of the *m/z* 30 signal is observed. Addition of TXPTS results in near complete quenching of this signal, indicating mostly HNO, but some NO production under these conditions. In absence of DTPA, the growth kinetics of the *m/z* 30 signal become much faster and the addition of TXPTS results in only modest quenching of the signal, indicating the production of more NO under these conditions. In contrast, analogous MIMS analysis of the decomposition of 2-BrPA, which has a much shorter half-life ($t_{1/2}$ = ca. 2 min at pH 7.4 and 37°C), indicates that 2-BrPA is less susceptible to oxidation and HNO production remains the dominant pathway regardless of the presence of a metal chelator or oxygen.[11]

13.6 Detection of HNO from HOCl-mediated oxidation of *N*-Hydroxyarginine (NOHA)

The positive pharmacological profile of HNO has led to interest in elucidating potential endogenous pathways to HNO generation. Oxidation of hydroxylamine

(NH_2OH), hydroxamic acids (RC(O)NHOH), and its derivatives (e.g., *N*-hydroxyguanidines) have been explored previously.[30–36] Recent work by Donzelli et al. has investigated the oxidation of NH_2OH by horseradish peroxidase (HRP) in the presence of hydrogen peroxide (H_2O_2).[34] Generation of HNO is observed from a variety of Fe^{III}-containing enzymes including HRP, the globins (metmyoglobin (metMb) and methemoglobin (metHb)), catalase, myeloperoxidase (MPO), lactoperoxidase, and hemin. HNO generation appears to be affected by the identity of the axial ligand, with histidine ligands producing the highest yields. Hydroxamic acid derivatives have their own history in applications as metal chelators,[37] and treatments for cancer.[38] Similar to the results of hydroxylamine, oxidation of hydroxamic acids by H_2O_2/Fe^{III} systems have the ability to generate HNO.[39–41] The amount of HNO that is able to avoid trapping in the enzyme pocket by the resulting Fe^{III} formed is dependent on the enzyme.

N-Hydroxyarginine (NOHA) is an established biosynthetic intermediate involved in the endogenous production of NO from arginine by nitric oxide synthase (NOS). NOS has also been shown to oxidize NOHA to HNO, rather than NO, in the absence of the biopterin cofactor.[30,31,42–45] The detection of an iron nitrosyl species in NOS, rather than the typical Fe^{III} resting state further suggests the generation of HNO.[43,46] Recently, it has been suggested that that the biopterin radical intermediate oxidizes the $\{FeNO\}^7$ by one electron to produce $\{FeNO\}^6$, which can then release NO. In the absence of the biopterin cofactor, HNO may be released from the $\{FeNO\}^7$ species.[47–49]

It is well established that *N*-hydroxyguanadines have the ability to produce NO and/or HNO from a variety of chemical oxidants.[32,35,44,50] This process is presumed to involve a nitroso intermediate that leads to the production of the corresponding cyanamide derivative (Scheme 13.2). A biologically relevant oxidant potentially capable of oxidizing NOHA to produce HNO is hypochlorous acid (HOCl). HOCl is a strong oxidant that is generated in vivo by MPO from the reaction of hydrogen peroxide (H_2O_2) with chloride ion. MPO is released by neutrophils, monocytes, macrophages, and has been proposed to play a role in cardiovascular disease.[51–53] Activated neutrophils have been shown to produce HOCl at concentrations up to 100 μM in vitro, which may be a concentration that is attainable at inflammatory sites in vivo.[54–56]

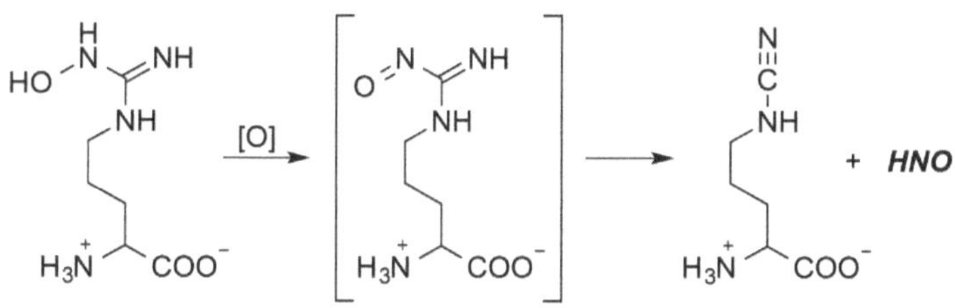

Scheme 13.2 Oxidation of *N*-hydroxyarginine to produce HNO and cyanamide derivative via a nitroso intermediate.

Investigation of the biologically relevant oxidation of NOHA has demonstrated that MIMS is a viable technique for the study of possible endogenous pathways for HNO production.[25] Consistent with the experiments performed by Donzelli et al. 500 μM NH_2OH or NOHA with 10 μM HRP and 100 μM H_2O_2 led to increases in MIMS signal intensities at *m/z* 30 and *m/z* 44 for NH_2OH, but not for NOHA. The oxidation of NOHA by HOCl was initially monitored by soluble gas chromatography (GC) headspace analysis to quantify the amount of N_2O produced.[25] NOHA (100 μM) was incubated with varying concentrations of HOCl at 37°C in 0.1 M PBS at pH 7.4. Significant N_2O production was observed only with excess HOCl, and a ratio of HOCl to substrate 5:1 was chosen for subsequent experiments. This ratio of oxidant to substrate is potentially plausible in a biological setting since HOCl can be generated up to concentrations of 100 μM and NOHA has been observed at concentrations of 15 μM in blood samples.[57,58]

MIMS experiments following the injection of 500 μM HOCl into a PBS solution of 100 μM NOHA resulted in observation of signals at *m/z* 30 and *m/z* 44.[25] The observed 44:30 intensity ratio was higher than that observed for authentic N_2O. A portion of the *m/z* 44 signal was determined to be due to CO_2 production that occurs via HOCl-mediated chlorination of the amine terminus to form ultimately the corresponding aldehyde and CO_2 (Scheme 13.3). Previous work has shown that HOCl can oxidize amino acids to produce CO_2 and the corresponding aldehydes,[59,60] and this was confirmed by MIMS experiments with L-arginine (CO_2 observed) and L-arginine methyl ester (no CO_2 observed).[25]

To confirm that the N_2O observed following HOCl oxidation of NOHA is due to HNO dimerization rather than from a direct N_2O-producing pathway, MIMS experiments were performed at lower concentrations (<5 μM NOHA).[25] Previous MIMS work has demonstrated that at HNO donor concentrations below 5 μM, N_2O dimerization is not observed and the *m/z* 30 signal observed under these conditions is derived from HNO. Reducing the concentration of NOHA to 2.5 μM or below results in the loss of the *m/z* 44 signal, while still maintaining the *m/z* 30 signal, corresponding to the direct detection of HNO. The possibility that this signal could be due to the parent ion of NO, rather than HNO fragmentation, was ruled out with the use of a liquid nitrogen cold trap. In the presence of this cold trap, a MIMS signal at *m/z* 30 is not observed, which indicated that NO is not produced.[25]

In addition to NOHA, the HOCl oxidation of NH_2OH, hydroxyurea, and acetohydroxamic acid were also examined.[25] All of these species have been shown capable of producing HNO upon oxidation (Scheme 13.4).[33,36] The oxidation of these substrates were monitored by MIMS and GC headspace analysis. MIMS experiments revealed the typical *m/z* 44 and *m/z* 30 signals for each substrate, albeit weaker ones for acetohydroxamic acid. GC headspace analysis was consistent with

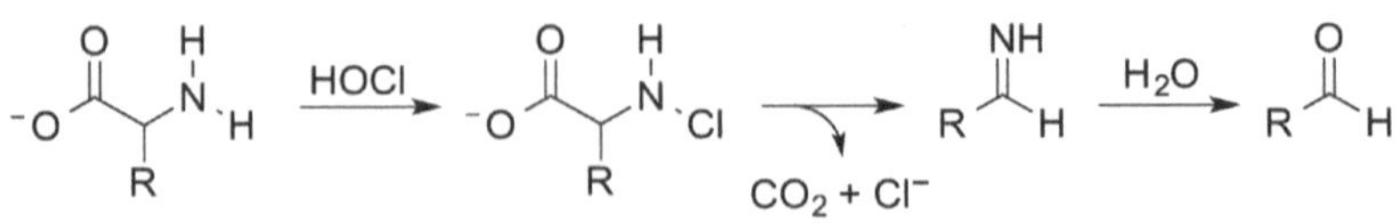

Scheme 13.3 Reaction of HOCl with amino acids.

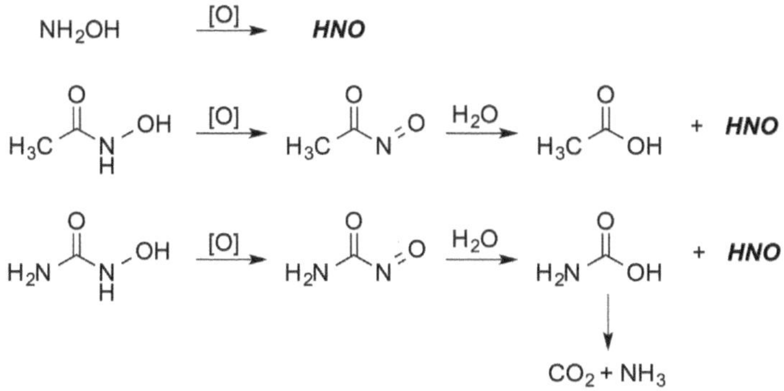

Scheme 13.4 HNO producing pathways for NH_2OH, acetohydroxamic acid, and hydroxyurea.

these MIMS observations, showing significant production of N_2O, relative to AS, from both NH_2OH (90%) and hydroxyurea (91 %), and relatively little from acetohydroxamic acid (8%).

The MPO-mediated HOCl oxidation of NOHA was examined to validate the biological relevance of this HNO producing pathway.[25] For comparison, oxidation of both NH_2OH and NOHA (100 μM) was monitored in the presence of 15 or 75 nM MPO with excess chloride ion (143 mM) and 500 μM H_2O_2. HOCl generation was initiated by the addition of H_2O_2 to a solution of MPO and chloride ion in the presence of substrate. The reactions were examined by GC headspace analysis of N_2O. With 15 nM MPO, only a small percentage of N_2O (25%, relative to AS) is observed for the oxidation of NH_2OH. Under the same conditions, NOHA produces very little N_2O (2%, relative to AS). Increasing the concentration of MPO to 75 nM led to much larger yields of N_2O from NH_2OH (73%, relative to AS); however, NOHA oxidation still produced only a small percentage of N_2O (8%, relative to AS). These results differ from those using HOCl itself where both NH_2OH and NOHA produce high yields of N_2O. It was suggested that this difference can be accounted for based on the slow production of HOCl from MPO and the reaction of *N*-hydroxyguanidines with HNO.[25,61]

13.7 Conclusions and future directions

The MIMS technique is a sensitive and selective detection method for HNO. It can be used to detect HNO under physiologically relevant conditions by both its parent ion (HNO^+, *m/z* 31) and its primary fragment ion (NO^+, *m/z* 30), as well as its dimerization product, N_2O (*m/z* 44). Reliable HNO detection down to precursor concentrations as low as 25 nM can be achieved. Distinguishing HNO from N_2O and NO is possible by the use of chemical (thiols or phosphines) or physical (liquid nitrogen) traps. The versatility of this technique allows for investigation of possible HNO producing pathways that have the potential to be physiologically relevant.

Future MIMS work should involve exploring other potential chemical reactions that can produce HNO. For example, recent work has shown that H_2S can react with *S*-nitrosothiols to generate thionitrous acid (HSNO), a potential precursor to HNO.[62] HSNO has also been proposed to be an intermediate in the reaction of H_2S and NO. Further, H_2S and NO have been shown to react to ultimately generate HNO.[63] Exploration of these reactions by MIMS could shed light on potential mechanisms of HNO formation from biologically relevant reactions.

Acknowledgment

J.P.T. gratefully acknowledges the National Science Foundation (CHE-1213438) and Cardioxyl Pharmaceuticals for generous support of this research.

References

1. Hoch G, Kok B. A mass spectrometer inlet system for sampling gases dissolved in liquid phases. *Arch Biochem Biophys* 1963;**101**:160–70.
2. Johnson RC, Cooks RG, Allen TM, Cisper ME, Hemberger PH. Membrane introduction mass spectrometry: trends and applications. *Mass Spectrom Rev* 2000;**19**:1–37.
3. Tu C, Swenson ER, Silverman DN. Membrane inlet for mass spectrometric measurement of nitric oxide. *Free Radic Biol Med* 2007;**43**:1453–7.
4. Tu C, Scafa N, Zhang X, Silverman DN. A comparison of membrane inlet mass spectrometry and nitric oxide (NO) electrode techniques to detect NO in aqueous solution. *Electroanalysis* 2010;**22**:445–8.
5. Paolocci N, Katori T, Champion HC, St John ME, Miranda KM, Fukuto JM, et al. Positive Inotropic and Lusitropic effects of HNO/NO^- in failing hearts: independence from beta-adrenergic signaling. *Proc Natl Acad Sci* 2003;**100**:5537–42.
6. Miranda KM, Paolocci N, Katori T, Thomas DD, Ford E, Bartberger MD, et al. A biochemical rationale for the discrete behavior of nitroxyl and nitric oxide in the cardiovascular system. *Proc Natl Acad Sci* 2003;**100**:9196–201.
7. Tocchetti CG, Wang W, Froehlich JP, Huke S, Aon MA, Wilson GM, et al. Nitroxyl improves cellular heart function by directly enhancing cardiac sarcoplasmic reticulum Ca^{2+} cycling. *Circ Res* 2007;**100**:96–104.
8. Paolocci N, Jackson MI, Lopez BE, Miranda K, Tocchetti CG, Wink DA, et al. The pharmacology of nitroxyl (HNO) and its therapeutic potential: not just the janus face of NO. *Pharmacol Ther* 2007;**113**:442–58.
9. Fukuto JM, Bartberger MD, Dutton AS, Paolocci N, Wink DA, Houk KN. The physiological chemistry and biological activity of nitroxyl (HNO): the neglected, misunderstood, and enigmatic nitrogen oxide. *Chem Res Toxicol* 2005;**18**:790–801.
10. Shafirovich V, Lymar SV. Nitroxyl and its anion in aqueous solutions: spin states, protic equilibria, and reactivities toward oxygen and nitric oxide. *Proc Natl Acad Sci* 2002;**99**:7340–5.
11. Cline MR, Tu C, Silverman DN, Toscano JP. Detection of nitroxyl (HNO) by membrane inlet mass spectrometry. *Free Radic Biol Med* 2011;**50**:1274–9.
12. Suárez SA, Bikiel DE, Wetzler DE, Martí MA, Doctorovich F. Time-resolved electrochemical quantification of azanone (HNO) at low nanomolar level. *Anal Chem* 2013;**85**:10262–9.

13. Lambert RM. Mass-spectrometric study of the system hydrogen + NO. *Chem Commun* 1966;**23**:850–1.
14. Wong PS, Hyun J, Fukuto JM, Shirota FN, DeMaster EG, Shoeman DW, et al. Reaction between S-Nitrosothiols and thiols: generation of nitroxyl (HNO) and subsequent chemistry. *Biochemistry* 1998;**37**:5362–71.
15. Reisz JA, Klorig EB, Wright MW, King SB. Reductive phosphine-mediated ligation of nitroxyl (HNO). *Org Lett* 2009;**11**:2719–21.
16. Pryor W, Church D, Govindan C. Oxidation of thiols by nitric oxide and nitrogen dioxide: synthetic utility and toxicological implications. *J Org Chem* 1982;**47**:156–9.
17. Folkes LK, Wardman P. Kinetics of the reaction between nitric oxide and glutathione: implications for thiol depletion in cells. *Free Radic Biol Med* 2004;**37**:549–56.
18. Wink DA, Nims RW, Darbyshire JF, Christodoulou D, Hanbauer I, Cox GW, et al. Reaction kinetics for nitrosation of cysteine and glutathione in aerobic nitric oxide solutions at neutral pH. Insights into the fate and physiological effects of intermediates generated in the NO/O_2 reaction. *Chem Res Toxicol* 1994;**7**:519–25.
19. Keshive M, Singh S, Wishnok JS, Tannenbaum SR, Deen WM. Kinetics of *S*-Nitrosation of thiols in nitric oxide solutions. *Chem Res Toxicol* 1996;**9**:988–93.
20. Goldstein S, Czapski G. Mechanism of the nitrosation of thiols and amines by oxygenated • NO solutions: the nature of the nitrosating intermediates. *J Am Chem Soc* 1996;**118**:3419–25.
21. Kuhn LP, Doali JO, Wellman C. The reaction of nitric oxide with triethyl phosphite. *J Am Chem Soc* 1960;**82**(18):4792–4.
22. Longhi BYR, Ragsdale R, Drago RS. Reactions of nitrogen(II) oxide with miscellaneous lewis bases. *Inorg Chem* 1962;**1**:768–70.
23. Lim MD, Lorkovic IM, Ford PC. Kinetics of the oxidation of triphenylphosphine by nitric oxide. *Inorg Chem* 2002;**41**:1026–8.
24. Bakac A, Schouten M, Johnson A, Song W, Pestovsky O, Szajna-Fuller E. Oxidation of a water-soluble phosphine and some spectroscopic probes with nitric oxide and nitrous acid in aqueous solutions. *Inorg Chem* 2009;**48**:6979–85.
25. Cline MR, Chavez TA, Toscano JP. Oxidation of *N*-Hydroxy-L-Arginine by hypochlorous acid to form nitroxyl (HNO). *J Inorg Biochem* 2013;**118**:148–54.
26. Zeller A, Wenzl MV, Beretta M, Stessel H, Russwurm M, Koesling D, et al. Mechanisms underlying activation of soluble guanylate cyclase by the nitroxyl donor angeli's salt. *Mol Pharmacol* 2009;**76**:1115–22.
27. Bonner F, Ko Y. Kinetic, isotopic, and ^{15}N NMR study of *N*-Hydroxybenzenesulfonamide decomposition: an nitrosyl hydride (HNO) source reaction. *Inorg Chem* 1992;**744**: 2514–19.
28. Zamora R, Grzesiok A, Weber H, Feelisch M. Oxidative release of nitric oxide accounts for guanylyl cyclase stimulating, vasodilator and anti-platelet activity of piloty's acid: a comparison with angeli's salt. *Biochem J* 1995;**312**:333–9.
29. Wilkins PC, Jacobs HK, Johnson MD, Gopalan AS. Mechanistic variations in the oxidation of piloty's acid by metal complexes. *Inorg Chem* 2004;**43**:7877–81.
30. Rusche KM, Spiering MM, Marletta MA. Reactions catalyzed by tetrahydrobiopterin-free nitric oxide synthase. *Biochemistry* 1998;**37**:15503–12.
31. Tejero J, Stuehr D. Tetrahydrobiopterin in nitric oxide synthase. *IUBMB Life* 2013;**65**: 358–65.
32. Fukuto JM, Stuehr DJ, Feldman PL, Bova MP, Wong P. Peracid oxidation of an *N*-Hydroxyguanidine compound: a chemical model for the oxidation of *N*-Hydroxyl-L-Arginine by nitric oxide synthase. *J Med Chem* 1993;**36**:2666–70.

33. Donzelli S, Espey MG, Flores-Santana W, Switzer CH, Yeh GC, Huang J, et al. Generation of nitroxyl by heme protein-mediated peroxidation of hydroxylamine but not *N*-Hydroxy-L-Arginine. *Free Radic Biol Med* 2008;**45**:578–84.
34. Donzelli S, Espey MG, Thomas DD, Mancardi D, Tocchetti CG, Ridnour LA, et al. Discriminating Formation of HNO from other reactive nitrogen oxide species. *Free Radic Biol Med* 2006;**40**:1056–66.
35. Fukuto JM, Wallace GC, Hszieh R, Chaudhuri G. Chemical oxidation of *N*-Hydroxyguanidine compounds. Release of nitric oxide, nitroxyl and possible relationship to the mechanism of biological nitric oxide generation. *Biochem Pharmacol* 1992; **43**:607–13.
36. Reisz JA, Bechtold E, King SB. Oxidative heme protein-mediated nitroxyl (HNO) generation. *Dalton Trans* 2010;**39**:5203–12.
37. Codd R. Traversing the coordination chemistry and chemical biology of hydroxamic acids. *Coord Chem Rev* 2008;**252**:1387–408.
38. Grant S, Easley C, Kirkpatrick P. Vorinostat. *Nat Rev Drug Discov* 2007;**6**:21–2.
39. Huang J, Sommers EM, Kim-Shapiro DB, King SB. Horseradish peroxidase catalyzed nitric oxide formation from hydroxyurea. *J Am Chem Soc* 2002;**124**:3473–80.
40. Huang J, Kim-Shapiro DB, King SB. Catalase-mediated nitric oxide formation from hydroxyurea. *J Med Chem* 2004;**47**:3495–501.
41. Samuni Y, Flores-Santana W, Krishna MC, Mitchell JB, Wink DA. The inhibitors of histone deacetylase suberoylanilide hydroxamate and trichostatin A release nitric oxide upon oxidation. *Free Radic Biol Med* 2009;**47**:419–23.
42. Adak S, Wang Q, Stuehr DJ. Arginine conversion to nitroxide by tetrahydrobiopterin-free neuronal nitric-oxide synthase: implications for mechanism. *J Biol Chem* 2000;**275**: 33554–61.
43. Schmidt HH, Hofmann H, Schindler U, Shutenko ZS, Cunningham DD, Feelisch M. No NO from NO synthase. *Proc Natl Acad Sci* 1996;**93**:14492–7.
44. Ishimura Y, Gao YT, Panda SP, Roman LJ, Masters BSS, Weintraub ST. Detection of nitrous oxide in the neuronal nitric oxide synthase reaction by gas chromatography-mass spectrometry. *Biochem Biophys Res Commun* 2005;**338**:543–9.
45. Clague MJ, Wishnok JS, Marletta MA. Formation of *N*-Cyanoornithine from *N*-Hydroxy-L-Arginine and hydrogen peroxide by neuronal nitric oxide synthase: implications for mechanism. *Biochemistry* 1997;**36**:14465–73.
46. Hobbs AJ, Fukuto JM, Ignarro LJ. Formation of free nitric oxide from L-Arginine by nitric oxide synthase: direct enhancement of generation by superoxide dismutase. *Proc Natl Acad Sci* 1994;**91**:10992–6.
47. Woodward JJ, Nejatyjahromy Y, Britt RD, Marletta MA. Pterin-centered radical as a mechanistic probe of the second step of nitric oxide synthase. *J Am Chem Soc* 2010;**132**: 5105–13.
48. Stoll S, Nejatyjahromy Y, Woodward JJ, Ozarowski A, Marletta MA, Britt RD. Nitric oxide synthase stabilizes the tetrahydrobiopterin cofactor radical by controlling its protonation state. *J Am Chem Soc* 2010;**132**:11812–23.
49. Doctorovich F, Bikiel D, Pellegrino J, Suárez SA, Larsen A, Martí MA. Nitroxyl (azanone) trapping by metalloporphyrins. *Coord Chem Rev* 2011;**255**:2764–84.
50. Yoo J, Fukuto JM. Oxidation of *N*-Hydroxyguanidine by nitric oxide and the possible generation of vasoactive species. *Biochem Pharmacol* 1995;**50**:1995–2000.
51. Nicholls SJ, Hazen SL. Myeloperoxidase and cardiovascular disease. *Arterioscler Thromb Vasc Biol* 2005;**25**:1102–11.

52. Reichlin T, Socrates T, Egli P, Potocki M, Breidthardt T, Arenja N, et al. Use of myeloperoxidase for risk stratification in acute heart failure. *Clin Chem* 2010;**56**:944–51.
53. Cook NL, Viola HM, Sharov VS, Hool LC, Schöneich C, Davies MJ. Myeloperoxidase-derived oxidants inhibit sarco/endoplasmic reticulum Ca^{2+}-ATPase activity and perturb Ca^{2+} homeostasis in human coronary artery endothelial cells. *Free Radic Biol Med* 2012;**52**:951–61.
54. Aruoma OI, Halliwell B. Action of hypochlorous acid on the antioxidant protective enzymes superoxide dismutase, catalase and glutathione peroxidase. *Biochem J* 1987; **248**:973–6.
55. Weiss SJ, Klein R, Slivka A, Wei M. Chlorination of taurine by human neutrophils. Evidence for hypochlorous acid generation. *J Clin Invest* 1982;**70**:598–607.
56. Agarwall R, Mohan RR, Ahmad N, Mukhtar H. Protection against malignant conversion in SENCAR mouse skin by all trans retinoic acid: inhibition of the RAS p21—processing enzyme farnesyltransferase and HA-RAS p21 membrane localization. *Mol Carcionog* 1996;**17**:13–22.
57. Hecker M, Schott C, Bucher B, Busse R, Stoclet J. Increase in Serum *N*-Hydroxy-L-Arginine in rats treated with bacterial lipopolysaccharide. *Eur J Pharmacol* 1995;**275**: 1–3.
58. Garlichs CD, Beyer J, Zhang H, Schmeisser A, Plötze K, Mügge A, et al. Decreased plasma concentrations of L-Hydroxy-Arginine as a marker of reduced NO formation in patients with combined cardiovascular risk factors. *J Lab Clin Med* 2000;**135**:419–25.
59. Hazen SL, Hsu FF, d'Avignon A, Heinecke JW. Human neutrophils employ myeloperoxidase to convert α-amino acids to a battery of reactive aldehydes: a pathway for aldehyde generation at sites of inflammation. *Biochemistry* 1998;**37**:6864–73.
60. Hazen SL, d'Avignon A, Anderson MM, Hsu FF, Heinecke JW. Human neutrophils employ the myeloperoxidase-hydrogen peroxide-chloride system to oxidize α-amino acids to a family of reactive aldehydes. *J Biol Chem* 1998;**273**:4997–5005.
61. Cho JY, Dutton A, Miller T, Houk KN, Fukuto JM. Oxidation of *N*-Hydroxyguanidines by copper(II): model systems for elucidating the physiological chemistry of the nitric oxide biosynthetic intermediate *N*-Hydroxyl-L-Arginine. *Arch Biochem Biophys* 2003; **417**:65–76.
62. Filipovic MR, Miljkovic JL, Nauser T, Royzen M, Klos K, Shubina T, et al. Chemical characterization of the smallest S-Nitrosothiol, HSNO; cellular cross-talk of H_2S and *S*-Nitrosothiols. *J Am Chem Soc* 2012;**134**:12016–27.
63. Eberhardt M, Dux M, Namer B, Miljkovic J, Cordasic N, Will C, et al. Cooperatively regulate vascular tone by activating a neuroendocrine HNO-TRPA1-CGRP signalling pathway. *Nat Commun* 2014;**5**:4381.

Spectroscopic NMR Characterizations of HNO Adducts of Ferrous Heme Proteins

14

M.R. Kumar and P.J. Farmer
Baylor University, Waco, TX, United States

Chapter Outline

14.1 Introduction

The small molecule HNO, variously termed "nitroxyl," "azanone," or "nitrosyl hydride," is the one-electron reduced and protonated form of nitric oxide. Many of the properties of this simple species have been previously misunderstood. For instance, in the last decade the oxidation potential for nitroxyl (as its protonated form, HNO) was revised from $+300$ to < -500 mV,[1] and likewise its pKa has been revised from 4.7 to >11.5.[2] Thus, nitroxyl is a powerful reductant that exists in aqueous solution almost exclusively in its protonated form.[3] A complication in studying HNO reactivity is that it cannot be purified or accumulated due its rapid dimerization producing N_2O between pH 2 to 11, Eq. (14.1).[4] Therefore, HNO must be generated in situ from precursor compounds that release it upon decomposition. The most widely used precursors are Angeli's Salt ($Na_2N_2O_3$), AS, which decomposes to give HNO at pH 4–8, Eq. (14.2),[5] and alkylsulfohydroxamic acids, e.g., Piloty's acid (PA), which releases HNO upon deprotonation at typically higher pH, Eq. (14.3).[6]

$$2HNO \rightarrow [H_2N_2O_2] \rightarrow N_2O_{(g)} + H_2O \tag{14.1}$$

$$HN_2O_3^- \rightleftarrows HNO + NO_2^- \text{ pH 4 to 8} \tag{14.2}$$

The Chemistry and Biology of Nitroxyl (HNO). DOI: http://dx.doi.org/10.1016/B978-0-12-800934-5.00014-1

$$RSO_2NHO^- \rightleftarrows RSO_2^- + HNO \quad >pH8 \tag{14.3}$$

$$NO\text{-}Fe^{II}Mb + X + H^+ \rightarrow HNO\text{-}Fe^{II}Mb + X^+ \quad X = Cr(II) \tag{14.4}$$

$$Mb\text{-}Fe^{II} + HNO \rightarrow HNO\text{-}Fe^{II}Mb \tag{14.5}$$

Although rare, a number of well-characterized complexes of HNO with transition metals have been identified by the groups of Hillhouse,[7] Sellman,[8] Caulton,[9] Richter-Addo,[10] and Olabe.[11] Several new complexes with anionic nitroxyl adducts have been recently characterized by Doctorovich[12] and Harrop[13] but this review will focus on the protonated HNO adducts which are low spin diamagnetic complexes.[14] Of the published routes to HNO-metal complexes, only the ferrous globin heme proteins directly trap HNO in solution to form an identifiable HNO-metal complexes.[15,16] We have reported extensively on the HNO adduct of myoglobin, HNO-Mb, including its synthesis[17,18] and characterization by ^{1}HNMR,[19] resonance Raman, and X-ray absorption spectroscopy.[20] HNO-Mb is quite stable once purified, in the absence of air and light.

There are two practical synthetic routes to HNO-Fe^{II} heme protein adducts; bulk reduction of NO-Fe^{II} adducts, Eq. (14.4). or trapping of HNO by the Fe^{II} species, e.g., deoxy-Mb, Mb-Fe^{II} in Eq. (14.5). Neither method is quantitative in producing the HNO adduct. Bulk reduction gives the highest yield, but the stoichiometry must be controlled to limit over-reduction. To this end, we have used a series of viologens that allow fine control of potential and concentration of reductant.[21] The radical anion of dimethyltrimethylene dipyridinium (DTDP, −670 mV), Fig. 14.1, is preferred for making high purity HNO-Mb.

As illustrated in Fig. 14.2, the nitroxyl N—H peak at c.15 ppm is in a relatively clean region of the ^{1}H NMR, and a direct proof of HNO-Fe^{II}Mb adduct formation. Comparable nitroxyl N—H resonances in small molecule HNO adducts are seen between 20 and 25 ppm, Table 14.1. Due to the strong ring current of the heme porphyrin, the effective chemical shifts of nuclei close to the center of the heme ring like the nitrosyl hydride are shifted upfield (lower ppm). A comparable shift is seen for the N—H resonance in the metalloporphyrin HNO-RuTPP adduct from Richter-Addo's group, at 13.64 ppm.[11] A similar upfield shift affects alkyl groups within the distal pocket to the region below 0 ppm. For HNO-Mb, the $\gamma 1$ and $\gamma 2$ methyl proton resonances of isovaline appear at −2.7 and −1.8 ppm; without the ring current these peaks are typically observed between 1 and 3 ppm. These fingerprint proton resonances of wild-type horse heart HNO-Mb are unaffected by paramagnetic impurities such as met-, deoxy-, or NO-Mb, and can be detected in 10 μM solutions with a signal to noise ratio c.18. Thus these alkyl heme pocket resonances are utilized in our lab as diagnostic of HNO adduct formation. Conversely, nuclei at the edge of the heme ring such as the meso-protons on the porphyrin ring experience a downfield shift to higher ppm; the meso C—H are comparable to vinylic or aromatic species which have resonances from 6 to 9 ppm, but are seen in diamagnetic porphyrin complexes between 10 and 12 ppm. Thus the strong heme ring current results in several distinctive regions of the ^{1}H NMR spectra of HNO heme

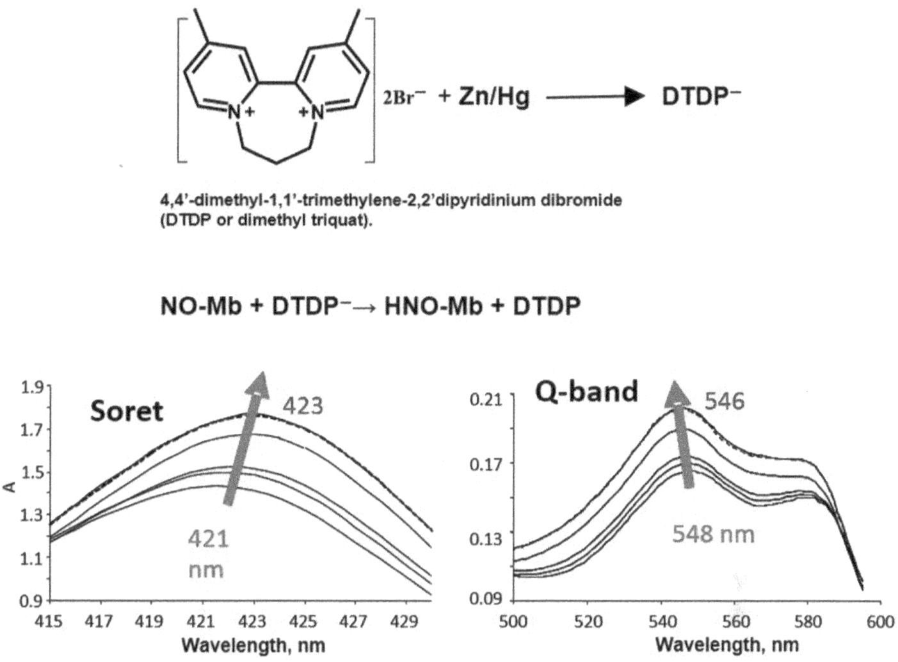

Figure 14.1 Top: reaction sequence for the DTDP method of HNO-Mb generation. Bottom: characteristic changes in spectra during the reaction.

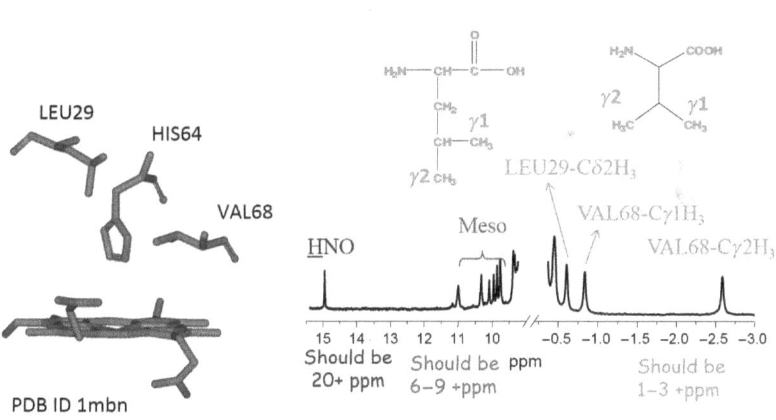

Figure 14.2 Left: crystal structure of deoxy-Mb. Right: ^{1}H NMR of HNO-Mb showing three characteristic regions, as described in the text.

protein adducts: the nitroxyl N—H region, the meso C—H region, and the distal pocket alkyl C—H region.

These diagnostic resonances are best demonstrated by comparing other stable HNO adducts of oxygen-binding globins such as clam hemoglobin, lpHb, from Lucinda

Table 14.1 **^{1}H NMR resonances for HNO-metal complexes**

HNO complex	δ, ppm
$Ru(H)COClL_2$	20.9
$OsCOCl_2L_2$	21.2
$Ir(H)Cl_2L_2$	22.75
$Re(CO)_2ClL_2$	22.66
$Os(CO)_2BrL_2$	20.7
$Re(CO)_3L_2$	21.66
$[Fe(CN)_5]^{3-}$	20.02
RuTPP(nMeIm)	13.64
Horse heart Mb	14.80

pectinata[22] and leghemoglobin, legHb, a well-studied globin from legumes.[23] HNO adducts of a series of sperm whale Mb mutants provided by Prof. John Olson are also provided. As seen in Fig. 14.3, the ^{1}H NMR spectra of these samples all show peaks c.15 ppm due to the HNO ligand. Of these, the HNO adduct of legHb is most illuminating. This H_2S-binding globin is found in the gills of a symbiote-harboring clam Lucina pectinata; the distal pocket above the heme is quite hydrophobic, with two Phe and a Gln residue at the positions of Leu29, Val68, and His64 in horse heart myoglobin. As a result of this unusual environment, the nitroxyl peak for HNO-lpHb is shifted upfield to 15.53 ppm, and the lack of alkyl residues close to the heme in the distal pocket is evident in the region below 0 ppm. An interesting comparison is obtained by examining the spectra of the HNO adduct of a triple mutant of sperm whale myoglobin, Phe29Gln64Phe68,[24] which mimics the distal pocket of the lpHb protein. Again, no peak is observed in the heme pocket region below 0 ppm, but the nitroxyl peak at 15.03 ppm is shifted downfield, but not as greatly as in the native l pHb. The spectra for the mutant myoglobins in Fig. 14.3 were obtained on low concentration samples, 10–100 μM, generated by HNO trapping, with spectra accumulation times of less than 1 h. As shown, the characteristic ring-current shifted signals are seen in the diagnostic regions even for these nonoptimal samples.

14.2 Solution structure determination

The unique characteristics of the NMR spectra of HNO-Mb allowed determination of its distal pocket structure.[19] In collaboration with Gerd La Mar at UC Davis, ^{1}H NMR, NOESY, and TOCSY data were obtained on HNO adduct of horse heart Mb and compared with that of the CO adduct. The 2D NOESY map, Fig. 14.4, showed over twenty NOE connectivities with the nitroxyl absorbance, indicative protons in the distal heme pocket. NOESY spectra were used to verify the spatial orientation of residues with respect to the heme and TOCSY spectra verified assignments of protons belonging to specific residues.

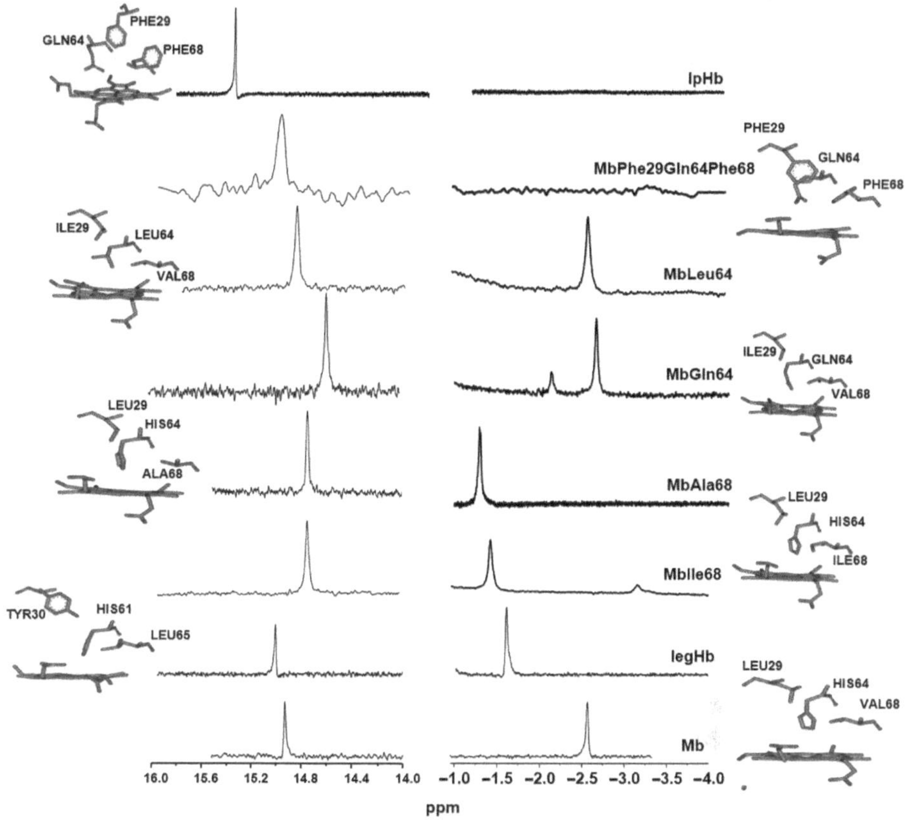

Figure 14.3 ^{1}H NMR spectra of HNO adducts of globins and myoglobin mutants (0.2–0.5 mM) in 50 mM phosphate buffer at pH 7. The samples were prepared by reacting deoxy states in carbonate buffer at pH 9.4 with fourfold excess PA for 40 min followed by purification on a G-25 column, preequilibrated with iP buffer at pH 7.

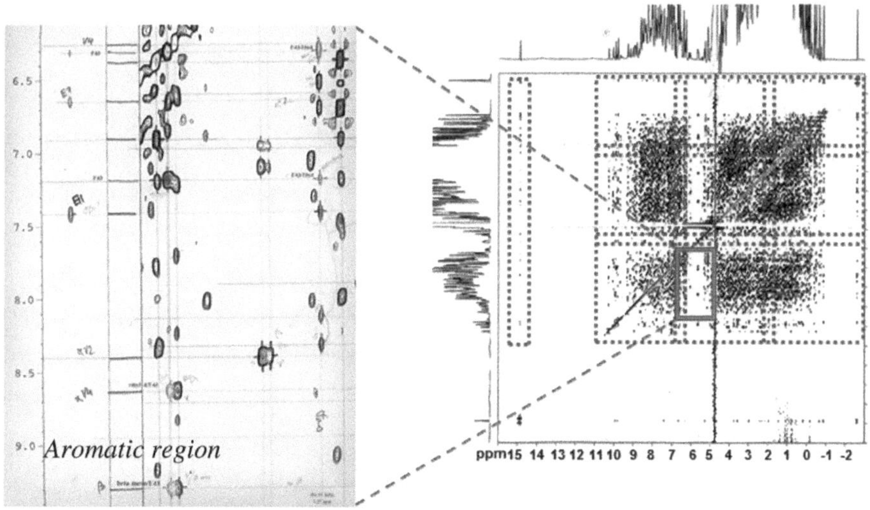

Figure 14.4 2D NOESY plot from 16 to −3 ppm of HNO-Mb, blue (black in print versions) box at left and bottom contain cross peaks of the nitroxyl N-H; inset on left shows identified cross peaks in the aromatic region due to the distal Phe43.

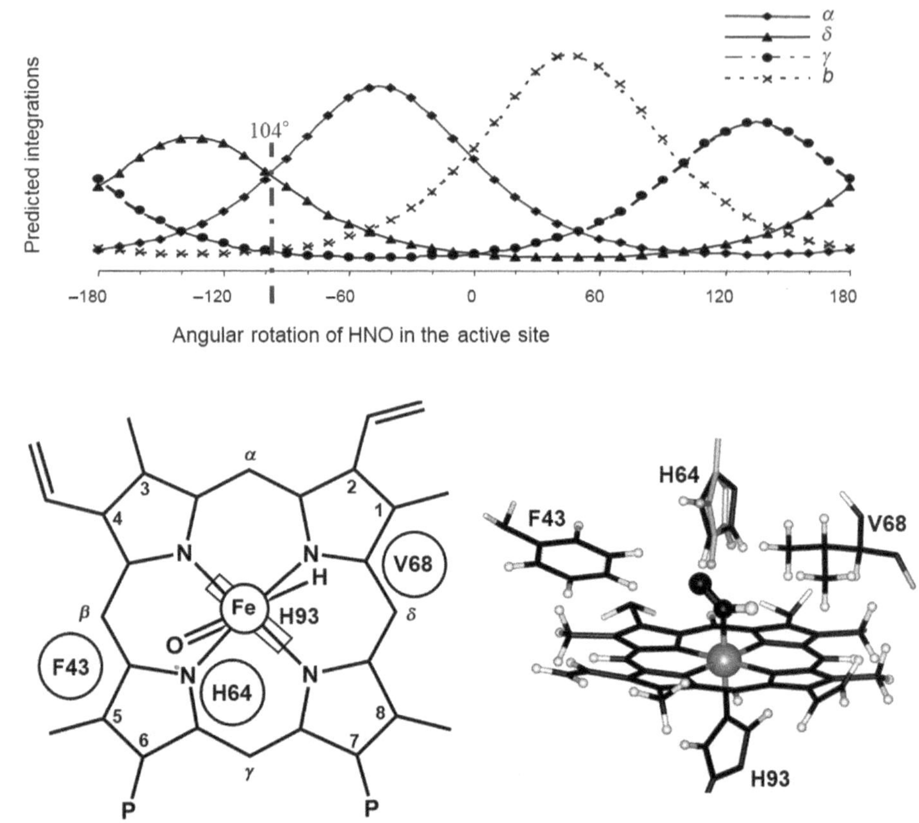

Figure 14.5 The orientation of the HNO ligand in Mb adduct. Left: axial overview of HNO orientation to the heme in HNO-Mb; circles indicate the position of particular protein residues. Right: stick model of HNO-Mb, showing nonbonding interactions with Val68 and His64.

The spatial orientation of the HNO ligand was determined by the cross peak intensities to heme meso-protons, and further supported by NOE intensities to other residues. A single orientation of the HNO was determined, c.$-104°$ to the N—Fe—N vector between α and β meso protons, Fig. 14.5. This orientation situates the HNO almost perpendicular to proximal His93 (at ~0°), suggestive of a π-back-bonding competition between the ligands. A computer model of HNO-Mb was generated by replacing CO with HNO in the crystal structure of CO-Mb and by adjusting position of residues and nitroxyl based on any differences in the position or the NOEs of the two adducts. The major difference between the residues of the two Mb adducts was movement of the distal His64, based on NOESY cross peak and chemical shift changes, by c.20° rotation to move within hydrogen bonding distance of the nitroxyl oxygen, Fig. 14.5.

14.2.1 ^{15}N labeling studies

The strong interaction of the H and N atoms in the nitroxyl adducts provides a route for enhancement of the NMR signals by spin-exchange. In ^{15}N-labeled samples of an HNO adduct, the nitroxyl N—H peak is split into a doublet due to the $S = ½$ spin of ^{15}N; syntheses of these ^{15}N-labeled species are relatively straight forward.[15,17] The splitting of the hydride resonance in such an ^{15}N-labeled sample is demonstrated in Fig. 14.6, for human hemoglobin, HNO-Hb; the natural abundance $H^{14}NO$-Mb adduct of has two hydride peaks at 14.6 and 14.8 ppm that split to four peaks for $H^{15}NO$-Hb samples.[16] Because of the spectral width and poor receptivity, the ^{15}N NMR analysis of $H^{15}NO$-Mb peak was first observed at 549.1 ppm (vs $^{15}NH_4^+$) by slow broad pulse spin-tickling experiments.[24] Subsequently, ^{15}N-^1H HSQC spectra allow determination of ^{15}N chemical shifts quickly (c.30 min) and at reasonable concentrations of a number of proteins (~500 μM protein), Table 14.2.

The combination of ^1H and ^{15}N NMR as well as vibrational data has generated interest from the computational community. The Zhang group has utilized these experimental data to test computational methods in modeling the active site structure of HNO-Mb.[25,26] In particular, their results suggest that a dual hydrogen-bond mode best matches the spectral data, Scheme 14.1. This suggests that, in addition to the previously described hydrogen bonding of the nitroxyl oxygen to the distal His68, a water molecule within the distal pocket interacts with the nitroxyl N—H. This possibility was subsequent investigated by the Spiro group, who suggested that the effect of H-bonding in metal nitrosyl complexes is to weaken both N—O and N—Fe bonds, thus promoting reduction to the HNO adduct, a putative intermediate in NO-reducing enzymes such as nitrite reductase.[27]

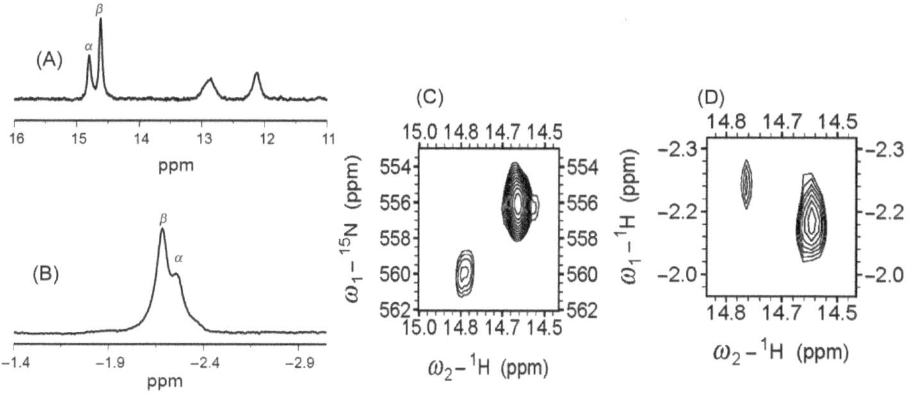

Figure 14.6 NMR of HNO adduct of human hemoglobin in iP buffer 7 (A) In the HNO region; (B) in the valine region; (C) ^1H-^{15}N HSQC spectrum of $H^{15}NO$ adduct; and (D) 2D ^1H-^1H NOESY spectrum showing the interaction between the nitroxyl and valine methyl protons.

Table 14.2 ^{1}H and ^{15}N NMR chemical shifts and coupling constants of $H^{15}NO$ adducts of globins

HNO-Fe^{II}	δ, ppm	
	^{1}H	$^{15}N^{a}$
WT Mb	14.80	549.1
Mb68A	14.74	
Mb68I	14.71	
Mb64Q	14.59	
Mb64L	14.84	
Mb29F64Q68F	15.03	
hHb	14.63	560.0
	14.80	556.2
lpHb	15.53	
legHb	15.00	553.0

[a]From ^{1}H-^{15}N HSQC experiments, ^{15}N referenced to $^{15}NH_4Cl$.

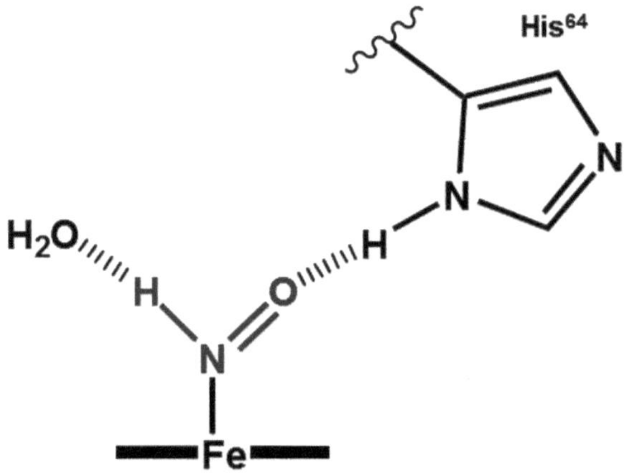

Scheme 14.1

14.2.2 H/D exchange in HNO-Mb

Deprotonated NO^- adducts of metallocenters have become of much interest, especially in regard to the nitroxyl intermediates of bacterial and fungal NO reductases.[28] The pKa of free HNO in solution was first estimated at 4.7, but subsequently dramatically revised to 11.6. Binding of the HNO to a cationic metal ion would suggest that HNO-Mb should have a lower pK_a, but deprotonation might lead to a spin change that would substantially affect pK_a. Similarly, the influence of an active site residues and hydrogen bonding to the HNO adduct within distal pocket likely has great effect.

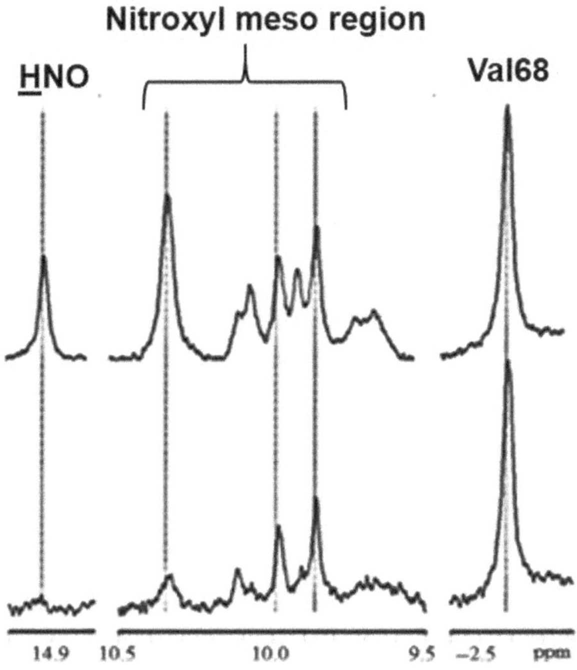

Figure 14.7 The ^{1}H NMR assignment of NMR spectra for HNO-Mb (top) and DNO-Mb (bottom) samples with small NO-Mb and Mb-Fe^{II} contaminations. The nitroxyl signal (left), the meso region (center), and the Val68 methyl (right) are shown.

For HNO-Mb, no change in the nitrosyl hydride resonance in the ^{1}H NMR spectrum is seen from pH 6.5−10, suggesting the pK_a to be well above the range of the well-known acid-alkaline transition for Fe^{III}Mb. NMR characterizations of HNO-Mb at higher pH is prohibited due to protein aggregation, but dilute samples at high pH do show characterizable changes in UV-vis spectra above pH 11, suggesting a pK_a well above 10.[29] On the other hand, H/D exchange in HNO-Mb is apparent by integration of the nitroxyl peak at 14.8 ppm (1 H) against that of the methyl group of Val68 (3 H) at −2.5 ppm in the ^{1}H-NMR spectra of samples containing >90% D_2O. Fig. 14.7 shows the full exchange of the nitroxyl H for D in a crude sample with signals from NO-Mb and Mb-Fe^{II} impurities also present, which also display exchangeable proton signals. After exchange, only the characteristic diamagnetic peaks remaining are due to the valine methyl c. −2.6 ppm and two meso peaks at ~9.75 and 10.0 ppm. The rate H/D exchange is quite slow at neutral pH but can be followed by time course ^{1}H NMR at more alkaline conditions. At pH 8 the $t_{1/2}$ of the exchange was ~5.5 h, but increased dramatically from pH 9.0, $t_{1/2}$~16 min, to pH 10, $t_{1/2}$~9 min. The rate of exchange is thus likely controlled by hydrogen bonding interactions within the distal pocket.

Analogous effects have been seen in small molecule complexes of HNO. It has long been suggested that such HNO-metal complexes are unstable in solution, due to coupling reactions that may generate N_2O or H_2. One exception is the highly reduced nitroprusside complex reported by Olabe, $[Fe(CN)_5(HNO)]^{3-}$, which is long-lived at low pH, but unstable at pH 10.[11] ^1H NMR of this species demonstrated a characteristic nitroxyl N–H resonance at 20.0 ppm, which split into two peaks for the ^{15}N-labeled species at pH 6.5, but the signal disappeared at pH above 8.5. This was initially interpreted as due to a pKa of c.7.7 for this species. This initial report has been reexamined by Wu,[30] using extensive labeling and multinuclear (^1H, ^{15}N, ^{17}O) NMR. The revised pKa of the metal-bound HNO was found to be greater than 11, using a unique ^{17}O signal as indication. The previously observed loss of the N–H resonance c. pH 7.7 was then characterized as a rapid hydrogen exchange with water, analogous to that previously described for HNO-Mb, which can be catalyzed by both acid and base. By comparison, the rate of hydrogen exchange for the Fe^{II}-bound HNO in the interior of globin proteins is a factor of 10^6 slower than in the reduced nitroprusside complex.

14.2.3 HNO trapping studies

Like the rate of H/D exchange, the trapping of free HNO by ferrous globins can be followed by time course ^1H NMR using slow-release HNO precursors such as methylsulfohydroxamic acid, MSHA, which has a half-life of ~160 min at pH 10.[16] The slow rate of reaction allows spectra to be binned over periods of 15 to 30 min, sufficient time to obtain spectra with good signal to noise resolution. As shown in Fig. 14.8, the time course spectra of such a reaction shows the expected growth in the nitroxyl N–H and heme pocket alkyl peak, as well as characteristic changes in the meso proton region. This analysis provided the first estimate of the rate constant for the bimolecular of HNO trapping by deoxy-Mb, giving a lower limit of $8 \times 10^4\ M^{-1}s^{-1}$, a subsequent global fitting analysis revised this to $2.8 \times 10^5\ M^{-1}s^{-1}$. This report also demonstrated a complicating side reaction, in which the HNO-Mb reacts with additional HNO in solution, producing NO-Mb and presumably H_2NO. This side reactivity limits the purity of HNO globin adduct samples generated by this method, and the rate of the following reaction appears quite dependent on the nature of the globin. But ^1H NMR characterization of products of the trapping reaction represents the easiest and most reliable method to determine if an HNO adduct has formed.

Similar time course analysis of the HNO trapping reaction with tetrameric human Hb is also very informative.[16] ^1H NMR spectra of HNO-Hb samples generated by various methods display two nitrosyl hydride peaks due to HNO binding at the α and β subunits. In spectra of samples derived from HNO trapping reactions, these peaks have a persistent 3:1 ratio during the trapping reaction, suggesting a kinetic binding difference between the subunits. There is also a difference in relative rates of reduction of the two subunits by the DTDP reduction method. The sequential reduction of NO-Hb with concurrent EPR analysis of these samples was used to assign changes in the concentrations of corresponding nitrosyl adducts of α and β subunits. As illustrated at right Fig. 14.9, the peaks at 14.6 and 14.8 ppm were assigned to the β and α subunits, respectively.

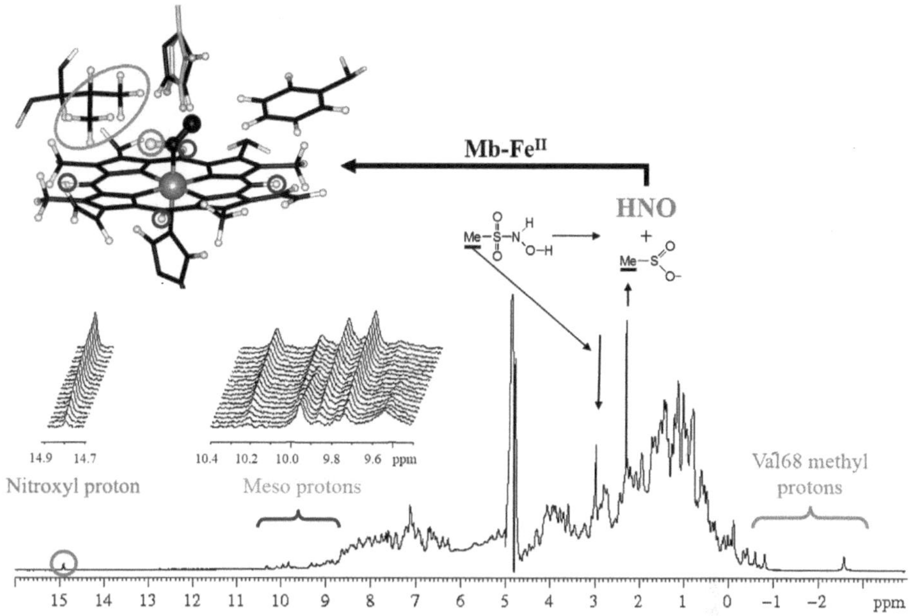

Figure 14.8 ^{1}H NMR spectra of anaerobic reaction mixture of MSHA and deoxy-Mb, labeled to identify the characteristic regions of the HNO-Mb spectra, with reaction sequence of MSHA decomposition illustrated above. Inset: time course spectra in the N-H and meso proton regions, color coded to the modeled distal pocket structure above left.

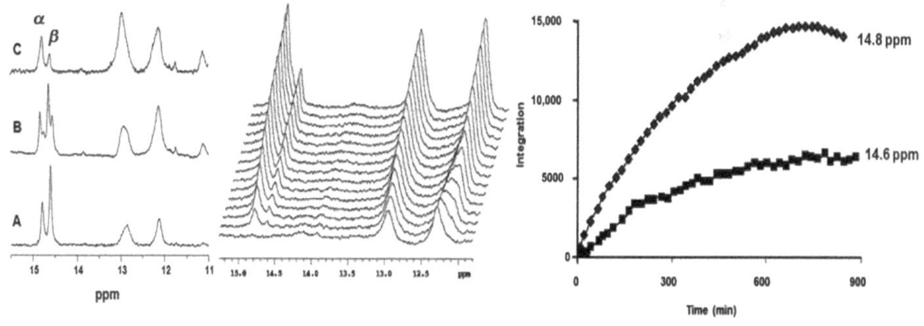

Figure 14.9 Left: ^{1}H NMR of HNO-Hb samples: (A) $H^{14}NO$-Hb by reduction of NO-Hb; (B) $H^{15}NO$-Hb by reduction of ^{15}NO-Hb; (C) $H^{14}NO$-Hb by trapping of HNO from PA decomposition at pH 9.4. Middle: ^{1}H NMR spectra of reaction of MSHA with deoxy-Hb. Right: peak height changes over the course of the reaction for N-H signals as labeled.

Overall, trapping of HNO by Hb-Fe^{II} gives yields of only ~20% HNO adduct, HNO-Hb, with the remainder the NO adduct. By comparison, yields of >80% HNO-Hb are obtained by the DTDP reduction of Hb-NO samples. The reason for the low trapping yield of HNO-Hb may be due to allosteric effects caused by HNO binding. In the ^{1}H NMR of HNO-Hb, left of Fig. 14.9, the two broad resonances between 12 and 13 ppm are due to protons at the α/β interface (His122 and His103) which form strong H-bonds between subunits; changes in these ^{1}H NMR signals have been linked to transitions between tensed and relaxed allosteric states. Examination of the time course spectra show that similar changes are observable during HNO trapping by deoxy-Hb, Fig. 14.9. Completion of the shift in the His103 band coincides with the maximum HNO signal, which may indicate that an allosteric structural change upon binding the first HNO inhibits further HNO binding.

14.3 Characterization of isoform mixtures

The characteristic HNO adduct resonances in the ^{1}H NMR may also be used to describe and quantitate mixtures of isoforms. In one example, we reported the ^{1}H NMR spectra of HNO adducts formed from samples of native legHb isoform mixtures isolated from soybeans.[16] As mentioned previously, this globin is generated to protect a symbiotic N_2-fixing bacteria which is sensitive to O_2; legHb has a very high O_2 affinity, thus maintains a pseudo anaerobic around the symbiote bacteria. The native legHb is isolated from red root nodules of soybean plants, and several isoforms are expressed over the life of a plant. Nanospray LC-Ms of the legHb sample, Fig. 14.10, showed it to be a mixture of 3 proteins in c.4:1:1 ratio, whose masses correlated to know isoforms of legHb. While the peaks in the nitroxyl N—H region were largely overlapped, in the distal pocket alkyl C—H region three distinct peaks were apparent. Gaussian fitting of the combined peak shape gave integration values of 2.7:1.8:8.2, in rough agreement with the LC-Ms ratios.

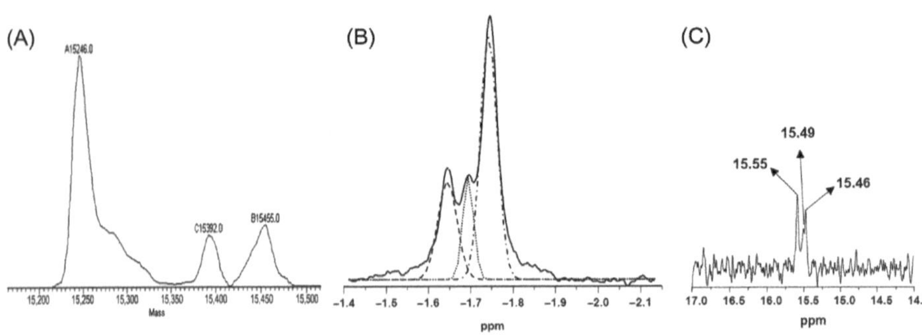

Figure 14.10 (A) An LC-MS trace of mixture of native legHb isoforms showing three distinct species. (B): ^{1}H NMR spectra in the distal pocket region showing three peaks due to Leu65 methyl groups in the respective HNO-legHb isoforms. (C) ^{1}H NMR of isoform mixture of lbHb in the HNO region.

Similarly, the mollusc *Lucina pectinata* expresses a mixture of 3 hemoglobin isoforms, termed HbI and HbII/III. Only HbI is involved in H_2S transport, as described previously, the other isoforms HbII/III are involved in O_2 metabolism despite having similar tertiary structures. A small sample of native lpHb mixture was obtained from Dr. Agustín Díaz from U. Puerto Rico, Rio Piedras, and its reaction with PA at pH 9.4 did indeed show the presence of three different species, as shown in Fig. 14.10, with HNO adduct nitroxyl resonances close to that of the expressed lpHb of 15.53 ppm.

14.3.1 Isomeric *HNO-Hb

The reaction of $NaBH_4$ and $NaNO_2$ produces HNO in situ, which can be trapped by deoxy-Mb as well as the other known HNO traps such as tetracyanatonickelate or ferric *N*-methyl D-glucaminedithiocarbamate.[31] While the overall yield of HNO-Mb by this convenient method is quite high, it yields an additional isomer, *HNO-Mb, in an apparent 1:1 mixture. This new isomeric form, *HNO-Mb, has slightly upfield shifted resonances in both ^{1}H and ^{15}N NMR spectra compared to the native HNO-Mb. As seen in Fig. 14.11, two distinct nitrosyl hydride peaks are observed, as well as splitting of several meso-heme peaks between 9 and 11 ppm and the proline methyl's c. −2 ppm. Heating of such a sample to 37°C for 2 days does not change the spectra, i.e., no equilibration occurs, nor does heating to 60°C over shorter periods. ESI-Ms analysis of isomer mixtures shows no evidence of protein alterations formed. But oxidation of HNO-Mb/*HNO-Mb mixtures to metMb, followed by reduction and trapping of HNO obtains an HNO-Mb spectra with a single hydride ^{1}H NMR peak at the expected chemical shift. ^{15}N-substituted substitution yields four peaks in the ^{1}H NMR and two in the HSQC spectra Fig. 14.11, with J_{N-H} larger that the isomeric splitting.

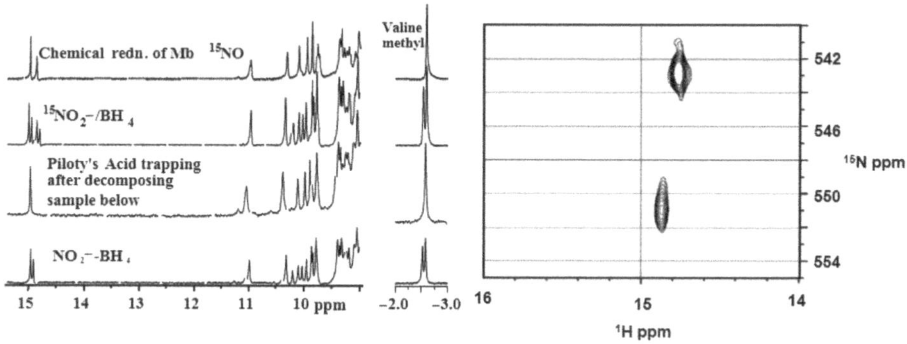

Figure 14.11 Left: ^{1}H NMR of HNO-Mb made by various routes, as described in the text. Middle: ^{1}H^{15}N HSQC (heteronuclear single quantum coherence) of isomeric mixture of *HNO-Mb showing the two distinct proton and nitrogen signals. Right: ^{1}H NMR spectra of HNO adducts of monomeric globins.

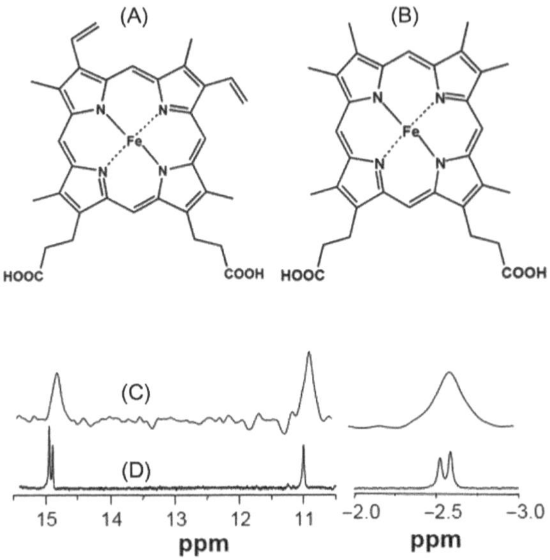

Figure 14.12 At top are structures of (A) hemin, iron protoporphyrin IX, (B) iron-2,4-dimethyldeuteroporphyrin. At bottom are ^{1}H NMR spectra of the HNO adducts of (C) WT Mb and (D) the iron-2,4-dimethyl deuteroporphyrin reconstituted Mb, both prepared by the nitrite-borohydride method.

A doubling of 1H NMR resonances had been previously observed in samples of reconstituted ferric metMb adducts and was attributed to heme-flip isomers, i.e., structural disorder due to a flipping of the native heme within the protein pocket that produces two species with different heme-orientations. To test this hypothesis, a sample of apomyoglobin was reconstituted with the iron complex of symmetrical 2,4-dimethyldeuteroporphyrin, and its HNO adduct was prepared by the nitrite/borohydride method, Fig. 14.12. The resulting 1H NMR spectrum of a low-concentration sample solution yielded broad but single nitroxyl N—H and valine C—H resonances, consistent with the heme orientational hypothesis.

An alternative explanation was proposed by the Zhang group, who modeled the observed ^{1}H NMR data computationally, and suggested the second species is an HNO rotational isomer, Fig. 14.13, in which the Fe-bound HNO is rotated c.170° from the original HNO-Mb isomer. The proposed dual H-bonding in *HNO-Mb invokes a different distal His conformation, which allows for it to act as a Lewis base in bonding to the nitroxyl N—H. This model predicts the slight upfield shift of *HNO-Mb NO ^{1}H and ^{15}N NMR resonances compared the original HNO-Mb isomer, and suggest that the N—H bond is shorter in *HNO-Mb, due to stronger hydrogen bonding interaction with the H_2O molecule. Further ^{1}H and ^{15}N NMR studies are underway to better characterize the structural differences of these isomeric forms.

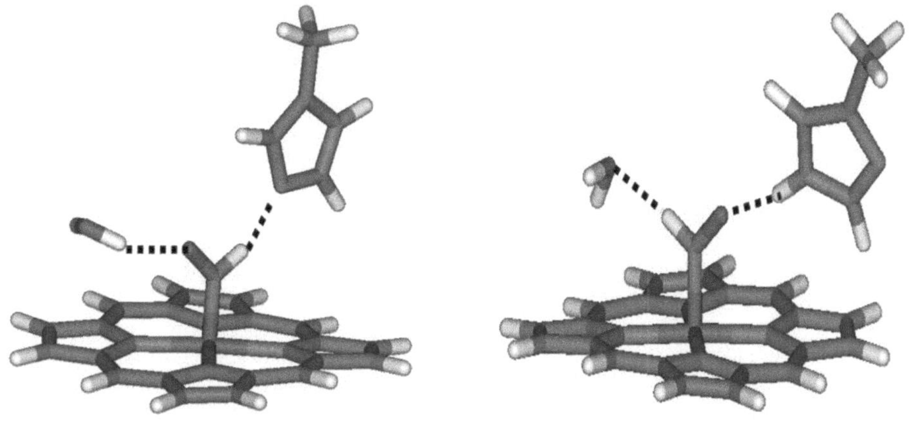

Figure 14.13 Illustrations of hydrogen bonding interactions of HNO with His64 and a water molecule in two different HNO orientations.

14.4 Conclusions

In this work, we have described the unique ^{1}H and ^{15}N NMR signals of HNO adducts of myoglobin and other ferrous heme proteins. As has been reported for other HNO metal ion complexes, HNO-Mb is diamagnetic with characteristic resonances upfield in ^{1}H NMR spectra due to the HNO ligand. For organometallic HNO complexes, the chemical shift of the HNO proton is at c.20 ppm; however, in HNO-Mb this proton signal is shifted upfield by c.5 ppm, due to the ring current effect of the porphyrin. A similar ring shift is observed for the γ methyl protons of valine 68 which lies above the heme in the distal pocket. These fingerprint proton resonances can be detected readily in low μM solutions, and thus allow differentiation of subunits and isoforms of proteins within mixtures. The combination of NMR characterizations have revealed several unique aspects of HNO-globin complexes, including H/D exchange kinetics and H-bonding within the distal pocket, and distinction of heme- or HNO-orientational isomers. Thus NMR characterization of HNO adducts provides a unique tool that is a sensitive to structural changes within the oxygen-binding cavity, and which may help define modes of oxygen binding and activation in other heme enzymes.

References

1. Bartberger MD, Liu W, Ford E, Miranda KM, Switzer C, Fukuto JM, et al. The reduction potential of nitric oxide (NO) and its importance to no biochemistry. *Proc Natl Acad Sci USA* 2002;**99**(17):10958–63.
2. Shafirovich V, Lymar SV. Nitroxyl and its anion in aqueous solutions: spin states, protic equilibria, and reactivities toward oxygen and nitric oxide. *Proc Natl Acad Sci USA* 2002;**99**(11):7340–5.
3. Lymar SV, Shafirovich V, Poskrebyshev GA. One-electron reduction of aqueous nitric oxide: a mechanistic revision. *Inorg Chem* 2005;**44**(15):5212–21.

4. Shafirovich V, Lymar SV. Spin-forbidden deprotonation of aqueous nitroxyl (HNO). *J Am Chem Soc* 2003;**125**(21):6547–52.
5. (a) Hughes MN, Cammack R. Synthesis, chemistry, and applications of nitroxyl ion releasers sodium trioxodinitrate or Angeli's salt and Piloty's acid. *Methods Enzymol* 1999;**301**:279–87.
   (b) Yi J, Namjou K, McCann P, Richter-Addo G. Simultaneous gas-phase detection of nitric oxide (NO) and nitrous oxide (N_2O) from the decomposition of Angeli's Salt ($Na_2N_2O_3$) at different pHs using tunable-diode laser absorption spectroscopy. *Am J Biomed Sci* 2009;38–46.
6. King SB, Nagasawa HT. Chemical approaches toward generation of nitroxyl. *Methods Enzymol* 1999;**301**:211–20.
7. (a) Melenkivitz R, Hillhouse GL. Synthesis, structure, and reactions of a nitroxyl complex of iridium(III), cis,trans-IrHC12(NH:O)(PPh3)2. *Chem Commun* 2002;660–1.
   (b) Southern JS, Green MT, Hillhouse GL, Guzei IA, Rheingold AL. Chemistry of coordinated nitroxyl. reagent-specific protonations of trans-$Re(CO)_2(NO)(PR_3)_2$ (R = Ph, Cy) that give the neutral nitroxyl complexes cis,trans-ReCl(CO)2(NH = O)$(PR3)_2$ or the cationic hydride complex [trans,trans-$ReH(CO)_2(NO)(PPh_3)_2{}^+$][$SO_3CF_3{}^-$]. *Inorg Chem* 2001;**40**:6039–46.
   (c) Southern JS, Hillhouse GL, Rheingold AL. Preparation of a nitroxyl (NH:O) complex of rhenium by selective oxidation of coordinated hydroxykamine. *J Am Chem Soc* 1997;**119**:12406–7.
   (d) Melenkivitz R, Southern JS, Hillhouse GL, Concolino TE, Liable-Sands LM, Rheingold AL. A new route to coordination complexes of nitroxyl (HN = O) via insertion reactions of nitrosonium triflate with transition-metal hydrides. *J Am Chem Soc* 2002;**124**:12068–9.
8. Sellmann D, Gottschalk-Gaudig T, Häussinger D, Heinemann FW, Hess BA. [Ru(HNO) ('py(bu)S4')], the first HNO complex resulting from hydride addition to a NO complex ('pybuS4'2-=2,6-Bis(2-Mercapto-3,5-Di-Tert-Butylphenylthio)dimethylpyridine(2−1)). *Chem Weinh Bergstr Ger* 2001;**7**(10):2099–103.
9. (a) Marchenko AV, Vedernikov AN, Dye DF, Pink M, Zaleski JM, Caulton KG, et al. An electron-excessive nitrosyl complex: reactivity of a ligand-centered radical leading to coordinated HNO. *Inorg Chem* 2002;**41**:4087–9.
   (b) Marchenko AV, et al. Reactivity of the hydrido/nitrosyl radical MHCl(NO)(CO) (PiPr3)2, M = Ru, Os. *Inorg Chem* 2004;(43):351–60.
10. Lee J, Richter-Addo GB. A nitrosyl hydride complex of a heme model [Ru(ttp)(HNO) (1-MeIm)] (ttp = tetratolylporphyrinato Dianion). *J Inorg Biochem* 2004;**98**(7):1247–50.
11. Montenegro AC, Amorebieta VT, Slep LD, Martin DF, Roncaroli F, Murgida DH, et al. Three redox states of nitrosyl: NO+, NO*, and NO-/HNO interconvert reversibly on the same pentacyanoferrate(II) platform. *Angew Chem Int Ed Engl* 2009;**48**:4213–16.
12. Pellegrino J, Bari SE, Bikiel DE, Doctorovich F. Successful stabilization of the elusive species {FeNO}8 in a heme model. *J Am Chem Soc* 2010;**132**(3):989–95.
13. (a) Patra AK, Dube KS, Sanders BC, Papaefthymiou GC, Conradie J, Ghosh A, et al. A thermally stable $\{FeNO\}^8$ complex: properties and biological reactivity of reduced MNO systems. *Chem Sci* 2012;**3**:364–9.
   (b) Sanders BC, et al. Synthesis, properties, and reactivity of a series of non-heme $\{FeNO\}^{7/8}$ complexes: implications for Fe-nitroxyl coordination. *J Inorg Biochem* 2013;**118**:115–27, 10230–10233.
   (c) Sanders BC, et al. Properties of $\{FeNO\}^8$ and $\{CoNO\}^9$ metal nitrosyls in relation to nitroxyl coordination chemistry. *Struct Bond* 2014;(160):57–88.

14. Farmer PJ, Sulc F. Coordination chemistry of the HNO ligand with hemes and synthetic coordination complexes. *J Inorg Biochem* 2005;**99**:166–84.
15. Sulc F, Immoos CE, Pervitsky D, Farmer PJ. Efficient trapping of HNO by deoxymyoglobin. *J Am Chem Soc* 2004;**126**(4):1096–101.
16. Kumar MR, Pervitsky D, Chen L, Poulos T, Kundu S, Hargrove MS, et al. Nitrosyl hydride (HNO) as an O_2 analogue: long-lived HNO adducts of ferrous globins. *Biochemistry (Mosc)* 2009;**48**(22):5018–25.
17. Lin R, Farmer PJ. The HNO adduct of myoglobin: synthesis and characterization. *J Am Chem Soc* 2000;**122**:2393–4, All Rights Reserved.
18. Kumar MR, Fukuto JM, Miranda KM, Farmer PJ. Reactions of HNO with heme proteins: new routes to HNO-heme complexes and insight into physiological effects. *Inorg Chem* 2010;**49**(14):6283–92.
19. Sulc F, Fleischer E, Farmer PJ, Ma D, La Mar GN. 1H NMR structure of the heme pocket of HNO-myoglobin. *J Biol Inorg Chem* 2003;**8**(3):348–52.
20. Immoos CE, Sulc F, Farmer PJ, Czarnecki K, Bocian DF, Levina A, et al. Bonding in HNO-myoglobin as characterized by X-ray absorption and resonance Raman spectroscopies. *J Am Chem Soc* 2005;**127**(3):814–15.
21. Guo M, Sulc F, Ribbe MW, Farmer PJ, Burgess BK. Direct assessment of the reduction potential of the [4Fe-4S](1+/0) couple of the Fe protein from *Azotobacter vinelandii*. *J Am Chem Soc* 2002;**124**(41):12100–1.
22. Navarro AM, Maldonado M, González-Lagoa J, López-Mejía R, López-Garriga J, Colón JL. Control of carbon monoxide binding states and dynamics in hemoglobin I of *Lucina pectinata* by nearby aromatic residues. *Inorganica Chim Acta* 1996;**243**(1–2):161–6.
23. Appleby CA. Leghemoglobin and rhizobium respiration. *Annu Rev Plant Physiol* 1984; **35**(1):443–78.
24. Nguyen BD, Zhao X, Vyas K, Mar GNL, Lile RA, Brucker EA, et al. Solution and crystal structures of a sperm whale myoglobin triple mutant that mimics the sulfide-binding hemoglobin from *Lucina pectinata*. *J Biol Chem* 1998;**273**(16):9517–26.
25. Ling Y, Mills C, Weber R, Yang L, Zhang Y. NMR, IR/Raman, and structural properties in HNO and RNO (R = alkyl and aryl) metalloporphyrins with implication for the HNO-myoglobin complex. *J Am Chem Soc* 2010;**132**(5):1583–91.
26. Yang L, Ling Y, Zhang Y. HNO binding in a heme protein: structures, spectroscopic properties, and stabilities. *J Am Chem Soc* 2011;**133**(35):13814–17.
27. Xu C, Spiro TG. Ambidentate H-bonding by heme-bound NO: structural and spectral effects of –O versus –N H-bonding. *J Biol Inorg Chem* 2008;**13**(4):613–21.
28. Speelman AL, Lehnert N. Heme versus non-heme iron-nitroxyl {FeN(H)O}8 complexes: electronic structure and biologically relevant reactivity. *Acc Chem Res* 2014;**47**(4): 1106–16.
29. Sulc F. HNO Adduct of deoxymyoglobin: structure, reactivity and biological importance of Mb-HNO, UCI: Irvine; 2006.
30. Gao Y, Toubaei A, Kong X, Wu G. Acidity and hydrogen exchange dynamics of iron(II)-bound nitroxyl in aqueous solution. *Angew Chem Int Ed Engl* 2014;**53**(43):11547–51.
31. (a) Bonner FT, Akhtar MJ. Formation of nitrosyltricyanonickelate ($NiNO(CN)_3^{2-}$) in a direct NO-displacement reaction. *Inorg Chem* 1981;**20**:3155–60.
    (b) Xia Y, Cardounel AJ, Vanin AF, Zweier JL. Electron paramagnetic resonance spectroscopy with *N*-methyl-D-glucamine dithiocarbamate iron complexes distinguishes nitric oxide and nitroxyl anion in a redox-dependent manner: applications in identifying nitrogen monoxide products from nitric oxide synthase. *Free Radic Biol Med* 2000;**29**(8):793–7.

Global Kinetic Analysis and Singular Value Decomposition Methods Applied to Complex Multicomponent Reactions of HNO

15

A.L. Zapata, M.R. Kumar and P.J. Farmer
Baylor University, Waco, TX, United States

Chapter Outline

15.1 Introduction

The physiological effects of nitric oxide (NO) and its redox congeners NO^+, nitrosonium, and NO^-, nitroxyl, have been largely attributed to reactivity with heme proteins and thiols.[1] The pharmacological activity of nitrosonium donors such as nitrite has garnered much interest, as they seem to play important, but different, biological roles from NO itself.[2] Likewise nitroxyl-releasing compounds have generated much interest in cardiovascular research for showing unique therapeutic effects on damaged hearts.[3] Such physiological investigations typically look for organismal effects

The Chemistry and Biology of Nitroxyl (HNO). DOI: http://dx.doi.org/10.1016/B978-0-12-800934-5.00015-3

that occur in response to treatment with NO*x* species, but typically cannot define fundamental reaction chemistries that underlie the observed behavior.

The biological targets for all three NO congeners are thought to be reactive thiols and ferric and ferrous hemes. Thus, an understanding the kinetics of individual reactions is essential to interpreting their individual effects within biological systems.[4] In a widely referenced review, the pharmaceutical effects of NO and HNO donor drugs were suggested to be orthogonal, with NO reacting mainly with ferrous hemes and HNO with ferric hemes, leading to distinctive cyclic guanosine monophosphate (cGMP) and cyclic adenosine monophosphate (cAMP) signaling cascades.[4] But the chemical basis for such a distinction is ambiguous and difficult to prove.

Chemical investigations of the reactivity of free HNO are inherently difficult and complex; nitroxyl has a short physiological lifetime due to its rapid reaction with oxygen[5] and thiols,[6] as well as a competitive dimerization reaction which limits its concentration in both aerobic and anaerobic solutions.[7] Since free nitroxyl is short-lived, precursor compounds like Angeli's salt (AS) and Piloty's acid (PA) are used as HNO sources. AS decomposes as its monoprotonated form, $HN_2O_3^-$, Scheme 15.1, to give HNO and NO_2^- with half-life of ca. 17 min from pH 4 to 8.[8,9] Similarly, PA decomposes via its the monoprotonated form to yield HNO and benzene sulfinate with a half-life of 33 min at pH 10.[10] Both decompositions rates are decreased with addition of by-products, nitrite or sulfinate, confirming the initial reversibility of these reactions.[11]

Our lab has long sought to define the kinetics of the reaction of HNO with deoxy-Mb (deoxy-myoglobin) which generates a unique HNO adduct, HNO-Mb.[12,13] Like the corresponding O_2 adduct, this species is diamagnetic and characterizable using ^{1}H-NMR.[14] The first kinetics measurements of this trapping reaction utilized a relatively long-lived HNO precursor, *N*-methanesulfonylhydroxyl-amine or MSHA ($t_{1/2}$ ~ 160 min, pH 10) in time course ^{1}H-NMR spectra. These initial experiments demonstrated the loss of paramagnetic deoxy-Mb and concurrent formation of signals indicative of HNO-Mb formation, Fig. 15.1. But midway during such a reaction, the diamagnetic HNO-Mb ^{1}H-NMR signal is seen to decrease; EPR experiments correlated this loss of HNO-Mb to the generation of the NO adduct, NO-Mb, via a secondary reaction with HNO. This secondary reactivity was verified

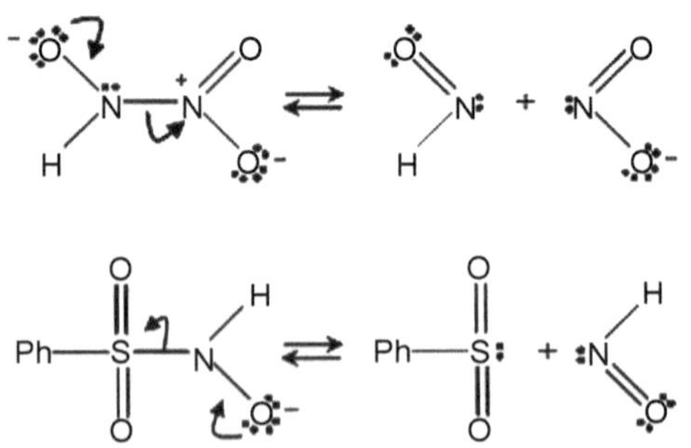

Scheme 15.1

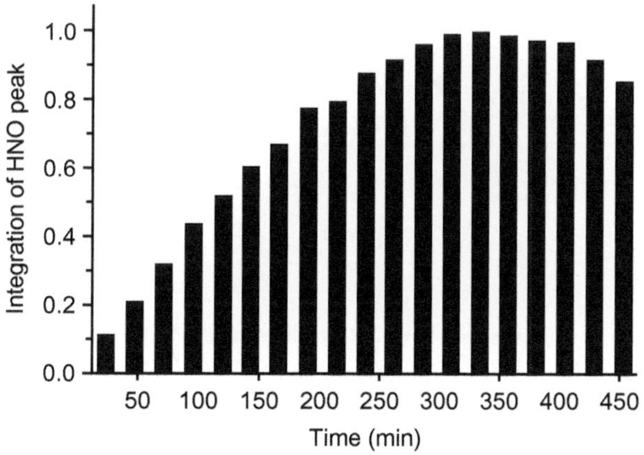

Figure 15.1 Sequential integration areas of HNO-Mb peak at 14.8 ppm over course of 1:1.5 reaction of deoxy-Mb with MSHA at 5.7 mM and pH 10. Integration area is represented as relative to maximum value, which corresponds to less than 70% conversion to Mb-HNO.

independently by reaction of a premade HNO-Mb sample with HNO, which resulted in NO-Mb generation.[15] Because HNO-Mb is both formed and depleted concurrently, and the rate constant was reported as a lower limit, $1.4 \times 10^4\ M^{-1}\ s^{-1}$, a value that is still commonly used in recent reviews.[16]

We also investigated flash photolysis of HNO-Mb, a method commonly used to determine the rate constants of association and dissociation, k_{on} and k_{off}, of small molecules like O_2, NO, and carbon monoxide (CO) to heme proteins.[17–19] But flash photolysis experiments on samples of HNO-Mb were dominated by redox reactivity. Instead of clean dissociation of HNO-Mb into free HNO and ferrous deoxy-Mb, photolysis instead generates a transient 5-coordinate ferric met-Mb species, and by inference HNO^-, the aminoxyl radical anion.[20] These two photo products were then seen to recombine on the unusually slow microsecond timescale.

The secondary reactivity of HNO-Mb with HNO also limits the purity of any HNO-Mb sample obtained in trapping reactions; isolated yields were reported from 60% to 80%, but such yields were difficult to determine as no single method (^{1}H-NMR, EPR, UV-vis) could be used to identify all the Mb species over the course of reaction.

The most readily accessible data for such an analysis is time course absorbance spectra. But the Soret absorbances of HNO-Mb, at 423 nm, and NO-Mb, at 421 nm, overlap and are difficult to resolve in mixtures. Initial attempts utilized pseudo-isosbestic points in the Q band region that correlated to concentration changes in NO-Mb (460 nm), deoxy-Mb (492 nm) and HNO-Mb (517 nm). The resulting concentration profiles over the course of the reaction, at top of Fig. 15.2, suggested a much faster trapping rate constant, ca. $k = 8 \times 10^5\ M^{-1}\ s^{-1}$, and gave an initial estimate for the secondary reaction of $k = 3 \times 10^4\ M^{-1}\ s^{-1}$ at pH 7.[16] But these analyses were flawed, as the two sequential reactions are too close in rate to distinguish by apparent isosbestic analysis.

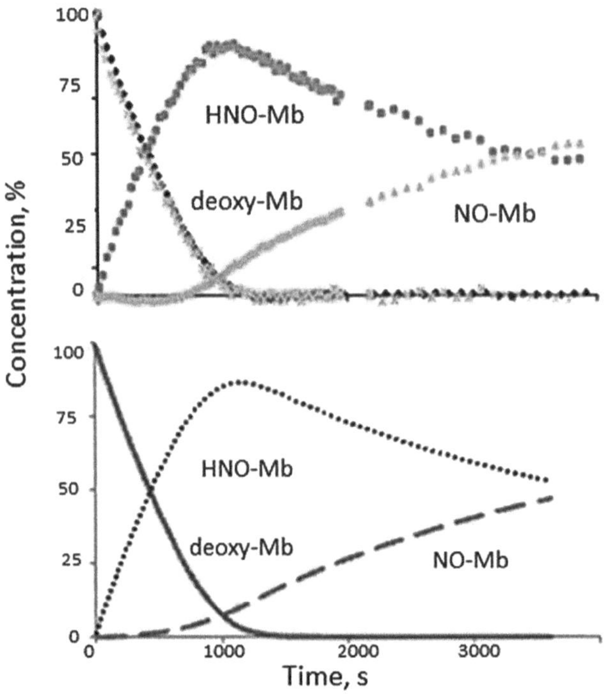

Figure 15.2 (Top) Concentration profiles derived from isosbestic spectral analysis of the absorbance spectra of the reaction of Mb-Fe^{II} with PA in 1:4 ratio at pH 10.[15] (Bottom) Concentration profiles from spectral global analysis of the reaction of Mb-Fe^{II} with PA in 1:4 ratio at pH 9.4.[21]

Recently, such shortcomings were overcome by use of global kinetic modeling methods to analyze these reactions. A critical component in applying this method to complex systems is singular value decomposition or SVD, a mathematical analysis used to separate a complex matrix into simpler, orthogonal component matrices.[21] This analysis is used in combination with global kinetic modeling to deconvolute components of mixed spectra using the evolution of time-based data.

By this method, time course spectra of the HNO trapping reactions were modeled, and the resulting speciation plots gave independent verification of the secondary reactivity producing NO-Mb, shown at the bottom of Fig. 15.2.[22] The determined bimolecular rate constant for the trapping of HNO by deoxy-Mb determined was $3.48 \times 10^5\ M^{-1}\ s^{-1}$, and a constant of $1.67 \times 10^4\ M^{-1}\ s^{-1}$ was found for the secondary reaction that generates NO-Mb; both rate constants were independent of pH and HNO donor (AS and PA). Importantly these methods also produce model-generated spectra for all colored species involved, which provided a derived spectrum for pure HNO-myoglobin, with a Soret maximum at 423 nm at an absorptivity of $1.66 \times 10^5\ M^{-1}\ cm^{-1}$. This work also determined the unimolecular HNO dissociation rate, found by competitive trapping with CO, which allowed calculation of the overall binding affinity, K_{eq}, of HNO to deoxy-Mb. At $3.9 \times 10^9\ M^{-1}$, this affinity constant is less than that of NO to deoxy-Mb, but orders of magnitude greater than those of O_2 or CO, Table 15.1.

In this report, we will demonstrate the application of this methodology step by step, in a more complex spectral analysis of the competitive trapping of HNO by mixtures of deoxy- and met-Mb, i.e., in a direct comparison of HNO trapping between ferrous (Fe^{II}) and ferric (Fe^{III}) hemes. The overall procedural strategy is shown in Scheme 15.2, which includes using SVD methods to determine the number of observed species and develop a working reaction mechanism, and subsequent iterative fitting of the data to the proposed mechanism.

15.2 Experimental

15.2.1 Materials

Lyophilized myoglobin was purchased from Sigma. AS ($Na_2N_2O_3$) was purchased from Cayman Chemical Company. All other chemicals used were of reagent grade

Table 15.1 Binding parameters for deoxy-Mb adducts[21]

Ligand	$k_{on} \times 10^6\ M^{-1}\ s^{-1}$	$k_{off}\ s^{-1}$	$K_{eq}\ M^{-1}$
NO[a]	22	1.0×10^{-4}	2.20×10^{11}
CO[a]	0.51	1.90×10^{-2}	26.80
O_2[a]	17	15	1.13
HNO[b]	0.35	8.90×10^{-5}	3.93×10^9

[a]For sperm whale Mb.
[b]For equine Mb.

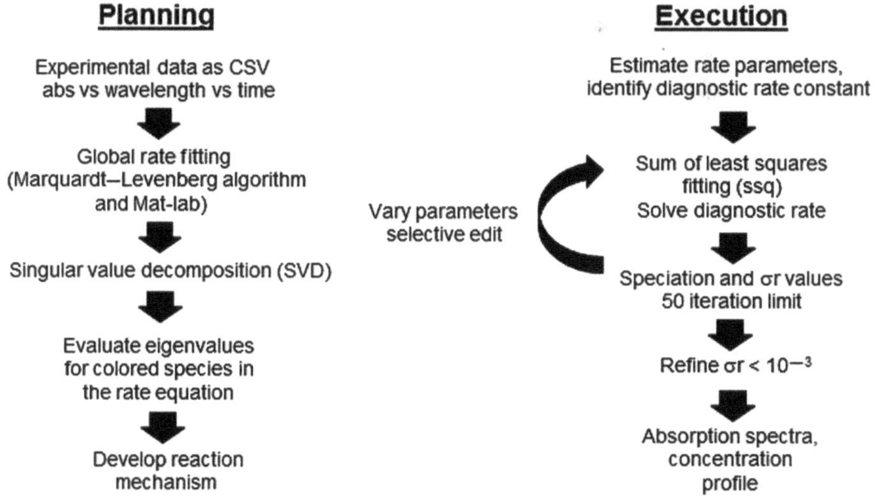

Scheme 15.2

and purchased from Sigma. Phosphate buffers (50 mM, pH 7.0) were prepared using monobasic and dibasic sodium phosphate salts purchased from Fisher. All solutions were degassed extensively before use.

15.2.2 Preparation of Mb samples

Samples of met-Mb were purified using a size exclusion filter as previously described.[15] Deoxy-Mb was prepared by the reduction with sodium dithionite,[13] and subsequent anaerobic purification on a Sephadex G25 column in a glovebox. All reaction mixtures were generated in an anaerobic glovebox which was maintained with O_2 levels below 5 ppm, and handled using strictly anaerobic procedures. Concentrations of Mb species were determined based on the protein band of Mb at 280 nm, with a molar absorptivity of 13,980 L $mol^{-1}cm^{-1}$.[23]

15.2.3 General protocol for reactions of HNO with Mb samples

The protein samples were diluted in 2 mL of pH 7 phosphate buffer to give solutions in the micromolar range, and transferred into airtight septa capped UV-vis cells in the glovebox. A standard solution of AS was dissolved in pH 9.4 carbonate buffer at a concentration of 0.1 mM and maintained anaerobically in the glovebox. A fourfold excess of the AS standard solution at pH 9.4 was placed as a drop in the cap of a sealed UV-vis cell above the mixture of deoxy-Mb and met-Mb. This cell was then carefully moved from the glovebox to the spectrometer, the sample shaken to initiate the reaction, and the spectra collected on a Hewlett Packard 8453 Diode Array spectrophotometer in the range of 280–650 nm, on the time scale of 3600 s at 27°C.

15.2.4 Kinetic simulations and data fitting

Time course absorbance data files were input into the Jplus ReactLab Kinetics modeling program, which utilizes Excel and Matlab operating procedures as previously described.[22] The number of Mb species, their concentration profiles, and fitted rate constant values for the modeled reaction sequence were determined as described below.

15.3 Results

15.3.1 Competitive trapping of HNO by met-Mb and deoxy-Mb

The reaction of AS with a mixture of deoxy-Mb and met-Mb induced characteristic heme Soret and Q band absorbance, as shown in Fig. 15.3. The initial spectrum shows two distinct Soret peaks corresponding to the two heme oxidation states, ferrous deoxy-Mb and ferric met-Mb. These spectra can be deconvoluted using authentic spectra to confirm the concentrations used; the total protein concentration

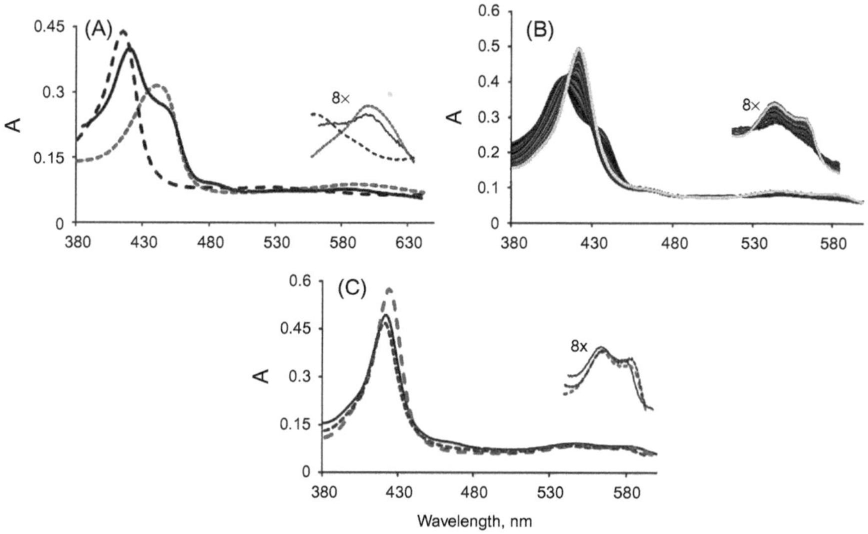

Figure 15.3 Reaction of met- and deoxy-Mb mixture (each at 8.4 μM) with AS (33 μM) in iP buffer, pH 7. (A) Initial spectra of met- and deoxy-Mb mixture (solid line), overlaid with the deconvoluted spectra of met-Mb (long dash) and deoxy-Mb (dash); *inset*: Q band region expanded eightfold. (B) Time course spectral data of reaction over 3600 s. (C) Final spectra of reaction (solid line), overlaid with deconvoluted spectra of HNO-Mb (long dash) and NO-Mb (dash); *inset*: Q band region expanded eightfold.

is given by absorbance at the protein band at 280 nm. The final spectra show one broad Soret peak at 422 nm, due to the mixture of NO-Mb and HNO-Mb products; this spectra can also be deconvoluted by similar methods. The mixtures seen in before and after spectra give evidence that the rate of HNO trapping by the two different heme oxidation states are of the same order of magnitude.

Free HNO cannot be stored in any practical way, therefore we utilize a commonly used precursor AS, which decomposes to release HNO within the range of pH 4 to 8, Eq. (15.1).[10] Once released, HNO does not accumulate due its rapid dimerization producing N_2O between pH 2 and 11, Eq. (15.2). The rate constants of these reactions are known, and will remain constant in the global fitting analysis. Previous reports for the rate constants of trapping of free HNO by deoxy-Mb and met-Mb are also available in the literature, Eqs. (15.3) and (15.4). In this work, these two constants as well as the secondary reaction, Eq. (15.5), will be solved independently in competitive reactions.

$$\text{AS} \rightarrow \text{HNO} \quad k = 6.75 \times 10^{-4}\text{s}^{-1} \quad \text{pH 4 to 8} \tag{15.1}$$

$$\text{HNO} + \text{HNO} \rightarrow \text{N}_2\text{O} \quad k = 8.00 \times 10^{6}\text{M}^{-1}\text{s}^{-1} \tag{15.2}$$

$$\text{HNO} + \text{deoxy} - \text{Mb} \rightarrow \text{HNO} - \text{Mb} \quad k = \textit{to be determined} \tag{15.3}$$

$$HNO + met - Mb \rightarrow NO - Mb \quad k = \textit{to be determined} \tag{15.4}$$

$$HNO + HNO - Mb \rightarrow NO - Mb \quad k = \textit{to be determined} \tag{15.5}$$

15.3.2 Global model fitting

The global kinetic fitting method applies matrix algebra to actual time course data using the Beer Lambert law, using a matrix of sequential time course absorbance data by wavelength over time. This collection of spectra should then match the matrix equation, Y = CA, where the rows of the matrix Y are the measured spectra over time, the columns of C are the concentration profiles of the reacting species changing over time and the rows of A are the molar absorptivities of each species.[24] The matrix C representing the concentration profiles of all components over time derives from the sequential reaction mechanism, which includes the rate constants and the initial concentration for all reacting species. The computation is explicit only for a relatively small number of simple reaction schemes; for more complex reaction sequences the computations involve numerical integration.[25] Nonlinear fitting is then applied to find the best set of rate constants for a proposed mechanism and the given data. Establishing the correct mechanism is more difficult, and requires the researcher to apply independent chemical knowledge of the system under investigation. There is no general algorithm for the determination of the correct mechanism, other than chemical plausibility.

The actual data used in the nonlinear fitting only accounts for the colored species, i.e., in this case only the Mb derivatives in Eqs. (15.3)–(15.5). All other species are considered colorless, which include AS, HNO, and N_2O, in Eqs. (15.1) and (15.2). The first iterations of fitting utilized previously determined rate constants for the independent reactions of deoxy-Mb and met-Mb with HNO, Eqs. (15.3) and (15.4). The data was plotted in dimensions of time, absorbance, and wavelength to generate a working 3D matrix (time, absorbance, wavelength) which was used for the decomposition analysis.

Factor analysis is a multivariate statistical method for extracting informational content from a data set. In this case we have a large matrix consisting of many sequential absorbance spectra as functions of both wavelength and time. SVD reduces these spectra to smaller matrices that contain all the information present in the data.[26] The kinetic information is contained in a kinetic eigenvector matrix.[24] Typically, only a small number of eigenvectors are needed to describe such data sets. The various rate processes are solved over a small number of eigenvectors and to which a kinetic model may be fit and the rate constants extracted. Similarly the spectral information is stored as a set of eigenvectors, which are derived from summations of the individual absorbances of species present.[24] Finally a vector of singular values is produced, which are listed from highest value (greatest likelihood) and fall off rapidly to zero. The initial analysis generates a table of possible eigenvalue solutions consistent with the time course absorbance changes. The number of high singular values is an indication of the number of colored species present and therefore useful for testing the chemical plausibility of various proposed reaction sequences.

In practice, the program generates a listing of eigenvector solutions ranked by their distance from zero, top of Fig. 15.4, and a plot of variation from the initial spectrum attributable to this possible eigensolution. The largest eigenvectors represent viable colored species present during the reaction, but chemical intuition must be used. In this reaction, the eigenvectors should represent four colored species: met-, deoxy-, NO-, and HNO-Mb but only three are observed. The initial SVD deconvolution defined the final products as NO-Mb and HNO-Mb as shown below in the Fig. 15.4. But examination of the dominant eigenvector spectra showed that it represents a mixture of both met- and deoxy-Mb absorbances as a single species; this was corrected in the kinetic model sequence used.

A preliminary modeled solution was obtained using only Eqs. (15.1)–(15.4), i.e., disregarding the secondary reaction of Eq. (15.5). The spectral combination of met- and deoxy-Mb was corrected by listing them as separate species within the working kinetic model, and defining their absorptivities and concentrations. To be

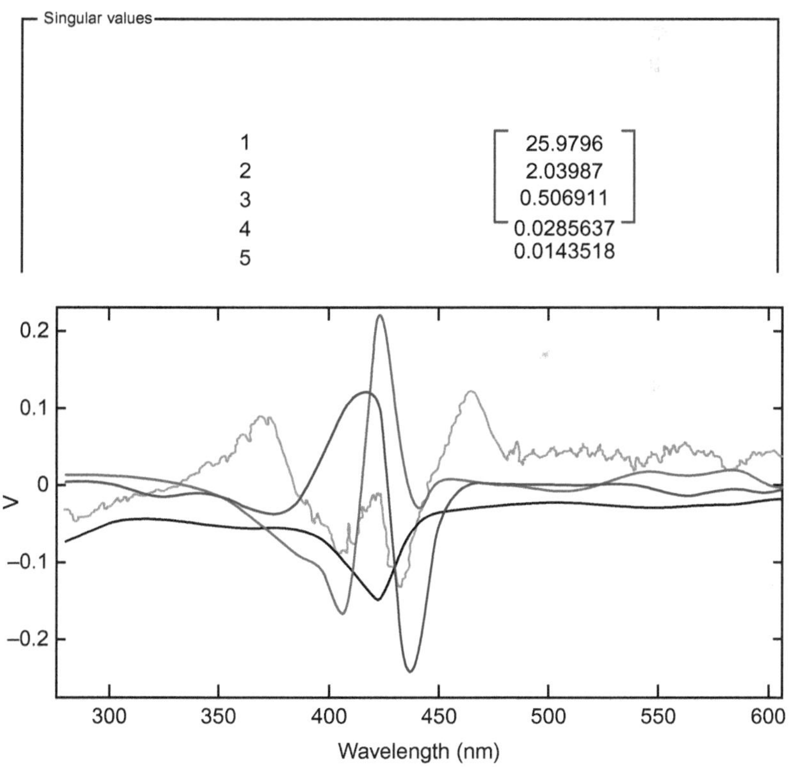

Figure 15.4 Screen shot from ReactLab program displaying at top, the table of possible SVD eigenvectors for absorbing species. At bottom are plots of spectral variations associated with the obtained individual SVD eigenvectors.

sure of initial concentrations, known spectra of both met- and deoxy-Mb were subtracted from the starting spectra to generate the difference spectra which matched with known spectra and absorptivities of the other component. The starting concentrations of both met- and deoxy-Mb were then introduced into the stepwise reaction model as well as the concentration of colorless AS used. All the products generated during the reaction, HNO, N_2O, NO-Mb, and HNO-Mb were included, thus giving a total of seven species whose concentrations would be solved within the multistep reaction mechanism. In agreement with our chemical intuition, the determined concentration profiles tracked initial met- and deoxy-Mb concentrations individually as two separate reactions occurring at distinctive rates and generating distinctive products.

From this point on, the Marquardt–Levenberg algorithm (MLA) is used to minimize least squares curve fitting errors, a softer method that allows variation in rate constants. With this method, all the data sets analyzed reached convergence using 12 or less iterations. The determined rate constants for the trapping reactions were close to previous modeled values, Table 15.2, but did not include the secondary reaction, Eq. (15.5). The generated spectra of the absorbing species were analyzed in tandem with speciation to determine the fit; in all cases they matched well with literature values for Soret wavelength and molar absorptivity.

The fitting error is determined using the matrices of the model-generated spectra and concentration profile, as seen at top of Fig. 15.5. First, the best fit spectra profile matrix is transposed onto a new spreadsheet along with the concentration profile matrix. Then the matrices are multiplied (e.g., using the Excel MMULT formula) to obtain the residual matrix, which contains model-generated UV-vis spectra of the entire reaction. The fit was evaluated by variance between expected and experimental total absorbances over time, σr, shown at bottom of Fig. 15.5, which was on the order of 10−3 for all three independent reaction data sets. Even though the fitting converged with low variance, the resulting concentration profile curves show unusual deviations around 1000 s, at the point where the effects of the secondary reaction, Eq. (15.5), would be apparent.

Table 15.2 Reported bimolecular rate constants, k_{on}, for reactions of ferrous and ferric Mb with HNO

	k_{on} values	
Previous report	**deoxy-Mb × 10^5 M^{-1} s^{-1}**	**met-Mb × 10^5 M^{-1} s^{-1}**
Miranda (2003)[a]		8.00
Sulc (2004)[b]	0.140	
Zapata (2013)[c]	3.71	2.73
This work	3.74	2.76

[a]From absorbance data at 37°C.[4]
[b]Lower limit from NMR data at room temperature.[13]
[c]From absorbance data at room temperature.[21]

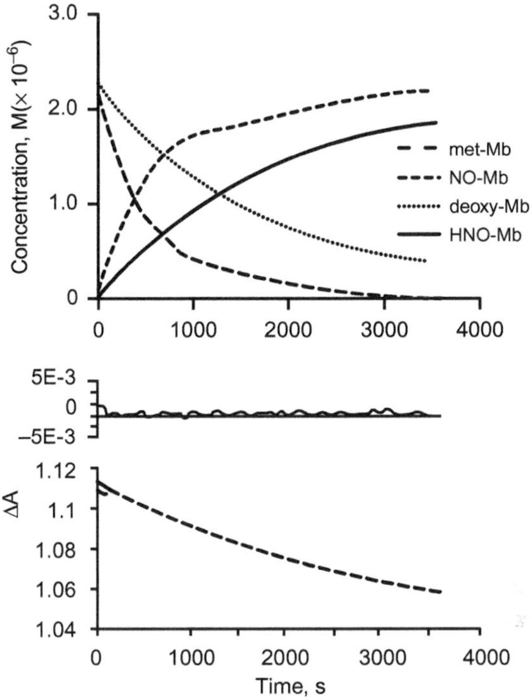

Figure 15.5 Initial data fitting without inclusion of Eq. (15.5). (Top) Concentration profile of the absorbing species as indicated. (Bottom) Plots of model and real data absorbance at 434 nm, corresponding to the loss of deoxy-Mb loss; the residual plot is shown at top, on a finer scale.

At the bottom of Fig. 15.5, the residual matrix data at 434 nm, which corresponds to deoxy-Mb loss, is plotted against the real data. Here the model and data values at this wavelength were subtracted to get a residual plot resembling a time course spectra. In this work, hand mixing was used to initiate the reactions; therefore, scattering noise is seen in the initial spectra of each reaction, as seen in Fig. 15.5, but rapidly falls within expected error. In previous work, initial spectra were omitted from the analysis if the final residual deviation exceeds $\pm 3 \times 10^{-3}$ absorbance; a rule of thumb is omit less than 1% of time course spectra used in the analysis. But in this initial solution, no omissions were necessary to obtain sufficient overall residual parameters.

15.3.3 The secondary reaction of HNO-Mb with HNO

Including the secondary reactivity of HNO-Mb with HNO, Eq. (15.5), within the global fitting presented several difficulties. The rate constant for this reaction is an order of magnitude slower than the trapping reactions in Eqs. (15.3) and (15.4), and thus the data fit reasonably well in its absence, even though examination of the speciation curves shows that the secondary reactivity is evident. When Eq. (15.5) was added to the global fitting model, convergent solutions were obtained but with a fitting errors $\sigma r >> 0.001$. Alternative attempts were made using the previously determined rate constant for Eq. (15.5) and solving for the constant for Eq. (15.3) or (15.4), but also resulted in large fitting errors, as well as giving concentration profiles inconsistent with known product yields.

Because the rates of all the reactions were determined independently, they were all input into the fitting program and a predicted spectral matrix for the reaction course was calculated. The resulting data is shown in Fig. 15.6, and does indeed fit the initial characteristics of the reaction speciation, up to ca. 1500 s, but deviates unrealistically afterward. The fitting error was allowable, at $\sigma r = 0.003$, but the apparent regeneration of deoxy- and met-Mb seen at latter stages of the reaction is counter to chemical plausibility.

The difficulty with the inclusion of Eq. (15.5) may be attributed to the linear dependence of the generation and loss of HNO-Mb to colorless HNO, as well as the generation of NO-Mb from two different sources, Eqs. (15.4) and (15.5). To circumvent this difficulty, the absorbance tolerance within the algorithm was decreased two orders of magnitude, from 10^{-6} to 10^{-8}, stiffening the MLA variation during fitting of nonlinear least squares. This resulted in a significantly better fit, shown in Fig. 15.7, with results that are self-consistent with the chemical transformations, e.g., at the end of the reaction [NO-Mb] is 25% larger than the initial [met-Mb], the difference being NO-Mb generated by the secondary reactivity. The residual error modeled at 409 nm is shown to have a $\pm 1.7 \times 10^{-3}$ difference in deviation, much better than previous fittings. As a final verification, the model-derived spectra for the individual species, also shown in Fig. 15.7, matched the known shapes and molar absorptivities of the colored Mb species.

The final global fitting analysis also provides concentration profiles for the nonabsorbing species: the nitroxyl donor AS, HNO, and N_2O over the course of the reaction, shown in Fig. 15.8. As illustrated, ca. 20% of HNO generated by AS

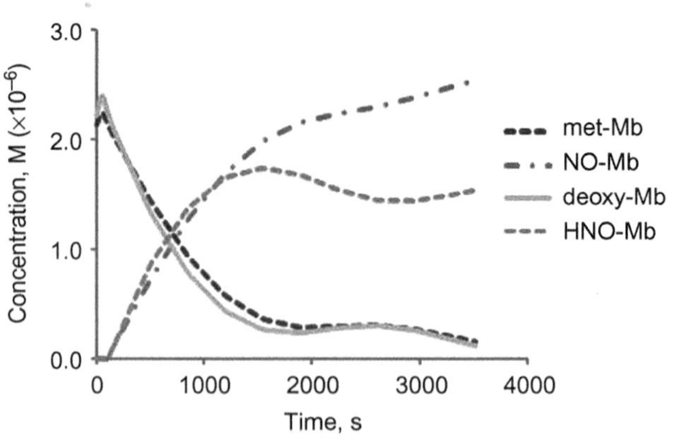

Figure 15.6 Concentration profile data calculated with inclusion of Eq. (15.5), using reported rates of Eqs. (15.1)–(15.5).

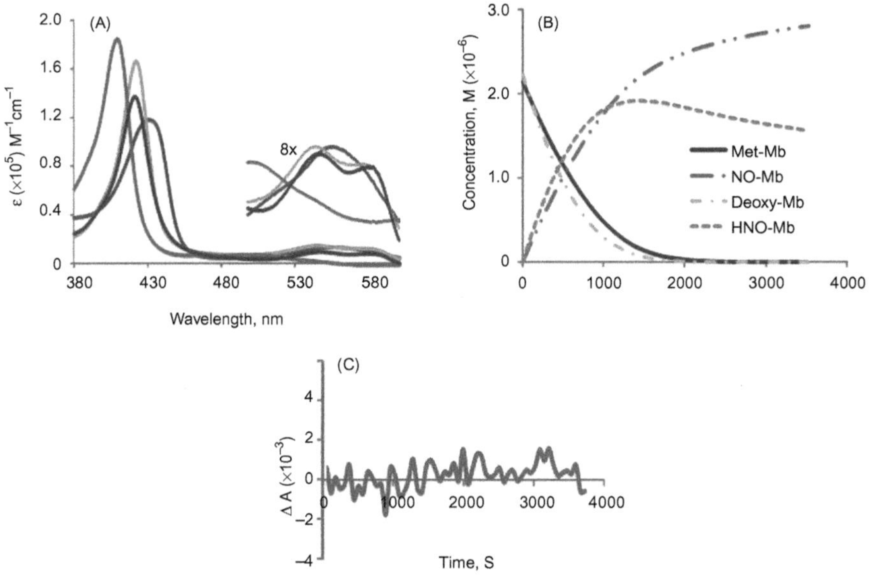

Figure 15.7 Final data fitting with inclusion of Eq. (15.5). (A) Plots of model-generated spectra of individual absorbing species (λ_{max}, absorptivity $M^{-1}cm^{-1}$): met-Mb (409 nm, 1.8×10^5), NO-Mb (421 nm, 1.4×10^5), deoxy-Mb (434 nm, 1.2×10^5), HNO-Mb (423 nm, 1.68×10^5); *inset*: Q band region expanded eightfold. (B) Concentration profiles of modeled species. (C) A residual plot of model versus data variance, showing a less than 2×10^{-3} deviation over the course of the reaction.

decomposition is lost in dimerization to N_2O. The global modeling solution shows a variance in HNO concentration of from 4 to 9 nM during the reaction, at the lowest range of current methods of electrochemical quantification.[27]

15.4 Discussion

In many reports on SVD modeling, a simple linear reaction model has been used, in synchronous order[28]; this is likely due to the limitations of commercial modeling packages as well as difficulties in handling concurrent reactivities. But such a simplified approach is not viable for reactions involving HNO, which rely on the relatively slow generation of HNO from its precursor and its rapid loss to dimerization. These two reactions have rate constants separated by 11 orders of magnitude, and thus are impossible to model as sequential reactions. Because of this, HNO trapping must be solved holistically with all reactions assumed to be occurring simultaneously. Here we have demonstrated how the global fitting methods provide self-consistent analysis of complex concurrent reactions. The analysis is verified by generation of modeled spectra of the absorbing species which match well with known myoglobin spectra. At each step, the results must be evaluated by and consistent with chemical intuition.

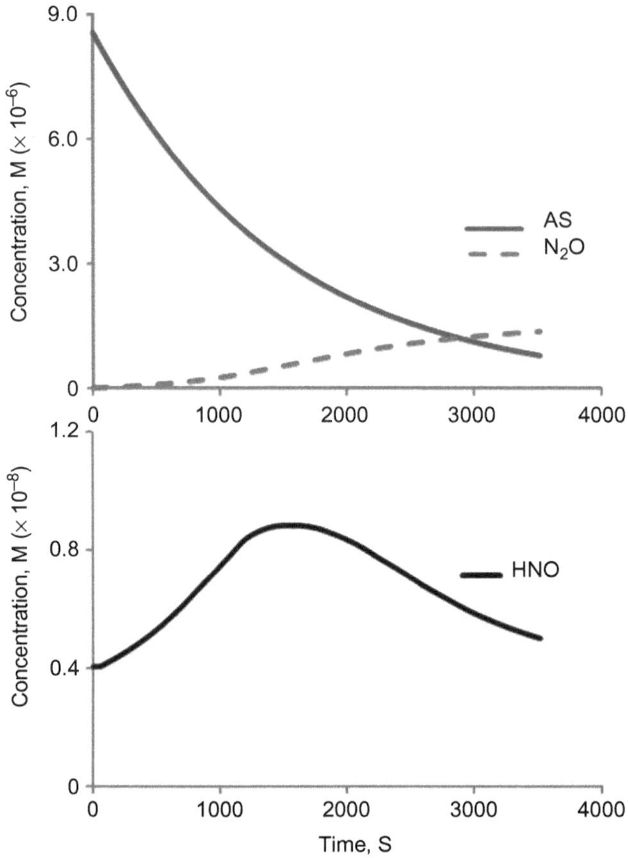

Figure 15.8 Plots of predicted concentration profiles for noncolored species. (Top) Concentration variation of AS (initially 33 μM at 100%) and byproduct N_2O. (Bottom) Predicted transient concentrations of free HNO during the reaction.

Physiological important reactivity is inherently complex due to the number and variability of such concurrent reactivity. In one recent paper on the NO dioxygenase reactivity of globins, SVD analysis was used to disprove a leading mechanistic hypothesis suggesting the generation of a ferryl heme intermediate.[29] The authors used the Runge–Kutta integration (RKI) method, which utilizes an adaptive step size that allows spectra to be viewed in steps along its time course. The sequential spectra were analyzed by comparison with known spectra of the proposed ferryl intermediate, which was but one of several species that might be present at any stage of the reaction. The authors were thus able to discount such an intermediate due to the lack of its buildup, using known rate constants to model several proposed reaction paths. Thus chemical intuition must be used to assess the modeling results throughout the process.

The competitive physiological trapping of HNO by different heme proteins has obvious importance to the proposed pharmacological use of HNO-releasing compounds. Of the various heme proteins in our bodies, the native states of hemoglobin, deoxy and oxy-Hb, dominate; they compose some 95% of the physiological Fe content. By contrast, normal individuals have less than 1% of Hb in the ferric met-Hb state,[30] thus the reactions of HNO with ferrous hemes should be of interest. Some 25 years ago, Doyle described the reaction of oxy-Hb with HNO to generate met-Hb, but the mechanism remains obscure.[31] More recently, HNO has been shown to accelerate the oxidation of oxy-Hb by nitrite, both being produced by the commonly used precursor AS.[32] A recent controversial report found that HNO precursors activate soluble guanylate cyclase in its native ferrous state.[33] Our results suggest that the formation of HNO adducts and their secondary reactivity with HNO, which have been largely ignored, may play a role in assays determining the physiological effects of HNO donors.

Acknowledgments

We thank Mr. Peter King from JPlus Consulting for help and advice. This research was supported by the National Science Foundation (PJF CHE-1057942).

Abbreviations

AS	angeli's salt or trioxodinitrate
cAMP	cyclic adenosine monophosphate
cGMP	cyclic guanosine monophosphate
CO	carbon monoxide
Mb	myoglobin
MLA	marquardt–levenberg algorithm
MSHA	*N*-methanesulfonylhydroxyl-amine
NO	nitric oxide
PA	piloty's acid or benzenesulfohydroxamic acid
RKI	runge–kutta integration
SVD	singular value decomposition

References

1. Fukuto JM, Bartberger MD, Dutton AS, Paolocci N, Wink DA, Houk KN. The physiological chemistry and biological activity of nitroxyl (HNO): the neglected, misunderstood, and enigmatic nitrogen oxide. *Chem Res Toxicol* 2005;**18**:790–801.
2. (a) Gladwin MT, Raat NJH, Shiva S, Dezfulian C, Hogg N, Kim-Shapiro DB, et al. Nitrite as a vascular endocrine nitric oxide reservoir that contributes to hypoxic signaling, cytoprotection, and vasodilation. *Am J Physiol Heart Circ Physiol* 2006;**291**: H2026–35.
   (b) Kim-Shapiro DB, Schechter AN, Gladwin MT. Unraveling the reactions of nitric oxide, nitrite, and hemoglobin in physiology and therapeutics. *Arterioscler Thromb Vasc Biol* 2006;**26**:697–705.

(c) Kim-Shapiro DB, Gladwin MT, Patel RP, Hogg N. The reaction between nitrite and hemoglobin: the role of nitrite in hemoglobin-mediated hypoxic vasodilation. *J Inorg Biochem* 2005;**99**:237–46.

3. Foresti R, Bains S, Sulc F, Farmer PJ, Green CJ, Motterlini R. The interaction of nitric oxide with distinct hemoglobins differentially amplifies endothelial heme uptake and heme oxygenase-1 expression. *J Pharmacol Exp Ther* 2006;**317**:1125–33.
4. Miranda KM, Nims RW, Thomas DD, Espey MG, Citrin D, Bartberger MD, et al. Comparison of the reactivity of nitric oxide and nitroxyl with heme proteins. A chemical discussion of the differential biological effects of these redox related products of NOS. *J Inorg Biochem* 2003;**93**:52–60.
5. Shafirovich V, Lymar SV. Nitroxyl and its anion in aqueous solutions: Spin states, protic equilibria, and reactivities toward oxygen and nitric oxide. *Proc Natl Acad Sci USA* 2002;**99**:7340–5.
6. Miranda KM, Paolocci N, Katori T, Thomas DD, Ford E, Bartberger MD, et al. A biochemical rationale for the discrete behavior of nitroxyl and nitric oxide in the cardiovascular system. *Proc Natl Acad Sci USA* 2003;**100**:9196–201.
7. Lymar SV, Shafirovich V, Poskrebyshev GA. One-electron reduction of aqueous nitric oxide: a mechanistic revision. *Inorg Chem* 2005;**44**:5212–21.
8. Bazylinski DA, Hollocher TC. Metmyoglobin and methemoglobin as efficient traps for nitrosyl hydride (nitroxyl) in neutral aqueous solution. *J Am Chem Soc* 1985;**107**:7982–6.
9. Hughes MN, Cammack R. Synthesis, chemistry, and applications of nitroxyl ion releasers sodium trioxodinitrate or Angeli's salt and Piloty's acid. *Methods Enzymol* 1999;**301**:279–87.
10. Bonner FT, Ko Y. Kinetic, isotopic, and nitrogen-15 NMR study of *N*-hydroxybenzenesulfonamide decomposition: a nitrosyl hydride (HNO) source reaction. *Inorg Chem* 1992;**31**:2514–19.
11. Liochev SI, Fridovich I. The mode of decomposition of Angeli's salt (Na_2 N_2O_3) and the effects thereon of oxygen, nitrite, superoxide dismutase, and glutathione. *Free Radic Biol Med* 2003;**34**:1399–404.
12. Lin R, Farmer PJ. The HNO adduct of myoglobin: synthesis and characterization [12]. *J Am Chem Soc* 2000;**122**:2393–4.
13. Sulc F, Immoos CE, Pervitsky D, Farmer PJ. Efficient trapping of HNO by deoxymyoglobin. *J Am Chem Soc* 2004;**126**:1096–101.
14. Sulc F, Fleischer E, Farmer PJ, Ma D, La Mar GN. 1H NMR structure of the heme pocket of HNO-myoglobin. *J Biol Inorg Chem* 2003;**8**:348–52.
15. (a) Sulc F. *HNO adduct of deoxymyoglobin: structure, reactivity and biological importance of Mb-HNO. Dissertation*. Irvine: UCI; 2006.
    (b) Kumar MR, Fukuto JM, Miranda KM, Farmer PJ. Reactions of HNO with heme proteins: new routes to HNO-heme complexes and insight into physiological effects. *Inorg Chem* 2010;**49**:6283–92.
16. Doctorovich F, Bikiel DE, Pellegrino J, Suárez SA, Martí MA. Reactions of HNO with metal porphyrins: underscoring the biological relevance of HNO. *Acc Chem Res* 2014;**47**:2907–16.
17. (a) Olson JS, Phillips Jr. GN. Kinetic pathways and barriers for ligand binding to myoglobin. *J Biol Chem* 1996;**271**:17593–6.
    (b) Gibson QH, Olson JS, McKinnie RE, Rohlfs RJ. A kinetic description of ligand binding to sperm whale myoglobin. *J Biol Chem* 1986;**261**:10228–39.

18. (a) Gibson QH, Ainsworth S. Photosensitivity of haem compounds. *Nature* 1951;**180**: 1416.
    (b) Hoshino M, Laverman L, Ford PC. Nitric oxide complexes of metalloporphyrins: an overview of some mechanistic studies. *Coord Chem Rev* 1999;**187**:75–102.
19. Kundu S, Snyder B, Das K, Chowdhury P, Park J, Petrich JW, et al. The leghemoglobin proximal heme pocket directs oxygen dissociation and stabilizes bound heme. *Proteins: Stuct Func Gen* 2002;**46**:268–77.
20. Pervitsky D, Immoos C, Van Der Veer W, Farmer PJ. Photolysis of the HNO adduct of myoglobin: transient generation of the aminoxyl radical. *J Am Chem Soc* 2007;**129**: 9590–1.
21. (a) Johnson KA. A century of Michaelis–Menten kinetics. *FEBS Lett* 2013;**587**: 2753–66.
    (b) Brereton RG. *Applied chemometrics for scientists*. John Wiley & Sons; 2007.
22. Zapata AL, Kumar MR, Pervitsky D, Farmer PJ. A singular value decomposition approach for kinetic analysis of reactions of HNO with myoglobin. *J Inorg Biochem* 2013;**118**:171–8.
23. Romberg RW, Kassner RJ. Nitric oxide and carbon monoxide equilibriums of horse myoglobin and (*N*-methylimidazole)protoheme. Evidence for steric interaction with the distal residues. *Biochem Mosc* 1979;**18**:5387–92.
24. http://olisweb.com/software/svd.html.
25. http://www.photophysics.com/tutorials/global-analysis.
26. http://jplusconsulting.com/files/ReactLab_Kinetics_Manual.pdf.
27. Suárez SA, Bikiel DE, Wetzler DE, Martí MA, Doctorovich F. Time-resolved electrochemical quantification of azanone (HNO) at low nanomolar level. *Anal Chem* 2013;**85**(21):10262–9.
28. (a) Silaghi-Dumitrescu R, Svistunenko DA, Cioloboc D, Bischin C, Scurtu F, Cooper CE. Nitrite binding to globins: linkage isomerism, EPR silence and reductive chemistry. *Nitric Oxide* 2014;**42**:32–9.
    (b) Hathazi D, Mahut SD, Scurtu F-V, Bischin C, Stanciu C, Attia AA, et al. Involvement of ferryl in the reaction between nitrite and the oxy form of globins. *J Biol Inorg Chem* 2014;**19**:1233–9.
29. Koebke KJ, Pauly DJ, Lerner L, Liu X, Pacheco AA. Does the oxidation of nitric oxide by oxymyoglobin share an intermediate with the metmyoglobin-catalyzed isomerization of peroxynitrite? *Inorg Chem* 2013;**52**:7623–32.
30. Umbreit J. Methemoglobin—it's not just blue: a concise review. *Am J Hematol* 2007; **82**:134–44.
31. Doyle MP, Mahapatro SN, Broene RD, Guy JK. Oxidation and reduction of hemoproteins by trioxodinitrate(II). The role of nitrosyl hydride and nitrite. *J Am Chem Soc* 1988;**110**:593–9.
32. Bellavia L, Dumond JF, Perlegas A, King SB, Kim-Shapiro DB. Nitroxyl accelerates the oxidation of oxyhemoglobin by nitrite. *Nitric Oxide Biol Chem* 2013;**31**:38–47.
33. Miller TW, Cherney MM, Lee AJ, Francoleon NE, Farmer PJ, King SB, et al. The effects of nitroxyl (HNO) on soluble guanylate cyclase activity: interactions at ferrous heme and cysteine thiols. *J Biol Chem* 2009;**284**:21788–96.

HNO as an Oxygen Substitute in Enzymes

16

X. Han, M.R. Kumar and P.J. Farmer
Baylor University, Waco, TX, United States

Chapter Outline

16.1 Introduction

HNO, termed nitroxyl or azanone, is the reduced and protonated congener of nitric oxide (NO). Recent work has demonstrated that HNO has distinct biological effects as compared with NO, such as the ability to elevate plasma levels of calcitonin gene-related peptide, thus serving as a positive cardiac inotrope.[1–4] HNO is thought to target metallo- and thiol- containing proteins, often as an inhibitor or modifier of enzyme or protein function. Evidence of HNO modulation of thiol-containing enzymes first emerged from studies with the antialcoholism drug cyanamide, which is an aldehyde dehydrogenase (ALDH) inhibitor used therapeutically as an alcohol-aversive agent (Temposil) in Europe, Canada, and Japan.[5] Cyanamide is oxidized by peroxidases to generate HNO, which subsequently reacts with the active-site cysteine thiolate in ALDH, inhibiting the enzyme reversibly through disulfide formation or irreversibly through sulfinamide formation.[6]

HNO has also been shown to act as an activator or signaling agent promoting enzyme activity for heme-containing soluble guanylyl cyclase, soluble guanylate

The Chemistry and Biology of Nitroxyl (HNO). DOI: http://dx.doi.org/10.1016/B978-0-12-800934-5.00016-5

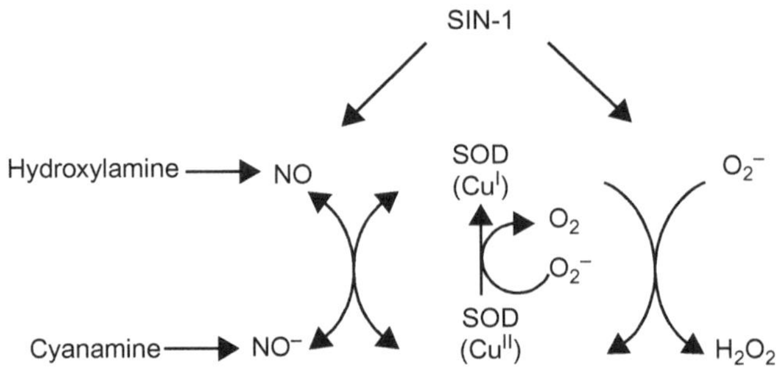

Scheme 16.1

cyclase (sGC), the primary receptor for NO, which stimulates cyclic GMP (cGMP, cyclic guanosine monophosphate) production.[7,8] As with NO, vasorelaxant responses to HNO are accompanied by an increase in cGMP and are impaired by the sGC inhibitor 1H-[1,2,4]oxadiazolo[4,3-a]quinoxalin-1-one (ODQ), indicating that HNO does target sGC.[9] It remains to be determined, however, whether HNO itself directly activates sGC or first requires oxidation to NO, or if it targets the oxidized, NO-insensitive sGC isoform, which is often found with disease.[10,11]

Rarer, though, are enzymes which utilize HNO as a substrate. The first reported example was superoxide dismutase (SOD), an enzyme whose native function is to rapidly disproportionate superoxide (O_2^-), generating H_2O_2 and O_2. Fridovich reported that SOD facilitates the interconversion of NO and HNO.[12] When cyanamide and catalase were used to generate HNO in the presence of oxidized SOD Cu(II), NO was measured by its conversion of oxyhemoglobin to methemoglobin. When reduced SOD Cu (I) was exposed to NO anaerobically, HNO was trapped by methemoglobin forming nitrosylmyoglobin. When NO was generated by 3-morpholinosydnonimine hydrochloride in the presence of SOD, HNO, or a similar reductant was formed, Scheme 16.1. Also, the activation of sGC by Angeli's salt (AS) ($Na_2N_2O_3$), a donor of HNO, implied that HNO reduces SOD to produce NO. In one study, AS had no significant effect on sGC activity in the absence of SOD. With SOD, AS caused biphasic sGC activation that was accompanied by the formation of NO.[13]

16.2 Nitroxygenase activity

HNO is isoelectronic with $^1O_2$ and it is efficiently trapped by O_2 binding proteins like myoglobin and hemoglobin.[14] Therefore, we hypothesized that HNO may inhibit or turnover oxygenases or dioxygenases in the place of O_2, and thus allow a unique approach to study the mechanism of such enzymes. Nonheme dioxygenases catalyze an amazing variety of complex oxidations, including oxidative cleavage of carbon–carbon bonds, monohydroxylation, and dihydroxylation reactions.[15] Among them, some of particular interest are dioxygenases which can catalyze the oxidative cleavage of aromatic substrates as part of bacterial aromatic degradation pathways.[16]

Quercetin 2,3-dioxygenase (QDO) is an important oxygenase in the metabolism of flavonols (Hflas). Its enzymatic turnover oxidatively cleaves the central ring of the Hfla quercetin producing the depside 2-protocatechuoylphloro-glucinolcarboxylic acid and carbon monoxide, as shown in Scheme 16.2.[17]

QDOs from *Apostichopus japonicus* and *Bacillus subtilis* have been crystallographically characterized,[18–20] and their metal-binding active sites have similar structures. The active site of the Fe(II) containing 2,3-QDO from *B. subtilis*, as shown in Fig. 16.1, has a distorted trigonal-bipyramidal geometry consisting of three histidine imidazoles (His62, His64, and His103), a water molecule, and one Glu69 at a 2.10 Å distance in the N-terminal cupin motif, and the other has a square pyramidal geometry consisting of three histidines (His234, His236, His275), a water molecule, and a weakly coordinating Glu241 at a 2.44 Å distance in the C terminal cupin motif. Metal-ion substitution experiments have shown that the catalytic activity of the Mn(II)- and Co(II)-containing enzymes were 35- and 24-fold more active, respectively, than the Fe(II)-containing native enzyme.[17] Thus, Mn(II) is proposed as the a preferred metal cofactor for 2,3-QDO from *B. subtilis*.

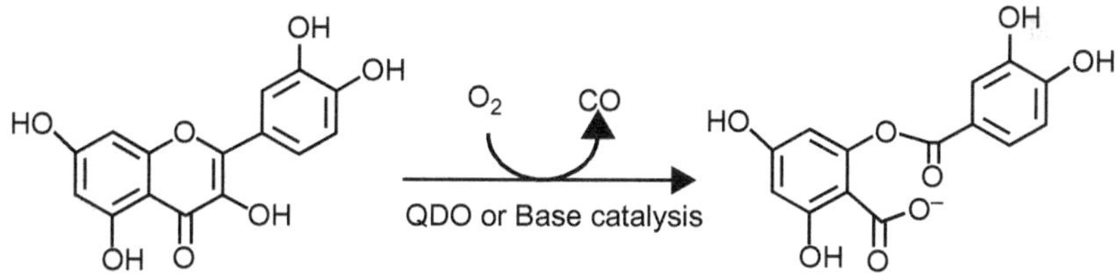

Scheme 16.2

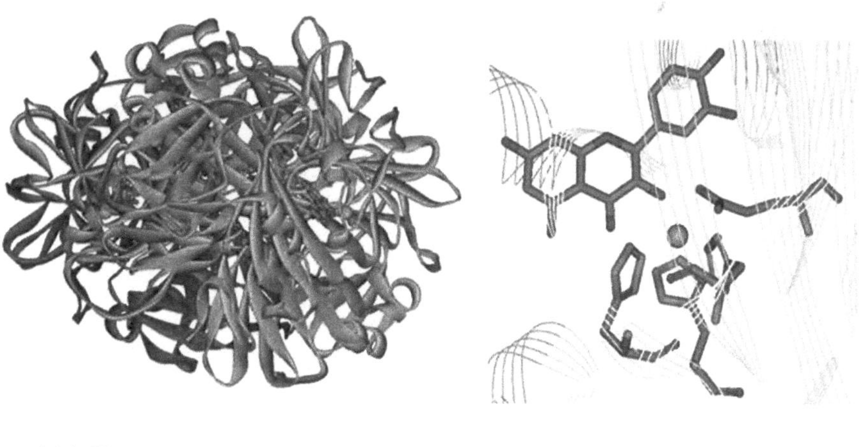

Figure 16.1 The crystal structures of QDO from *Bacillus subtilis*. The Fe atom is shown as red (dark gray in print versions) sphere. Also shown are the substrate quercetin, three histidines and Glu69 which all bind to the Fe(II) within the active site.

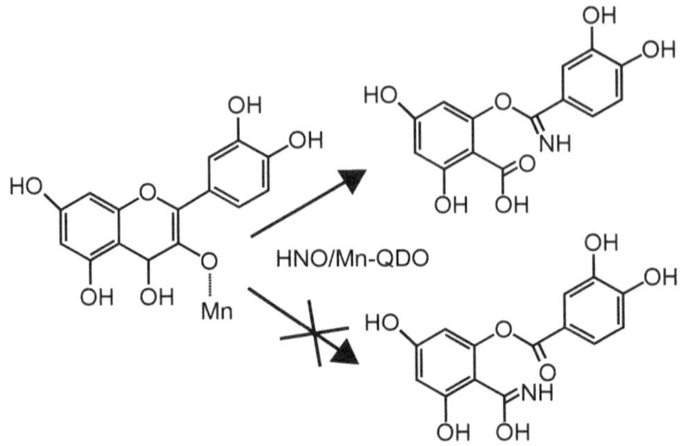

Scheme 16.3

The Cu(II) active site of resting 2,3-QDO from *A. japonicus* (homo bicupin glycoprotein) also exhibits two distinct geometries: a distorted trigonal-bipyramidal geometry consisting of three histidine imidazoles (His66, His68, and His112), one water molecule, and a carboxylate group of Glu73 (minor conformation 30%), and a distorted tetrahedral geometry without the direct coordinative interaction between copper(II) ion and the carboxylate group of Glu73 (major conformation 70%). In its principal conformation, Glu73 acts as a hydrogen bond acceptor for the metal-bound water molecule. The substrate quercetin is bound to the copper(II) ion through the deprotonated 3-hydroxy group of flavonolate (Fla^-), with displacement of the water molecule, forming an ES (enzyme−substrate) complex with a distorted square pyramidal geometry under anaerobic conditions.

We demonstrated that HNO can replace O_2 in the catalytic turnover of the Mn(II)-substituted QDO from *B. subtilis*; this "nitroxygenase activity" results in the incorporation of both N and O into products derived from quercetin, as is shown in Scheme 16.3.[21] An analogous noncatalytic reaction of quercetin and HNO also occurs under base-catalyzed conditions in the absence of enzyme, producing the same products, which can be rationalized from a depside intermediate similar to that generated in the reaction of quercetin and dioxygen, Scheme 16.2. This is the first demonstration of the nitroxygenase activity using HNO as a substrate, and we believe it may have significance in the pharmacology of HNO. In this report, we further analyze this reactivity by comparing the nitroxygenase activity of the Mn(II)-substituted QDO with a series of flavonoid substrates.

16.3 Results

Six flavonoid antioxidants were tested for the nitroxygenase reactivity, the Hflas quercetin, myricetin, galangin, the flavone luteolin, the flavanonol taxifolin, and the

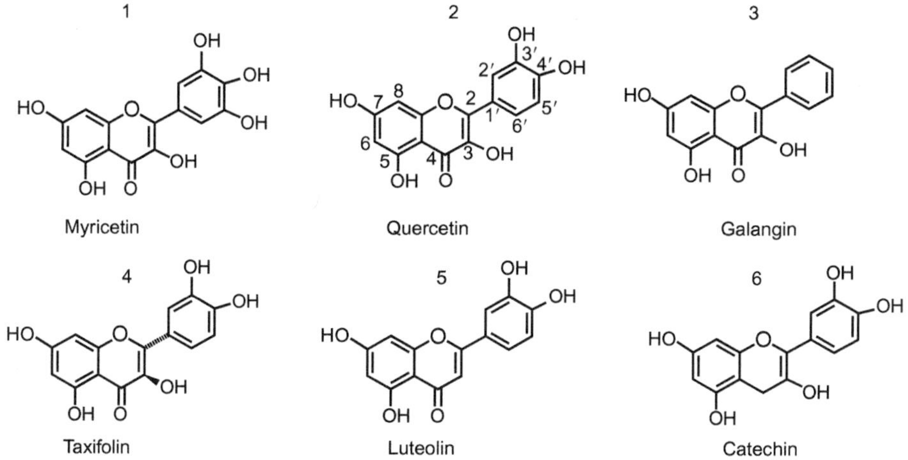

Scheme 16.4

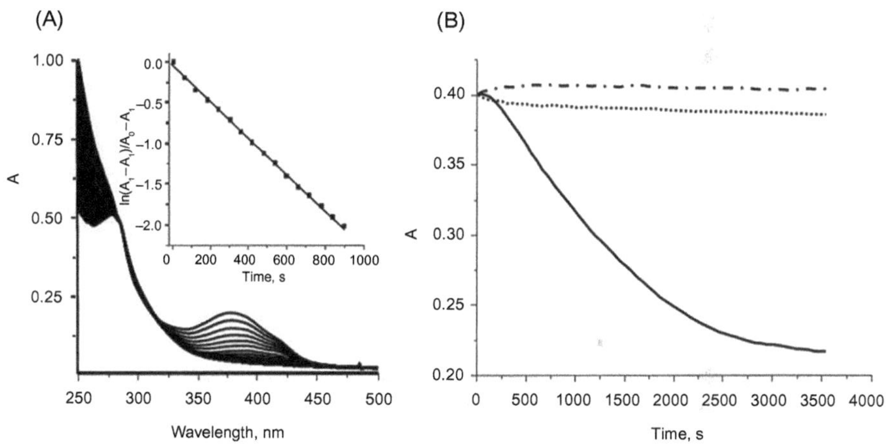

Figure 16.2 (A) Absorbance spectrum of reaction of HNO (137 μM) with Mn-QDO (8.5 μM) showing the decrease of quercetin (13.8 μM) absorbance at 380 nm over time. Inset shows a plot of $\ln(A_t - A_f)/(A_0 - A_f)$ versus time data (■), with a best fit line through data points. (B) Time course UV-vis spectra of reactions of quercetin with HNO (solid), NO_2^- anion (dash-dot) and NH_2OH (dot) in the presence of enzyme.

flavanol catechin, Scheme 16.4. In typical anaerobic enzymatic assays, premade solutions of the substrate Hfla in DMSO and the Mn-QDO in sodium phosphate buffer at pH 7 were mixed. The reaction was initiated by the addition of a stock AS at high pH. The reaction of substrate was quantitated by the loss of its absorbance ca. 380 nm, as shown in Fig. 16.2A. After completion, the reaction solutions were analyzed by liquid chromatography mass spectrometry (LC-MS). A series of

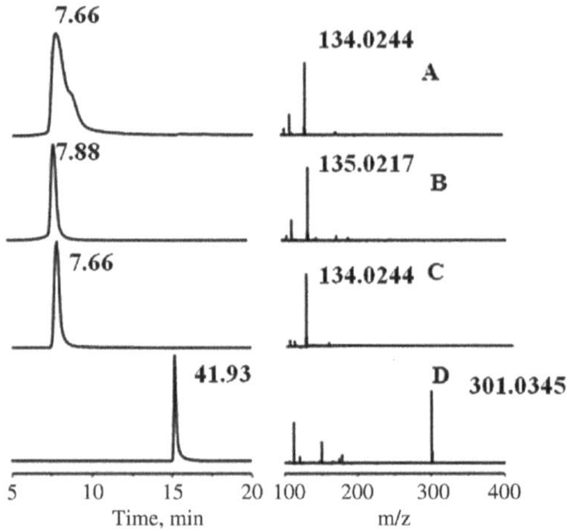

Figure 16.3 LC-MS analysis of reaction mixtures. Total ion chromatograms (left) and mass spectra (right) of product mixtures from: (A) authentic 3,4-dihydroxybenzonitrile; (B) the reaction of labeled angeli's salt (AS), $Na_2{}^{15}NONO_2$ with quercetin; (C) the reaction of unlabeled AS, $Na_2N_2O_3$, with quercetin; (D) the substrate quercetin.

analogous nonenzymatic anaerobic base-catalyzed assays were performed to compare with enzyme-catalyzed reactions, and analyzed in the same fashion.

Of the tested flavonoids, only the true Hflas myricetin, quercetin, and galangin react with AS. As a control, we tested the reactivity of the Hflas with NO_2^- and NH_2OH, possible decomposition by-products of AS, but no reactions were observed on the timescale of the HNO-based activity, Fig. 16.2B.

For the three reactive substrates, analysis of negative ion LC-MS data of the product mixtures from enzymatic and nonenzymatic assays gives similar results. The substrate is completely converted to a new product ion, 134 m/z, that matches a $C_7H_4O_2N^-$ anion; LC-MS retention time and fragmentation pattern is consistent with 3,4-dihydroxybenzonitrile, as confirmed with authentic sample of the product. The use of ^{15}N-labeled AS (specifically $Na_2[O^{15}N^{14}NO_2{}^{-2}]$) in which only the HNO precursor is labeled results in ^{15}N-labeled product with 135 m/z peak observable in LC-MS, Fig. 16.3. The generation of CO during these reactions was confirmed by the conversion of a solution of deoxymyoglobin to its ferrous CO adduct upon exposure to the head gas above the assay mixture, Fig. 16.4. Although not observed in the MS, it is assumed that an initial depside-like product, 2-iminomethoxy-3,4-dihydroxyphenyl-4,6-dihydroxybenzoate, undergoes a 1,3proton transfer to generate the observed nitrile and phenolic products. An analogous hydrolytic cleavage of the depside ester produced under typical QDO turnover is seen at basic pH, resulting in carboxylic and phenolic products Scheme 16.5.

Kinetic analyses of the enzymatic and nonenzymatic reactions were performed and analyzed by using the sequential reactions described in Eqs. (16.1)–(16.6). Rate analysis of the reaction between Hfla and HNO is complicated due to the slow decomposition rate of the HNO-donor AS and the competitive dimerization of free

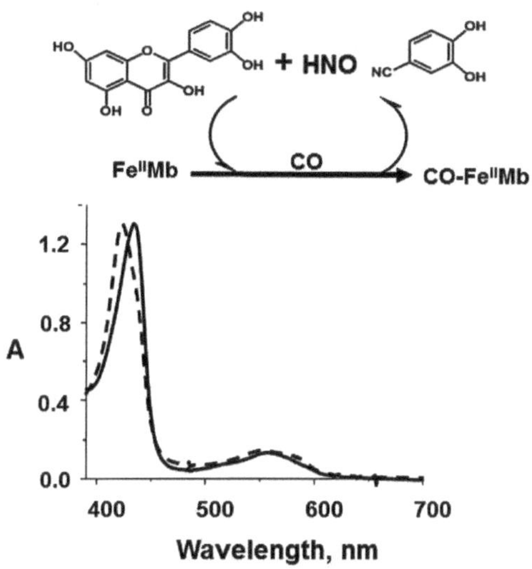

Figure 16.4 Absorbance spectra illustrating the formation of CO-Fe^{II}Mb (dotted line) by trapping of CO by deoxymyoglobin (solid line) released in the reaction of quercetin (0.033 mM) with AS (0.327 mM); a simplified reaction scheme is shown at top.

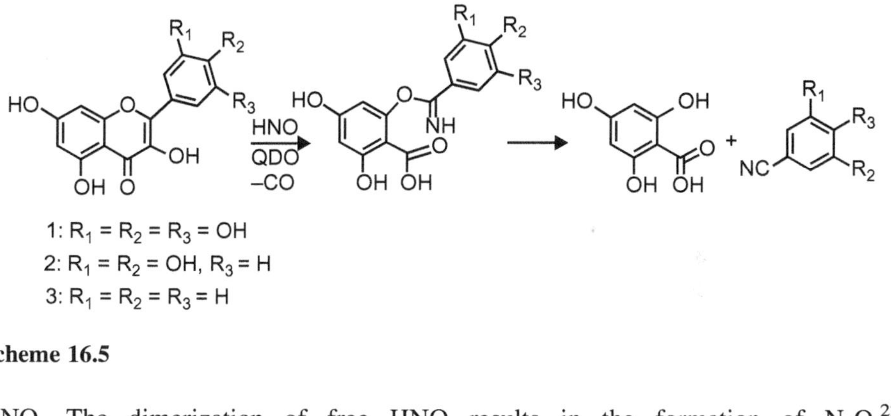

Scheme 16.5

HNO. The dimerization of free HNO results in the formation of N_2O.[22] Eqs. (16.1)–(16.4) describe the nonenzymatic reaction between Hfla and HNO, while Eqs. (16.1), (16.2) and (16.5), (16.6) relate to the enzyme-catalyzed reaction.

$$HN_2O_3^- \rightarrow HNO + NO_2 \quad (16.1)$$

$$HNO + HNO \rightarrow N_2O + H_2O \quad (16.2)$$

$$Hfla \rightleftharpoons fla^- + H^+ \quad (16.3)$$

$$fla^- + HNO \rightarrow Product \quad (16.4)$$

$$\text{Hfla} + \text{QDO} \rightleftharpoons \text{fla}^-/\text{QDO complex} \quad (16.5)$$

$$\text{fla}^-/\text{QDO complex} + \text{HNO} \rightarrow \text{Product} + \text{QDO} \quad (16.6)$$

A first-order rate constant at 25°C of $7.0 \times 10^{-4}\ s^{-1}$ for Eq. (16.1) was calculated by Guggenheim's method, derived from plots of the first-order rate constants versus the concentrations of AS in controlled temperature bath reactions. This value is somewhat larger than the published constant for this reaction of $6.75 \times 10^{-4}\ s^{-1}$ at room temperature, but was derived under conditions equivalent to those of the subsequent determinations. The widely reported bimolecular rate constant of the dimerization in Eq. (16.2) of $8.00 \times 10^{6}\ M^{-1}\ s^{-1}$ was used.[22]

The dependence of the nonenzymatic reactivity on basic conditions suggests that the substrate must be deprotonated, Eq. (16.3), preceding its reaction with HNO. All of the substrates used have reported pK_a of Hflas in the range of the onset of HNO reactivity, shown in Table 16.1.[23] There is some variability in the reported pK_as of Hflas and related polyhydroxylic antioxidants,[24] the values used are mid-range within those reported. Likewise, there is much variability in reported oxidation potentials of the substrate Hflas; selected values used here are for general comparison, given in Table 16.1.[25]

The Henderson–Hasselbalch equilibrium was used to estimate the percentages of deprotonated fla^- in pH 8 buffer: 96% for myricetin, 93% for quercetin, and 87% for galangin. These percentages were used to adjust substrate flavonate concentrations when solving for the second-order rate constants (k_4) of Eq. (16.4) from kinetic data using the software, REACT for Windows,[26] as illustrated in Fig. 16.5 and reported in Table 16.1 as an average of three trials. For comparison, the rate constant for the nonenzymatic dioxygenation of quercetin, Eq. (16.7), was determined under equivalent conditions as $k_7 = 0.46\ M^{-1}\ s^{-1}$. The rate of oxygenations of Fla^-s are dramatically slower than that of nitroxygenations. Likewise, a difference of ca. 10^5 is found between pseudo-bimolecular rates reported for the enzymatic dioxygenase and nitroxygenase reaction of Mn-QDO.[21]

Table 16.1 Enzymatic and nonenzymatic reaction rate constants between flavonols (Hfla) and AS at room temperature, as well as reported pK_a and oxidation potentials of each Hfla

Flavonol	Enzyme catalyzed[a] $k_6(M^{-1}s^{-1}) \times 10^4$	Base catalyzed[a] $k_4(M^{-1}s^{-1}) \times 10^4$	pK_a[b]	E_{ox}[c] (V vs Ag/AgCl)
1 (m)	1.97 (±0.09)	1.79 (±0.03)	6.5	−0.030
2 (q)	3.32 (±0.12)	1.10 (±0.05)	7.1	0.020
3 (g)	5.60 (±0.11)	0.86 (±0.04)	7.2	0.280

[a]Average of three trials (error in parenthesis).
[b]Data from Ref. [23].
[c]Oxidation potentials determined by column electrolysis, from Ref. [25].

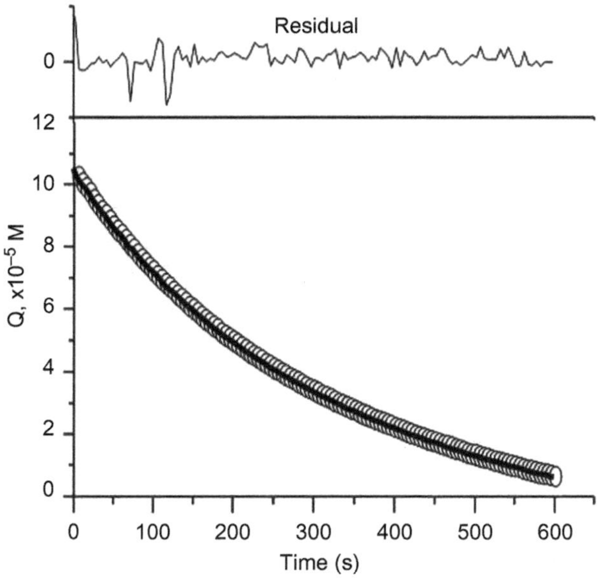

Figure 16.5 A plot of best fit kinetic model (line) versus data absorbance at 380 nm (circles). A residual plot of (model minus data) over time is shown at top.

$$Fla^- + O_2 \rightarrow Product \quad (16.7)$$

For the Mn-QDO catalyzed reactions (Eqs. 16.1, 16.2 and 16.5, 16.6), the rate constants of the Eqs. (16.1) and (16.2) were obtained by the same method as in the nonenzymatic reactions, following the loss of absorbance at 380 nm at 25°C. Previous studies had determined a Michealis constant, K_M, of ca. 4 μM for the formation of the quercetin/QDO complex, Eq. (16.5).[17,21] To determine the bimolecular rate constants given, quantitative formation of the Fla^-/QDO complex is assumed under conditions equivalent to that used for quercetin, i.e., at concentrations of substrate in >100-fold excess to the enzyme. Using the same modeling REACT program, the rate constants for Eq. (16.6) was obtained for each substrate, listed in Table 16.1 as an average of three trials.

16.4 Discussion

The substrates chosen in Scheme 16.1 were intended to test for structural dependences that might distinguish nitroxygenase reactivity from the corresponding dioxygenase reactivity. As with native QDO activity, only the true Hflas react with AS either enzymatically or nonenzymatically; the other flavonoids luteolin, taxifolin and catechin do not react. Thus the structural requirements for dioxygenase and nitroxygenase reactivities are to have an alpha hydroxy-ketone functionality at the three and four position, as well as a double bond between C2 and C3.

Both dioxygenase and nitroxygenase reactivity also have similar bimolecular rate constants for enzyme-catalyzed reactions at neutral or acidic pH as compared with the corresponding nonenzymatic reactions at high pH.[21] But there is a large difference in rates between the two reactivities, the nitroxygenase reactivity is ca. 10^5 times faster that of the dioxygenase reactivity both enzymatically and nonenzymatically. This is to be expected, as HNO is much more electrophilic than dioxygen, and also reacts from a singlet ground state rather than the triplet ground state of dioxygen; the required spin-flip results in a substantial kinetic barrier to dioxygenation reactions.[21,27]

In the nonenzymatic reactions, the interaction of Fla^- and HNO may be initiated by an initial electron transfer,[28,29] or by nucleophilic attack of the anionic substrate on HNO,[30] as have been proposed for the corresponding dioxygenation reactions. In both scenarios, the reaction rate should be affected by the pK_a of Hfla and thermodynamic potential of the flavonate anion HOMO. The determined rate constants for Eq. (16.4) do appear to follow to the substrate's assigned pK_a and oxidation potential, but the enzymatic rates run counter to both trends which may suggest a different mechanistic path is involved.

In the enzymatic reaction, the Mn-QDO and Hfla must first combine to form a complex, which assumes prior coordination of the Fla^- with the Mn(II) within the QDO active site. Steric hindrance within the binding pocket might favor a less bulky Hfla; the data seems most consistent with this hypothesis. Galangin, **3**, having the least number of hydroxyl groups undergoes the enzymatic nitroxygenase reaction almost six times faster than the nonenzymatic reaction. Myricetin, **1**, having the highest number of hydroxyl groups, reacts at comparable rates under both conditions tested.

For Mn-QDO catalyzed mechanism, it remains unclear whether HNO binds to the Mn cofactor in QDO enzyme prior to its insertion into the substrate. In most previously proposed mechanisms for QDO native activity, the β-O atom of a metal-bound superoxide attacks the C2 position which initiates the ring-opening reaction sequence, Scheme 16.6. All known metal-ion complexes of HNO are N-bound; thus an analogous nitroxygenase reaction mechanism would predict the β-O atom of a metal-bound aminoxyl radical attacks this site. But the observed products derive from N-atom incorporation at the C2 position, counter to the prediction and perhaps more consistent with a direct reaction of the activated substrate with HNO.

Importantly, only the Mn-substituted QDO enzyme undergoes nitroxygenase reactivity; samples of Fe(II) and Co(II) QDO did not show demonstrable activity.[21] Such selectivity argues for a metal-mediated mechanism, in which both the HNO

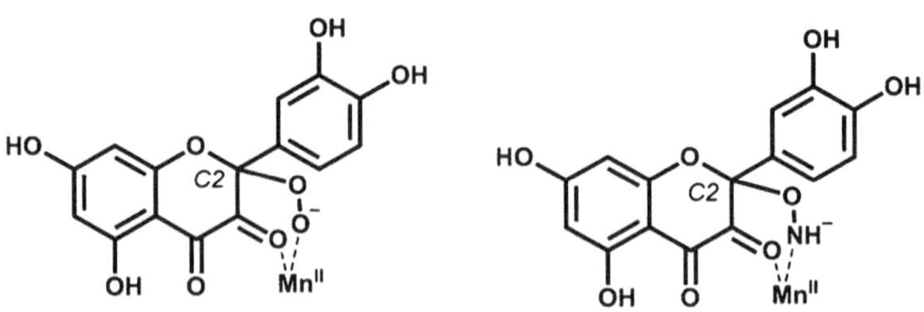

Scheme 16.6

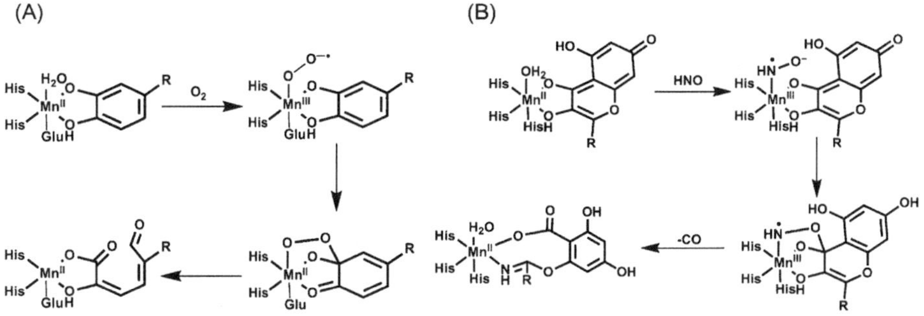

Scheme 16.7

and the substrate Fla⁻ are bound to the metal prior to product formation, rather than a metal-activation of the substrate, in which the metal-coordinated Fla⁻ reactions directly with HNO.

Similar questions of substrate versus metal ion initiation have been raised in the native QDO reactivity, where reactivity is seen for several metal ions including Fe(II), Co(II), Mn(II), and Cu(II).[17] A true tertiary complex of substrate/metal ion/dioxygen has recently been characterized in a similar nonheme dioxygenase homoprocatechuate 2,3-dioxygenase (HPCD).[16] The substrate 4-nitrocatechol and dioxygen were characterized bound to the active-site Mn ion, and distinct intermediate states observed by both EPR and Mossbauer spectroscopy; these radical intermediate states are represent sequential oxidation and reduction of the metal ion during enzymatic turnover. An important distinction in the proposed HPCD mechanism is that the Mn-bound superoxide attacks via a 4-centered cyclic transition state, rather than the 6-atom cyclic intermediate proposed for QDO, as shown in Scheme 16.7A.

For such a mechanistic sequence to apply to the nitroxygenase reaction of Mn-QDO, a tautomeric quinone methide must be formed in ring A of the Hfla, as shown in Scheme 16.7B, which would be favored by metal binding at the two oxygen appended to positions C4 and C3. The substrate quinone methide allows attack of the Mn(III)–aminoxyl radical adduct at the C4 position, analogous to that of the dioxygenase mechanism. The plausibility of such a tautomeric substrate intermediate is supported by recent a nuclear magnetic resonance (NMR) study which suggests that the C7 hydroxyl is the most acidic site in flavonoids, even in the presence of additional hydroxyls at positions C3, C5, or C6.[24] Thus the reactive Fla⁻ anionic species likely has quinone methide character in both the enzymatic and nonenzymatic reactions. To our knowledge, such an intermediate has not been previously proposed, and ongoing tests of its viability in both the dioxygenase and nitroxygenase reactions are in progress.

16.5 Conclusions

The unprecedented substitution of HNO for dioxygen in the enzymatic turnover of Mn-QDO has been described, termed "nitroxygenase activity," which results in the regioselective incorporation of both N and O atoms into the Hfla-derived products, as well as CO release. A similar nonenzymatic reaction of the Fla^{-}s with HNO is observed at high pH. Rate analysis studies imply that nitroxygenase reactivity of Mn-QDO depends mainly on steric hindrance of Hfla under conditions tested, while the base-catalyzed reaction appears to depend strongly on the pK_a and oxidation potential of Hfla. The possible involvement of a quinone methide tautomer of the Hfla substrates was used to rationalize the site of nitroxyl N-atom incorporation into the product.

16.6 Materials and methods

16.6.1 Materials

AS was purchased from Cayman chemicals and used as received. Quercetin, myricetin, galangin, and other Hflas shown in the text and 3,4-dihydroxy benzonitrile were purchased from Tokyo Chemical Industry America and checked for purity by ^{1}H NMR and ^{13}C NMR.

16.6.2 Enzyme expression and purification

Mn-substituted QDO was prepared by growing *Escherichia coli* BL21 (DE3) carrying plasmid pQuer4, grown in M9 minimal medium. Each culture was induced to express protein by the addition of 50 mg L^{-1} isopropyl-β-D-thiogalactopyranoside. The Mn-QDO was selectively generated by supplementation of the media with 1 mM $MnCl_2$. Cells were harvested and the protein purified, except the protein was eluted from the DEAE-Sephacel column stepwise with 125 mM NaCl in 50 mM Tris buffer, pH 7.5 (wash buffer), and 175 mM NaCl in 50 mM Tris buffer, pH 7.5 (elution buffer). Also, the DEAE-Sepharose column was omitted from the purification procedure. Enzymatic activity for isolated batches was assayed at 380 nm in a standard spectroscopic assay using quercetin as a substrate. Active samples were analyzed for purity by SDS-PAGE electrophoresis. The purest, most active samples were pooled and concentrated and the buffer was exchanged to 50 mM phosphate buffer (pH 7.0) by centrifugation in a Vivaspin 15R centrifugal filter unit (10,000 molecular weight cutoff). Protein concentration was determined by a Bradford assay, and enzyme was stored in 10% glycerol at −20 °C.

16.6.3 CO trapping by deoxymyoglobin

A home-built long-necked quartz cuvette with a side arm attached to a 25 mL round-bottom flask was used to trap the CO by deoxymyoglobin. In the cuvette was added 11 μM deoxymyoglobin in IP buffer, pH 7.0. For the nonenzymatic reaction, quercetin (33 μM) in IP buffer, pH 8 was added to the round-bottom flask. The reaction was initiated by adding AS (327 μM) to the quercetin. The progress of the reaction was monitored by the shift in Soret absorbance from 434 to 423 nm, confirming formation of CO-Fe^{II}Mb.

16.7 Kinetic measurements

16.7.1 The nonenzymatic reaction of HNO with flavonols

Assay reactions between Hfla and HNO-precursor AS were carried out in a screw-capped UV cuvette and monitored by following the decrease of the substrate absorbance at 380 nm. An initial estimate of the rate of decomposition of AS was determined by a natural log of concentration versus time plot, and this value was used as a starting point in the subsequent simulations using REACT for Windows, Version 1.2.

16.7.2 Assays of Mn-QDO nitroxygenase activity

Assay reactions with varying amounts of Hfla, Mn-QDO, and the HNO-precursor AS were carried out in a screw-capped UV cuvette and monitored by following the decrease of the substrate absorbance at 380 nm. In a typical assay, 13.7 μM Hfla is mixed with 8.5 nM of enzyme in deareated IP buffer at pH 7; the reaction is initiated by the addition of a stock AS solution to give final concentration of 137 μM. The reaction was initiated by gently shaking the cuvette before placing it in the spectrometer, followed by observing the loss of absorbance at 380 nm.

Acknowledgments

This material is based upon work supported by the National Science Foundation Graduate Research Fellowship under Grant No. CHE – 1057942 and funding from Baylor University.

Abbreviations

AS angeli's salt
Fla^- flavonolate
Hfla flavonol
HPCD homoprocatechuate 2,3-dioxygenase
QDO quercetin 2,3-dioxygenase
SOD superoxide dismutase

References

1. Paolocci N, Saavedra WF, Miranda KM, Martignani C, Isoda T, Hare JM, et al. Nitroxyl anion exerts redox-sensitive positive cardiac inotropy *in vivo* by calcitonin gene-related peptide signaling. *Proc Natl Acad Sci* 2001;**98**:10463–8.
2. Paolocci N, Katori T, Champion HC, St John ME, Miranda KM, Fukuto JM, et al. Positive inotropic and lusitropic effects of HNO/NO^- in failing hearts: independence from β-adrenergic signaling. *Proc Natl Acad Sci* 2003;**100**:5537–42.
3. Paolocci N, Jackson MI, Lopez BE, Miranda K, Tocchetti CG, Wink DA, et al. The pharmacology of nitroxyl (HNO) and its therapeutic potential: not just the Janus face of NO. *Pharmacol Ther* 2007;**113**:442–58.

4. Dai T, Tian Y, Tocchetti CG, Katori T, Murphy AM, Kass DA, et al. Nitroxyl increases force development in rat cardiac muscle. *J Physiol* 2007;**580**:951–60.
5. Kappaka V, Thomposon DC, Chen Y, Vasiliou V. Aldehyde dehydrogenase inhibitors: a comprehensive review of the pharmacology, mechanism of action, substrate specificity, and clinical application. *Pharmacol Rev* 2012;**64**:520–39.
6. DeMaster EG, Redfern B, Nagasawa HT. Mechanisms of inhibition of aldehyde dehydrogenase by nitroxyl, the active metabolite of the alcohol deterrent agent cyanamide. *Biochem Pharmacol* 1998;**55**:2007–15.
7. Irvine JC, Ritchie RH, Favaloro JL, Andrews KL, Widdop RE, Kemp-Harper BK. Nitroxyl(HNO): the cinderella of the nitric oxide story. *Trends Pharmacol Sci* 2008;**29**: 601–8.
8. Irvine JC, Cao N, Gossain S, Alexander A, Love JE, Qin C, et al. HNO/cGMP-dependent antihypertrophic actions of isopropylamine-NONOate in neonatal rat cardiomyocytes: potential therapeutic advantages of HNO over NO. *Am J Physiol Heart Circ Physiol* 2013;**305**:H365–77.
9. Chin KY, Qin C, Kemp-Harper BK, Woodman OL, Ritchie RH. The concomitant coronary vasodilator and positive inotropic actions of the nitroxyl donor angeli's salt in the intact rat heart: contribution of soluble guanylyl cyclase-dependent and -independent mechanisms. *Br J Pharmacol* 2014;**7**:1722–34.
10. Miller TM, Cherney ME, Franco N, Farmer PJ, King SB, Hobbs AJ, et al. The effects of nitroxyl (HNO) on soluble guanylate cyclase activity: interactions at ferrous heme and cysteine thiols. *J Biol Chem* 2009;**284**:21788–96.
11. Irvine JC, Favaloro JL, Kemp-Harper BK. NO^- activates soluble guanylate cyclase and K_v channels to vasodilate resistance arteries. *Hypertension* 2003;**41**:1301–7.
12. Liochev SI, Fridovich I. Nitroxyl (NO^-): a substrate for superoxide dismutase. *Arch Biochem Biophys* 2002;**402**:166–71.
13. Zeller A, Wenzl M, Beretta M, Stessel H, Russwurm M, Koesling D, et al. Mechanism underlying activation of soluble guanylate cyclase by the nitroxyl donor angeli's salt. *Mol Pharmacol* 2009;**76**:1115–22.
14. (a) Kumar RM, Pervisky D, Chem L, Poulos T, Kundu S, Hargrove MS, et al. Nitrosyl hydride (HNO) as an O_2 analogue: long-lived HNO-adducts of ferrous globins. *Biochemistry* 2009;**48**:5018–25.
    (b) Reisz JA, Bechtold E, King SB. Oxidative heme protein-mediated nitroxyl (HNO) generation. *Dalton Trans* 2010;**39**:5203–12.
15. Bugg T. Dioxygenase enzymes: catalytic mechanisms and model chemistry. *Tetrahedron* 2003;**59**:7075–101.
16. Bugg T, Ramaswamy S. Non-heme iron-dependent dioxygenases: unraveling catalytic mechanisms for complex enzymatic oxidations. *Curr Opin Chem Biol* 2008;**12**:134–40.
17. Schaab MR, Barney BM, Francisco WA. Kinetic and spectroscopic studies on the quecetin 2,3-dioxygenase from *Bacillus subtilis*. *Biochemistry* 2006;**45**:1009–16.
18. Fusetti F, Schroter KH, Steiner RA, Noort P, Pijing T, Rozeboom HJ, et al. Crystal structure of the copper-containing quercetin 2,3-dioxygenase from *Aspergillus japonicus*. *Structure* 2002;**10**:259–68.
19. Gopal B, Madan LL, Betz SF, Kossiakoff AA. The crystal structure of a quercetin 2,3-dioxygenase from *Bacillus subtilis* suggests modulation of enzyme activity by a change in the metal ion at the active site. *Biochemistry* 2005;**44**:193–201.
20. Sun YJ, Huang QQ, Zhang JJ. Series of structural and functional models for the ES (enzyme-substrate) complex of the Co(II)-containing quercetin 2,3-dioxygenase. *Inorg Chem* 2014;**53**:2932–42.

21. Kumar RM, Zapata A, Ramirez A, Bowen S, Francisco W, Farmer JP. Nitrosyl hydride (HNO) replaces dioxygen in nitroxygenase activity of manganese quercetin dioxygenase. *Proc Natl Acad Sci* 2011;**108**:18926–31.
22. Shafirovich V, Lymar SV. Nitroxyl and its anion in aqueous solutions: Spin states, protic equilibria, and reactivities toward oxygen and nitric oxide. *Proc Natl Acad Sci* 2002; **99**:7340–5.
23. Lunte C, Sanli S. Determination of eleven flavonoids in chamomile and linden extracts by capillary electrophoresis. *Anal Methods* 2014;**6**:3858–64.
24. Musialik M, Kuzmicz R, Litwinienko G. Acidity of hydroxyl groups: an overlooked influence on antiradical properties of flavonoids. *J Org Chem* 2009;**74**:2699–709.
25. Yang B, Kotani A, Arai K, Kusu F. Estimation of the antioxidant activities of flavonoids from their oxidation potentials. *Anal Sci* 2001;**17**:599–604.
26. (a) REACT for Windows, Version 1.2*;* Manka MJ, editor. Chapel, FL: Alchemy Software, Wesley; 2001.
    (b) Braun W, Herron JT, Kahaner DK. Acuchem: a computer program for modeling complex chemical reaction systems. *Int J Chem Kinet* 1988;**20**:51–62.
27. Costas M, Mehn MP, Jensen MP, Que L. Dioxygen activation at mononuclear nonheme iron active sites: enzymes, models, and intermediates. *Chem Rev* 2004;**104**:939–86.
28. Kaizer J, Speier G. Radical-initiated oxygenation of flavonols by dioxygen. *J Mol Catal A: Chem* 2001;**171**:33–6.
29. Balogh-Hergovich E, Speier G. Kinetics and mechanism of the base-catalyzed oxygenation of flavonol in DMSO-H_2O solution. *J Org Chem* 2001;**66**:7974–8.
30. Nishinaga A, Tojo T, Tomita H, Matsuura T. Base-catalyzed oxygenolysis of 3-hydroxy-flavones. *J Chem Soc Perkin Trans* 1979;**10**:2511–25.

The Reactions of HNO With Nonheme Proteins: An Emphasis on Thiol-Containing Proteins

17

J.M. Fukuto and R.J. Millikin
Sonoma State University, Rohnert Park, CA, United States

Chapter Outline

17.1 Introduction

Azanone (HNO, also referred to as nitroxyl, hydrogen oxonitrate(I) or nitrosyl hydride) is a simple three atom species with fascinating chemistry and important biological and/or pharmacological activity. Although HNO has been of interest to chemists in the past,[1] more recent reports regarding its unique biological activity have prompted a rebirth of investigative activity regarding its fundamental chemical biology (for a review of the important biological activity of HNO, please see Ref. [2]). Moreover, recent developments in HNO detection and monitoring have also been forthcoming,[3,4,5] motivated in part by its potential biological relevance and pharmacological utility. In spite of the possible importance of HNO as a physiological and/or pharmacological agent, the intimate details of the chemical mechanism(s) of its biological function are not firmly established. That is, although some biological targets have been proposed or identified, in many cases the specific details of the interaction of HNO with these targets have not been unequivocally determined. For example, HNO is considered to be a novel and potent agent for the

The Chemistry and Biology of Nitroxyl (HNO). DOI: http://dx.doi.org/10.1016/B978-0-12-800934-5.00017-7

treatment of heart failure since it simultaneously increases heart contractility (inotropy), heart relaxation (lusitropy), myofilament sensitivity to Ca^{2+} while not increasing metabolic demand or heart rate (vide infra). Further studies on the physiological mechanisms for these effects have proposed protein targets that are involved in myocardial Ca^{2+} cycling and binding. However, the intimate details of HNO interactions with these targets have not been, as yet, determined.

Perusal of the recent HNO literature indicates that a likely biological functional group that stands out as a potential target by which HNO can elicit its biological effects are thiols and thiol-containing proteins. Although this revelation does not narrow down significantly the field of possible biological targets, it is noteworthy that many of the proteins that regulate/modulate the physiology of heart function are thiol-containing proteins (vide infra). To be sure, there are numerous thiol-containing proteins, and it is unlikely that HNO interacts significantly with all of them since the actions of HNO, e.g., as a treatment for heart failure, appear to be very specific and are without significant side effects. As is always the case for biological or pharmacological signaling agents, issues of target specificity, and reversibility are paramount. That is, if HNO is generally reactive with thiols and thiol-containing proteins, what is the reason that HNO can react with specific thiol-containing proteins in a literal "sea of thiols"? Also, if HNO chemically modifies a thiol protein to elicit a biological response, how is this modification reversed to bring the protein back to its original state? These questions remain unanswered and considering the potential therapeutic utility of HNO, it will be important to address these issues in the future. Regardless, as mentioned immediately above, one of the most established targets for HNO are thiols/thiol-containing proteins and the discussion herein will focus on the chemistry of the interactions of HNO with these proteins.

17.2 The chemistry of HNO with thiols

Although covered elsewhere in this book and reviewed previously,[6] it is worthwhile to briefly present some of the basic chemistry between HNO and thiols as a preface to discussing the interaction of HNO with thiol-containing proteins. One of the first reports describing the ability of HNO to react with thiols was by Doyle and coworkers[1] who reported a reaction of HNO with thiophenol to give the corresponding disulfide via the presumed intermediacy of an *N*-hydroxysulfenamide (Reactions 17.1 and 17.2).

$$RSH + HNO \rightarrow RSNHOH \quad (17.1)$$

$$RSNHOH + RSH \rightarrow RSSR + NH_2OH \quad (17.2)$$

This report indicated the electrophilicity of HNO and its ability to modify thiol groups. The thiophilicity of HNO was further established computationally when

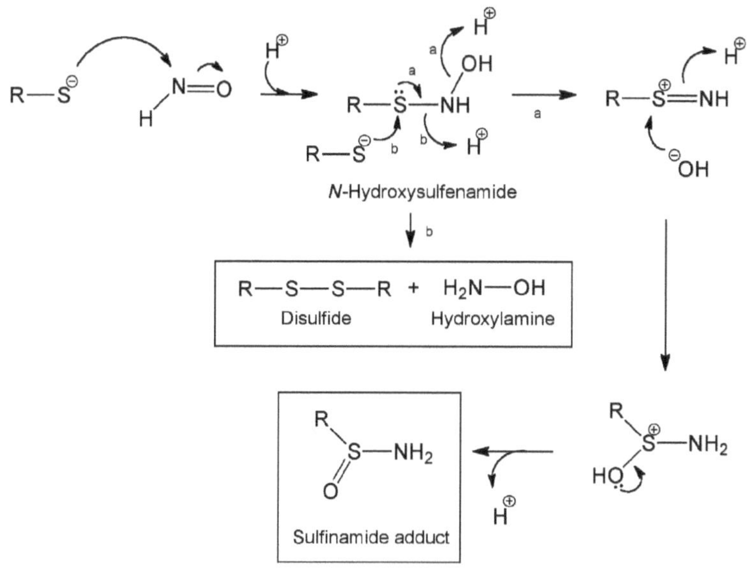

Figure 17.1 HNO reaction with thiols to give disulfide or sulfinamide.

Bartberger and coworkers found that this reaction was thermodynamically favorable.[7] Interestingly, HNO was calculated not to react appreciably with water, indicating a lack of electrophilicity toward "hard" oxygen nucleophiles (as opposed to "soft" sulfur nucleophiles). Wong et al.[8] also confirmed the thiol reactivity of HNO and further proposed that another product from this reaction could be formed. In the absence of a large excess of thiol, the intermediate *N*-hydroxysulfenamide was capable of rearranging to give a sulfinamide (Reaction 17.3).

$$RSNHOH \rightarrow RS(O)NH_2 \qquad (17.3)$$

The details of the HNO-thiol reaction and rearrangement of the *N*-hydroxysulfenamide to the corresponding sulfinamide have been examined computationally and found to be highly dependent on the nature of the reaction environment, the thiol involved and the presence/absence of base.[9] Fig. 17.1 schematically shows the generation of the two products, disulfide and sulfinamide, for the reaction of a thiol with HNO.

The generation of two possible products from the reaction of a thiol with HNO is important since the two products can have different biological fates/lifetimes. It is well established that disulfides can be readily reduced in biological systems back to the corresponding thiols by numerous reductive processes (Reaction 17.4).

$$R - S - S - R + \text{Reductant} \rightarrow 2RSH \qquad (17.4)$$

It has also been reported that sulfinamides can be converted back to the corresponding thiol in the presence of excess reductant.[10] However, this appears to be an extremely slow process indicating that sulfinamide formation is, at the very least, a modification that is not easily or very slowly reversible (and considered to be essentially irreversible compared to disulfide formation). Thus, it is important to recognize that HNO-mediated modification/oxidation of a thiol/thiol protein can lead to a biologically reversible disulfide species or an irreversible (or at least a slowly reversible) sulfinamide adduct. It is also worth noting that sulfinamides can be slowly hydrolyzed to the corresponding sulfinic acid/sulfinate (Reaction 17.5).

$$RS(O)NH_2 + H_2O \rightarrow RS(O)O^- + NH_4^+ \quad (17.5)$$

Biological sulfinate generation is known to occur via other distinct processes in biological systems. Interestingly, protein sulfinates can be converted back to the corresponding thiols by the ATP-dependent enzyme sulfiredoxin.[11] However, to date, it is found that this is not a general process and appears to be very specific for oxidized peroxiredoxins.

Clearly, many physiological/pharmacological agents are known to react with and affect thiol protein activity. For example, cell-signaling associated with hydrogen peroxide (H_2O_2) and other electrophilic agents such as peroxynitrite ($ONOO^-$) is thought to involve modification of thiol-containing proteins.[12] One of the most important factors in determining the specificity associated with signaling is the relative kinetics of the reaction of the signaling effector species (i.e., HNO) with its protein target. Importantly, the reaction of HNO with thiols is considerably faster than with most other electrophiles. For example, the second-order rate constants for the reaction of glutathione (GSH) with HNO, H_2O_2, and $ONOO^-$ have been reported to be $2\text{–}8 \times 10^6$, 0.87, and $1.3 \times 10^3\ M^{-1}\ s^{-1}$, respectively.[6,13–15] Therefore, from both kinetic and thermodynamic perspectives, HNO is extremely reactive with thiols, which is why thiol-containing proteins are currently considered to be primary targets for HNO-mediated signaling.

17.3 Examining HNO in biological systems

Structurally, HNO possesses a slight dipole with nitrogen having a partial positive charge, making it somewhat analogous to an aldehyde in terms of reactivity. HNO is thus vulnerable to nucleophilic attack at the nitrogen atom. The nitrogen atom possesses a lone pair of electrons that are orthogonal to the pi system (the N = O bond), making HNO also somewhat nucleophilic. At a qualitative level, it may not be surprising that HNO rapidly dimerizes with a second-order rate constant of $8 \times 10^6\ M^{-1}\ s^{-1}$ [16] since it possesses both nucleophilic and electrophilic properties. The initial product of this reaction has long been understood to be hyponitrous acid (HONNOH), which is known to degrade via dehydration. The final decomposition products of the HNO dimerization reaction are thus N_2O and water (Reaction 17.6).[17]

$$2HNO \rightarrow H_2N_2O_2 \rightarrow N_2O + H_2O \quad (17.6)$$

These rapid and nearly unavoidable series of reactions makes studying HNO inherently complex and difficult, especially in biological systems. Because HNO cannot be concentrated without rapid self-consumption occurring (Reaction 17.6), pure HNO cannot be isolated and stored for use in biological studies. Thus, HNO-donating molecules are employed in research in lieu of pure HNO. However, the use of HNO-donors (and all donors for that matter) always presents fundamental problems that need to be considered.[18] For example, if donors require biological activation (i.e., reduction or oxidation) prior to release of the species of interest, it may be the activation process that actually elicits the biological effect. There is also the possibility that the donor itself and not the released species is the actual biological effector. Moreover, since most donors typically release other products as well as the intended molecule, it is always possible that these other released species can either antagonize, increase or be responsible for the biological effect of interest. Chemical impurities in the donor may also behave similarly. In these scenarios, it will be difficult to determine the identity or potency of the biological effector. A common way to avoid some of the caveats associated with the use of donor molecules is to use several chemically and structurally distinct donors. Multiple structurally distinct donors are unlikely to have the same impurities, the same coproducts, the same inherent biological activities or the same activation processes. However, even this method for alleviating the problems with donor molecule usage is problematic since the kinetics of the release of the species of interest are often different. In the case of HNO donors, it is also always important to test completely decomposed donor since HNO is fleeting and in this way the other products of donor decomposition (which typically are not fleeting) can be examined and controlled for.

Angeli's salt ($Na_2N_2O_3$) is the mainstay of HNO donors among researchers due to the fact that its chemistry is well established, it is relatively easy to synthesize or commercially available and it is water soluble.[18] Indeed, many of the studies discussed below rely on Angeli's salt as the source of HNO and for many years it has been considered the "gold standard" of HNO donors. It should be realized, however, that Angeli's salt releases HNO very quickly (often leading to high initial concentrations of HNO), releases an equivalent of nitrite (NO_2^-), and is very basic, requiring significant buffering. Regardless, most of the studies discussed below utilize HNO donors and, for the most part, had proper controls when Angeli's salt was used.

17.4 Reaction of HNO with thiol-containing proteins

The brief discussion above indicates that thiol-containing proteins are likely biological targets for the actions of HNO. Discussed below are several examples of studies focused on the effect of HNO on thiol protein activity. Hopefully these few examples will serve to illustrate the range of possibilities associated with HNO-thiol protein interactions.

17.4.1 Aldehyde dehydrogenase

Pioneering work by the Nagasawa lab on the mechanism by which the antialcoholic drug cyanamide (H_2NCN) inhibits aldehyde dehydrogenase represents the first example of HNO-mediated interaction (and inhibition) of a thiol protein. In a series of seminal studies, Nagasawa and coworkers demonstrated that oxidation of H_2NCN by, e.g., catalase/H_2O_2, leads to HNO generation and subsequent inhibition of aldehyde dehydrogenase.[19–25] In this case, H_2NCN can be considered to be a prodrug for HNO since HNO is the pharmacological agent. Indeed, this represents the first clinical use of an HNO donor species. The use of H_2NCN as an antialcoholic drug in humans is significant since it allows for an evaluation of the general biology/physiology of HNO administration. For example, due to the high levels of GSH in cells (as high as 1–10 mM) and the above-mentioned thiophilicity of HNO, it may be expected that H_2NCN administration (and subsequent HNO generation) would alter intracellular GSH. However, it has been reported that aldehyde dehydrogenase inhibition occurs without a significant change in the GSH status in cells,[26] indicating that HNO can be very specific with respect to its biological target (at least at therapeutic levels).

17.4.2 Glyceraldehyde-3-phosphate dehydrogenase (GAPDH)

When examining the mechanisms of the toxicity of HNO in the yeast *Saccharomyces cerevisiae*, it was found that HNO is primarily toxic while yeast were glycolytic and considerably less toxic when yeast were respiring.[27] Further study revealed that HNO inhibited yeast growth via interaction and inhibition of glyceraldehyde-3-phosphate dehydrogenase (GAPDH), a crucial thiol-dependent protein involved in glycolysis. Significantly, as was the case for HNO-mediated inhibition of aldehyde dehydrogenase, inhibition of GAPDH occurred at HNO levels that did not alter the intracellular GSH status of yeast, providing further evidence that HNO can have significant specificity toward certain thiols/thiol-containing proteins. For both aldehyde dehydrogenase and GAPDH, the enzyme substrates are both aldehydes. From a purely chemical perspective, one can envision a similarity between the structures and chemical properties of HNO and an aldehyde. Furthermore, any catalytic entities that promote reaction of the active-site thiol (or thiolate) with an aldehyde should do the same thing with HNO (i.e., Lewis acid activation, general acid–base catalysis, etc.). Thus, it is proposed that at least some of the specificity of HNO toward reaction with the thiol of a dehydrogenase can be explained by the similarity between HNO and the endogenous substrates (Fig. 17.2). As discussed above, HNO is capable of either reversibly or irreversibly modifying thiols. In the case of GAPDH inhibition by HNO, GAPDH activity in yeast remained depressed long after HNO was present, possibly indicating an irreversible modification of the protein.

Mass spectral analysis of the HNO-GAPDH adduct confirmed the generation of irreversible sulfinamide adducts on cysteine residues.[28] However, sulfinamide formation was not observed on the active-site cysteine but rather on other,

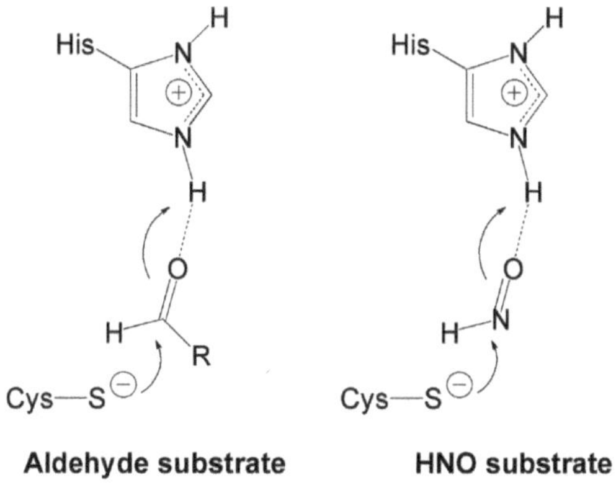

Figure 17.2 Similarity between an aldehyde and HNO as possible substrates for dehydrogenases and other reactive cysteine proteins.

nonactive-site cysteines. The active-site cysteine was, instead, shown to form a disulfide with another proximal cysteine. Thus, it was speculated that sulfinamide formation on the peripheral cysteines may have led to dissociation of the homotetrameric protein, thus causing irreversible loss of activity. Mass spectral analysis of another thiol protein treated with HNO, HCalB, also showed sulfinamide formation and cysteine mixed disulfides when cysteine was present. When serum albumin, which contains only one free cysteine residue, was exposed to HNO the predominant product exhibited an M + 14 mass. Although speculative at this time, it was proposed that an intermediate *N*-hydrosulfenamide species could react with a proximal lysine amine group, releasing NH_2OH (Reaction 17.7), in a reaction analogous to the reaction of a proximal thiol to give disulfide and NH_2OH (Reaction 17.2).

$$\text{RSNHOH} + \text{R}'\text{-NH}_2 \rightarrow \text{RS-NH-R}' + \text{NH}_2\text{OH} \quad (17.7)$$

Subsequent oxidation of the RS-NH-R′ species to give the corresponding sulfinamide, RS(O)-NH-R′, would be consistent with the M + 14 mass observed. A similar effect was also observed by Keceli and Toscano[29] who found that HNO-mediated modification of a terminal cysteine could generate a sulfenic acid (and other further oxidized products), possibly via initial attack of a free carboxylate on an *N*-hydroxysulfenamide, generating a cyclic intermediate followed by hydrolysis (Fig. 17.3). Overall, this work rigorously confirms the idea of protein sulfinamide (and disulfide) formation in thiol-containing proteins treated with HNO. These studies also indicate that other possible species can be formed, presumably via reaction of a proximal nucleophile with the intermediate *N*-hydroxysulfenamide. Thus, it is important to note, as these authors point out, that the nature of HNO-mediated modification is highly dependent on the protein and cysteine microenvironment.

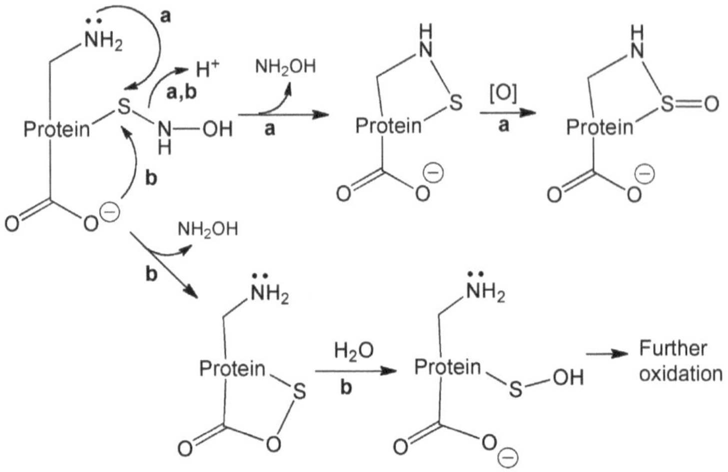

Figure 17.3 Possible and proposed reactions of the *N*-hydroxysulfenamide intermediate with proximal protein nucleophiles such as an amine[28] or a carboxylate.[29]

17.4.3 The yeast transcription factor ACE1

The transcription factor ACE1 is a thiol protein that serves to "sense" copper (Cu^I) levels in yeast by binding it via protein cysteines. Thus, in the presence of high intracellular copper, ACE1 binds copper and becomes activated to elicit gene expression of proteins that sequester copper.[30] Due to the versatility of the yeast system[31] and the presence of a thiol-dependent and well-defined transcription factor (with readily available transcriptional activity read-outs), yeast represent an ideal model for examining the chemical details of the interaction of HNO with thiol-containing proteins (such as ACE1).[32] HNO was found to potently inhibit ACE1 activation by copper, presumably by modifying the otherwise copper-binding cysteines. This effect was also examined under both aerobic and anaerobic conditions (since yeast can survive under both) and found to be independent of O_2. Importantly, nitric oxide (NO) inhibition of ACE1 has been found to be entirely O_2-dependent[33] indicating that HNO-mediated inhibition of this thiol protein was not due to simple oxidation of HNO to NO. Moreover, anaerobic inhibition of ACE1 by HNO excluded the possibility that other HNO/O_2-derived molecules were responsible for the observed inhibition. Thus, all data from this study was consistent with the idea that HNO directly modified the ACE1 cysteine thiols, inhibiting its ability to bind and "sense" intracellular copper.

17.4.4 Cysteine proteases papain and cathepsin B

Papain and cathepsin B are both cysteine proteases and have been used as model proteins to examine the effect various nitrogen oxides have on thiol protein activity.[34,35] Papain was found to be irreversibly inhibited (as evidenced by the inability of dithiothreitol to bring back enzyme activity) in the presence of the HNO-donor compound Angeli's salt.[34] Interestingly, high-level exposure of papain to HNO under aerobic conditions resulted in protein carbonyl formation, consistent with the idea that HNO can react with O_2 to generate a potent oxidant. Although the chemical nature of this oxidant is not known, it has been reported that the reaction of HNO with O_2 can create a species capable of oxidation chemistry.[6] Regardless, it is generally thought that the primary HNO–protein interaction (and the major mechanism of HNO-mediated effects on thiol-containing proteins) is not the generation of potent oxidizing species, but rather involves either reversible of irreversible cysteine thiol modification.[35] In a later study using macrophages, the inhibition of lysosomal cathepsin B by HNO was found to be similar to that observed for papain.[35] This study also examined the effect of altered intracellular GSH levels on HNO-mediated cathepsin B inactivation and found that inhibition was relatively independent of GSH levels. Thus, as was reported for inhibition of GAPDH in yeast (vide supra), it appears that the inhibition of this cysteine protease was not due to changes in the GSH/GSSG (glutathione/glutathione disulfide) redox status and, possibly, that HNO is able to exhibit some level of specificity with regards to biological thiol modification. As was proposed earlier for GAPDH, the reaction of HNO with these proteases was thought to be analogous to that of the normal substrates. That is, the reactivity of HNO as an electrophile appears to be analogous to that of carbonyl compounds, allowing it to substitute for aldehydes and amides as substrates for cysteine proteins. Moreover, due to their structural and chemical similarities, the catalytic mechanisms that enhance the reactivity of the carbonyl compounds should similarly enhance reactivity with HNO.

17.4.5 Other proteins

The above examples should suffice in indicating the propensity for HNO to react with, modify and alter the activity of thiol-containing proteins. These studies also provide mechanistic insight into the nature of the possible HNO-thiol protein interactions. There are numerous other reports of HNO-mediated effects on thiol-containing proteins, and some of these are worth mentioning here since several of these putative targets may be responsible for the observed biological/physiological effects associated with HNO.

The effects of HNO on the cardiovascular system represent one of its most important pharmacological applications.[2] Expectedly, HNO has been reported to affect several crucial proteins involved in heart function and regulation. For example, HNO has been shown to interact with proteins involved in the regulation of Ca^{2+} release and uptake by the sarcoplasmic reticulum (SR). The ryanodine receptor RyR2 is responsible for Ca^{2+} release from the SR and shown to be significantly activated by HNO.[36,37] Calcium reuptake into the SR occurs via the actions of the Ca^{2+} pump sarcoplasmic reticulum calcium ATPase (SERCA), which is regulated by the protein phospholamban (PLN) and PLN has been shown to be oxidized by HNO resulting in an increase in activity of the SERCA reuptake pump.[38] It is also reported that HNO can modify SERCA directly, leading to an increase in the maximal activation.[39] Thus, the simultaneous activation of RyR2 and SERCA (via PLN or directly) by HNO results in an increase in Ca^{2+} cycling and, therefore, an increase in both cardiac contractility and relaxation. It has also been reported that HNO can directly modify the cardiac myofilaments leading to an increase in contractility.[40] Taken altogether, it is not difficult to envision why HNO has been touted as an ideal potential treatment for heart failure[41] since simultaneous activation of both Ca^{2+} release and reuptake in the SR will lead to both an increased inotropy and lusitropy of the heart. Also, sensitization of myofilaments to Ca^{2+} will mean that increased Ca^{2+}-mediated contraction will occur at all concentrations of Ca^{2+}. These factors coupled to the fact that there is no increase in heart rate or metabolic demand indicate a near ideal therapeutic scenario for treating heart failure.

Importantly, all of the protein targets discussed immediately above, cardiac myofilaments, RyR2 and SERCA/PLN, contain regulatory cysteine thiols and the actions of HNO in affecting protein activity have all been proposed to occur via thiol modification. It is interesting to note that these specific target sites all combine physiologically to have a positive effect on cardiac function and that interaction of HNO with other thiol protein targets, that may negatively affect this system, have not been reported. Considering the many thiol-containing proteins that could have been affected, this appears to be either a remarkable pharmacological coincidence or, more likely, a coordinated system by which some endogenous thiol modifying species (possibly HNO) can positively regulate cardiac function. Related to this idea, it has been shown that HNO exhibits a reactivity profile that is nearly identical to that of a peroxide.[42] That is, the proteins that are known to interact with H_2O_2 (i.e., thiols, selenoproteins, heme proteins) are likely the same as those that have the potential to interact with HNO. This may be partially evidenced by the finding that exposure of cells to HNO can elevate H_2O_2 levels, presumably by inhibiting H_2O_2 degrading systems (i.e., HNO interacts with the same proteins that can interact with and degrade H_2O_2).[42] This raises the possibility that some of the effects associated with HNO can be, in part, due to HNO-mediated increases in peroxide and peroxide signaling. However, considering the above-mentioned kinetics (i.e., the relatively rapid reaction of HNO with thiols compared to H_2O_2), it seems more likely that HNO biology is due to direct reaction with thiol-containing proteins and that the increases in, e.g., H_2O_2 represent a less significant overall effect.

HNO has also been shown to affect other thiol-containing proteins as well. For example, HNO appears to be capable of activating glucose uptake by oxidizing cysteine residues in the glucose transporter GLUT1.[43] Indeed, the preponderance of the literature regarding the possible physiological targets for HNO involves thiol-containing proteins. Based on the reported chemistry of HNO, this is not at all surprising (vide supra). Importantly, however, the nature of the interaction of HNO with protein thiol residues appears to be highly dependent on the cysteine environment, with several pathways possible (Figs. 17.1–17.3).

It is clear that thiol-containing proteins are primary (although not exclusive) targets for the biological actions of HNO. Based on the above discussions, it has been reasonably established that:

1. The reaction of HNO with thiols and thiol-containing proteins is fast. Among other established thiol modifiers (e.g., H_2O_2, peroxynitrite), HNO is an extremely facile thiol modifier.
2. The reaction of HNO with thiol-containing proteins can be very specific. For example, inhibition of thiol-containing proteins can occur without significant changes in the GSH status of cells. However, the mechanism(s) for this specificity is(are) unknown.
3. HNO-mediated thiol modification can lead to numerous possible products, some representing reversible protein modifications and others representing irreversible (or at least slowly reversible) protein modification.

17.5 HNO versus NO

It is worthwhile to briefly address the chemical and possible biological relationship between HNO and NO with regards to interactions with thiol-containing proteins. As alluded to above, thiol-containing proteins are undoubtedly possible (if not likely) targets responsible for the biological activity of HNO. It has also been proposed that thiol protein modification by NO (or more likely NO-derived species) represents a fundamental pathway in NO signaling.[44] Due to the close chemical relationship between HNO and NO and the large volume of published work indicating a role for NO in regulating thiol protein function, it is intriguing for some to speculate that the biological actions of HNO can, at least in part, be attributed to conversion to NO. One-electron reduction and protonation of NO gives HNO. The reduction potential of the $NO,H^+/HNO$ couple is -0.55 V (vs a normal hydrogen electrode (NHE), pH 7)[45] indicating that one-electron reduction of NO would be a very difficult process in biological systems. This value also indicates that oxidation of HNO to NO is favorable. Moreover, the H–NO bond is very weak (46–47 kcal mol^{-1} [46]) indicating that HNO can be readily oxidized to NO via simple H-atom abstraction by a relatively weak oxidant. Thus, HNO can be relatively easily oxidized to NO but the reverse process will be considerably more difficult in biological systems. However, somewhat in opposition to this view is more recent work indicating that NO can be reduced by, e.g., H_2S and aromatic alcohols to give HNO.[47,48] The biological relevance of these reactions remains to be established. Regardless, it is not difficult to envision that some of the biological activity of HNO can be attributed to conversion to NO (while the reverse is highly unlikely). Thus, it is reasonable to speculate that the overall biological effect associated with pharmacological administration of HNO can be due to both HNO and NO. However, with regard to effects on thiol-containing proteins, numerous lines of evidence indicate that conversion of HNO to NO is not responsible for the actions of HNO. For example, one way to distinguish between the biological actions of HNO versus NO is to add a thiol to the experiment, which is reported to selectively block HNO-mediated activity while having little effect on NO-mediated activities.[49] This effect relies on the fact that HNO readily reacts with thiols, and therefore can be "intercepted" by the addition of exogenous thiols, while NO does not directly react with thiols and therefore cannot be efficiently trapped. From a study mentioned earlier, it was found that HNO-mediated inhibition of the thiol-containing transcription factor ACE1 was actually attenuated by the addition of NO.[32] Furthermore, HNO-mediated inhibition of the cysteine dehydrogenase GAPDH is expectedly independent of O_2, whereas NO-dependent inhibition requires the presence of O_2, indicating that HNO itself can inhibit GAPDH.[50] The effect of O_2 on NO-mediated inhibition of thiol-containing proteins is not surprising since NO will not react directly with thiols but requires oxidation (in this case by O_2) to other thiol-reactive species.[51] On the other hand, the reaction of HNO with thiols is direct (not requiring any oxidation or reduction) and fast. Thus, HNO-mediated thiol modification does not require HNO conversion to NO. Indeed, the reactivity of HNO with thiols/thiol-containing proteins is straightforward (except for the possible multitude of products) and well-established. NO-mediated thiol modification, however, requires NO to be oxidized prior to reaction with the thiol, a process that is not necessarily facile.

17.6 Summary

A primary target for HNO in biological systems appears to be thiol-containing proteins. This is not surprising since the thiophilicity of HNO is well-documented and the reaction of HNO with thiols can be both kinetically facile and thermodynamically favorable. Importantly, the nature of the interaction of HNO with thiol-containing proteins can differ significantly, depending on the protein cysteine thiol environment and function. HNO-mediated modification of thiol-containing proteins can follow several distinct chemical paths, some involving reversible and possibly regulatory processes and others involving irreversible modifications. The factors governing these distinct pathways are currently a matter of some speculation and further work will be required to establish reactivity motifs that generally explain the propensity for proteins to react in the differing ways.

References

1. Doyle MP, Mahapatro SN, Broene RD, Guy JK. Oxidation and reduction of hemoproteins by trioxodinitrate(II). The role of nitrosyl hydride and nitrite. *J Am Chem Soc* 1988;**110**:593–9.
2. Irvine JC, Ritchie RH, Favaloro JL, Andrews KL, Widdop RE, Kemp-Harper BK. Nitroxyl (HNO): the cinderella of the nitric oxide story. *Trends Pharmacol Sci* 2008;**29**:601–8.
3. Wrobel AT, Johnstone TC, Liang AD, Lippard SJ, Rivera-Fuentes P. A fast and selective near-infrared fluorescent sensor for multicolor imaging of biological nitroxyl. *J Am Chem Soc* 2014;**136**:4697–705.

4. Suarez SA, Bikiel DE, Wetzler DE, Marti MA, Doctorovich F. Time-resolved electrochemical quantification of azanone (HNO) at low nano-molar level. *Analyt Chem* 2013;**85**:10262–9.
5. Johnson GM, Chozinski TJ, Salmon DJ, Moghaddam AD, Chen HC, Miranda KM. Quantitative detection of nitroxyl upon trapping with glutathione and labeling with a specific fluorogenic reagent. *Free Radic Biol Med* 2013;**63**:476–84.
6. Miranda KM. The chemistry of nitroxyl (HNO) and implications in biology. *Coord Chem Rev* 2005;**249**:433–55.
7. Bartberger MD, Fukuto JM, Houk KN. On the acidity and reactivity of HNO in aqueous solution and biological systems. *Proc Natl Acad Sci USA* 2001;**98**:2194–8.
8. Wong PSY, Hyun J, Fukuto JM, Shiroda FN, DeMaster EG, Nagasawa HT. The reaction between nitrosothiols and thiols: generation of nitroxyl (HNO) and subsequent chemistry. *Biochemistry* 1998;**37**:5362–71.
9. Sherman MP, Grither WR, McCulla RD. Computational investigation of the reaction mechanisms of nitroxyl and thiols. *J Org Chem* 2010;**75**:4014–25.
10. Kecili G, Toscano JP. Reactivity of nitroxyl-derived sulfinamides. *Biochemistry* 2012;**51**:4206–16.
11. Woo HA, Jeong W, Chang T-S, Park KJ, Park SJ, Yang JS, et al. Reduction of cysteine sulfinic acid by sulfiredoxin is specific to 2-cys peroxiredoxins. *J Biol Chem* 2004;**280**:3125–8.
12. Forman HJ, Fukuto J, Torres M. Redox signaling—chemistry defines which reactive oxygen and nitrogen species can act as second messengers. *Am J Physiol (Cell Physiology)* 2004;**287**(2):C246–56.
13. Jackson MI, Han TH, Dutton A, Ford E, Miranda KM, Houk KN, et al. Kinetic feasibility of nitroxyl (HNO) reduction by physiological reductants and biological implications. *Free Radic Biol Med* 2009;**47**:1130–9.
14. Winterbourne CC, Metodiewa D. Reactivity of biologically important thiol compounds with superoxide and hydrogen peroxide. *Free Radic Biol Med* 1999;**27**:322–9.
15. Koppenol WH, Moreno JJ, Pryor WA, Ishiropoulos H, Beckman JS. Peroxynitrite, a cloaked oxidant formed by nitric oxide and superoxide. *Chem Res Toxicol* 1992;**5**:834–42.
16. Shafirovich V, Lymar SV. Spin-forbidden deprotonation of aqueous nitroxyl (HNO). *J Am Chem Soc* 2003;**125**:6547–52.
17. Bazylinski DA, Hollocher TC. Evidence from the reaction between trioxodinitrate(II) and ^{15}NO that trioxodinitrate(II) decomposes into nitrosyl hydride and nitrite in neutral aqueous solution. *Inorg Chem* 1985;**24**:4285–8.
18. Fukuto JM, Jackson MI, Kaludercic N, Paolocci N. Examining nitroxyl (HNO) in biological systems. *Methods Enzymol* 2008;**440**:411–31.
19. DeMaster EG, Kaplan E, Shirota FN, Nagasawa HT. Metabolic activation of cyanamide by liver mitochondria, a requirement for the inhibition of aldehyde dehydrogenase enzymes. *Biochem Biophys Res Commun* 1982;**107**:1333–9.
20. DeMaster EG, Nagasawa HT, Shirota FN. Metabolic activation of cyanamide to an inhibitor of aldehyde dehydrogenase in vitro. *Pharmacol Biochem Behav* 1983;**18**:273–7.
21. DeMaster EG, Shirota FN, Nagasawa HT. The metabolic activation of cyanamide to an inhibitor of aldehyde dehydrogenase is catalyzed by catalase. *Biochem Biophys Res Commun* 1984;**122**:358–65.
22. DeMaster EG, Shirota FN, Nagasawa HT. Catalase mediated conversion of cyanamide to an inhibitor of aldehyde dehydrogenase. *Alcohol* 1985;**2**:117–21.

23. DeMaster EG, Redfern B, Nagasawa HT. Mechanisms of inhibition of aldehyde dehydrogenase by nitroxyl, the active metabolite of the alcohol deterrent agent cyanamide. *Biochem Pharmacol* 1998;**55**:2007–15.
24. Nagasawa HT, DeMaster EG, Redfern B, Shirota FN, Goon DJW. Evidence for nitroxyl in the catalase-mediated bioactivation of the alcohol deterrent agent cyanamide. *J Med Chem* 1990;**33**:3120–2.
25. Nagasawa HT, Lee MJC, Kwon CH, Shirota FN, DeMaster EG. An *N*-hydroxylated derivative of cyanamide that inhibits yeast aldehyde dehydrogenase. *Alcohol* 1992;**9**:349–53.
26. Hammond AH, Fry JR. Effect of cyanamide on toxicity and glutathione depletion in rat hepatocyte cultures: differences between two dichloropropanol isomers. *Chem Biol Interact* 1999;**122**:107–15.
27. Lopez BE, Rodriguez CE, Pribadi M, Cook N, Shinyashiki M, Fukuto JM. Selective inhibition of yeast glycolysis by nitroxyl (HNO): implications to HNO pharmacology/physiology. *Arch Biochem Biophys* 2005;**442**:140–8.
28. Shen B, English AM. Mass spectrometric analysis of nitroxyl-mediated protein modification: comparison of products formed with free and protein-based cysteines. *Biochemistry* 2005;**44**:14030–44.
29. Keceli G, Toscano JP. Reactivity of C-terminal cysteines with HNO. *Biochemistry* 2014;**53**:3689–98.
30. Winge DR, Graden JA, Posewitz MC, Martins LJ, Jensen LT, Simon JR. Sensors that mediate copper-specific activation and repression of gene expression. *J Biol Inorg Chem* 1997;**2**:2–10.
31. Shinyashiki M, Lopez BE, Rodriguez CE, Fukuto JM. Yeast model systems for examining nitrogen oxide biochemistry/signaling. *Methods Enzymol* 2005;**396**:301–16.
32. Cook NM, Shinyashiki M, Jackson MI, Leal FA, Fukuto JM. Nitroxyl (HNO)-mediated disruption of thiol proteins: inhibition of the yeast transcription factor ACE1. *Arch Biochem Biophys* 2003;**410**:89–95.
33. Shinyashiki M, Chiang KT, Switzer CH, Gralla EB, Valentine JS, Thiele DJ, et al. The interaction of nitric oxide (NO) with the yeast transcription factor ace1: a model system for NO-protein thiol interactions with implications to metal metabolism. *Proc Natl Acad Sci USA* 2000;**97**(6):2491–6.
34. Vaananen AJ, Kankuri E, Rauhala P. Nitric oxide-related species-induced protein oxidation: reversible, irreversible and protective effects on enzyme function of papain. *Free Rad Biol Med* 2005;**38**:1102–11.
35. Vaananen AJ, Salmenpera P, Hukkanen M, Rauhala P, Kankuri E. Cathepsin B is a differentiation-resistant target for nitroxyl (HNO) in THP-1 monocyte/macrophages. *Free Rad Biol Med* 2006;**41**:120–31.
36. Cheong E, Tumbev V, Abramson J, Salama G, Stoyanovsky DA. Nitroxyl triggers Ca^{2+} release from skeletal and cardiac sarcoplasmic reticulum by oxidizing ryanodine receptors. *Cell Calcium* 2005;**37**:87–96.
37. Tochetti CG, Wang W, Froelich JP, Huke S, Aon MA, Wilson GM, et al. Nitroxyl improves cellular heart function by directly enhancing cardiac sarcoplasmic reticulum Ca^{2+} cycling. *Circ Res* 2007;**100**:96–104.
38. Froehlich JP, Mahaney JE, Keceli G, Pavlos CM, Goldstein R, Redwood AJ, et al. Phospholamban thiols play a central role in activation if the cardiac muscle sarcoplasmic reticulum calcium pump by nitroxyl. *Biochemistry* 2008;**47**:13150–2.

39. Lancel S, Zhang J, Evangelista A, Trucillo MP, Tong XY, Siwik DA, et al. Nitroxyl activates SERCA in cardiac myocytes via glutathionylation of cysteine 674. *Circ Res* 2009;**104**:720–3.
40. Gao WD, Murrayu CI, Tian Y, Zhong X, DuMond JF, Shen X, et al. Nitroxyl-mediated disulfide bond formation between cardiac myofilament cysteines enhances contractile function. *Circ Res* 2012;**111**:1002–11.
41. Feelisch M. Nitroxyl gets to the heart of the matter (Commentary). *Proc Natl Acad Sci USA* 2003;**100**:4978–80.
42. Jackson MI, Fields HF, Lujan TS, Cantrell MM, Lin J, Fukuto JM. The effects of nitroxyl (HNO) on H_2O_2 metabolism and possible mechanisms of HNO signaling. *Arch Biochem Biophys* 2013;**538**:120–9.
43. Salie MJ, Oram DS, Kuipers DP, Scripture JP, Chenge J, MacDonald GJ, et al. Nitroxyl (HNO) acutely activates the glucose uptake activity of GLUT1. *Biochimie* 2012;**94**:864–9.
44. Smith BC, Marletta MA. Mechanisms of S-nitrosothiol formation and selectivity in nitric oxide signaling. *Curr Opin Chem Biol* 2012;**16**:498–506.
45. Shafirovich V, Lymar SV. Nitroxyl and its anion in aqueous solutions: Spin states, protic equilibria, and reactivities toward oxygen and nitric oxide. *Proc Natl Acad Sci USA* 2002;**99**:7340–5.
46. Gomes JRB, Ribeiro da Silva MDMC, Ribeiro da Silva MAV. Solvent and structural effects in the N–H bond hemolytic dissociation energy. *J Phys Chem A* 2004;**108**: 2119–30.
47. Suarez SA, Neuman NI, Munoz M, Alvarez L, Bikiel DE, Brondino CD, et al. Nitric oxide is reduced to HNO by proton—coupled nucleophilic attack by ascorbate, tyrosine and other alcohols. A new route to HNO in biological media? *J Am Chem Soc* 2015;**137**:4720–7.
48. Eberhardt M, Dux M, Namer B, Miljkovic J, Cordasic N, Will C, et al. H_2S and NO cooperatively regulate vascular tone by activating a neuroendocrine HNO-TRPA1-CGRP signaling pathway. *Nat Commun* 2014;**5**.
49. Pino RZ, Feelisch M. Bioassay discrimination between nitric oxide (NO) and nitroxyl (NO^-) using L-cysteine. *Biochem Biophys Res Commun* 1994;**201**:54–62.
50. Lopez BE, Wink DA, Fukuto JM. The inhibition of glyceraldehyde-3-phosphate dehydrogenase by nitroxyl (HNO). *Arch Biochem Biophys* 2007;**465**:430–6.
51. Fukuto JM, Cisneros CJ, Kinkade RL. A comparison of the chemistry associated with the biological signaling and actions of nitroxyl (HNO) and nitric oxide (NO). *J Inorg Biochem* 2013;**118**:201–8.

Is Azanone Endogenously Produced in Mammals?

18

M.A. Marti, L. Álvarez, S.A. Suarez and F. Doctorovich
Universidad de Buenos Aires and CONICET, Buenos Aires, Argentina

Chapter Outline

18.1 The newest small molecule signaling agent

Nitric oxide was the first small reactive molecule shown to be produced in vivo under particular circumstances to act as a signaling agent, as evidenced by the presence of a specific physiological response. Its discovery was later followed by the recognition of other endogenously generated small molecule species that in principle could also act as signaling agents, namely carbon monoxide (CO) and hydrogen sulfide (H_2S) among others.[1] These small agents changed the paradigm in cell signaling and opened up a whole new world of bioinorganic chemistry and pharmacology. Although HNO has been shown to have relevant pharmacological properties,[2–6] the question of whether it actually can be considered, or not, a specific signaling molecule, i.e., produced in vivo under given circumstances and with a certain purpose, is still a matter of debate.[1,7,8] Key for answering this interrogation is an understanding of how, when, and where HNO can be endogenously (i.e., not derived from an HNO-releasing molecule) produced.

The Chemistry and Biology of Nitroxyl (HNO). DOI: http://dx.doi.org/10.1016/B978-0-12-800934-5.00018-9

18.2 Possible routes for enzymatic HNO production

18.2.1 Incomplete arginine oxidation by nitric oxide synthase

Possibly, the oldest suggested mechanism for endogenous HNO production is derived from experiments performed on the heme-containing nitric oxide synthase (NOS) under tetrahydrobiopterin (BH_4)-deficient conditions.[9–12] In the accepted mechanism for NO production by NOS, the substrate arginine is oxidized in two NOS oxidation cycles using molecular oxygen and electrons derived from nicotinamide adenine dinucleotide reduced (NADH), according to the following mechanism: (Scheme 18.1).

The first cycle comprises the simple oxidation of L-arginine to yield *N*-hydroxy-L-arginine (NOHA), which occurs similarly to what is observed in aliphatic or aromatic carbon hydroxylating reactions mediated by P450s. As shown in Scheme 18.1, molecular oxygen is first activated (shuffling two electrons) leading to an oxo-ferryl species (Compound I), which subsequently hydroxylates one of the two guanidinium nitrogen atoms. Interestingly, as shown by Marletta and coworkers[13,14] during the second reaction cycle, the same mechanism would lead to citrulline and HNO, instead of NO. This is evidenced, e.g., when a second reaction cycle with NOHA, as the substrate, is started using the peroxide shunt (see the following section). This shunt produces Compound I through the reaction of the enzyme ferric resting state with hydrogen peroxide (i.e., without the need of an internal electron transfer), leading in this case to the formation of a ferrous nitrosyl instead of the ferric nitrosyl required for NO release. The key to NO (and not HNO) production by NOS is provided by BH_4, which at the beginning of each cycle provides one electron to activate molecular oxygen, leading to the formation of a pterin-centered radical ($BH_4^{+\bullet}$), which was experimentally observed using ENDOR spectroscopy.[13,14] This radical would be responsible for removing one electron on the last step of the reaction leading to NO, or a ferric nitrosyl that releases NO, instead of HNO or the observed stable ferrous nitrosyl.

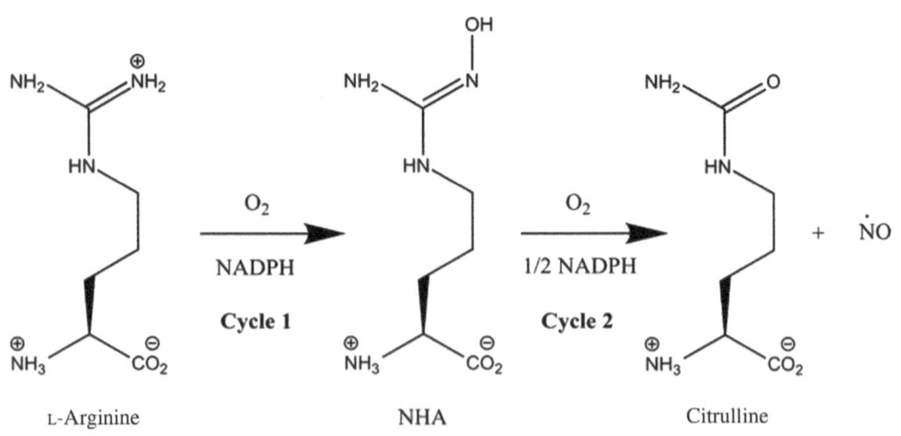

Scheme 18.1 Two cycles of NOS reaction mechanism leading to citrulline and NO.

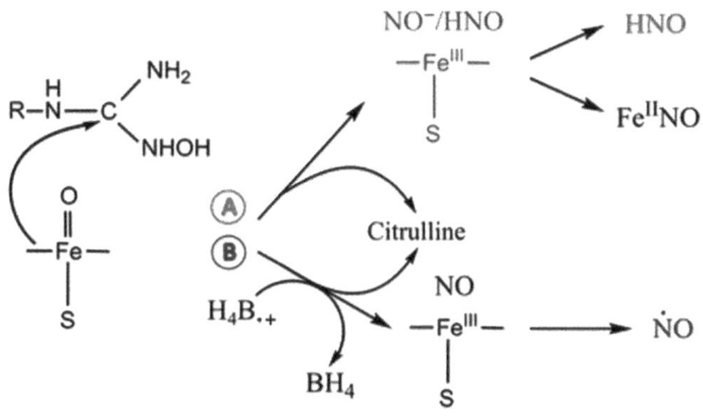

Scheme 18.2 Pathway A: In the absence of $BH_4^{+\bullet}$ the reaction produces HNO. The species in gray were not experimentally detected. Pathway B: if the electron is provided by $BH_4^{+\bullet}$, then NO is produced.

Although the electron donating/withdrawing role of BH_4 is consistent with all available experimental observations, and explains why deficiency of BH_4 would lead NOS to produce HNO, several issues remain to be addressed to confirm that it is a physiologically relevant azanone source. First, it should be noted that evidence concerning NOS production of HNO is indirect, based on the deficiency of "NO effects" in the absence of BH_4, and the requirement of the presence of superoxide dismutase (SOD), which presumably converts HNO to NO, to recover these effects. The most convincing evidence for its presence comes from the observation concerning the formation of the ferrous nitrosyl during the second NOS cycle. Also, the mechanism of HNO release from the heme active site is still a matter of debate (Scheme 18.2).

18.2.2 Heme protein-mediated oxidation of hydroxylamines

Another well-established potential endogenous HNO source, which is directly related to the described NOS mechanism, relies on the generic heme protein-mediated oxidation of hydroxylamines (HAs), such as HA itself, or hydroxyurea. The reaction has been shown to occur in vitro with myoglobin, horse radish peroxidase, and myeloperoxidase, using hydrogen peroxide as the oxidizing agent.[15] HNO production was detected using a selective assay which consisted on adding glutathione, GSH,[16] and quantifying by HPLC (high pressure liquid chromatography) the $GS(O)NH_2$ product formed through reaction (Eq. 18.1).

$$HNO + GSH \rightarrow GSNHOH \rightarrow GS(O)NH_2 \qquad (18.1)$$

The yields of $GS(O)NH_2$ observed were higher in proteins in which the heme displays a histidine coordinated as the proximal ligand, rather than a tyrosine or a cysteine. These results indicate that the protein structure controls the efficiency of

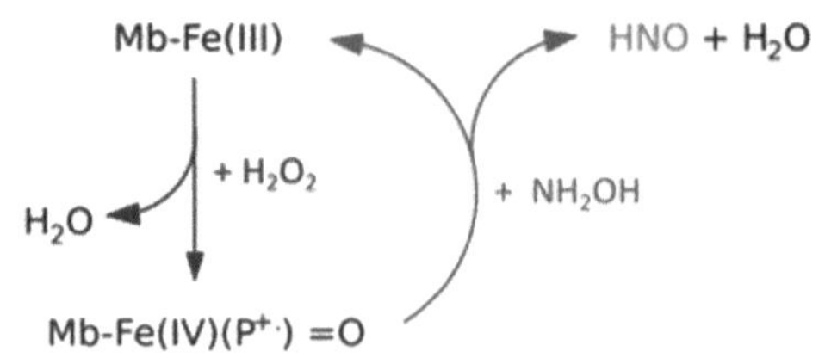

Scheme 18.3 Reaction mechanism proposed in the case of myoglobin.

the reaction since it may affect differentially the stability of the oxo-ferryl oxidant and/or the substrate binding and product release steps.

The underlying mechanism is assumed to be similar to the NOS peroxide shunt mechanism (described above) which involves first the reaction of the heme group with H_2O_2 to form an oxo-ferryl adduct (known as Compound I) which subsequently reacts with HA to yield HNO (Scheme 18.3). It is interesting to note how the absence of possible oxidation of the produced nitroxyl, which in NOS is mediated by BH_4, results in HNO instead of NO production.

Concerning the physiological relevance of the presented mechanism leading to azanone, it is clear that the odds for its occurrence will be directly related to the presence of HAs and active peroxidases. In mammals, although HA is not produced endogenously by enzymatic means, it is used or produced pharmacologically and interestingly, it has been shown to induce significant cardiovascular effects which could hint to the presence of HNO.[17–19]

18.2.3 Enzymatic NO reduction

The third proposal for enzymatic production of HNO relies on NO reduction as performed by either SOD[20] or xanthine oxidase (XO),[21] both of which are able to reduce O_2 to superoxide (Scheme 18.4). SODs are a group of proteins that catalyze the dismutation of superoxide into molecular oxygen and hydrogen peroxide. There are two families of SODs: CuZnSODs, present in cytosol of eukaryotic organisms, and Mn/FeSODs, present in prokaryotes and mitochondria. They perform a critical role in all organisms exposed to oxygen by preventing superoxide-induced cell damage. Early studies on the SOD reaction with NO showed that while CuZnSOD was not inhibited by NO, Mn/FeSODs were sensitive to NO.[20] Murphy et al.[22] also suggested that CuZnSOD was able to convert NO to nitroxyl anion as judged by the formation of nitrosylmyoglobin. Furthermore, other studies also suggested that Mn/FeSODs were able to disproportionate NO to NO^+ and NO^-/HNO. Although it should be noted that this reaction is substantially unfavorable. In this case, evidence supporting azanone production relied on the formation of HA from supposed reaction of HNO with thiols. More recent studies[23] analyzed in detail the reaction of *Escherichia coli* MnSOD with NO. The authors clearly showed that MnSOD catalyzes NO decomposition, and becomes inactivated during the process (especially in aerobic conditions) in part due to active site tyrosine nitration. Tyrosine nitration in MnSOD is known to be mediated by peroxynitrite and enhanced by direct

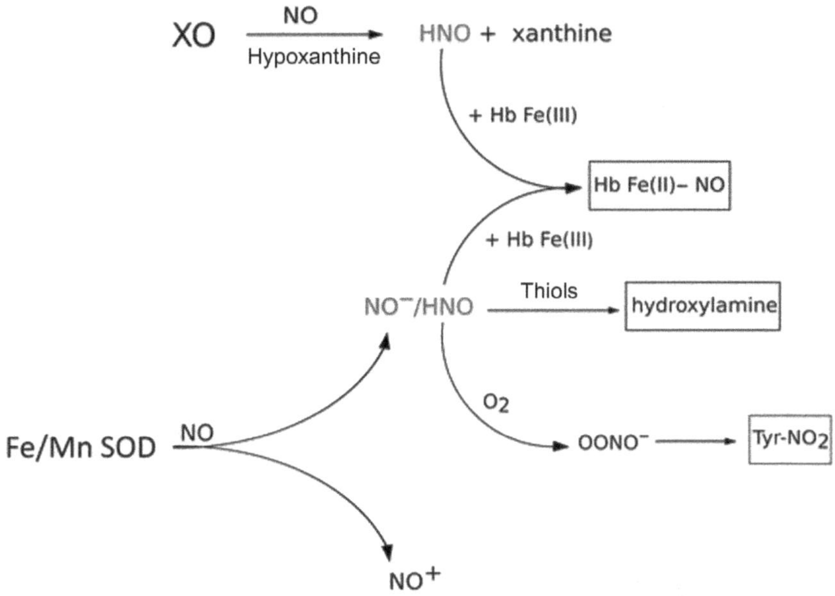

Scheme 18.4 Mechanisms and evidence for enzymatic reduction of NO to HNO. The squared species represent the ones that were experimentally detected.

Mn-peroxynitrite interaction; the authors thus suggested that reaction of HNO with oxygen would lead to peroxynitrite production and the observed results.[24] However, other authors have pointed out that HNO autoxidation does not produce peroxynitrite.[25] Again, final evidence for MnSOD-mediated NO to HNO conversion is derived from the observed formation of nitrosyl hemoglobin from met-hemoglobin.

XO is a metalloflavoprotein involved in the metabolism of purines and pyrimidines, where it catalyzes the oxidation of hypoxanthine to xanthine, and that of xanthine to uric acid, with the concomitant reduction of molecular oxygen to superoxide. As for MnSOD, exposure of XO to NO and hypoxanthine as the substrate, was shown to result in enzyme inactivation and conversion of met-hemoglobin to nitrosyl hemoglobin, again suggesting NO to HNO conversion.[21]

Although both the above-described reactions are potential endogenous sources of HNO, and the reactions make perfect biochemical sense, it is important to note: first, both reactions were shown to occur in vitro, and thus lacked a proper biological context; and second, and perhaps more important, in both cases evidence of HNO production is indirect and inconclusive. Particularly relevant is careful interpretation of the evidence presented using met-myoglobin. Met-myoglobin—as well as most ferric heme proteins—is known to react with NO leading to ferric nitrosyls, which react with nucleophiles, like hydroxide ion, yielding a ferrous heme. This in turn can react with a second NO molecule to yield the observed ferrous nitrosyl. This process, called reductive nitrosylation, could be responsible for the observed results, even in the absence of HNO production. It is also important to know that there is an absence of follow up studies on these reactions, which could not only provide definite proof of their HNO production capacity (e.g., using recent developments in HNO detection chemistry) but also analyze them in a physiologically relevant context.

18.3 Biochemical spontaneous "reducing" pathways leading to HNO

Chemical reduction of $NO^{\bullet}$ to HNO under physiological conditions was commonly thought not to be significant, or even possible, mainly due to the estimated negative reduction potential of $NO^{\bullet}$ which results in an $E^{\circ\prime} \approx -0.5$ V at pH 7.[26,27] Nonetheless, in the last decades several works, which studied reactions of $NO^{\bullet}$ (g), or NO donors with common biological mild reductants bearing alcohol functional groups (such as ascorbic acid, ubiquinol, or thiols), showed indirect evidence of the presence of azanone. Although none of them proved HNO production unequivocally or analyzed the underlying mechanism, they kept the idea of NO to HNO conversion alive, allowing to research this venue once better methodologies were developed, as will be described below.

18.3.1 NO reaction with hydrogen sulfide and thiols

Common proposals concerning possible "biological compatible" spontaneous HNO production are based on the decomposition (or reaction) of nitrosothiols (RSNOs) with other thiols. In vitro experiments show, e.g., that the reaction of RSNO with excess thiol (RSH), leads to disulfide and HNO formation.[28] Interestingly, this potential azanone source relies on NO (in aerobic environments), or HNO itself, to firstly yield the nitrosothiol.[29] More appealing are thus proposals based on the direct reaction of NO with thiols or H_2S (as shown below in Scheme 18.6). Already back in 1982, Pryor and coworkers had found that both aliphatic and aromatic thiols react with NO under anaerobic conditions to yield N_2O and disulfide, clearly suggesting HNO as an intermediate.[30] More than 10 years later, it was also shown by Nagasawa and coworkers that albumin, which displays several exposed CysSH residues, also produces N_2O after reaction with NO or NO-donors.[31] Although these results (Scheme 18.5) show the potential of NO and thiol reactions as physiological

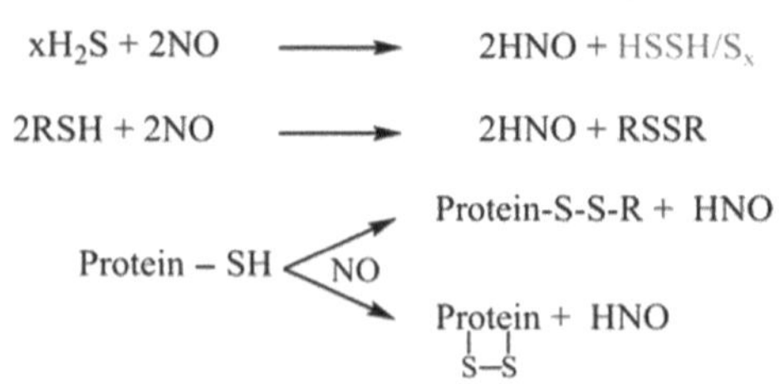

Scheme 18.5 NO reaction with thiols (or thiol-containing proteins) as a potential in vivo HNO source. The species in gray are not well known.

HNO sources, they lack proper biological context. Also important to note, is that the reaction of NO with a thiol is expected to lead to some RSN(H)O* radical intermediate, whose S–N bond is quite strong, and thus its HNO-releasing capacity may be limited, leading instead directly to N_2O.

More recently, works showing endogenous, physiologically relevant, HNO production have appeared.[32] The authors first showed that HNO derived from Angeli's salt (AS) is able to activate transient receptor potential channel A1 (TRPA1) in sensory neurons, leading to a release of calcitonin gene-related peptide that results in positive inotropic and lusitropic effects.[2,33] Subsequently, using a combination of chemical site-directed mutagenesis and mass-spectrometry analysis, they showed that the activation was mediated by HNO-induced disulfide formation on TRPA1.

The possible physiological origin of HNO in the studied system came from previous studies of the NO reaction with H_2S.[34,35] The authors therefore set out to show that the reaction of NO with H_2S leads to HNO, which was accomplished in vitro using an HNO-selective electrode (see chapters: Electrochemical detection of Azanone, Detection of HNO by Membrane Inlet Mass Spectrometry and references therein) and in cells by using the HNO fluorescent detector CuBOT (see Chapter 10 on Fluorescent Probes for HNO Detection and references therein). Although use of CuBOT as a selective HNO detection method has been questioned, authors also showed that neither NO nor H_2S alone resulted in TRPA1 activation (or chemical modifications), but their combination produced even more efficient activation than AS, showing again the formation of key specific disulfide bridges. Thus it was clear that H_2S and NO worked together to produce a species (probably HNO) that activates the channel.

Further proof of endogenous HNO generation from this reaction came from cell culture studies, where the authors demonstrated that neurons that expressed both the H_2S-producing enzyme cystathionine beta synthase (CBS), and NO-producing iNOS, showed basal levels of HNO production that were inhibited by selectively blocking either CBS or iNOS. Last but not least, the authors showed that previously known H_2S-evoked vasodilator effects are dependent on NO production and the HNO-TRPA1 pathway, providing a potential specific physiological role for endogenously produced azanone. These results are summarized in Scheme 18.6. Recently, Feelisch and coworkers[36] further studied the reaction products and biological effects

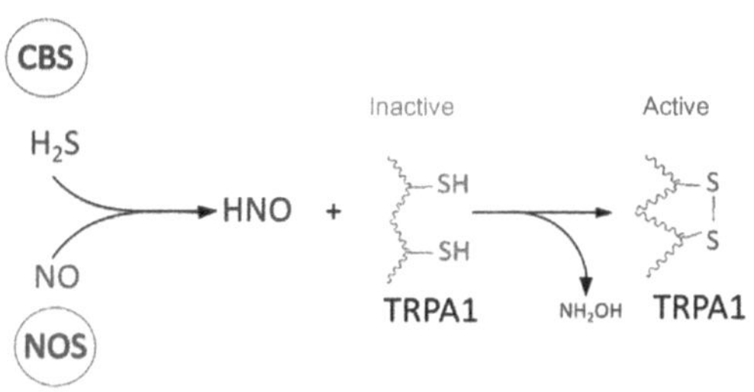

Scheme 18.6 Neuronal HNO production derived from reaction of NO with H_2S, and subsequently activation of TRPA1 channels.

in NO/H_2S systems. Although they reported nitroxyl as one of the products, they also showed evidence of the presence of polysulfides and anionic S/N-hybrid species with apparently distinct biological activities.

18.3.2 NO reaction with vitamin C and aromatic alcohols

The last, quite unsuspected, and a physiological compatible HNO source relies on the reaction of NO with mono- or di-alcohols, like ascorbic acid (vitamin C), and aromatic ones such as hydroquinone or even tyrosine, among other common biochemical compounds of this type (Scheme 18.7).[37,38] As shown in Chapter 1, HNO: Redox Chemistry and Interactions With Small Inorganic Molecules, the first evidence showing that this reaction produces azanone was obtained in vitro. Detailed mechanistic studies show that the reactions occur through a proton-coupled nucleophilic attack of the alcohol to $NO^{\bullet}$, which yields an intermediate RO-N(H)$O^{\bullet}$ species, which subsequently decomposes and releases HNO. The analysis also showed that the reaction can be described as bimolecular—first order both in NO and alcohol—and characterized by an effective bimolecular reaction rate constant (k_{eff}) displaying values in the $0.1-10\ M^{-1}\ s^{-1}$ range (Table 18.1). The rate is related to the stability of the alcohol-derived phenoxyl radical, and thus its redox potential. Bi-alcohols such as ascorbic acid, that allow for two HNO production cycles, are more effective.[37,38]

Having shown and studied the reaction in vitro, the question was whether it can happen in a real biological context. To answer this, two different situations were studied. In the first, bovine arterial endothelial cells were pretreated with 1 mM ascorbate for 1 h. These cells, as well as neurons and immune cells, are known to accumulate vitamin C to concentrations higher than millimolar and to produce basal amounts of NO.[49] After washing, cells were exposed to the HNO fluorescence sensor CuBOT1[50] (described in the chapter: Fluorescent Probes for HNO Detection). The results, presented in Fig. 18.1, showed a clear increase of the fluorescence of

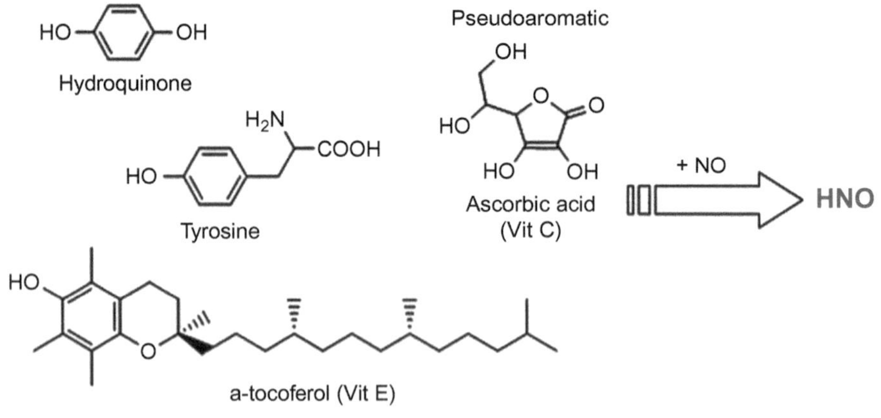

Scheme 18.7 Vitamins and other biological relevant compounds produce HNO from NO.

Table 18.1 HNO producing reactions comparative kinetics and plasmatic concentration

Compound	E°′ (pH 7) $RX^{\bullet},H^{+}/RXH$ (volt)	PEC[a] ($\mu g\ mL^{-1}$) or (μM)	k_{app} ($M^{-1}s^{-1}$)	References (E°′)	References (PEC)
Ascorbate	0.28	8.8–440 or 50–2500	8.0 ± 0.5	[39]	[40]
Hydroquinone	0.10	–	6.1 ± 0.4	[39]	–
Phenol	0.91	–	3.2 ± 0.4	[39]	–
Tyrosine	0.97	–	0.9 ± 0.4	[39]	–
Tocopherol	0.48	129 or 300	3.0 ± 0.4	[41]	[42]
Salicylic acid	1.10	110–160 or 800–1600	0.34 ± 0.06	[43]	[44]
Acetaminophen	0.70	10–20 or 70–130	1.4 ± 0.6	[45]	[46]
Piroxicam	0.80	4.5 or 10	0.26 ± 0.04	[47]	[48]

[a]Plasmatic effective concentration (PEC): Concentration of a drug in plasma required to produce a desired pharmacological effect in most patients.

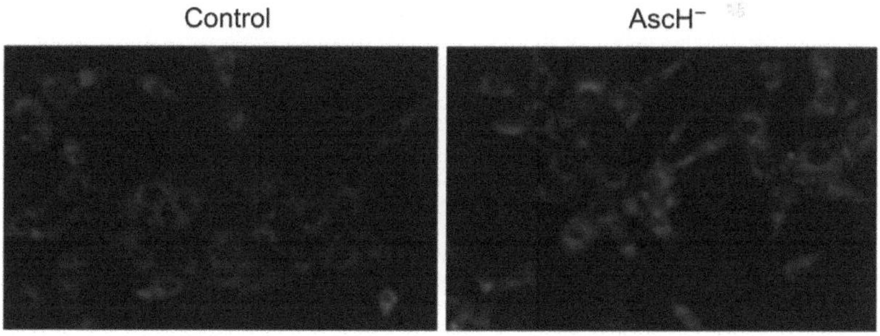

Figure 18.1 Intracellular HNO formation in bovine arterial endothelial cells as revealed by the HNO fluorescence sensor, CuBOT1.

ascorbate pretreated cells. In the second case, RAW macrophages were stimulated with lipopolysaccharide and interferon gamma, which are known to induce iNOS and subsequent NO production. Addition of 1 mM ascorbic acid to the extracellular medium resulted in an immediate rise of the HNO signal as detected by an HNO-sensing electrode. Although data from CuBOT should be taken with care, since Cu-based systems are now known to be susceptible to reduction by compounds other than HNO, together these experiments showed that NO conversion to HNO through the reaction with ascorbic acid and other aromatic alcohols is possible under physiological concentrations of both reactants and in complex biological environments such as inside a living cell.

The above-described reactions become relevant in light of physio-, patho-, and pharmacological conditions were circulating levels of these type of compounds are high (e.g., when under consumption of aspirin as an analgesic) and combined with metabolic or pharmacologically induced NO production. It is important to realize that aromatic alcohols are found in many trendy antioxidants or vitamins, such as tocopherol, (consumed in certain foods or as dietary supplements) or in over-the-counter extensively used drugs like salicylic acid (active form of common aspirin) or acetaminophen, several of which are able to convert NO to HNO (Table 18.1). Several works have reported nonadditive unexpected effects when L-Arg, which augments endogenous NO production, was co-administered with some of these compounds.[38] A retrospective view of these results suggests HNO formation and its specific biological actions as the most likely source of the unexpected effects. Also interesting, are recent comparisons between NO-aspirin and HNO-aspirin which showed distinct pharmacological effects and promising therapeutic potential.[51,52]

In any case, the alcohol mediated NO to HNO conversion seems to occur, and the rates, although relatively slow (which is the main drawback for this mechanism), allow for estimating possible HNO physiological concentration ranges. For example, assuming that NO concentrations in inflammatory conditions are in the 10–100 μM range,[53–55] and that circulating concentrations of NO-reducing alcohols can be as high as 2500 μM, HNO concentrations could easily reach low nM levels, which could under the adequate circumstances, produce a biological response.[3,56,57] What remains ahead is to show that the reaction actually occurs in vivo, to determine under which conditions, and what is the resulting physiological-specific outcome of its presence.

A final note on the NO to HNO conversion mechanisms is related to the NO, H^+/HNO reduction potential. It is important to note that the reported redox potential has not, and possibly cannot, be determined directly, since it is an irreversible process. The value is thus estimated and relies on computed/estimated HNO and NO formation energies in solution, and given the observed evidence of NO reduction, it should probably be revisited.

18.4 Physiological context and effects of endogenous HNO production

The results reviewed above show that there are several potential routes for endogenous, physiological relevant HNO production, with some of them showing solid evidence of their occurrence in cells and even in vivo. Having thus established some of the biological pathways that could lead to HNO, the questions that are now brought forward are:

1. What are the most likely conditions and places for physiological HNO production, or in other words, when, where, and most important, why is HNO produced in vivo?
2. What are the consequences or effects of endogenous HNO production? What is it produced for?
3. Given the prevalence of NO to HNO interconversion, which effects usually attributed to NO could indeed be mediated by HNO?

To begin answering these questions, the first important thing to notice is that HNO production routes are of two kinds, those relying on the oxidation of amines (Sections 2.1 and 2.2), and those relying on NO reduction. In the first case, either if HNO is produced by BH_4-free NOS or by other heme protein acting as a peroxidase, it is clear that an oxidizing or even oxidatively stressed environment is needed. Molecular oxygen is required for NOS function and the presence of hydrogen peroxide, superoxide or perhaps peroxynitrite are needed to produce significant amounts of active Compound I in heme peroxidases. Moreover, the environment should also favor formation of HAs, such as NOHA, hydroxyurea, HA, or other as-yet unidentified species. Of course, any increase in HA concentration induced pharmacologically would promote HNO production by these means.

The problem with an oxidizing environment is that oxygen is an efficient HNO scavenger and thus formation of HNO could be masked by its reaction with oxygen, leading to peroxynitrite, although the kinetic viability of this reaction has been suggested to be minimal. In these situations, its main effects would be to produce nitro-oxidative damage. Also, in an oxidizing environment, conversion of HNO to NO, either chemically or enzymatically, is also a likely event. Since its main effects in this situation would also be to produce oxidative damage, it could be very difficult if not impossible to distinguish which one is responsible for the observed effects.

On the other hand, for the NO to HNO interconversion routes, it is clear that a hypoxic reducing environment is needed, since elevated concentrations of oxygen would not only result in thiol/H_2S/aromatic alcohols/MnSOD and XO oxidation, but would also significantly react with HNO itself. However, since NO is also strictly required, either nonoxidative NO production should be looked for, or a situation where NOS-derived NO production is confronted with excess reductants, or NO transported away from NOS by a metal center or in the form of a nitrosothiol.

A possible situation for HNO production could be in mitochondria-rich tissues under hypoxic conditions. This situation leads to not only NO production through nitrite reductase activity of globins, a widely accepted NO source,[58–60] but also to an accumulation of respiratory chain reduced intermediates (like reduced quinones) which could react with NO, leading to HNO. Also interesting to note is that NO has been shown to have significant effects in mitochondrial respiration, mainly through the partial inhibition of cytochrome c oxidase (CCO).[58] Inhibition of CCO could also lead to accumulation of the reduced quinone pool, creating an ideal situation for NO to HNO conversion, especially if oxygen tension is low. Another ideal condition for HNO production by these means could be induced pharmacologically by the joint incorporation of an NO promoting or releasing compounds (like L-Arg) with common aromatic alcohol bearing drugs (i.e., aspirin) or foods rich in antioxidants (like vitamin C). These combinations are not uncommon, and there are several reports of unsuspected effects, which retrospectively could be attributed to HNO formation.[61]

Concerning the reaction with H_2S, it is clear that primary sites suspected for HNO production are those where H_2S and NO concentrations are high enough, like those where NOS and CBS are coexpressed and active, although due to their diffusive nature, they do not need to be produced inside the same cell. This mechanism is also so far the only one where a direct link between endogenous azanone production and its effect is shown to occur. In neurons expressing TRPA1, combination of enzymatically produced H_2S and NO (either by the same neurons or nearby endothelial cells) may produce HNO, which would react with cysteine targets present in the channel, triggering formation of disulfide bridges and channel activation. Since TRPA1 activation is related to release of calcitonin gene-related peptide that results in positive inotropic and lusitropic effects, TRPA1 activation emerges as the possible underlying mechanism of several HNO observed pharmacological effects.

In summary, in the last decade several possibilities of endogenous HNO production mechanisms have emerged. Their complex and sometimes unsuspected nature suggest that they could be operative under different conditions and environments and more important, that a revision of many of the effects believed to be mediated by NO is needed in light of the recently presented findings.

References

1. Fukuto JM, Carrington SJ, Tantillo DJ, Harrison JG, Ignarro LJ, Freeman BA, et al. Small molecule signaling agents: the integrated chemistry and biochemistry of nitrogen oxides, oxides of carbon, dioxygen, hydrogen sulfide, and their derived species. *Chem Res Toxicol* 2012;**25**(4):769–93.
2. Paolocci N, Katori T, Champion HC, St John ME, Miranda KM, Fukuto JM, et al. Positive inotropic and lusitropic effects of HNO/NO−in failing hearts: independence from beta-adrenergic signaling. *Proc Natl Acad Sci USA* 2003;**100**(9):5537–42.
3. Miranda KM, Paolocci N, Katori T, Thomas DD, Ford E, Bartberger MD, et al. A biochemical rationale for the discrete behavior of nitroxyl and nitric oxide in the cardiovascular system. *Proc Natl Acad Sci USA* 2003;**100**(16):9196–201.
4. Moncada S, Palmer RM, Higgs EA. Nitric oxide: physiology, pathophysiology, and pharmacology. *Pharmacol Rev* 1991;**43**(2):109–42.
5. Yuill KH, Yarova P, Kemp-Harper BK, Garland CJ, Dora KA. A novel role for HNO in local and spreading vasodilatation in rat mesenteric resistance arteries. *Antioxid Redox Signal* 2011;**14**(9):1625–35.
6. Bullen ML, Miller AA, Andrews KL, Irvine JC, Ritchie RH, Sobey CG, et al. Nitroxyl (HNO) as a vasoprotective signaling molecule. *Antioxid Redox Signal* 2010;**14**(9):1675–86.
7. Martin W. Nitroxyl anion—the universal signalling partner of endogenously produced nitric oxide? *Br J Pharmacol* 2009;**157**(4):537–9.
8. Kemp-Harper BK. Nitroxyl (HNO): a novel redox signaling molecule. *Antioxid Redox Signal* 2011;**14**(9):1609–13.
9. Adak S, Wang Q, Stuehr DJ. Arginine conversion to nitroxide by tetrahydrobiopterin-free neuronal nitric-oxide synthase: implications for mechanism. *J Biol Chem* 2000;**275**(43):33554–61.
10. Schmidt HHHW, Hofmann H, Schindler U, Shutenko ZS, Cunningham DD, Feelisch M. NO from NO synthase. *Proc Natl Acad Sci USA* 1996;**93**(December):14492–7.
11. Fukuto JM, Wallace GC, Hszieh R, Chaudhuri G. Chemical oxidation of *N*-hydroxyguanidine compounds. *Biochem Pharmacol* 1992;**43**(3):607–13.

12. Tantillo DJ, Fukuto JM, Hoffman BM, Silverman RB, Houk KN. Theoretical studies on NG-hydroxy-L-arginine and derived radicals: implications for the mechanism of nitric oxide synthase. *J Am Chem Soc* 2000;**122**(3):536–7.
13. Stoll S, Nejatyjahromy Y, Woodward JJ, Ozarowski A, Marletta MA, Britt RD. Nitric oxide synthase stabilizes the tetrahydrobiopterin cofactor radical by controlling its protonation state. *J Am Chem Soc* 2010;**132**(33):11812–23.
14. Woodward JJ, Nejatyjahromy Y, Britt RD, Marletta MA. Pterin-centered radical as a mechanistic probe of the second step of nitric oxide synthase. *J Am Chem Soc* 2010; **132**(14):5105–13.
15. Donzelli S, Espey MG, Flores-Santana W, Switzer CH, Yeh GC, Huang J, et al. Generation of nitroxyl by heme protein-mediated peroxidation of hydroxylamine but not *N*-hydroxy-L-arginine. *Free Radic Biol Med* 2008;**45**:578–84.
16. Johnson GM, Chozinski TJ, Gallagher ES, Aspinwall CA, Miranda KM. Glutathione sulfinamide serves as a selective, endogenous biomarker for nitroxyl after exposure to therapeutic levels of donors. *Free Radic Biol Med* 2014;**76**:299–307.
17. Huang Y. Hydroxylamine-induced relaxation inhibited by K+ channel blockers in rat aortic rings. *Eur J Pharmacol* 1998;**349**(1):53–60.
18. Booth BP, Tabrizi-Fard MA, Fung HL. Calcitonin gene-related peptide-dependent vascular relaxation of rat aorta. An additional mechanism for nitroglycerin. *Biochem Pharmacol* 2000;**59**(12):1603–9.
19. Zhang Y, Kang SA, Mukherjee T, Bale S, Crane BR, Begley TP, et al. Crystal structure and mechanism of tryptophan 2,3-dioxygenase, a heme enzyme involved in tryptophan catabolism and in quinolinate biosynthesis. *Biochemistry* 2007;**46**(1):145–55.
20. Niketić V, Stojanović S, Nikolić A, Spasić M, Michelson AM. Exposure of Mn and FeSODs, but not Cu/ZnSOD, to NO leads to nitrosonium and nitroxyl ions generation which cause enzyme modification and inactivation: an in vitro study. *Free Radic Biol Med* 1999;**27**(9–10):992–6.
21. Saleem M, Ohshima H. Xanthine oxidase converts nitric oxide to nitroxyl that inactivates the enzyme. *Biochem Biophys Res Commun* 2004;**315**(2):455–62.
22. Murphy SH, Suzuki K, Downes M, Welch GL, De Jesus P, Miraglia LJ, et al. Tumor suppressor protein (p)53, is a regulator of NF-κB repression by the glucocorticoid receptor. *Proc Natl Acad Sci* 2011;**108**(41):17117–22.
23. Filipović MR, Stanić D, Raicević S, Spasić M, Niketić V. Consequences of MnSOD interactions with nitric oxide: nitric oxide dismutation and the generation of peroxynitrite and hydrogen peroxide. *Free Radic Res* 2007;**41**(1):62–72.
24. Moreno DM, Martí MA, De Biase PM, Estrin DA, Demicheli V, Radi R, et al. Exploring the molecular basis of human manganese superoxide dismutase inactivation mediated by tyrosine 34 nitration. *Arch Biochem Biophys* 2011;**507**(2):304–9.
25. Jorolan JH, Buttitta LA, Cheah C, Miranda KM. Comparison of the chemical reactivity of synthetic peroxynitrite with that of the autoxidation products of nitroxyl or its anion. *Nitric Oxide* 2015;**44**:39–46.
26. Lymar SV, Shafirovich V, Poskrebyshev GA. One-electron reduction of aqueous nitric oxide: a mechanistic revision. *Inorg Chem* 2005;**44**(15):5212–21.
27. Shafirovich V, Lymar SV. Nitroxyl and its anion in aqueous solutions: Spin states, protic equilibria, and reactivities toward oxygen and nitric oxide. *Proc Natl Acad Sci USA* 2002;**99**(11):7340–5.
28. Wong PS-Y, Hyun J, Fukuto JM, Shirota FN, DeMaster EG, Shoeman DW, et al. Reaction between S-nitrosothiols and thiols: generation of nitroxyl (HNO) and subsequent chemistry. *Biochemistry* 1998;**37**(16):5362–71.

29. Fukuto JM, Cisneros CJ, Kinkade RL. A comparison of the chemistry associated with the biological signaling and actions of nitroxyl (HNO) and nitric oxide (NO). *J Inorg Biochem* 2013;**118**:201−8.
30. Pryor W, Church D, Govindan C. Oxidation of thiols by nitric oxide and nitrogen. *J Org Chem* 1982;**47**:156−9.
31. DeMaster EG, Quast BJ, Redfern B, Nagasawa HT. Reaction of nitric oxide with the free sulfhydryl group of human serum albumin yields a sulfenic acid and nitrous oxide. *Biochemistry* 1995;**34**(36):11494−9.
32. Eberhardt M, Dux M, Namer B, Miljkovic J, Cordasic N, Will C, et al. H_2S and NO cooperatively regulate vascular tone by activating a neuroendocrine HNO-TRPA1-CGRP signalling pathway. *Nat Commun* 2014;**5**:4381.
33. Paolocci N, Saavedra WF, Miranda KM, Martignani C, Isoda T, Hare JM, et al. Nitroxyl anion exerts redox-sensitive positive cardiac inotropy in vivo by calcitonin gene-related peptide signaling. *Proc Natl Acad Sci USA* 2001;**98**(18):10463−8.
34. Filipovic MR, Eberhardt M, Prokopovic V, Mijuskovic A, Orescanin-Dusic Z, Reeh P, et al. Beyond H(2)S and NO interplay: hydrogen sulfide and nitroprusside react directly to give nitroxyl (HNO). A new pharmacological source of HNO. *J Med Chem* 2013;1−10.
35. Filipovic MRM, Miljkovic JLJ, Nauser T, Royzen M, Klos K, Shubina T, et al. Chemical characterization of the smallest S-nitrosothiol, HSNO, cellular cross-talk of H(2)S and S-nitrosothiols. *J Am Chem Soc* 2012;**134**(29):12016−27.
36. Cortese-Krott MM, Kuhnle GGC, Dyson A, Fernandez BO, Grman M, DuMond JF, et al. Key bioactive reaction products of the NO/H_2S interaction are S/N-hybrid species, polysulfides, and nitroxyl. *Proc Natl Acad Sci* 2015;**112**(34):E4651−60.
37. Suarez SA, Neuman NI, Muñoz M, Álvarez L, Bikiel DE, Brondino CD, et al. Nitric oxide is reduced to HNO by proton-coupled nucleophilic attack by ascorbate, tyrosine, and other alcohols. A new route to HNO in biological media? *J Am Chem Soc* 2015;**137** (14):4720−7.
38. Hamer M, Suarez SA, Neuman NI, Alvarez L, Muñoz M, Marti MA, et al. Discussing endogenous NO•/HNO interconversion aided by phenolic drugs and vitamins. *Inorg Chem* 2015;**54**(19):9342−50.
39. Armstrong DA, Huie RE, Koppenol WH, Lymar SV, Merényi G, Neta P, et al. Standard electrode potentials involving radicals in aqueous solution: inorganic radicals. *Pure Appl Chem* 2015;**87**(11−12).
40. Frei B, England L, Ames BN. Ascorbate is an outstanding antioxidant in human blood plasma. *Proc Natl Acad Sci USA* 1989;**86**(16):6377−81.
41. Laranjinha J, Cadenas E. Redox cycles of caffeic acid, alpha-tocopherol, and ascorbate: implications for protection of low-density lipoproteins against oxidation. *IUBMB Life* 1999;**48**:57−65.
42. Vogel S, Contois JH, Tucker KL, Wilson PWF, Schaefer EJ, Lammi-Keefe C. Plasma retinol and plasma and lipoprotein tocopherol and carotenoid concentrations in healthy elderly participants of the Framingham Heart Study. *Am J Clin Nutr* 1997; **66**(4):950−8.
43. Enache TA, Fatibello-Filho O, Oliveira-Brett AM. Electrochemical behavior of triflusal, aspirin and their metabolites at glassy carbon and boron doped diamond electrodes. *Comb Chem High Throughput Screen* 2010;**13**:569−77.
44. Gearba RI, Bondar AI, Goderis B, Bras W, Ivanov DA. Tailoring the thermotropic behavior of tetra-substituted phthalocyanines via the lateral chains architecture. *Chem Mater* 2005;**17**(11):2825−32.

45. Bisby RH, Tabassum N. One-electron oxidation of acetaminophen—a pulse radiolysis study. *Biochem Pharmacol* 1988;**37**(14):2731–8.
46. Van Der Westhuizen J, Kuo PY, Reed PW, Holder K. Randomised controlled trial comparing oral and intravenous paracetamol (acetaminophen) plasma levels when given as preoperative analgesia. *Anaesth Intensive Care* 2011;**39**(2):242–6.
47. Ghaempanah A, Jameh-Bozorghi S, Darvishpour M, Fekri MH. Electrochemical calculations of some non-steroidal anti-inflammatory drugs: solvent effect and antioxidant activity. *Int J Electrochem Sci* 2012;**7**:6127–33.
48. Twomey TM, Bartolucci SR, Hobbs DC. Analysis of piroxicam in plasma by high-performance liquid chromatography. *J Chromatogr* 1980;**183**(1):104–8.
49. May JM. How does ascorbic acid prevent endothelial dysfunction? *Free Radic Biol Med* 2000;**28**(9):1421–9.
50. Rosenthal J, Lippard SJ. Direct detection of nitroxyl in aqueous solution using a tripodal copper(II) BODIPY complex. *J Am Chem Soc* 2010;**132**(16):5536–7.
51. Basudhar D, Cheng RC, Bharadwaj G, et al. Chemotherapeutic potential of diazeniumdiolate-based aspirin prodrugs in breast cancer. *Free Radic Biol Med* 2015;**83**: 101–14.
52. Basudhar D, Bharadwaj G, Cheng RY, et al. Synthesis and chemical and biological comparison of nitroxyl- and nitric oxide-releasing diazeniumdiolate-based aspirin derivatives. *J Med Chem* 2013;**56**(20):7804–20.
53. Coleman JW. Nitric oxide in immunity and inflammation. *Int Immunopharmacol* 2001; **1**(8):1397–406.
54. Wink DA, Kasprzak KS, Maragos CM, Elespuru RK, Misra M, Dunams TM, et al. DNA deaminating ability and genotoxicity of nitric oxide and its progenitors. *Science* 1991;**254**(5034):1001–3.
55. Ridnour LA, Thomas DD, Mancardi D, Espey MG, Miranda KM, Paolocci N, et al. The chemistry of nitrosative stress induced by nitric oxide and reactive nitrogen oxide species. Putting perspective on stressful biological situations. *Biol Chem* 2004;**385**: 1–10.
56. Ma XL, Gao F, Liu G-L, Lopez BL, Christopher TA, Fukuto JM, et al. Opposite effects of nitric oxide and nitroxyl on postischemic myocardial injury. *Proc Natl Acad Sci USA* 1999;**96**(25):14617–22.
57. Pagliaro P, Mancardi D, Rastaldo R, Penna C, Gattullo D, Miranda KM, et al. Nitroxyl affords thiol-sensitive myocardial protective effects akin to early preconditioning. *Free Radic Biol Med* 2003;**34**(1):33–43.
58. Schumacker PT. Nitric oxide quenches the fire in heart mitochondria. *Nat Med* 2013; **19**(6):666–7.
59. Bykov D, Neese FJ. Reductive activation of the heme iron-nitrosyl intermediate in the reaction mechanism of cytochrome c nitrite reductase: a theoretical study. *J Biol Inorg Chem* 2012;**17**:741–60.
60. Averill BA. Dissimilatory nitrite and nitric oxide reductases. *Chem Rev* 1996; **96**(7):2951–64.
61. Burgaud J-L, Ongini E, Del Soldato P. Nitric oxide-releasing drugs: a novel class of effective and safe therapeutic agents. *Ann NY Acad Sci* 2002;**962**:360–71.

From Heaven to Heart: Nitroxyl (HNO) in the Cardiovascular System and Beyond

19

N. Paolocci[1,2], *G. Keceli*[1], *D.A. Wink*[3] *and D.A. Kass*[1]
[1]Johns Hopkins Medical Institutions, Baltimore, MD, United States, [2]University of Perugia, Perugia, Italy, [3]National Cancer Institute, Frederick, MD, United States

Chapter Outline

"Natura operatur breviori modo quo potest"
(Nature operates in the shortest way possible)

Roger Bacon—Opus Maius II

The Chemistry and Biology of Nitroxyl (HNO). DOI: http://dx.doi.org/10.1016/B978-0-12-800934-5.00019-0

"There are more things in heaven and earth, Horatio, Than are dreamt of in your philosophy"

William Shakespeare—Hamlet.

19.1 Introduction

In 1901, Angelo Angeli published his first study on nitroxamic acids showing their ability of resolving in nitrate and nitroxyl (HNO). Presumably, he never imagined that, almost a century later, his new HNO releasing salt (Angeli's salt, AS) revealed pharmacological properties quite distinct from those of nitrite (NO_2^-), whose effects were already tested in humans by Sobrero, Guthrie, and Brunton,[1] or of the yet to be discovered, nitric oxide ($NO^•$). Moreover, it is not entirely clear, at least to us, what the primary input was that led Angeli to undertake the generation of nitroxamic acids, generating the salt that now goes under his name.

As reported by Fontani and Orna, Angelo Angeli was the symbolic figure of the 19th century scientist, who devoted his entire existence to science.[2] What is more, his nature was extremely shy and suspicious of any human relationship. So much so that, at the end of World War I, he was awarded a gold medal for his efforts in developing new antigas masks for the *Regio Esercito* (Italian Royal Army). Yet he declined this prestigious acknowledgment, formulating instead an unusual request to the Italian Ministry of University and Public Education. The reply of the Ministry was sent to the *Istituto di Studi Pratici e Perfezionamento*: "*Due to his excessive shyness, it is recommended that Professor Angeli be released from all his academic duties, such as teaching or public speeches.*"[2]

Surprisingly enough, the life trajectory of the scientist mirrored the research interest around his creature: studies on HNO—in any field have remained rather cursory for many decades. Indeed, it was only in the mid-70s that Dr. Snyder's group identified in HNO the most, if not the only one N=O bond present in the interstellar space.[3] Few years later, the seminal work by Drs. Jon Fukuto, Gautam Chaudhuri, and their colleagues ushered HNO into the realm of Pharmacology.[4] These authors were the first to demonstrate that HNO has vasorelaxing properties akin to those displayed by $NO^•$, and they also proposed HNO as a possible endothelium-derived relaxing factor (EDRF), a possibility still entertained by investigators nowadays.[5] Few years later, Martin Feelisch and his group demonstrated that L-cysteine can differentiate HNO from NO-induced vascular relaxation, thus highlighting for the first time the redox sensitivity of HNO in biological settings.[6] Work from one of the present authors, Dr. David A. Wink, showed that AS is cytotoxic at high (mM) concentrations to Chinese hamster lung fibroblast cells, in a manner sensitive to intracellular glutathione (GSH) concentration and relatively insensitive to the presence of low molecular weight metal complexes.[7] These studies provided the very first proof that the formation of reactive oxygen species (ROS) is not involved in HNO actions, regardless of harmful or beneficial nature of the latter. Moreover, Wink's group discovered that AS/HNO can be more cytotoxic under aerobic conditions, thus suggesting that oxygen can dramatically enhance AS-mediated cytotoxicity.[7] This landmark study started to delineate in great detail the unique chemical repertoire of HNO, while providing an important yardstick for parsing AS/HNO pharmacological effects out from those of other reactive nitrogen species (RNS) or ROS.[7] In essence, they delineated the orthogonality between $NO^{\bullet}$ and HNO[8] showing that these two species have very different in vivo properties and that signaling to cytotoxicity is often inverse.[8,9] Examination of the unique chemistry of HNO revealed that its reactivity was unique and selective. Its reaction with O_2 to form potentially toxic species occurred at much higher level than those required for HNO to exert its cardiovascular effects. The inability of HNO to spontaneously generate $NO^{\bullet}$ through oxidation and the inability of $NO^{\bullet}$ to readily generate HNO in contrast to $O_2/O_2^{-\bullet}$ made this a unique system. The reactivity of $NO^{\bullet}$ is toward ferrous heme and other radicals while HNO mainly "prefers" ferric heme and thiols. Differentiation of these two alternative reactivity patterns between $NO^{\bullet}$ and HNO open the possibility for pharmacological uniqueness and possibly to different signaling pathways.

Meanwhile, Dr. Nagasawa's group demonstrated that HNO could be the second product of the decomposition of cyanamide (cyanide being the first), and that HNO can ultimately be the active inhibitor of aldehyde dehydrogenase (ALDH).[10,11] Therapies centered on ALDH inhibition for treating alcoholism still remain important nowadays. In fact, the inhibition of this enzyme triggers extremely unpleasant effects, spanning from headache to unconsciousness, in individuals treated with cyanamide (or its derivatives) consuming ethanol after the medication. Likely, the contribution of Dr. Nagasawa's group was the first to propose some clinical applicability for HNO donors.

Altogether this pioneering work, along with studies on the cardiovascular system that we shall discuss here, finally sparkled interest on this RNS and its donors. HNO field now counts dozens of papers published each year. Here, we will review the current status and outlook of the most advanced research on HNO pharmacological (biological) actions particularly in the cardiovascular system. We will discuss HNO properties in a clinical perspective, particularly as a possible therapy for acute decompensated heart failure (ADHF). ADHF is a special form of heart failure (HF) that still represents a major therapeutic challenge, an enormous economic burden, and, needless to say, a highly invalidating disease condition. We will conclude suggesting new directions for HNO research that, at least in our opinion, would foster more translational work, from bench-to-bedside, for HNO donors.

19.2 Possible pathways of endogenous HNO production in the cardiovascular system

It is worth remembering that the endogenous generation of HNO remains speculative (see Chapter 18 on this book for further details). This actual gap in our knowledge still forces us to present and discuss HNO biological actions in a pharmacological rather than in a physiological or endogenous signaling perspective.

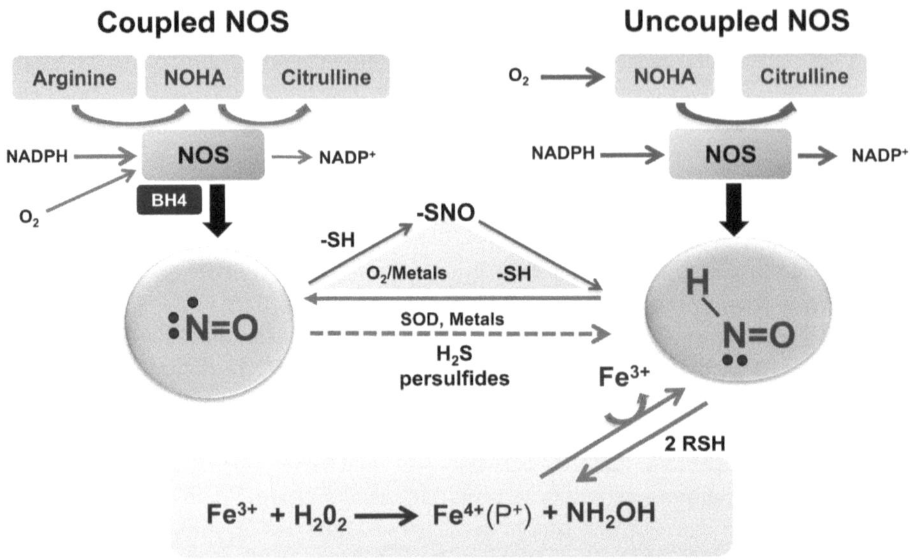

Figure 19.1 Possible NOS-dependent and NOS-independent pathways leading to the endogenous formation of HNO in the cardiovascular system.

Notwithstanding, several putative routes accounting for in vivo HNO generation have been proposed over the last two decades or so. Importantly, all of them may be relevant when cardiac or vascular basal or disease-altered function is concerned. One initial way to categorize them is according to their dependence on the enzymatic activity of nitric oxide synthase (NOS). As shown in Fig. 19.1, NOS function involves the conversion of L-arginine to $NO^{\bullet}$ and L-citrulline, in a process that consumes O_2 and dihydronicotinamide-adenine dinucleotide phosphate (NADPH), and requires the cofactor tetrahydrobiopterin (BH_4).[12] These reactions however also produce the intermediate compound, N^{γ}-hydroxy-L-arginine (NOHA). When NOS is uncoupled, due to the absence of either arginine or BH_4, the direct oxidation of NOHA can lead to the formation of peroxide and superoxide ($O_2^{-}\bullet$), along with that of HNO. It has been suggested that HNO can also be produced by the catalysis of arginine to citrulline in the absence of BH_4[13] (Fig. 19.1). Among the numerous possibilities for an NOS-independent formation of HNO, we can include the decomposition of nitrosothiols[14] that arguably remain the most physiologically relevant way for endogenous HNO formation to date, and its generation by a mechanism involving the interaction of H_2S with $NO^{\bullet}$. Decomposition of RSNO to HNO has been demonstrated through nitrosonium transfer to a vicinal dithiol dithiothreitol (DTT)[15] and thioredoxin, TRX.[16] Also, an endogenously produced species, presumably from the reaction of H_2S and $NO^{\bullet}$, is attributed to HNO based on the in vitro observations (isolated rat ventriculocytes) showing that the combination of H_2S donor, NaHS, and a $NO^{\bullet}$ donor (sodium nitroprusside, SNP) increases contraction of the isolated cells in a manner highly reminiscent of AS/HNO effects (vide infra).[17] This possibility has been suggested even more recently by Filipovic and colleagues in the mouse heart.[18] Other mechanisms that imply HNO formation from H_2S and $NO^•$ may be via the intermediacy of dinitrososulfite (SULFI/NO)[19] or perthionitrite ($SSNO^-$).[20] These intermediates may be an interface between H_2S and nitrosative stress where the formation of HNO may have important consequences. One possible mechanism by which this could occur is that HS^- acts as a nucleophile, attacking the Fe nitrosyl bond of SNP forming an $S-NO^•$ intermediate that decomposes through the reaction with SH- to give HNO. However, all these NOS-independent mechanisms suffer from the limitation that unlike $O_2^-\bullet$ and O_2, HNO cannot be formed through the direct reduction of $NO^•$ to HNO due to energetic barriers; in fact the conversion of HNO to $NO^•$ requires the use of metal catalysts. Studies have also proposed the formation of HNO from the reaction of $NO^•$ with reduced Cu/Zn,[21,22] Fe superoxide dismutase (SOD),[23] cytochrome C,[24] and xanthine oxidase.[25] Despite the possible mechanisms these reactions are uphill since the reduction potential of $NO^•$ to HNO is >0.9 V,[26] and suffer from the back reaction with the metal center. Interestingly, recent work has shown that $NO^•$ can be reduced to HNO in the presence of aromatic and "pseudoaromatic" alcohols such as vitamin E, hydroquinone, and ascorbate, potentially via a mechanism involving proton-coupled nucleophilic attack.[27,28] Similar to the studies with nitrogenase a decade earlier, N_2O can be generated via HNO dimerization or through a metal dimerization of $NO^•$.[29,30]

Importantly, there are additional pathways that do not implicate the generation and interconversion of nitrogen oxides in HNO formation. One is through the heme protein–mediated oxidation of hydroxylamine (NH_2OH) in the presence of peroxide.[31] This is a particularly intriguing biomediator since it implicates a possible HNO cycle involving peroxide, thiols, and NH_2OH. When HNO is initially formed, it can react with thiols such as GSH, which results in NH_2OH. In the cell, there is between 2 and 5 mM GSH, making it a primary target of HNO, and consequently denoting NH_2OH both as a primary product and a decomposition product. As stated above, NH_2OH can form HNO in the presence of peroxide and heme protein. This forms a cycle that under oxidative conditions can generate HNO, providing a recycled supply of this molecule. However, this mechanism would be limited if there was a breakdown in disulfide reduction. This possible route to HNO formation is especially interesting because it suggests a possible endogenous pathway by which signaling molecules such as H_2O_2 and the oxygenated derivative of ammonia may lead to HNO formation.

19.3 The cardiovascular actions of HNO donors

In 2001, more than a century after Angeli discovered his salt, it was first administered to an intact conscious mammal while cardiovascular function was determined. Given the anticipated complexity of effects on vascular tone and potentially cardiac function, the studies employed pressure–volume relationships to dissect its influences.[32,33] In conscious fully instrumented dogs, AS/HNO

($10\ \mu g\ kg^{-1} = 0.082\ \mu mol\ kg^{-1}$ for 10–20 min, i.v.) increased peak systolic chamber stiffness (end-systolic elastance, a load-independent index of contractility). HNO also hastened ventricular relaxation and reduced the filling volume (preload) of the normal heart.[32] In failing hearts, HNO also lowered arterial vascular tone.[33] When the latter was offset by volume expansion, and autonomic reflexes concomitantly suppressed, we still observed the positive contractility change. The effect on contractility was rapid and was accompanied by faster myocardial relaxation (lusitropy). These effects were neither reproduced by administering equimolar doses of $NO^{\bullet}$-releasing compounds such as diethylamine NONOate (DEA/NO) or nitrate, which ranged from neutral to negatively impacting contractility.[33] The effects of AS/HNO were not suppressed by β-adrenergic blockade. The combination of positive inotropy/lusitropy and prominent venodilation was not mimicked by equimolar amounts of DEA/NO, nitrate, or synthetic peroxynitrite.[34] The latter is interesting as prior work had suggested that peroxynitrite can decompose to yield HNO[35] and conversely, HNO should react with molecular oxygen to form peroxynitrite.[36] The HNO-evoked rise in cardiac function was fully prevented by coadministering a thiol-donating compound such as *N*-acetyl-L-cysteine (NAC), an effect that distinguished the vasodilatory effect of HNO from that of $NO^{\bullet}$. This further suggested that when the intracardiac environment is enriched with free floating thiols, the ability of HNO to interact with its thiol target is reduced, thus its pharmacological actions are markedly blunted. Conditions with oxidative stress that might compromise free thiols might thus enhance HNO efficacy.

HNO donors in the normal cardiovascular system

- Cardiac positive inotropy and lusitropy
- Prominent venodilation
- Insensitivity to β-AR blockade
- No tachyphylaxis

HF can be defined as the inadequacy of the heart to pump enough blood to match the perfusion need of the different organs and tissues in the body. The failing heart presents with enlarged left ventricle (LV) dimensions (reflecting volume overload and pump inefficiency) and thinning of the LV walls, due to an adverse remodeling process. Both events contribute to reduce myocardial performance. The failing heart generates excessive ROS and RNS, accompanied by the downregulation of enzymatic and nonenzymatic antioxidant defenses.[37] This alters cardiac redox balance status, prompting in turn oxidative modifications of cardiac proteins and/or quenching bio-available $NO^{\bullet}$.[38] The latter eventuality can abrogate the cardioprotective, antihypertrophic effects by NO-cGMP/PKG (NO-cyclic guanosine monophosphate/protein kinase G) signaling.[39] This inadequately controlled ROS/RNS generation can worsen: (1) excitation–contraction coupling; (2) maladaptive myocyte hypertrophy; (3) extracellular matrix remodeling; (4) mitochondrial function resulting in impaired energetics and increased ROS release; (5) capillary density and endothelial function; and (6) stimulate proinflammatory states with dysregulated immunological reactions. All these events are ultimately responsible for loss of viable myocytes and/or reduction in their function, both hallmarks of all forms of human congestive heart failure (CHF). Furthermore, we should consider that the β-adrenergic system whose activation is set to adjust cardiac force development to increased work demand is downregulated in failing hearts.[40] This is due to βAR-desensitization and uncoupling to their signaling molecules,[41] although increased ROS/RNS emission can also factor in blunting βAR-agonist evoked rise in contractility and relaxation, either directly via oxidative modification of βAR-agonist structure per se,[34] or indirectly, via negative interaction with their associated signaling pathways.[42] For instance, endogenous $NO^{\bullet}$ and its oxidized product, nitrate, have been reported to attenuate β-adrenergic response, both in experimental and human CHF,[43] with mechanisms that remain poorly understood. As we shall see later, this negative interaction limits the benefits afforded by combining β-agonists and nitrate, a therapeutic regimen that is often required in HF patients to provide inotropic support and relieve congestion.[44]

In the early 2000s, a series of studies performed in vivo and in vitro revealed that HNO donors provided both inotropic/lusitropic and vasodilating effects in CHF, avoided tolerance and suppression of efficacy by ROS/RNS, and induced signaling that prevented pathways known to be harmful. Fig. 19.2 (panel A) shows a schematic representation of P−V loops as they could be obtained at baseline and

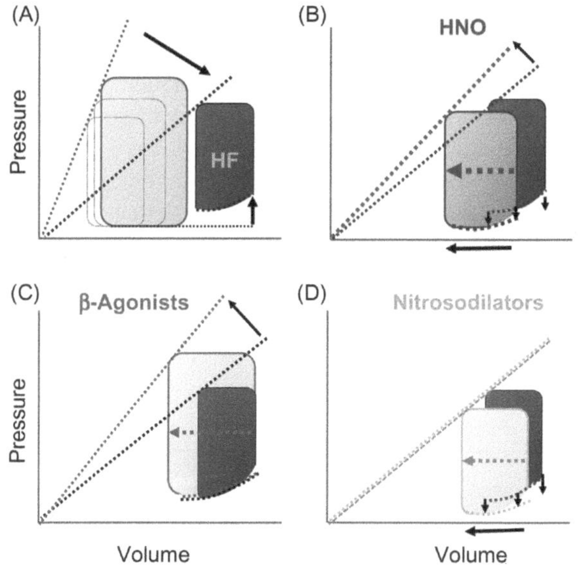

Figure 19.2 Schematic examples of pressure−volume relationships (P−V loops) obtained under (A) control conditions and after the induction of HF; in HF preparations after (B) acute infusion (20 min) of a HNO donor such as AS; (C) a synthetic β1 agonist such as dobutamine; (D) a nitrosovasodilator such as nitroglycerin.

after the induction of HF. Compared to normal, the failing condition is shifted rightward, reflecting chamber dilation, reduced cardiac work (loop area), and depressed systolic and diastolic function [two boundaries at the upper left, end-systolic PV relation (ESPVR), and lower, end-diastolic PV relation (EDPVR)]. The ESPVR is shallower with failure reflecting decreased contractility, while lower EDPVR is steeper, indicating reduced distensibility in the relaxed state. The latter limits the ability of a heart to fill at low pressures. As in normal hearts, HNO enhanced contractility (steeper ESPVR) and reduced diastolic pressure (left-shift; venodilation), with a concomitant widening due to arterial vasodilation (Fig. 19.2, panel B). Panels C and D show schematic examples of P–V loops from the β-receptor agonist, dobutamine, and organonitrate. The differences are that dobutamine has no effect on venous or arterial loading, while nitrite has minimal effect on contractility but does reduce loading much like AS/HNO. When dobutamine and AS/HNO were combined, their impact on contractility was additive, whereas an $NO^{•}$ donor blunted dobutamine-evoked inotropic action.[33] This revealed a major difference between HNO and $NO^{•}$. Importantly, repeated systemic AS/HNO infusions displayed no tachyphylaxis for positive inotropy or load reduction.[32] Tolerance to $NO^{•}$ or nitrate-induced vasodilation in isolated rat aorta is also not observed with AS/HNO.[45] Other HNO donors such as isopropylamine-NONOate (IPA/NO) reproduced these findings, so they are not AS specific.[46] Finally, a very recent paper from Ritchie's group have tested the impact of the HNO donor 1-nitrosocyclohexylacetate (1-NCA) on LV diastolic dysfunction induced by diabetes mellitus in vivo.[47] In this study, the authors also determined whether 1-NCA would limit cardiomyocyte hypertrophy in this mouse model of metabolic syndrome. They found that 1-NCA therapy (83 mg kg^{-1} per day IP for 4 weeks) prevented diabetes-induced diastolic dysfunction as well as the increase in LV cardiomyocyte size, hypertrophic gene expression, and the upregulation of superoxide production found in cells from diabetic animals. These studies represent the first evidence that administering HNO donors can be useful also to treat diabetic cardiomyopathy, particularly diabetes-induced diastolic dysfunction, and they also suggest that HNO donors can have a favorable impact on HF even when chronically administered. This evidence, in addition to previous findings obtained by our group, attests that HNO donors can improve diastolic function in failing hearts, although the possible enhancement of LV compliance still awaits a more direct and load-independent demonstration via P–V loops or other approaches.

HNO in experimental and human heart failure

- Cardiac positive inotropy and lusitropy
- Modest reduction in systemic blood pressure and marked unloading effects
- Additive effects to inotropes such as β-AR agonists
- No tachyphylaxis

19.4 Studies with the novel HNO donors, CXL-1020 and CXL-1427

A major limitation of AS as an HNO donor concerns its chemical instability, being largely degraded within 5−10 min in aqueous solution. As a result, several novel stable HNO donors were developed, and two in particular, CXL-1020 and CXL-1427, have been tested in both animals and humans. CXL-1020 converts solely to HNO and an inactive metabolite CXL-1051, avoiding the generation of NO_2^-, and has a half-life similar to that of AS (approximately 2 min). Sabbah and colleagues tested CXL-1020 in an ischemic model of CHF, and found it reduced LV end-diastolic pressure and myocardial oxygen consumption while increasing ejection fraction from 27% to 40% and maximal power index (another index of contractility) by 42%.[48] Similarly, in conscious dogs with tachypacing-induced HF, CXL-1020 enhanced contractility as assessed by the rise in ESPVR, i.e., end-systolic elastance, and provided venoarterial dilation. Importantly, heart rate was minimally affected by CXL-1020. In humans with systolic HF, a 6-h up-titration study showed that CXL-1020 reduced pulmonary capillary wedge pressure and systemic vascular resistance (SVR), accompanied by an increase in cardiac index (CI) and stroke volume index. Higher doses revealed a local inflammatory response at the injection site,[48] and this was removed by developing HNO donors with longer dissociation kinetics such as CXL-1427. In a dose-finding, ascending dose cohort Phase I trial in 70 healthy volunteers, CXL-1427 improved CI with some decreased blood pressure without increases in heart rate, at the maximally tolerated dose of 10 $\mu g\ kg^{-1}$ per min.[49] Phase IIa dose-ranging studies were completed in 2015, indicating efficacy in a HF population. Phase IIb trials are currently being developed.

19.5 HNO and systemic and coronary vasodilation

Fukuto and colleagues were the first to report that, in both rabbit aorta and bovine intrapulmonary artery, AS/HNO relaxes vessels by a soluble guanylyl cyclase (sGC)-dependent mechanism.[50] Since then, several studies have explored the modalities by which AS, Piloty's acid, and its derivatives relax vessels, both in vitro and in vivo. The total or partial sGC-dependency of HNO vasodilation, and more in general the way by which HNO vasorelaxes both capacitance and resistance vessels as well as the possibility that HNO embodies the endothelial-derived hyperpolarization factors has been nicely reviewed by Bullen and colleagues.[51] However, a recent paper from our group clearly documents that mice and isolated arteries lacking the sGC β-subunit (sGC-KO mice resulting in full sGC deficiency) or expressing solely a redox-dead C42S mutant were fully insensitive to the pure HNO donors, CXL-1020. In stark contrast, these mice exhibited fully maintained contractile response.[52] Interestingly, the CXL-1020 dose reversing 50% of preconstricted force in aortic rings was ≈400-fold greater in sGC-KO than controls. Cyclic-GMP and cyclic adenosine monophosphate (cAMP) levels were unaltered in myocardium exposed to

CXL-1020, despite its inotropic-vasodilator activity. In protein kinase G-1α(C42S) mice, CXL-1020 induced identical vasorelaxation in vivo and in isolated aortic and mesenteric vessels as in littermate controls. In both groups, dilation was near fully blocked by pharmacologically inhibiting sGC. Based on these findings, we concluded that sGC- and cGMP-dependent signaling are necessary and sufficient to explain HNO-induced vasodilation in vivo, but these mechanisms are not implicated in HNO-evoked positive inotropy, thus validating in full our previous observations.[53] Moreover, they demonstrate that redox modulation of protein kinase G-1α is not a required mechanism for HNO-mediated vasodilation.

19.6 HNO's mechanisms of action in the heart and in the vasculature: "nitroxylation"

The next question is how HNO works in the heart and the cardiovascular system. Previous studies have shown that NO-related species such as nitroglycerin (NTG)[54] can release nonadrenergic/noncholinergic (NANC)-related peptides such as calcitonin gene-related peptide (CGRP). CGRP is a neuropeptide that experimental and human studies have shown to induce both vasodilation[55] and positive inotropy.[56,57] We first sought to determine whether CGRP release could account for in vivo HNO inotropy and found that the latter can be blocked by the CGRP receptor antagonist ($CGRP_{8-37}$).[32] Moreover, subsequent studies revealed that, following AS/HNO in vivo infusion, CGRP levels raise both in veins and arteries.[33] However, CGRP-induced inotropy is essentially indirect, because it is mediated by norepinephrine released from the sympathetic efferent fibers. Therefore, it is not surprising that, in contrast to HNO donors, CGRP inotropic action can be fully prevented by β-blockers such as timolol.[58] What is more, CGRP stimulatory action is lost in CHF preparations, confirming its dependency on β-adrenergic receptor signaling, which is notoriously downregulated in human and experimental HF.[40] Finally, CGRP does not reduce LV filling volumes, thus it does not unload the heart as AS/HNO does. Hence, while some yet to be defined off-target effects of the CGRP receptor antagonist $CGRP_{8-37}$ might have a bearing on this apparent discrepancy, together these findings lend further support to the notion that CGRP release can at least in part explain HNO-induced vasodilation.[59] Studies from Filipovic's group suggest that CGRP remains an important component of HNO signaling capacity. These authors showed that H_2S and $NO^{\bullet}$ production colocalizes with transient receptor potential channel A1, and the ability of HNO donors to activate these chemoreceptor channels via formation of amino-terminal disulfide bonds, resulting in sustained Ca^{2+} influx.[60] This event prompts CGRP release, thus accounting, at least in part, for local and systemic vasodilation.

So, how does HNO increase inotropy and lusitropy in the heart? To gain more definitive insights into mechanisms of HNO action in the heart, we studied AS/HNO impact on single myocyte function, indexing sarcomere contraction and relaxation and whole Ca^{2+} transient kinetics. We found that HNO modulates the activity of the sarcoplasmic reticulum (SR) Ca^{2+} pump (SERCA2a) and the ryanodine receptor 2 (RyR2), leading to increased Ca^{2+} uptake and release from the SR, and thus resulting in an enhancement of Ca^{2+} cycling within the myocytes.[53] Our findings were preceded by a study from Cheong and colleagues who investigated the impact of HNO on skeletal ryanodine receptor 1 (RyR1) and cardiac (RyR2) ryanodine receptors.[61] These authors found that, at steady-state concentrations of nanomolar AS, HNO induces a rapid release of Ca^{2+} from the SR, an effect reversed by the reducing agent DTT. When HNO was applied to the *cis*-side of RyR1 reconstituted in planar bilayers, it increased the open probability of the channels. They also demonstrated that the hydroxyl radical scavenger, phenol did not alter AS/HNO-induced SR Ca^{2+} release, and concluded that HNO is more potent than $NO^{\bullet}$ in activating RyR1.[61] Extracellular Ca^{2+} enters the cardiac cells via activation of L-type Ca^{2+} channels, and this is an essential step to trigger the release of higher amounts of Ca^{2+} from the SR via RyR2 in order to activate myofilaments, thus resulting in myocardial contraction. This Ca^{2+}-induced SR Ca^{2+} release is typically induced by β-AR agonists and other positive inotropic agents.[62] However, we found that, differently from the latter agents, HNO donors such as AS do not activate L-type Ca^{2+} channels.[63] Moreover, altogether these studies demonstrated that HNO increases in vitro positive inotropy/lusitropy without rising diastolic Ca^{2+} levels, and that HNO modulation of myocardial contractility does not involve either cAMP/protein kinase A or cGMP/protein kinase G signaling.[53] Importantly, unlike β-AR agonists whose effects are blunted in isolated failing myocytes,[48] HNO donated by CXL-1020 remains effective in failing hearts,[48] thus confirming earlier observations demonstrating that HNO preserved inotropy in in vivo failing hearts.

HNO avidly and readily reacts with thiols in vitro,[64,65] and likely HNO has fast reactivity toward its targets in biological systems too. Thiol(-SH)-based switches, mainly via posttranslational modification of cysteine residues, are at the core of many cell functions' modulation.[66] They typically consist of two redox-sensitive cysteine residues and a disulfide bond between them. The possibility of oxidizing these cysteine residues to form a disulfide bond depends upon the topology, temperature- and pH-dependent redox potential of cysteine-cystine.[67] HNO can modify cysteines in two distinct ways: following the formation of the initial *N*-hydroxysulfenamide (RSNHOH) intermediate with reactive thiols in cysteine residues, this species can either rearrange to generate a sulfinamide ($RS(O)NH_2$) or, when in the presence of a vicinal thiol, may form a disulfide bond (RSSR).[68,69] Additionally, in the atypical case of C-terminal cysteines, the formation of a sulfenic acid is reported in vitro.[70] Recent studies on the reversibility and hydrolysis of HNO-derived sulfinamides utilizing mass spectrometry techniques and a new ^{15}N-edited nuclear magnetic resonance (NMR) method for sulfinamide detection demonstrate that this species can be reverted back to thiols, albeit slowly compared with disulfides.[71,72] Furthermore, these reports point to the importance of local environment on the reactivity of HNO-derived modifications.[72] While the signaling potential of sulfinamide formation remains to be ascertained in full, at least in the cardiovascular system, over the last few years, we and others have been collecting experimental evidence that, under mildly oxidizing conditions, HNO can induce the formation of inter- and/or intramolecular bonds. By doing so, it likely induces a conformational change in the resident protein, modulating its function. This new type of posttranslational modifications that we can call "nitroxylation" may or may not implicate an oligomer transition.[67]

But, where are these HNO-sensitive redox switches situated in cardiomyocytes? First, we observed that HNO actions on myocyte contractility can be prevented, at least in part, by pretreating isolated ventricular myocytes with cell-permeable GSH to enrich the intracellular environment with free floating thiols. Then, we showed that treating RyR2 (reconstituted in lipid bilayers) with the reducing agent DTT resulted in no enhancement of RyR2 activity after HNO, so confirming early findings in RyR1.[61] In separate studies, we found that the activation of SERCA2a by HNO is a phospholamban (PLN)-dependent process, and that cysteine residues in the PLN transmembrane domain, namely, cys 36, 41, and 46, play an essential role in this event since their replacement with alanines fully abolished HNO's ability to increase SERCA2a-mediated Ca^{2+} reuptake into the SR.[73] Moreover, we reported that HNO affects PLN/SERCA2a interaction, by stabilizing PLN in an oligomeric (likely disulfide bond-dependent) configuration, ultimately resulting in a functional uncoupling of PLN from SERCA2a via reducing the amount of available monomeric PLN, the ultimate inhibitory species for SERCA2a.[74] However, primary actions of HNO on SERCA2a per se can also account in HNO-induced enhancement of SERCA2a function. Colucci's group nicely showed that HNO increases the maximal rate of thapsigargin-sensitive Ca^{2+} uptake mediated by SERCA2a in sarcoplasmic vesicles via reversible modifications of thiols present in the pump.[75] More in detail, HNO appears to increase *S*-glutathionylation of SERCA2a. In fact, adenoviral overexpression of glutaredoxin-1 prevented both HNO-stimulated oxidative modifications of SERCA2a and its activation, as did overexpression of a mutated SERCA2a in which cysteine 674 was mutated to serine. These authors concluded that HNO enhances the maximal activation of SERCA2a via *S*-glutathionylation at cysteine 674. These two possibilities, i.e., HNO-induced oligomerization of PLN versus direct effects on SERCA2a, are not mutually exclusive. Rather, it is very plausible that they cooperate, perhaps reinforcing each other, to enhance the catalytic efficiency of the SERCA2a pump. Differently from studies on SERCA2a/PLN interaction in which specific cysteine residues targeted by HNO have already been identified, the same cannot be said for RyR2 whose tetramers harbor a large number of free cysteines.[76] Identifying the "cysteine hot-spots" targeted by HNO in RyR2 represents important studies in perspective.

In addition and in parallel to changes in Ca^{2+} handling, HNO donors also modify the function of contractile and regulatory proteins resident in myofilaments, which in response to these SR Ca^{2+} waves, develop cardiac force followed by relaxation of the cardiac muscle. In an initial study, we found that, in rat right ventricle trabeculae, AS/HNO increased in a dose-dependent manner both whole Ca^{2+} transient and force generation, resulting in an overall enhancement of myofilament responsiveness to Ca^{2+}.[77] These effects were thiol-sensitive because administering DTT either in concomitance with AS or after HNO effect onset prevented or reversed HNO-induced enhancement in force generation, respectively. In a following study, we used both AS and 1-NCA, a pure HNO donor, and a new mass spectrometry capture technique based on a modified biotin switch assay to identify cysteine residues whose modification could eventually contribute to HNO-induced enhancement in cardiac force generation. We found that these HNO donors can induce the formation of DTT-reversible disulfide bonds between actin-tropomyosin (likely involving cys190 on tropomyosin and cys257 on actin) and myosin heavy chain (possibly Cys37)-myosin light chain 1 (cys81).[78] These interactions were also independently correlated with the increased Ca^{2+} sensitivity and force generation, respectively. Intriguingly, in contrast to the cardiac one, the skeletal muscle that lacks the candidate cys81 in MLC1 did not show any enhancement in force generation after HNO treatment. By revealing for the first time that HNO exerts a direct effect on myofilament proteins, increasing myofilament Ca^{2+} responsiveness via promoting disulfide bond formation between critical cysteine residues, this study provided one of the very first evidence that redox-cysteine-based modifications of myofilament proteins by RNS, such as HNO, can positively modulate cardiac contractility.[78]

19.7 Myocardial ischemia and HNO

Myocardial ischemia-reperfusion (I/R) injury occurs when blood supply to the heart is reduced (due to a critical coronary artery obstruction) and then restored. At reperfusion, a burst of ROS occurs, mainly originated from mitochondria. While the exact modalities by which ROS production occurs during I/R is still highly debated,[79] decades of bench work has clearly demonstrated that preventing or limiting this ROS burst at reperfusion can salvage the myocardium. Yet clinical trials with several, pharmacologically different antioxidants have been so far unsuccessful in preventing I/R injury in humans so far.[79] One of the first in vivo studies, if not the very first one employing AS/HNO is a contribution by Ma and colleagues. In that study, the authors compared the impact of AS and the $NO^{\bullet}$ donor *S*-nitrosoglutathione (GSNO) in rabbits subjected to 45 min of regional myocardial ischemia.[80] The administration of GSNO 5 min before reperfusion resulted in a markedly attenuated reperfusion injury, as evidenced by an improvement in myocardial function, reduced infarct size, and decreased myocardial activity of creatine kinase (CK) and myeloperoxidase (MPO). In stark contrast, the administration of hemodynamically equieffective doses of AS (3 μmol kg^{-1} for 20–30 min) aggravated reperfusion injury as indicated by increased myocardial necrotic size, CK release, and end-diastolic pressure. Data in the heart are in agreement with findings in the brain where HNO has been reported to worsen I/R injury by increasing oxidative stress and decreasing brain perfusion.[81] As far as the heart is concerned, the ability of HNO to recruit neutrophils in the ischemically injured area may have had some part in AS/HNO detrimental impact on reperfused myocardium,[82] but the mechanistic underpinnings of these HNO deleterious effects remain unclear, along with the full dose-dependency of AS/HNO. In fact, a lower dose of AS, i.e., 1 μmol kg^{-1}, did not exacerbate I/R injury.

The AS/HNO concentrations used in this in vivo I/R injury study widely exceed those used later for hemodynamic studies in conscious dogs.[32,33] Further to this, subsequent studies by us and other groups unequivocally demonstrated the preconditioning potential of AS/HNO. In fact, in isolated perfused rat hearts, AS given prior to ischemia afforded a robust protection against I/R injury whose effects are akin to those produced by classical ischemic preconditioning (IPC).[83] Intriguingly, the magnitude of AS/HNO-afforded protection was significantly more than that achieved via equimolar amounts of the NO donor, DEA/NO. Of note, DEA/NO has a half-life and in vitro kinetics identical to those of AS.[46] AS/HNO beneficial effects were fully abrogated by NAC (4 mM), suggesting once again that HNO myocardial actions are sensitive to the local concentrations of free floating thiols. Unpublished data from our group (obtained in isolated rat hearts) suggest that HNO protection, conveyed by either AS or IPA/NO, is not prevented by the mitochondrial K(ATP) channel (mK_{ATP}) blocker, 5-hydroxydecanoate.[84] However, another study by Queliconi and coworkers has documented that in isolated mitochondria AS/HNO inhibits complex II and opens mK_{ATP}, showing that HNO protects cardiomyocyte against I/R injury in an mK_{ATP}-dependent manner.[85] Additional studies are required to resolve this apparent discrepancy, and to provide further mechanistic insights on the routes followed by HNO to afford protection against I/R injury in the heart. Using a proteomics approach, an extensive modification of mitochondrial protein thiols by AS/HNO has been reported,[86] suggesting that HNO interacts with mitochondria via a mechanism that is distinct from those of either $NO^\bullet$ or peroxynitrite ($ONOO^-$). However, similar to $NO^\bullet$,[87] HNO appears to inhibit mitochondrial respiration through the inhibition of complexes I and II, most probably via modification of specific cysteine residues in these proteins. The meaning of these findings remains to be contextualized in I/R injury and CHF settings, where HNO may exert important influences on energy production and/or ROS emission.

We anticipate that the myocardial protective potential exhibited by HNO donors against I/R injury as well as the maintained efficacy in elevating myocardial contraction and relaxation in failing, oxidatively stressed hearts stem to large extent from the unique chemistry of nitroxyl. In fact, as sketched in Fig. 19.3, given its lower intrinsic reactivity toward O_2, HNO will be substantially unaffected by either hypoxic or hyperoxic cardiac tissue conditions. Further to this, HNO is rather insensitive to ROS inactivation.[65] Therefore, when, and if endogenously formed, or when exogenously administered, HNO is likely to preserve its signaling ability and/or pharmacological power. Finally, its PKG- and protein kinase A (PKA)-independency suggests that HNO unlikely interferes with posttranslational modifications induced by the second messengers cGMP and cAMP, and vice versa its signaling should be unaffected by altered patterns of PKG- or PKA-induced phosphorylation. More research is needed to identify those myocardial redox-sensitive sites targeted by HNO to confer protection to hearts subjected to in vitro or in vivo I/R injury (or to increased myocardial performance), and to elucidate the chemical nature of these modifications. We should bear in mind that these sites should be rather insensitive to all oxidizing agents present in the failing and/or I/R injured heart. Last, but not least, it would be of interest to establish if, and to what extent, once generated or

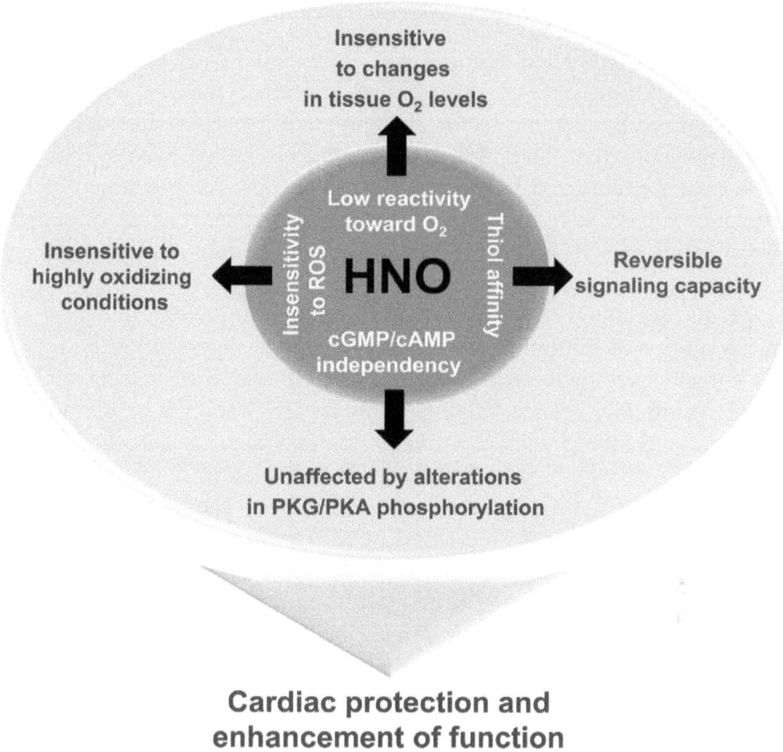

Figure 19.3 Basic chemical properties of HNO that may explain HNO preserved ability of positively modulating myocardial function and to afford cardiac protection against ischemic-reperfusion injury, acutely or in chronic settings.

delivered, HNO interacts with other gas transmitters such as CO, H_2S, and $NO^{\bullet}$ to modulate the myocardial response to I/R injury or myocardial contractility and relaxation.

19.8 Additional pharmacological effects of HNO in the cardiovascular system and on other pathophysiological conditions

HNO donors possess several additional pharmacological properties that can nicely complement its peculiar actions on cardiac and vascular function. Altogether, these effects can contribute to arrest the progression of HF syndrome, lower the risk of recurrent cardiovascular accidents such as ischemic attacks, and ultimately improve expectancy and quality of life in HF patients. Here we will provide also a brief summary of HNO impact on other processes such as inflammation, pain and cancer growth.

19.8.1 HNO and platelets

One aspect of HNO pharmacology that was appreciated early on was the ability of HNO donors to inhibit platelet aggregation. Impaired platelet responsiveness to $NO^{\bullet}$ is a recognized pathogenic factor of many cardiovascular disorders, and represents an independent risk factor for acute cardiac ischemic events, thus mortality for cardiovascular events and sickle cell disease (SCD). As mentioned earlier, vascular resistance to $NO^{\bullet}$ effects can be due to its direct scavenging of superoxide[38] and/or impairment of sGC activity.[88] However, HNO could still signal in the presence of excessive ROS production. Mondoro and colleagues were the first to recognize that HNO, generated from AS, has some antiaggregation effects on platelets, reporting its ability of inhibiting ADP-induced platelet aggregation both in normal and SCD subjects, although with a reduced magnitude with respect to DEA/NO, and without specifically acknowledging AS as an HNO and not a $NO^{\bullet}$ donor.[89] In a following study, Bermejo and coworkers demonstrated that both AS and SNP (tested as a $NO^{\bullet}$ releaser) significantly inhibited platelet aggregation and ATP release induced by different agonists and adrenaline. These authors showed that both AS and SNP significantly increased cGMP accumulation in a 1H-[1,2,4]oxadiazolo[4,3-a]quinoxalin-1-one (ODQ)-sensitive manner. As in other settings, L-cysteine reduced AS effects, whereas it increased the potency of SNP on some of the parameters tested. This study demonstrated HNO's ability to inhibit aggregation in human platelets. Later on, using a comprehensive tandem mass spectrometric analysis including MS/MS by CID and electron capture dissociation as well as an MS^3 analysis, Hoffman and colleagues identified 10 proteins in human platelets that are modified by HNO, in a dose-dependent manner, and whose specific functions range from metabolism and cytoskeletal rearrangement to signal transduction. Hence, these Authors provided for the first time a possible mechanistic link between HNO-induced modifications and the physiological effects of HNO in platelets.[90] Importantly, Bullen and colleagues reported that, differently from GTN, IPA/NO's ability to inhibit platelet aggregation is preserved in hypercholesterolemic mice.[91]

From a different but complementary angle, it is well documented that plasma hemoglobin (Hb) released during intravascular hemolysis carries many potential adverse effects, which range from increased $NO^{\bullet}$ scavenging to enhanced ROS production and platelet activation.[92] Solomon and coworkers tested the novel and pathophysiologically relevant idea that AS/HNO could ameliorate hemolysis-associated pathology via its preferential reactivity with plasma Hb, as opposed to red cell–encapsulated Hb, and through its intrinsic vasodilatory capacity.[93] Using a canine model, these Authors found that coinfusion of AS/HNO and cell-free Hb leads to preferential conversion of plasma Hb to metHb, but the extent of this conversion is lower than anticipated based on the in vivo AS concentration relative to plasma Hb, suggesting that this lower metHb yield is likely due to reactions of HNO with plasma thiol-containing components. Interestingly, in this study, the infusion of Hb alone led to a marked rise in mean arterial pressure and SVR, whereas AS alone lowered all these parameters. Importantly, when AS and Hb were infused together, AS countered the negative hemodynamic effects of plasma Hb. Likely, these effects are partially attributable to inactivation of NO$^{\bullet}$ scavenging through conversion to metHb and/or, more even likely to the overriding vasodilatory effects of HNO opposing the vasoconstrictive action of cell-free Hb. It is worth noting that AS/HNO's ability to maintain its action in the presence of large amounts of plasma Hb, without oxidizing the plasma Hb, could be very instrumental to develop Hb-based oxygen carriers. In stark contrast, a similar approach, i.e., coinfusion of Hb and HNO in the pulmonary system, led to significantly less impact of AS/HNO on pressure, suggesting that inactivation of NO$^{\bullet}$ scavenging with AS reduced the direct vasodilatory effects of AS on the vasculature. In a following study, Donzelli and coworkers found that another HNO donor, NCA, significantly reduces contractile response and platelet aggregation mediated by the thromboxane A(2) mimetic 9,11-dideoxy-11α-9α-epoxymethanoprostaglandin F(2)(α) in a cGMP-dependent manner.[94] More recently, Dautov and colleagues tested whether HNO can circumvent NO$^{\bullet}$ resistance in platelets isolated from human subjects with or without ischemic heart disease.[95] These authors reported that IPA/NO was more effective than SNP in inhibiting platelet aggregation in normal subjects. Importantly, in NO-resistant individuals the IPA/NO: SNP response ratio was markedly increased, consistent with partial circumvention of NO$^{\bullet}$ resistance by HNO. cGMP accumulation was greater with IPA/NO than with SNP stimulation, and the use of the NO$^{\bullet}$ scavenger carboxy-PTIO revealed that IPA/NO effects are mostly NO$^{\bullet}$-independent since only a very small portion of them was inhibited by carboxy-PTIO. However, ODQ inhibited IPA/NO response by $36 \pm 8\%$, consistent with a contribution of sGC/heme to its antiplatelet aggregation effects.

Altogether, the studies reported above document the capacity of HNO donors such as AS, IPA/NO, and NCA to inhibit platelet aggregation, circumventing platelet NO$^{\bullet}$ resistance while acting, and with mechanisms that, at least in part, involve a heme-mediated sGC activator. Moreover, HNO donors could serve as therapeutic agents to counteract the negative vasoconstrictive consequences of hemolysis that occur in hemolytic anemias, transfusion of stored blood, and other diseases. However, the study by Solomon and colleagues suggest that the eventual formation of metHb should be monitored when infusing HNO donors in vivo.

19.8.2 HNO impact on cardiac hypertrophy

Differently from exercise, i.e., physiological stress that induces a condition of concentric hypertrophy with compensated LV function, chronic hemodynamic stress such as that occurring during hypertension or other chronic cardiovascular disorders triggers a status of eccentric hypertrophy characterized by marked reduction in both systolic and diastolic function.[96] This condition, frequently referred as maladaptive or pathological hypertrophy, is an independent risk for HF and death, and measures directed to either prolong a status of physiological, compensated hypertrophy or to counter the onset/progression of pathological hypertrophy, or both are among the most sought-after strategies to prevent and/or cure HF. NO$^{\bullet}$/cGMP/PKG signaling is a well-known negative regulator of maladaptive hypertrophy.[39] Recent in vitro reports suggest that HNO can also counter this adverse event. Using neonatal rat cardiomyocytes, Lin and colleagues were the first to report that AS inhibits Ang II-induced hypertrophic response in cardiomyocytes, including increases in cardiomyocyte size, de novo protein synthesis and β-myosin heavy chain expression, all well-known markers of pathologic hypertrophy.[97] In this setting, AS also suppressed Ang II-evoked upregulation in NADPH oxidase, both in terms of Nox2 expression itself and superoxide generation, along with that of p38 mitogen-activated protein kinase that is another marker of maladaptive hypertrophy. These protective effects by HNO were coupled to elevated levels of cGMP and mimicked by brain natriuretic peptide. In this study as in previous settings, the specific involvement of HNO (and not of NO• or nitrite) was demonstrated by the fact that AS-induced antihypertrophic effects with subsequent activation of cardiomyocyte sGC and cGMP signaling were inhibited by L-cysteine but not by carboxy-PTIO. This study attesting that HNO prevents in vitro Ang-II cardiomyocyte hypertrophy via mechanisms that, at least in part, involves cGMP-dependent suppression of NADPH oxidase activity was followed by another contribution from the same group of investigators showing that IPA/NO elicits a concentration-dependent inhibition of endothelin-1 (ET1)-induced increase in cardiomyocytes size, accompanied by suppression of prohypertrophic genes.[98] In aggregate, all this evidence shows that HNO donors acutely limit cardiac hypertrophy, independently and/or in addition to NO• signaling, and under conditions of elevated superoxide production. Hence, by limiting or suppressing Nox2, and possibly other oxidase activity, longer-lasting HNO donors can be helpful in countering the adverse consequences of renin-angiotensin system (RAAS) activation that takes places during acute and chronic conditions of hemodynamic stress (see section 19.10).

19.8.3 In vitro impact of HNO donors on oxidative stress, vascular proliferation, inflammation, pain, and tumor growth

Dr. Fukuto's group was the first to recognize that HNO can act as an antioxidant via facile hydrogen atom donation to oxidizing radical species—an event similar to what happens with tocopherols—and subsequent generation of NO• that is a well-established ROS quencher.[99] They used a yeast *Saccharomyces cerevisiae* model system that is highly resistant to oxidative stress conditions that would lead to lipid peroxidation and would be toxic to mammalian cells. However, these authors rendered yeasts susceptible to oxidative stress by letting them incorporating polyunsaturated fatty acids (PUFAs) into their membranes and deleting the membrane antioxidant Q. In this model only membranes are expected to be oxidized, and both aerobic and anaerobic conditions can be applied (because yeasts can grow under both). With this tool in hand, the authors tested the direct impact of HNO on membrane integrity and evaluated the O_2 dependence of the effects of various conditions on membrane lipids. They found that PUFA supplementation and Q deletion were toxic to yeast only when O_2 was present, thus implicating lipid peroxidation as the damaging/toxicity mechanism. HNO protected yeast cells against PUFA- and Q deletion-induced toxicity, both in terms of cell viability and malondialdehyde content, with effects that were sensitive to GSH, but not to the metal chelator diethylenetriaminepentaacetic acid (DTPA). Thus, AS/HNO-mediated protection against toxicity and lipid peroxidation of PUFA-supplemented and Q-deleted yeast in the presence of O_2 can be attributed to its action as an antioxidant, while the insensitivity to DTPA actions clearly demonstrates that any eventual metal-catalyzed production of $NO^{\bullet}$ from HNO is not involved in the protective effects exerted by the latter. Previously Naughton and colleagues have shown that AS/HNO is able to induce the expression of mRNA for heme-oxygenase-1 (HO-1), in a concentration- and time-dependent manner, which translated into increased antioxidant activities by HO-1 in rat H9c2 cells.[100] Consistent with previous in vivo and in vitro reports, these authors reported that multiple additions of AS/HNO always promoted an increase in HO-1 activity. HO-1 is a well-known controller of heme and iron distribution, therefore playing a crucial role in cell survival. In fact, HO-1 and its product CO and bilirubin play a major protective role against endothelial cell damage, atherogenesis, and oxidation of low-density lipoproteins,[101] all desirable effects in patients suffering from acute or chronic coronary syndrome, eventually culminating in HF. Very recently, Miller and colleagues reported that HNO, donated by both AS and IPA/NO, suppresses the activity of vascular Nox2 oxidase triggered by AngII in mouse cerebral arteries. The administration of IPA/NO resulted in the suppression of both superoxide and H_2O_2 in an HNO-dependent manner because inhibitable by L-cysteine and via an ODQ-sensitive, thus cGMP-dependent mechanism.[102] The molecular intricacies of HNO-induced antagonism of Nox2 activity remain to be determined; however, in aggregate, the evidence reported above clearly confirm that HNO not only is not significantly affected by ROS signaling, but also that its donors can act as effective antioxidant agents.

In the same vein, another potentially relevant and yet poorly explored aspect of HNO pharmacology is the possibility that HNO such as IPA/NO inhibit neo-intimal hyperplasia, by inhibiting vascular smooth muscle cell (VSMC) proliferation and macrophage infiltration. In fact, Tsihlis and colleagues demonstrated that IPA/NO inhibits VSMC and endothelial cell proliferation in vitro without inhibiting migration or inducing cell death.[103] Using an in vivo carotid artery balloon injury model, these investigators also demonstrated IPA/NO's ability to cause a significant reduction in neo-intimal area and medial area, compared with injury alone, along with reduced macrophage infiltration. Interestingly, IPA/NO also prevented reendothelization. However, in this model, a 50% mortality rate was observed in the IPA/NO administered carotid injured rats. The latter outcome comes as no surprise because we and others have demonstrated that high levels of HNO may have prooxidant effects,[7] as demonstrated by its exacerbating effects on I/R injury in the heart[80] or in the brain.[81] More studies are definitely needed to determine whether the latter outcome is due to dosing and/or modalities of administration, i.e., whether lower doses of IPA/NO or different HNO donors (with respect to those used by Tsihlis and colleagues) can achieve the same beneficial effects without affecting survival.

Importantly from a cardiovascular therapeutic perspective are also the following very recent findings. First, we showed that HNO donor, NCA, inhibits leukemia-inhibiting factor (LIF)-induced IL-6 type cytokine activation of JAK-STAT signaling,[104] previously shown to be redox-sensitive.[105] This signaling pathway plays a central role in cardiac pathophysiology as shown by its implication in pressure overload–induced cardiac hypertrophy and remodeling, IPC, and I/R-induced cardiac dysfunction. Different STAT family members expressed in cardiac myocytes appear to be linked to different, and sometimes, orthogonal responses, such as cell growth/survival and apoptosis. Thus, differential activation and/or selective inhibition of the STAT proteins by agonists for G protein–coupled receptors, such as angiotensin II, may contribute to cardiac dysfunction during either I/R injury of HF. Pretreatment of human microvascular endothelial cells (HMEC-1) or neonatal cardiomyocytes with two different HNO donors, AS and NCA, inhibited LIF-induced STAT3 activation. Importantly, NCA pretreatment also blocked the induction of downstream proinflammatory genes such as intercellular adhesion molecule 1 (ICAM-1) and CCAAT/enhancer binding protein delta. The related 1-nitrosocyclohexyl pivalate (NCP; not an HNO donor) was equally effective in inhibiting STAT3 activation, suggesting that these compounds act as thiolate targeting electrophiles. Thus, these findings suggest that the intriguing possibility of targeting STAT3 signaling via a redox-active agent, i.e., HNO, in the context of cardiac and noncardiac diseases with prominent activation of proinflammatory signaling cascades. In the same vein, recent evidence from Basudhar and colleagues demonstrates the ability of diazeniumdiolate-based HNO-releasing aspirin to inhibit, among other enzymatic activity, cyclooxygenase-2 that is major mediator of inflammation.[106] Inflammatory responses frequently accompany the development of pain, in particular the neuropathic one,[107] and several lines of evidence suggest that ROS/RNS, particularly $NO^{\bullet}$, plays complex and often diverse role in the pathophysiology of pain as well as analgesia.[108] Zarpelon and colleagues were the first to report that local pretreatment with AS in rats inhibits the carrageenan-induced mechanical hyperalgesia in a dose- and time-dependent fashion, with a maximum inhibition of 97%, and the consequent rise in cytokine production.[109] Moreover, these authors also tested AS/HNO ability to inhibit hyperalgesia due to lipopolysaccharide, tumor necrosis factor-α, and prostaglandin E2, all inhibited by AS. These salutary actions of HNO were prevented by the soluble guanylate cyclase inhibitor, ODG, the PKG inhibitor KT5823, and by the ATP-sensitive K^+ channel blocker glibenclamide, but not by the opioid receptor blocker, naloxone. As in many other, different experimental settings, L-cysteine fully prevented the antinociceptive actions of AS/HNO. In a following study, the same group of investigators tested the possible antinociceptive potential of AS/HNO on phenyl-*p*-benzoquinone (PBQ)- and acetic acid-induced writhings and via the formalin test in mice. They found that the nociceptive responses were significantly reduced, in a dose-dependent manner, by subcutaneous pretreatment with AS/HNO. As in the previous study, ODQ, KT5823, and glibenclamide as well as L-cysteine prevented HNO's ability to act as an analgesic agent.

In addition to this evidence, there is the fact that HNO can irreversibly inhibit cathepsin B,[110] a cysteine protease involved in intracellular protein turnover and implicated, among other processes, in tumor invasion and metastasis.[111] When cancer growth is concerned, Norris and colleagues were the first to hypothesize that, given its ability to inhibit glyceraldehyde-3-phosphate dehydrogenase (GAPDH),[112,113] an enzyme essential for the anaerobic (glycolytic) metabolism of many malignant cancer cell growth, HNO could adversely affect tumor growth. Using in vivo and in vitro models of breast cancer, these investigators showed that AS derived HNO, suppresses the proliferation of both estrogen receptor (ER)-positive and ER-negative human breast cancer cell lines, in a dose-dependent manner.[114] In vivo, they observed that mice treated with HNO, either injected into the tumor itself or via the intraperitoneal approach had smaller xenograft tumor size. In addition to significantly decreased blood vessel density in the HNO-treated tumors, Norris and colleagues also observed lower levels of circulating serum vascular endothelial growth factor and a decrease in total hypoxia-inducible factor (HIF-1α) protein in HNO-treated human breast cancer cell lines and in HNO-treated tumor tissue derived from xenografts. They argued that the multiplicity of actions observed after HNO treatment could be explained by the initial inhibition of GAPDH, thus providing a potential therapeutic opportunity based on blocking glycolysis that then results in decreased HIF-1α expression/activity, ultimately leading to angiogenesis inhibition. This study was the first to suggest HNO's potential as an anticancer therapy, acting via mechanisms different from those of existing breast cancer drugs, thus rendering HNO donors potential candidates to overcome known and emerging drug resistance pathways. The anticancer therapeutic potential of diazeniumdiolate-based aspirin prodrugs (the same showing the anti-inflammatory properties of aspirin that however protect against stomach ulceration) have recently evaluated by Basudhar and colleagues who prepared two new diazeniumdiolate-NSAID adducts (NONO-adducts) by derivatizing both a primary and a secondary amine-based diazeniumdiolate with aspirin to produce O^2-(acetylsalicyloyoxymethyl)-1-(*N*-isopropylamino)-diazen-1-ium-1,2-dilate IPA/NO-aspirin and O^2-(acetylsalicyloyloxymethyl)-1-(*N,N*-diethylamino)-diazen-1-ium-1,2-diolate (DEA/NO-aspirin).[115] They demonstrated increased cytotoxicity of the newly generated NONO-adducts toward human lung carcinoma cells compared to either aspirin or the parent nitrogen oxide (either HNO or NO) donor, reporting that these agents were more effective in reducing growth of breast adenocarcinoma cells than the parent compounds while not being appreciably cytotoxic in a related nontumorigenic cell line (MCF-10A). Importantly, the diazeniumdiolate-based aspirin prodrug that releases HNO (i.e., the IPA/NO-aspirin) was more cytotoxic than the related $NO^{\bullet}$ donating agent (DEA/NO-aspirin). Thus HNO donors have a significant pharmacological potential for preventing cancer growth and metastasis, and in particular for the treatment of breast cancer.

19.9 HNO impact on the central nervous system

The current knowledge about how exogenous HNO affects the central nervous system, particularly with respect to neuronal cells and glia, under normal and disease conditions, has been nicely reviewed by Choe and colleagues.[116] In this review, the authors provide a detailed account on HNO donors' ability to cross the blood–brain barrier, on their impact of neuronal and glial cell function and on the possible additional molecular targets of HNO in nerve cells, with particular attention to NOS. Moreover, the authors thoroughly reviewed possible sources of endogenous HNO in the brain taking into account that this organ has unique properties such as the expression of brain-specific enzymes (e.g., nNOS), high iron content, and increased (highly oxidative) energy metabolism.

Findings in all these different body compartments reiterate the same take-home-message: *different levels of ROS/RNS elicit different thiol chemistry and ultimately functional outcomes*. In the end, at low levels, HNO is a mild oxidizing agent, and its interaction with highly reactive groups unlikely exceeds the formation of reversible disulfide bonds. This is in contrast to stronger oxidants (or high fluxes of ROS/RNS) that may lead to an overriding oxidative chemistry, including the formation of Cys-sulfinic and sulfonic acid[5] and/or the modification of additional protein residues, such as tyrosine. It should be noted that although HNO has been shown to react with tryptophan in the presence of oxygen,[117,118] this reaction is approximately 1–2 orders of magnitude slower than the corresponding reaction with cysteine.[119] Therefore, HNO-induced irreversible modifications at this tryptophan residue levels are very unlikely to occur in the presence of thiols under normal physiological conditions. Significantly, the studies demonstrating the prooxidant effects were performed at high levels of HNO, whereas the antioxidant properties were observed at much lower concentrations. Future studies should address whether the presence of endogenous or exogenously applied HNO donors can result in beneficial effects when high tissue oxidative stress is present in the context of pathophysiologically relevant conditions, in which, for instance proinflammatory, prohypertrophic signaling pathways, or conditions of enhanced tissue proliferation/neoangionesis are at play.

19.10 HNO donors and human ADHF

HF affects millions of individuals worldwide, and each year millions of HF patients are admitted to hospitals for ADHF. This disease status can be defined as either the rapid onset of signs and symptoms of HF or their gradual worsening up to the point that intensive, intravenous therapy is required.[120] Fig. 19.4 provides a brief overview of ADHF pathophysiology and therapeutic treatment. In patients with preexisting chronic HF, due to acute ischemic attack, chronic hypertension, arrhythmias, or metabolic disorders such as diabetes and obesity, a condition of ADHF can be precipitated by a new episode of acute coronary syndrome, pneumonia, hypertensive bouts, recurrent arrhythmias, infections, high salt intake as well as nonadherence to therapeutic regimen.[120] Under these conditions, a status of acute LV failure becomes evident, with reduced cardiac output and volume overload. Patients refer fatigue, marked exercise intolerance, anorexia, and cognitive impairment. In turn, this condition of low cardiac output (thus inadequate perfusion of peripheral organs and tissues) triggers the activation of the RAAS at the renal level, promoting Na^+ and H_2O retention and renal vasoconstriction. Initially,

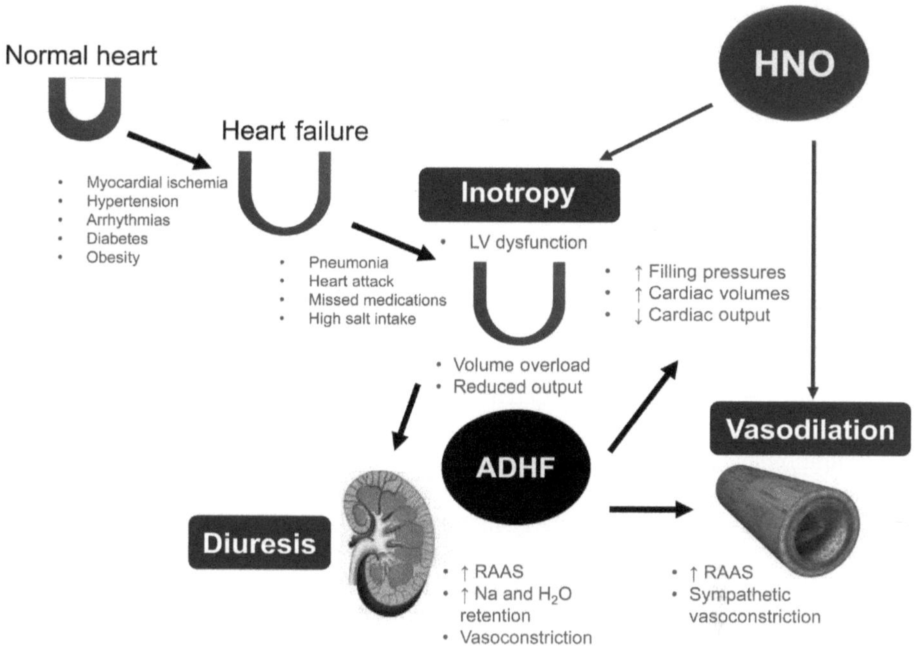

Figure 19.4 Brief overview of the causes and key pathophysiological events leading to acute decompensated heart failure (ADHF) and the potential therapeutic opportunities offered by HNO donors in this clinical situation.

this is a compensatory/adaptive measure whose scope is to maintain perfusion of vital organs in two main ways: (1) maintenance of systemic pressure by vasoconstriction, resulting in redistribution of blood flow to vital organs; (2) restoration of cardiac output by increasing myocardial contractility and LV filling pressure and volume. RAAS activation also increases heart rate in order to maintain cardiac output. However, with time a vicious cycle ensues that, as indicated in Fig. 19.4, ultimately ends with the exacerbation of LV dysfunction due to excessive LV volumes and filling pressure, ultimately reducing cardiac output.

ADHF still represents a major therapeutic challenge for clinicians, and current therapeutic options are basically limited to provide short-term relief from symptoms, although other important goals are to reduce postdischarge readmission and mortality rates. As indicated in Fig. 19.4, the forefront therapies for ADHF remain: (1) diuretics to relieve congestion; (2) vasodilators to reduce the load on the heart; (3) inotropic agents to sustain cardiac contractility, thus tissue perfusion. The latter are particularly needed in ADHF to treat patients who are in cardiogenic shock and those with signs and symptoms of end-organ dysfunction and hypoperfusion.[120] Further to this, as recently stated,[120] ultrafiltration was proposed as a potential mechanical solution to achieve greater volume and rate of fluid removal with the potential advantage of reducing neurohormonal activation as compared to diuretic therapy.[121] However, it is still unclear whether this procedure can be really useful in those ADHF patients who are failing diuretic therapy.[120] Excellent recent review articles provide a detailed and up-to-date account of contemporary treatment of ADHF, discussing in great detail old and new diuretics, agents with vasodilator properties as well as inotropic agents.[44,120] However, it is worth briefly recalling here that some of the current therapies have major drawbacks and/or side effects that limit their long-term use.

In the case of nitrovasodilators such as nitroglycerin, isosorbide mononitrate, isosorbide dinitrate, and SNP that, at low doses, predominantly affect venous circulation, there is still a substantial lack of evidence demonstrating beneficial effects on clinical outcomes[44]; moreover, these agents induce tolerance. Another serious drawback of vasodilating agents (when used at high dosage) is the induction of marked hypotension, and multiple clinical trials have demonstrated that a relationship exists between hypotension and subsequent adverse events and/or poor clinical outcomes.[122–124] This evidence calls for a careful dose selection, protocol-defined down-titration and stopping rules to limit these hypotensive episodes.[44] When inotropic agents are concerned, despite the fact that they remain an important adjunct to current ADHF treatment (particularly in those subjects with signs of shock or organ hypoperfusion), their long-term use is associated with increased mortality. The effect on mortality of old and novel inotropic agents such as the Na-K pump inhibitor digoxin, the synthetic β-agonist dobutamine, PDE inhibitors such as milrinone, Ca^{2+} sensitizers such as levosimendan, and other hormone- or SR-based interventions apt to increase intracellular Ca^{2+} levels have been expertly reviewed by Goldhaber and Hamilton.[125] Almost invariably, all these old and new inotropic avenues raise diastolic Ca^{2+} levels, either in a cAMP/PKA-dependent or independent manner. In a long term, this inotropic agent-induced rise in cytosolic Ca^{2+} and the concomitant impairment of Ca^{2+} sequestration into the SR due to altered expression/activity of SERCA2a (another hallmark of decreased contractile performance in human and experimental HF[126]) can ignite arrhythmias, promote and/or sustain impairment of cardiac relaxation, alter mitochondrial function, and ultimately induce myocyte cell death due to the activation of enzymatic activities such as caspases. Notwithstanding, loss in contractile performance, i.e., systolic HF, remains a leading cause of death and disability. Therefore, therapeutic interventions that directly target the central feature of systolic HF remain an important objective. Cardiac myosin activators such as omecamtiv mecarbil (formerly CK-1827452) directly activate the enzymatic pathway that within the cardiac myocytes leads to ventricular contraction.[127] Preclinical and clinical studies have shown that new inotropic agents are effective in improving cardiac contractility by increasing systolic time without the unwanted effects of the currently available indirect inotropic drugs, and without increasing cardiac O_2 consumption. Hence, these new agents may represent a safer and more therapeutic approach for the treatment of systolic HF.[127] However, by prolonging the systolic period, they may impair diastolic function, i.e., relaxation and coronary flow.[128]

Why HNO donors are a good addition/alternative to current ADHF therapeutic armamentarium? As shown in Fig. 19.4, HNO donors positively modulate cardiac contraction and relaxation; moreover, they induce an easy-to-titrate and balanced venous and arterial dilation. Whether or not HNO donors have diuretic properties,

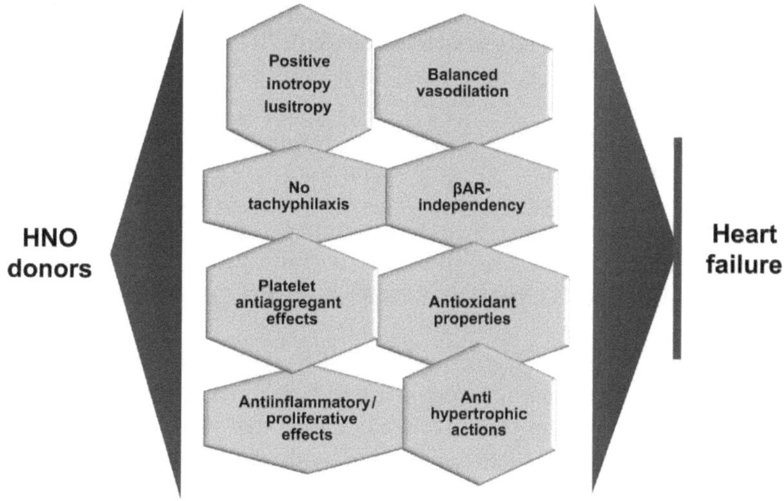

Figure 19.5 Synopsis of the in vivo and in vitro cardiovascular actions of HNO donors that support their use as an effective therapy during the course of acute or chronic heart failure and other cardiovascular disorders.

and eventually how these effects occur, remain totally unknown at the moment, but an important question to answer in future studies. Importantly, as compared to other inotropic agents such as dobutamine or milrinone, HNO inotropy and lusitropy is mild, fully β-AR- and PKA-independent and not accompanied by accumulation of diastolic Ca^{2+} at cellular (myocyte) level.[53,77] Unpublished observations from our group also suggest that HNO donors such as AS or IPA/NO (at least when infused acutely) do not prolong the Q–T interval. Long QT syndrome is a heart rhythm disorder that can cause serious arrhythmias. These characteristics single HNO donors out from the vast majority of currently employed inotropic agents.

When coronary or systemic vasodilation is concerned, our in vivo studies have repeatedly demonstrated that HNO-induced balanced venous and arterial dilation has a magnitude, that is substantially less than that obtained with equimolar amounts of pure $NO^{\bullet}$ donors, but similar to that induced by nitrite/nitrate. Importantly, as shown by us in vivo[32] and by Kemp-Harper's group in vitro,[45,129] HNO does not induce tolerance, thus representing a significant therapeutic departure from well-known nitrosovasodilators. As summarized in Fig. 19.5, all these acute effects dovetail nicely with the additional, potentially beneficial effects of HNO donors represented by their ability to counter platelet aggregation, to exert antioxidant, antiinflammatory, antiproliferative, and antihypertrophic effects, at least in vitro and under certain experimental conditions and with certain doses of HNO donors. While these intriguing and therapeutically promising findings await to be fully validated in vivo, using more long-lasting donors and more chronic administration regimens, all these findings give hope that HNO donors can be helpful not only to treat ADHF, but also chronic HF and other cardiac disorders. In the latter conditions, in fact, besides reduced cardiac contractility and altered vascular conditions, several additional pathological features must be corrected. For instance, whether and how HNO donors can counter chronic LV remodeling, modulate sympatho-vagal imbalance, influence cardiac energy metabolism, and counter cardiac starvation that is a prominent adverse event in chronic HF[130] remain to be tested, along with any possible impact of HNO donors chronic HF-related activation of proinflammatory response and immunological alterations.

19.11 Questions to ponder and future perspectives

HNO remains a somewhat elusive and certainly worth-studying molecule. Many aspects of its biology and pharmacology warrant further, more in-depth investigation. First and foremost, there is no definitive evidence yet that HNO is endogenously produced, under which conditions this may happen, and what are the factors required for HNO generation and cessation of production, and likely signaling capacity. This current gap in our knowledge is presumably technical in nature and may not be a true biological limitation. In fact, currently there are no methods available to unequivocally detect the "footprints" of nitroxyl in any in vivo biological system. HNO is different from $NO^{\bullet}$ and from some ROS that, the latter being a radical, can be directly trapped by EPR-based methods even in biological systems. This actual inadequacy, and its innate capacity to dimerize into N_2O render HNO a fleeting, hard-to-catch species. Second, given these basic limitations, and the somehow still limited availability of long-lasting HNO donors able to mimic persistent HNO generation over time, little-to-nothing is still known about the chronic effects produced by chronic exposure to HNO either in the cardiovascular system or in other tissues, organs or systems. In this regard, it would be very important, both from a conceptual and practical point of view, to validate in vivo the intriguing in vitro findings showing that HNO donors may counter oxidative stress and maladaptive hypertrophy, for instance. By doing this, the use of HNO donors can be conveniently extended to the treatment of chronic HF since they may counter adverse cardiac remodeling, another major negative factor accounting for HF progression and worsening of symptoms in HF patients. Third, setting aside in vitro evidence showing that HNO can affect the function of GAPDH that is involved in anaerobic glycolysis,[112,113] nothing is known on how HNO affects the metabolism of organs with high oxidative metabolism such as the heart and the brain. Fourth, the mapping of HNO-targetable cysteines has just begun. However, in our opinion, it is essential to expand the identification of these HNO "hot-spots" in the heart and vessels as well as to start mapping HNO-targetable (or targeted) residues in other body compartments, especially in the pulmonary and renal systems because alterations at these levels are crucially involved in the onset and progression of both acute and chronic cardiac diseases. Informatics approach would certainly facilitate these tasks. In essence, undertaking biochemical and functional investigations at these levels will likely disclose additional pharmacological, or potentially biological effects of HNO, helping to clarify where and under which conditions HNO actions are more likely important, and whether they are salutary or deleterious.

19.12 Concluding remarks

Two decades ago, HNO was still only a vague shape on the horizon of biologically relevant RNS. Now, HNO donors are currently in clinical trials for treating patients with ADHF.[48,49] Proof that HNO is formed in vivo remains elusive, but if obtained, it would alter our understanding of HNO profoundly, changing HNO from purely a drug to a physiological gaso-transmitter with signaling capacity, akin to O_2, $NO^•$, CO, and H_2S. Meanwhile, the unique and clinically promising cardiac and vascular properties exhibited by HNO donors so far are moving it forward for clinical trial testing in human HF. We have just begun to explore the implications of the prior 15 years of pharmacological and physiological studies, and still have much to learn about HNO's biology and pharmacological potential.

19.13 Disclosures

Nazareno Paolocci and David A. Kass are scientific founders and stock owners of Cardioxyl Pharmaceuticals, Inc.

19.14 Sources of funding

This work was supported by the NIH (R01HL075265 to NP, by the American Heart Association (BGA−0265435U, SDG−0435154N, and GIA−0855242E) and by Cardioxyl Pharmaceuticals, Inc. to NP.

Abbreviations

AS	Angeli's salt
BH_4	tetrahydrobiopterin
cAMP	cyclic adenosine monophosphate
cGMP	cyclic guanosine monophosphate
CGRP	calcitonin gene-related peptide
CHF	congestive heart failure
CuZnSOD	copper/zinc superoxide dismutase
DEA/NO	diethylamine/NO complex
DTT	mononitrated dithiothreitol
EDRF	endothelium-derived relaxing factor
GAPDH	glyceraldehyde-3-phosphate dehydrogenase
GSNO	*S*-nitrosoglutathione
HNO	nitroxyl, azanone, or nitrosyl hydride
IPC	ischemic preconditioning
MPO	myeloperoxidase
NANC	nonadrenergic/noncholinergic
NH_2OH	hydroxylamine
NO	nitric oxide

NOHA	N^{ω}-hydroxy-L-arginine
NOS	nitric oxide synthase
pK_a	acid dissociation constant
PKA	protein kinase A
PLN	phospholamban
RNS	reactive nitrogen species
ROS	reactive oxygen species
RyR2	ryanodine receptor 2
SERCA2a	Ca^{2+}-ATPase
sGC	soluble guanylyl cyclase
SNP	sodium nitroprusside
SR	sarcoplasmic reticulum
TRX	thioredoxin

References

1. Pagliaro P, Gattullo D, Penna C. Nitroglycerine and sodium trioxodinitrate: from the discovery to the preconditioning effect. *J Cardiovasc Med (Hagerstown)* 2013; **14**:698–704.
2. Fontani M, Orna MV. The shy angel who missed the Nobel prize. *Chem Today* 2012;**30**:58–60.
3. Ulich BL, Hollis JM, Snyder LE. Radiodetection of nitroxyl (HNO): the first interstellar NO bond. *Astrophys J* 1977;**217**:L105–8.
4. Fukuto JM, Chiang K, Hszieh R, Wong P, Chaudhuri G. The pharmacological activity of nitroxyl: a potent vasodilator with activity similar to nitric oxide and/or endothelium-derived relaxing factor. *J Pharmacol Exp Ther* 1992;**263**:546–51.
5. Andrews KL, Irvine JC, Tare M, Apostolopoulos J, Favaloro JL, Triggle CR, et al. A role for nitroxyl (HNO) as an endothelium-derived relaxing and hyperpolarizing factor in resistance arteries. *Br J Pharmacol* 2009;**157**:540–50.
6. Pino RZ, Feelisch M. Bioassay discrimination between nitric oxide (NO•) and nitroxyl (NO^-) using L-cysteine. *Biochem Biophys Res Commun* 1994;**201**:54–62.
7. Wink DA, Feelisch M, Fukuto JM, Chistodoulou D, Jourd'heuil D, Grisham MB, et al. The cytotoxicity of nitroxyl: possible implications for the pathophysiological role of NO. *Arch Biochem Biophys* 1998;**351**:66–74.
8. Wink DA, Miranda KM, Katori T, Mancardi D, Thomas DD, Ridnour L, et al. Orthogonal properties of the redox siblings nitroxyl and nitric oxide in the cardiovascular system: a novel redox paradigm. *Am J Physiol* 2003;**285**:H2264–76.
9. Flores-Santana W, Salmon DJ, Donzelli S, Switzer CH, Basudhar D, Ridnour L, et al. The specificity of nitroxyl chemistry is unique among nitrogen oxides in biological systems. *Antioxid Redox Signal* 2011;**14**:1659–74.
10. Demaster EG, Redfern B, Nagasawa HT. Mechanisms of inhibition of aldehyde dehydrogenase by nitroxyl, the active metabolite of the alcohol deterrent agent cyanamide. *Biochem Pharmacol* 1998;**55**:2007–15.
11. Nagasawa HT, DeMaster EG, Redfern B, Shirota FN, Goon DJW. Evidence for nitroxyl in the catalase-mediated bioactivation of the alcohol deterrent agent cyanamide. *J Med Chem* 1990;**33**:3120–2.
12. Forstermann U, Sessa WC. Nitric oxide synthases: regulation and function. *Eur Heart J* 2012;**33**:829–37.

13. Adak S, Wang Q, Stuehr DJ. Arginine conversion to nitroxide by tetrahydrobiopterin-free neuronal nitric-oxide synthase. Implications for mechanism. *J Biol Chem* 2000; **275**:33554–61.
14. Arnelle DR, Stamler JS. NO^+, NO, and NO^- donation by *S*-nitrosothiols: implications for regulation of physiological functions by *S*-nitrosylation and acceleration of disulfide formation. *Arch Biochem Biophys* 1995;**318**:279–85.
15. Donzelli S, Espey MG, Thomas DD, Mancardi D, Tocchetti CG, Ridnour LA, et al. Discriminating formation of HNO from other reactive nitrogen oxide species. *Free Radic Biol Med* 2006;**40**:1056–66.
16. Stoyanovsky DA, Tyurina YY, Tyurin VA, Anand D, Mandavia DN, Gius D, et al. Thioredoxin and lipoic acid catalyze the denitrosation of low molecular weight and protein *S*-nitrosothiols. *J Am Chem Soc* 2005;**127**:15815–23.
17. Yong QC, Hu LF, Wang S, Huang D, Bian JS. Hydrogen sulfide interacts with nitric oxide in the heart: possible involvement of nitroxyl. *Cardiovasc Res* 2010;**88**:482–91.
18. Filipovic MR, Eberhardt M, Prokopovic V, Mijuskovic A, Orescanin-Duric Z, Reeh P, et al. Beyond H_2S and NO interplay: hydrogen sulfide and nitroprusside react directly to give nitroxyl (HNO). A new pharmacological source of HNO. *J Med Chem* 2013;**56**:1499–508.
19. Cortese-Krott MM, Kuhnle GGC, Dyson A, Fernandez BO, Grman M, DuMond JF, et al. Key bioactive reaction products of the NO/H_2S interaction are S/N-hybrid species, polysulfides, and nitroxyl. *Proc Natl Acad Sci USA* 2015;**112**:E4651–60.
20. Wedmann R, Zahl A, Shubina TE, Durr M, Heinemann FW, Eberhardt M, et al. Does perthionitrite ($SSNO^-$) account for sustained bioactivity of NO? A (bio)chemical characterization. *Inorg Chem* 2015;**54**:9367–80.
21. Liochev SI, Fridovich I. Copper, zinc superoxide dismutase as univalent NO^- oxidoreductase and as a dichlorofluorescein peroxide. *J Biol Chem* 2001;**276**:35253–7.
22. Murphy ME, Sies H. Reversible conversion of nitroxyl anion to nitric oxide by superoxide dismutase. *Proc Natl Acad Sci USA* 1991;**88**:10860–4.
23. Niketic V, Stojanovic S, Nikolic A, Spasic M, Michelson AM. Exposure of Mn and FeSODs, but not Cu/ZnSOD, to NO leads to nitrosonium and nitroxyl ions generation which cause enzyme modification and inactivation: an in vitro study. *Free Radic Biol Med* 1999;**27**:992–6.
24. Sharpe MA, Cooper CE. Reactions of nitric oxide with mitochondrial cytochrome C: a novel mechanism for the formation of nitroxyl anion and peroxynitrite. *Biochem J* 1998;**332**:9–19.
25. Ohshima H, Saleem M. Xanthine oxidase converts nitric oxide to nitroxyl that inactivates the enzyme. *Biochem Biophys Res Commun* 2004;**315**:455–62.
26. Bartberger MD, Liu W, Ford E, Miranda KM, Switzer CH, Fukuto JM, et al. The reduction potential of nitric oxide (NO) and its importance to NO biochemistry. *Proc Natl Acad Sci USA* 2002;**99**:10958–63.
27. Hamer M, Suarez SA, Neuman NI, Alvarez L, Munoz M, Marti MA, et al. Discussing endogenous NO/HNO interconversion aided by phenolic drugs and vitamins. *Inorg Chem* 2015;**54**:9342–50.
28. Suarez SA, Neuman NI, Munoz M, Alvarez L, Bikiel DE, Brondino CD, et al. *J. Am. Chem. Soc.* 2015;**137**:4720–7.
29. Ye RW, Averill BA, Tiedje JM. Denitrification: production and consumption of nitric oxide. *Appl Environ Microbiol* 1994;**60**:1053–8.
30. Shafirovich V, Lymar SV. Nitroxyl and its anion in aqueous solutions: Spin states, protic equilibria, and reactivities toward oxygen and nitric oxide. *Proc Natl Acad Sci USA* 2002;**99**:7340–5.

31. Donzelli S, Espey MG, Flores-Santana W, Switzer CH, Yeh GC, Huang J, et al. Generation of nitroxyl by heme protein-mediated peroxidation of hydroxylamine but not *N*-hydroxy-L-arginine. *Free Radic Biol Med* 2008;**45**:578–84.
32. Paolocci N, Saavedra WF, Miranda KM, Martignani C, Isoda T, Hare JM, et al. Nitroxyl anion exerts redox-sensitive positive cardiac inotropy in vivo by calcitonin gene-related peptide signaling. *Proc Natl Acad Sci USA* 2001;**98**:10463–8.
33. Paolocci N, Katori T, Champion HC, St John ME, Miranda KM, Fukuto JM, et al. Positive inotropic and lusitropic effects of HNO/NO^- in failing hearts: Independence from beta-adrenergic signaling. *Proc Natl Acad Sci USA* 2003;**100**:5537–42.
34. Katori T, Donzelli S, Tocchetti CG, Miranda KM, Cormaci G, Thomas DD, et al. Peroxynitrite and myocardial contractility: in vivo versus in vitro effects. *Free Radic Biol Med* 2006;**41**:1606–18.
35. Khan AU, Kovacic D, Kolbanovskiy A, Desai M, Frenkel K, Geacintov NE. The decomposition of peroxynitrite to nitroxyl anion (NO^-) and singlet oxygen in aqueous solution. *Proc Natl Acad Sci USA* 2000;**97**:2984–9.
36. Smulik R, Debski D, Zielonka J, Bartosz A, Adamus J, Marcinek A, et al. Nitroxyl (HNO) reacts with molecular oxygen and forms peroxynitrite at physiological pH. Biological implications. *J Biol Chem* 2014;**289**:35570–81.
37. Casas AI, Dao VT, Daiber A, Maghzal GJ, Di Lisa F, Kaludercic N, et al. Reactive oxygen-related diseases: therapeutic targets and emerging clinical indications. *Antioxid Redox Signal* 2015;**23**:1171–85.
38. Paolocci N, Biondi R, Bettini M, Lee C-I, Berlowitz CO, Rossi R, et al. Oxygen radical-mediated reduction in basal and agonist-evoked NO release in isolated rat heart. *J Mol Cell Cardiol* 2001;**33**:671–9.
39. Tsai EJ, Kass DA. Cyclic GMP signaling in cardiovascular pathophysiology and therapeutics. *Pharmacol Ther* 2009;**122**:216–38.
40. Lymperopoulos A, Rengo G, Koch WJ. Adrenergic nervous system in heart failure: pathophysiology and therapy. *Circ Res* 2013;**113**:739–53.
41. Dorn GW. Adrenergic signaling polymorphisms and their impact on cardiovascular disease. *Physiol Rev* 2010;**90**:1013–62.
42. Haenen GRMM, Veerman M, Bast A. Reduction of B-adrenoceptor function by oxidative stress in the heart. *Free Radic Biol Med* 1990;**9**:279–88.
43. Hare JM, Colucci WS. Role of nitric oxide in the regulation of myocardial function. *Prog Cardiovasc Dis* 1995;**38**:155–66.
44. Singh A, Teerlink JR, Laribi S, Mebazaa A. Agents with vasodilator properties in acute heart failure. *Eur Heart J* 2016. [Epub ahead of print.]
45. Irvine JC, Favaloro JL, Widdop RE, Kemp-Harper BK. Nitroxyl anion donor, angeli's salt, does not develop tolerance in rat isolated aortae. *Hypertension* 2007;**49**:885–92.
46. Miranda KM, Katori T, Torres de Holding CL, Thomas L, Ridnour LA, McLendon WJ, et al. Comparison of the NO and HNO donating properties of diazeniumdiolates: primary amine adducts release HNO in vivo. *J Med Chem* 2005;**48**:8220–8.
47. Cao N, Wong YG, Rosli S, Kiriazis H, Huynh K, Qin C, et al. Chronic administration of the nitroxyl donor 1-nitrosocyclo hexyl acetate limits left ventricular diastolic dysfunction in a mouse model of diabetes mellitus in vivo. *Circ: Heart Fail* 2015;**8**:572–81.
48. Sabbah HN, Tocchetti CG, Wang M, Daya S, Gupta RC, Tunin RS, et al. Nitroxyl (HNO): a novel approach for the acute treatment of heart failure. *Circ: Heart Fail* 2013;**6**:1250–8.
49. Cowart D, Venuti R, Guptill J, Noveck R, Foo S. A phase 1 study of the safety and pharmacokinetics of the intravenous nitroxyl prodrug, CXL-1427. *J Am Coll Cardiol* 2015;**65**:A876.

50. Fukuto JM, Gulati P. Involvement of nitroxyl (HNO) in the cyanamide-induced vasorelaxation of rabbit aorta. *Biochem Pharmacol* 1994;**47**:922–4.
51. Bullen ML, Miller AA, Andrews KL, Irvine JC, Ritchie RH, Sobey CG, et al. Nitroxyl (HNO) as a vasoprotective signaling molecule. *Antioxid Redox Signal* 2011;**14**:1675–86.
52. Zhu G, Groneberg D, Sikka G, Hori D, Ranek MJ, Nakamura T, et al. Soluble guanylate cyclase is required for systemic vasodilation but not positive inotropy induced by nitroxyl in the mouse. *Hypertension* 2015;**65**:385–92.
53. Tocchetti CG, Wang W, Froehlich JP, Huke S, Aon MA, Wilson GM, et al. Nitroxyl improves cellular heart function by directly enhancing cardiac sarcoplasmic reticulum Ca^{2+} cycling. *Circ Res* 2007;**100**:96–104.
54. Booth BP, Nolan TD, Fung H-L. Nitroglycerin-inhibited whole blood aggregation is partially mediated by calcitonin gene-related peptide—a neurogenic mechanism. *Br J Pharmacol* 1997;**122**:577–83.
55. Russell F, King R, Smillie S-J, Kodji X, Brain SD. Calcitonin gene-related peptide: physiology and pathophysiology. *Physiol Rev* 2014;**94**:1099–142.
56. Franco-Cereceda A, Lundberg JM. Calcitonin gene-related peptide (CGRP) and capsaicin-induced stimulation of heart contractile rate and force. *Naunyn Schmiedebergs Arch Pharmacol* 1985;**331**:146–51.
57. Opgaard S, Hasbak P, Vries R, Saxena PR, Edvinsson L. Positive inotropy mediated via CGRP receptors in isolated human myocardial trabeculae. *Eur J Pharmacol* 2000;**397**:373–82.
58. Katori T, Hoover DB, Ardell JL, Helm RH, Belardi DF, Tocchetti CG, et al. Calcitonin gene-related peptide in vivo positive inotropy is attributable to regional sympatho-stimulation and is blunted in congestive heart failure. *Circ Res* 2005;**96**:234–43.
59. Favaloro JL, Kemp-Harper BK. The nitroxyl anion (HNO) is a potent dilator of rat coronary vasculature. *Cardiovasc Res* 2007;**73**:587–96.
60. Eberhardt M, Dux M, Namer B, Miljkovic J, Cordasic N, Will C, et al. H_2S and NO cooperatively regulate vascular tone by activating a neuroendocrine HNO-TRPA1-CGRP signalling pathway. *Nat Commun* 2014;**5**:4381.
61. Cheong E, Tumbev V, Abramson J, Salama G, Stoyanovsky DA. Nitroxyl triggers Ca^{2+} release from skeletal and cardiac sarcoplasmic reticulum by oxidizing ryanodine receptors. *Cell Calcium* 2005;**37**:87–96.
62. Viatchenko-Karpinski S, Gyorke S. Modulation of the Ca^{2+}-induced Ca^{2+} release cascade by β-adrenergic stimulation in rat ventricular myocytes. *J Physiol* 2001; **533**:837–48.
63. Kohr MJ, Kaludercic N, Tocchetti CG, Gao WD, Kass DA, Janssen PML, et al. Nitroxyl enhances myocyte Ca^{2+} transients by exclusively targeting SR Ca^{2+}-cycling. *Front Biosci, Elite Ed* 2010;**E2**:614–26.
64. Fukuto JM, Jackson MI, Kaludercic N, Paolocci N. Examining nitroxyl in biological systems. *Methods Enzymol* 2008;**440**:411–31.
65. Miranda KM, Paolocci N, Katori T, Thomas DD, Ford E, Bartberger MD, et al. A biochemical rationale for the discrete behavior of nitroxyl and nitric oxide in the cardiovascular system. *Proc Natl Acad Sci USA* 2003;**100**:9196–201.
66. Wouters MA, Fan SW, Haworth NL. Disulfides as redox switches: from molecular mechanisms to functional significance. *Antioxid Redox Signal* 2010;**12**:53–91.
67. Nagahara N. Intermolecular disulfide bond to modulate protein function as a redox-sensing switch. *Amino Acids* 2011;**41**:59–72.
68. Wong PSY, Hyun J, Fukuto JM, Shirota FN, DeMaster EG, Shoeman DW, et al. Reaction between *S*-nitrosothiols and thiols: generation of nitroxyl (HNO) and subsequent chemistry. *Biochemistry* 1998;**37**:5362–71.

69. Doyle MP, Mahapatro SN, Broene RD, Guy JK. Oxidation and reduction of hemoproteins by trioxodinitrate(II). The role of nitrosyl hydride and nitrite. *J Am Chem Soc* 1988;**110**:593–9.
70. Keceli G, Toscano JP. Reactivity of C-terminal cysteines with HNO. *Biochemistry* 2014;**53**:3689–98.
71. Keceli G, Toscano JP. Reactivity of nitroxyl-derived sulfinamides. *Biochemistry* 2012;**51**:4206–16.
72. Keceli G, Moore CD, Labonte JW, Toscano JP. NMR detection and study of hydrolysis of HNO-derived sulfinamides. *Biochemistry* 2013;**52**:7387–96.
73. Froehlich JP, Mahaney JE, Keceli G, Pavlos CM, Goldstein R, Redwood AJ, et al. Phospholamban thiols play a central role in activation of the cardiac muscle sarcoplasmic reticulum calcium pump by nitroxyl. *Biochemistry* 2008;**47**:13150–2.
74. Sivakumaran V, Stanley BA, Tocchetti CG, Ballin JD, Caceres V, Zhou L, et al. HNO enhances SERCA2a activity and cardiomyocyte function by promoting redox-dependent phospholamban oligomerization. *Antioxid Redox Signal* 2013;**19**:1185–97.
75. Lancel S, Zhang J, Evangelista A, Trucillo MP, Tong XY, Siwik DA, et al. Nitroxyl activates SERCA in cardiac myocytes via glutathionylation of cysteine 674. *Circ Res* 2009;**104**:720–3.
76. Meissner G. Molecular regulation of cardiac ryanodine receptor ion channel. *Cell Calcium* 2004;**35**:621–8.
77. Dai T, Tian Y, Tocchetti CG, Katori T, Murphy AM, Kass DA, et al. Nitroxyl increases force development in rat cardiac muscle. *J Physiol* 2007;**580**:951–60.
78. Gao WD, Murray CI, Tian Y, Zhong X, DuMond JF, Shen X, et al. Nitroxyl-mediated disulfide bond formation between cardiac myofilament cysteines enhances contractile function. *Circ Res* 2012;**111**:1002–11.
79. Choucani ET, Pell VR, James AM, Work LM, Saeb-Parsy K, Frezza C, et al. A unifying mechanism for mitochondrial superoxide production during ischemia-reperfusion injury. *Cell Metab* 2016;**23**:254–63.
80. Ma XL, Gao F, Liu G-L, Lopez BL, Christopher TA, Fukuto JM, et al. Opposite effects of nitric oxide and nitroxyl on postischemic myocardial injury. *Proc Natl Acad Sci USA* 1999;**96**:14617–22.
81. Choe CU, Lewerenz J, Fischer G, Uliasz TF, Espey MG, Hummel FC, et al. Nitroxyl exacerbates ischemic cerebral injury and oxidative neurotoxicity. *J Neurochem* 2009;**110**:1766–73.
82. Vanuffelen BE, Van Der Zee J, De Koster BM, Vansteveninck J, Elferink JG. Intracellular but not extracellular conversion of nitroxyl anion into nitric oxide leads to stimulation of human neutrophil migration. *Biochem J* 1998;**330**:719–22.
83. Pagliaro P, Mancardi D, Rastaldo R, Penna C, Gattullo D, Miranda KM, et al. Nitroxyl affords thiol-sensitive myocardial protective effects akin to early preconditioning. *Free Radic Biol Med* 2003;**34**.
84. Mancardi D, Ridnour L, Donzelli S, Miranda KM, Thomas DD, Katori T, et al. The nitroxyl donors angeli's salt and IPA/NO afford equal cardiac early preconditioning-like effect that is independent of mitochondrial K-ATP channel activation. *Free Radic Biol Med* 2004;**37**:S86–7.
85. Queliconi BB, Wojtovich AP, Nadtochiy SM, Kowaltowski AJ, Brookes PS. Redox regulation of the mitochondrial K(ATP) channel in cardioprotection. *Biochim Biophys Acta, Mol Cell Res* 2011;**1813**:1309–15.
86. Shiva S, Crawford JH, Ramachandran A, Ceaser EK, Hillson T, Brookes PS, et al. Mechanisms of the interaction of nitroxyl with mitochondria. *Biochem J* 2004;**379**:359–66.

87. Erusalimsky JD, Moncada S. Nitric oxide and mitochondrial signaling: from physiology to pathophysiology. *Arterioscler Thromb Vasc Biol* 2007;**30**:643–7.
88. Stasch JP, Schmidt PM, Nedvetsky PI, Nedvetskaya TY, Arun KHS, Meurer S, et al. Targeting the heme-oxidized nitric oxide receptor for selective vasodilatation of diseased blood vessels. *J Clin Invest* 2006;**116**:2552–61.
89. Mondoro TH, Ryan BB, Hrinczenko BW, Schechter AN, Vostal JG, Alayash AI. Biological action of nitric oxide donor compounds on platelets from patients with sickle cell disease. *Br J Haematol* 2001;**112**:1048–54.
90. Hoffman MD, Walsh GM, Rogalski JC, Kast J. Identification of nitroxyl-induced modifications in human platelet proteins using a novel mass spectrometric detection method. *Mol Cell Proteomics* 2009;**8**:887–903.
91. Bullen ML, Miller AA, Dharmarajah J, Drummond GR, Sobey CG, Kemp-Harper BK. Vasorelaxant and antiaggregatory actions of the nitroxyl donor isopropylamine NONOate are maintained in hypercholesterolemia. *Am J Physiol* 2011;**301**:H1405–14.
92. Schaer DJ, Buehler PW, Alayash AI, Belcher JD, Vercellotti GM. Hemolysis and free hemoglobin revisited: exploring hemoglobin and hemin scavengers as a novel class of therapeutic proteins. *Blood* 2013;**121**:1276–84.
93. Solomon SB, Bellavia L, Sweeney D, Piknova B, Perlegas A, Helms CC, et al. Angeli's salt counteracts the vasoactive effects of elevated plasma hemoglobin. *Free Radic Biol Med* 2012;**53**:2229–39.
94. Donzelli S, Fischer G, King SB, Niemann C, DuMond JF, Heeren J, et al. Pharmacological characterization of 1-nitrosocyclohexyl acetate, a long-acting nitroxyl donor that shows vasorelaxant and antiaggregatory effects. *J Pharmacol Exp Ther* 2013;**344**:339–47.
95. Dautov RF. The nitric oxide redox sibling nitroxyl partially circumvent impairment of platelet nitric oxide responsiveness. *Nitric Oxide* 2013;**35**:72–8.
96. Dorn GW. The fuzzy logic of physiological cardiac hypertrophy. *Hypertension* 2007;**49**:962–70.
97. Lin EQ, Irvine JC, Cao AH, Alexander AE, Love JE, Patel R, et al. Nitroxyl (HNO) stimulates soluble guanylyl cyclase to suppress cardiomyocyte hypertrophy and superoxide generation. *PLoS ONE* 2012;**7**:e34892.
98. Irvine JC, Cao N, Gossain S, Alexander AE, Love JE, Qin C, et al. HNO/cGMP-dependent antihypertrophic actions of isopropylamine-NONOate in neonatal rat cardiomyocytes: potential therapeutic advantages of HNO over NO. *Am J Physiol* 2013;**305**: H365–7.
99. Lopez BE, Shinyashiki M, Han TH, Fukuto JM. Antioxidant actions of nitroxyl (HNO). *Free Radic Biol Med* 2007;**42**:482–91.
100. Naughton P, Foresti R, Bains S, Hoque M, Green C, Motterlini R. Induction of heme oxygenase I by nitrosative stress. A role for nitroxyl anion. *J Biol Chem* 2002; **277**:40666–74.
101. Czibik G, Derumeaux G, Sawaki D, Valen G, Motterlini R. Heme oxygenase-1: an emerging therapeutic target to curb cardiac pathology. *Basic Res Cardiol* 2014;**109**:1–15.
102. Miller AA, Maxwell KF, Chrissobolis S, Bullen ML, Ku JM, Michael De Silva T, et al. Nitroxyl (HNO) suppresses vascular Nox2 oxidase activity. *Free Radic Biol Med* 2013;**60**:264–71.
103. Tsihlis ND, Murar J, Kapadia MR, Ahanchi SS, Oustwani CS, Saavedra JE, et al. Isopropylamine NONOate (IPA/NO) moderates neointimal hyperplasia following vascular injury. *J Vasc Surg* 2010;**51**:1248–59.

104. Zgheib C, Kurdi M, Zouein FA, Gunter BW, Stanley BA, Zgheib J, et al. Acyloxy nitroso compounds inhibit LIF signaling in endothelial cells and cardiac myocytes: evidence that STAT3 signaling Is redox-sensitive. *PLoS ONE* 2012;**7**:e43313.
105. Smith JK, Patil CN, Patlolla S, Gunter BW, Booz GW, Duhe RJ. Identification of a redox-sensitive switch within the JAK2 catalytic domain. *Free Radic Biol Med* 2012;**52**:1101–10.
106. Ricciotti E, Fitzgerald GA. Prostaglandins and inflammation. *Arterioscler Thromb Vasc Biol* 2011;**31**:986–1000.
107. White FA, Bhangoo SK, Miller RJ. Chemokines: integrators of pain and inflammation. *Nat Rev Drug Discov* 2005;**4**:834–44.
108. Gamper N, Ooi L. Redox and nitric oxide-mediated regulation of sensory neuron ion channel function. *Antioxid Redox Signal* 2015;**22**:486–504.
109. Zarpelon AC, Souza GR, Cunha TM, Schivo IRS, Marchesi M, Casagrande R, et al. The nitroxyl donor, angeli's salt, inhibits inflammatory hyperalgesia in rats. *Neuropharmacology* 2013;**71**:1–9.
110. Vaananen Antti J, Selmenpera P, Hukkanen M, Miranda KM, Harjula A, Rauhala P, et al. Persistent susceptibility of cathepsin B to irreversible inhibition by nitroxyl (HNO) in the presence of endogenous nitric oxide. *Free Radic Biol Med* 2008;**45**: 749–55.
111. Gondi CS, Rao JS. Cathepsin B as a cancer target. *Expert Opin Ther Targets* 2013;**17**:281–91.
112. Lopez BE, Rodriguez CE, Pribadi M, Cook NM, Shinyashiki M, Fukuto JM. Inhibition of yeast glycolysis by nitroxyl (HNO): a mechanism of HNO toxicity and implications to HNO biology. *Arch Biochem Biophys* 2005;**442**:140–8.
113. Lopez BE, Wink DA, Fukuto JM. The inhibition of glyceraldehyde-3-phosphate dehydrogenase by nitroxyl (HNO). *Arch Biochem Biophys* 2007;**465**:430–6.
114. Norris AJ, Sartippour MR, Lu M, Park T, Rao JY, Jackson MI, et al. Nitroxyl inhibits breast tumor growth and angiogenesis. *Int J Cancer* 2008;**122**:1905–10.
115. Basudhar D, Bharadwaj G, Cheng RY, Jain S, Shi S, Heinecke JL, et al. Synthesis and chemical and biological comparison of nitroxyl- and nitric oxide-releasing diazeniumdiolate-based aspirin derivatives. *J Med Chem* 2013;**56**:7804–20.
116. Choe CU, Lewerenz J, Gerloff C, Magnus T, Donzelli S. Nitroxyl in the central nervous system. *Antioxid Redox Signal* 2011;**14**:1699–711.
117. Suzuki T, Mower HF, Friesen MD, Gilibert I, Sawa T, Ohshima H. Nitration and nitrosation of *N*-acetyl-L-tryptophan and tryptophan residues in proteins by various reactive nitrogen species. *Free Radic Biol Med* 2004;**37**:671–81.
118. Peyrot F, Fernandez BO, Bryan NS, Feelisch M, Ducrocq C. *N*-Nitroso products from the reduction of indoles with angeli's salt. *Chem Res Toxicol* 2006;**19**:58–67.
119. Keceli G, Moore CD, Toscano JP. Comparison of HNO reactivity with tryptophan and cysteine in small peptides. *Bioorg Med Chem Lett* 2014;**24**:3710–13.
120. Hsiao R, Greenberg B. Contemporary treatment of acute heart failure. *Prog Cardiovasc Dis* 2016;**58**:367–78.
121. Bart BA, Goldsmith SR, Lee KL, Givertz MM, O'Connor CM, Bull DA, et al. Ultrafiltration in decompensated heart failure with cardiorenal syndrome. *N Engl J Med* 2012;**367**:2296–304.
122. Packer M, Colucci WS, Fisher L, Massie BM, Teerlink JR, Young J, et al. Effect of levosimendan on the short-term clinical course of patients with acutely decompensated heart failure. *JACC Heart Fail* 2013;**1**:103–11.

123. Voors AA, Davison BA, Felker GM, Ponikowski P, Unemori E, Cotter G, et al. Early drop in systolic blood pressure and worsening renal function in acute heart failure: renal results of Pre-RELAX-AHF. *Eur J Heart Fail* 2011;**13**:961–7.
124. Mebazaa A, Longrois D, Metra M, Mueller C, Richards AM, Roessig L, et al. Agents with vasodilator properties in acute heart failure: how to design successful trials. *Eur J Heart Fail* 2015;**17**:652–64.
125. Goldhaber JI, Hamilton MA. Role of inotropic agents in the treatment of heart failure. *Circulation* 2010;**121**:1655–60.
126. Kranias EG, Hajjar RJ. Modulation of cardiac contractility by the phospholamban/SERCA2a regulatome. *Circ Res* 2012;**110**:1646–60.
127. Malik FI, Hartman JJ, Elias KA, Morgan BP, Rodriguez H, Brejc K, et al. Cardiac myosin activation: a potential therapeutic approach for systolic heart failure. *Science* 2011;**331**:1439–43.
128. Greenberg BH, Chou W, Saikali KG, Escandón R, Lee JH, Chen MM, et al. Safety and tolerability of omecamtiv mecarbil during exercise in patients with ischemic cardiomyopathy and angina. *JACC Heart Fail* 2015;**3**:22–9.
129. Irvine JC, Kemp-Harper BK, Widdop RE. Chronic administration of the HNO donor angeli's salt does not lead to tolerance, cross-tolerance, or endothelial dysfunction: comparison with GTN and DEA/NO. *Antioxid Redox Signal* 2011;**14**:1615–24.
130. Gupta A, Akki A, Wang Y, Leppo MK, Chacko VP, Foster DB, et al. Creatine kinase-mediated improvement of function in failing mouse hearts provides causal evidence the failing heart is energy starved. *J Clin Invest* 2012;**122**:291–302.

Index

Note: Page numbers followed by "*f*" and "*t*" refer to figures and tables, respectively.

G

T

FSC
www.fsc.org
MIX
Papier | Fördert
gute Waldnutzung
FSC® C083411

Kolibri 360 GmbH
Ferdinand-Jühlke-Straße 7
99095 Erfurt, Deutschland
produktsicherheit@kolibri360.de